More information about this series at http://www.springer.com/series/7409

Lecture Notes in Computer Science　10816

Commenced Publication in 1973
Founding and Former Series Editors:
Gerhard Goos, Juris Hartmanis, and Jan van Leeuwen

Editorial Board

John Krogstie · Hajo A. Reijers (Eds.)

Advanced Information Systems Engineering

30th International Conference, CAiSE 2018
Tallinn, Estonia, June 11–15, 2018
Proceedings

 Springer

Editors
John Krogstie
Department of Computer Science
Norwegian University of Science and
  Technology
Trondheim
Norway

Hajo A. Reijers
Vrije Universiteit Amsterdam
Amsterdam
The Netherlands

ISSN 0302-9743   ISSN 1611-3349   (electronic)
Lecture Notes in Computer Science
ISBN 978-3-319-91562-3   ISBN 978-3-319-91563-0   (eBook)
https://doi.org/10.1007/978-3-319-91563-0

Library of Congress Control Number: 2018942346

LNCS Sublibrary: SL3 – Information Systems and Applications, incl. Internet/Web, and HCI

Printed on acid-free paper

This Springer imprint is published by the registered company Springer International Publishing AG
part of Springer Nature
The registered company address is: Gewerbestrasse 11, 6330 Cham, Switzerland

Preface

CAiSE 2018, the 30th International Conference on Advanced Information Systems Engineering was held in Tallinn, Estonia, during June 11–15, 2018.

The goal of the CAiSE conference series is to bring together the R&D community working on the models, methods, techniques, architecture, and technologies that address the design, engineering, operation, and evolution of information systems. The conference theme of CAiSE 2018 was "Information Systems in the Big Data Era," which acknowledges the disruptions brought about by the abundance of big data sources on government and business services, their users and customers, and their environments. This data abundance creates new opportunities to develop smart and personalized information systems. Concomitantly, it raises new challenges for information systems engineers, for example, in the areas of scalable data cleaning, integration and processing, real-time and predictive data analytics, and the cognification of information systems engineering.

The three invited keynotes of CAiSE 2018 provided three important perspectives on the big data era:

- Prof. Jan Recker (University of Cologne): "Information Systems Engineering in the Digital Age"
- Prof. Frank van Harmelen (Vrije Universiteit Amsterdam): "Information Systems in the Linked Data Era"
- Dr. Dan Bogdanov (Cybernetica): "Embedding Privacy by Design in Information Systems, One Business Process at a Time"

The accepted research papers and the CAiSE Forum address facets related to the theme of the conference, as well as the core topics associated with information systems design, engineering, and operation. The program included the following paper sessions:

- Process Execution
- User-Oriented IS Development
- Social Computing and Personalization
- The Cloud and Data Services
- Process Discovery
- Decisions and the Blockchain
- Process and Multi-level Modeling
- Data Management and Visualization
- Big Data and Intelligence
- Data Modeling and Mining
- Quality Requirements and Software

For CAiSE 2018, we received 179 full paper submissions. This year, we followed up with the selection process that was put in place in 2017. Each paper was initially reviewed by at least two Program Committee members. At the end of the first phase, we rejected papers with consistent negative evaluations. In the second phase of the evaluation, all papers with at least one positive evaluation were reviewed by a member of the Program Board. All reviewers then engaged in an online discussion of these papers. Finally, during the physical meeting of the Program Board, the final decision was made about the acceptance or rejection of each paper. The Program Board meeting took place in Amsterdam during February 20–21, 2018. The overall evaluation process of the papers resulted in the selection of 37 high-quality papers, which amounts to an acceptance rate of 21%. The final program of CAiSE 2018 was complemented by workshops, co-located working conferences, and a PhD consortium. For each of these events, separate proceedings were published.

We would like to thank the general chairs, Marlon Dumas and Andreas Opdahl, as well as their organization team, including Fabrizio Maggi, Fredrik Milani, Alexander Nolte, and Eva Pruusapuu, for all their efforts with respect to organizing CAiSE 2018. We warmly thank the workshop chairs (Remco Dijkman and Raimundas Matulevičius), the forum chairs (Jan Mendling and Haralambos Mouratidis), the tutorial chairs (Massimo Mecella and Selmin Nurcan), the doctoral consortium chairs (Marite Kirikova, Audrone Lupeikiene, and Ernest Teniente), and the publicity chairs (Jennifer Horkoff, Ingo Weber, and Liang Zhang) for their excellent work and contributions. We are grateful, too, to Richard van der Stadt, who helped run the review process and prepare the proceedings. Henrik Leopold and Han van der Aa contributed greatly to a smooth execution of the Program Board meeting in Amsterdam. As editors of this volume, we also offer our sincere thanks to the members of the Program Committee, as well as the external reviewers involved, for their dedication in providing fair and constructive evaluations. Our final thanks are to the members of the Program Board, who played a great role in the reviewing process to which they devoted much of their time and wisdom.

April 2018   John Krogstie
                                                                                    Hajo A. Reijers

Organization

General Chairs

Marlon Dumas   University of Tartu, Estonia
Andreas Opdahl   University of Bergen, Norway

Program Chairs

John Krogstie   Norwegian University of Science and Technology,
     Norway
Hajo A. Reijers   VU Amsterdam, The Netherlands

Workshop Chairs

Remco Dijkman   Eindhoven University of Technology, The Netherlands
Raimundas Matulevičius   University of Tartu, Estonia

Forum Chairs

Jan Mendling   Vienna University of Business and Economics, Austria
Haralambos Mouratidis   University of Brighton, UK

Tutorial/Panel Chairs

Massimo Mecella   University of Rome La Sapienza, Italy
Selmin Nurcan   Université Paris 1 Panthéon-Sorbonne, France

Doctoral Consortium Chairs

Marite Kirikova   Riga Technical University, Latvia
Audrone Lupeikiene   Vilnius University, Lithuania
Ernest Teniente   Universitat Politècnica de Catalunya, Spain

Industry Chairs

Dirk Draheim   Tallinn University of Technology, Estonia
Alex Norta   Tallinn University of Technology, Estonia

Publicity Chairs

Jennifer Horkoff   Chalmers University of Technology, Sweden
Ingo Weber   CSIRO Data61 Lab, Australia
Liang Zhang   Fudan University, China

Web and Social Media Chair

Alexander Nolte   University of Pittsburgh and Carnegie Mellon University,
                          USA

Sponsorship Chair

Fredrik P. Milani   University of Tartu, Estonia

Organization Chair

Fabrizio Maggi   University of Tartu, Estonia

Conference Steering Committee Chairs

Johann Eder   Alpen Adria Universität Klagenfurt, Austria
John Krogstie   Norwegian University of Science and Technology,
                          Norway
Barbara Pernici   Politecnico di Milano, Italy

Advisory Committee

Janis Bubenko Jr   Royal Institute of Technology, Sweden
Arne Sølvberg   Norwegian University of Science and Technology,
                          Norway
Colette Rolland   Université Paris 1 Panthéon Sorbonne, France
Óscar Pastor   Universitat Politècnica de València, Spain

Programe Committee Board

Sjaak Brinkkemper, The Netherlands
Eric Dubois, Luxembourg
Johann Eder, Austria
Xavier Franch, Spain
Paolo Giorgini, Italy
Matthias Jarke, Germany
Pericles Loucopoulos, UK
Heinrich C. Mayr, Austria
John Mylopoulos, Canada
Selmin Nurcan, France
Andreas L. Opdahl, Norway

Oscar Pastor, Spain
Henderik A. Proper, Luxembourg
Jolita Ralyté, Switzerland
Manfred Reichert, Germany
Stefanie Rinderle-Ma, Austria
Pnina Soffer, Israel
Janis Stirna, Sweden
Yannis Vassiliou, Greece
Barbara Weber, Denmark
Roel Wieringa, The Netherlands

Program Committee

Wil van der Aalst, The Netherlands
Daniel Amyot, Canada
João Araújo, Portugal
Marko Bajec, Slovenia
Jörg Becker, Germany
Boualem Benatallah, Australia
Nacer Boudjlida, France
Albertas Caplinskas, Lithuania
Maxime Cordy, France
Valeria De Antonellis, Italy
Oscar Díaz, Spain
Schahram Dustdar, Austria
João Falcão e Cunha, Portugal
Ulrich Frank, Germany
Avigdor Gal, Israel
Jānis Grabis, Latvia
Giancarlo Guizzardi, Italy
Jennifer Horkoff, Sweden
Marta Indulska, Australia
Marite Kirikova, Latvia
Agnes Koschmider, Germany
Lea Kutvonen, Finland
Marcello La Rosa, Australia
Julio Cesar Leite, Brazil
Henrik Leopold, The Netherlands
Lin Liu, China
Fabrizio Maria Maggi, Estonia
Raimundas Matulevičius, Estonia

Florian Matthes, Germany
Jan Mendling, Austria
Isabelle Mirbel, France
Boris Novikov, Russia
Jeff Parsons, Canada
Anna Perini, Italy
Günther Pernul, Germany
Pierluigi Plebani, Italy
Geert Poels, Belgium
Naveen Prakash, India
Gil Regev, Switzerland
Manuel Resinas, Spain
Antonio Ruiz Cortés, Spain
Kurt Sandkuhl, Germany
Flavia Maria Santoro, Brazil
Guttorm Sindre, Norway
Kari Smolander, Finland
Monique Snoeck, Belgium
Ernest Teniente, Spain
Irene Vanderfeesten, The Netherlands
Olegas Vasilecas, Lithuania
Panos Vassiliadis, Greece
Matthias Weidlich, Germany
Hans Weigand, The Netherlands
Mathias Weske, Germany
Eric Yu, Canada
Yijun Yu, UK

Additional Reviewers

Okhaide Akhigbe
Sanaa Alwidian
Amal Anda
Kevin Andrews
Ada Bagozi
Malak Baslyman
Kimon Batoulis
Jan Hendrik Betzing
Devis Bianchini
Alexander Bock
Veronica Burriel
Fabian Böhm

Kristof Böhmer
José Carlos Camposano
José Miguel Cañete-Valdeón
Yossi Dahari
Mohammad Danesh
Marietheres Dietz
Monica Dragoicea
Najah Mary El-Gharib
Ludwig Englbrecht
Montserrat Estañol
Siamak Farshidi
Walid Fdhila

Pablo Fernandez

Manuel Gall

José María García

Mahdi Ghasemi

Catarina Gralha

Avihai Greenvald

Sebastian Groll

Jens Gulden

Marcin Hewelt

Moritz von Hoffen

Conrad Indiono

Amin Jalali

Slinger Jansen

Monika Kaczmarek

Klaus Kammerer

Fitsum Meshesha Kifetew

Sybren De Kinderen

Julius Köpke

Tony Leclercq

Armel Lefebvre

Gururaj Maddodi

Alfonso E. Marquez-Chamorro

Nicolas Mayer

Michele Melchiori

Florian Menges

Markus Monhof

Stefan Nastic

Adriatik Nikaj

Kestutis Normantas

Nadine Ogonek

Xavier Oriol

Michiel Overeem

Ute Paukstadt

Alexander Puchta

Luise Pufahl

Ivan S. Razo-Zapata

Dennis M. Riehle

Marcela Ruiz

Titas Savickas

Ognjen Scekic

Johannes Schobel

Hendrik Scholta

Arik Senderovich

Vladimir Shekhovtsov

Michael Stach

Sebastian Steinau

Irene Teinemaa

Arava Tsoury

Slim Turki

Ilya Verenich

Manfred Vielberth

Jan Martijn van der Werf

Karolin Winter

Slavko Žitnik

Contents

Big Data and Intelligence

Data Modelling and Mining

Quality Requirements and Software

CAiSE 2018 Tutorials

Process Execution

Association Rules for Anomaly Detection and Root Cause Analysis in Process Executions

Kristof Böhmer(✉) and Stefanie Rinderle-Ma

Faculty of Computer Science, University of Vienna, Vienna, Austria
{kristof.boehmer,stefanie.rinderle-ma}@univie.ac.at

Abstract. Existing business process anomaly detection approaches typically fall short in supporting experts when analyzing identified anomalies. Hereby, false positives and insufficient anomaly countermeasures might impact an organization in a severely negative way. This work tackles this limitation by basing anomaly detection on association rule mining. It will be shown that doing so enables to explain anomalies, support process change and flexible executions, and to facilitate the estimation of anomaly severity. As a consequence, the risk of choosing an inappropriate countermeasure is likely reduced which, for example, helps to avoid the termination of benign process executions due to mistaken anomalies and false positives. The feasibility of the proposed approach is shown based on a publicly available prototypical implementation as well as by analyzing real life logs with injected artificial anomalies.

Keywords: Anomaly detection · Process · Root cause · Rule mining

1 Introduction

Process *anomaly detection* enables to reveal anomalous process execution behavior which can indicate fraud, misuse, or unknown attacks, cf. [3,4]. Typically, whenever anomalous behavior is identified an *alarm* is sent to a security expert. Subsequently, the expert determines the alarm's root cause to choose an appropriate anomaly *countermeasure*, such as, terminating an anomalous process, ignoring a false alarm, or manually correcting process execution behavior, cf. [4].

Analyzing anomaly detection alarms and choosing countermeasures is *challenging*, cf. [13]. This applies also to the process domain as processes operate in *flexible open environments*, cf. [4]. Hence, thousands of alarms must be carefully analyzed as they could be false positives that report benign behavior as anomalous [4,13] (e.g., because of incorrectly interpreted noise or ad hoc changes).

Existing work, cf. [3,4,6], reports only if an execution is *anomalous or not*. However, we assume that additional information, e.g., which behaviour motivated the (non-) anomalous decisions or the anomaly severity, are a necessity. Without such information anomalies, likely, cannot be fully understood and it

© Springer International Publishing AG, part of Springer Nature 2018
J. Krogstie and H. A. Reijers (Eds.): CAiSE 2018, LNCS 10816, pp. 3–18, 2018.
https://doi.org/10.1007/978-3-319-91563-0_1

becomes hard to differentiate between harmful anomalies and false positives but also to choose appropriate countermeasures as anomalies vary in effect and form.

Further on, existing process focused work frequently applies *monolithic* anomaly detection signatures, cf. [3,6], to identify anomalies. Such signatures, compress all the expected execution behavior into a complex interconnected structure. Thus, they must be recreated from scratch whenever a process changes, are hard to understand, and can be computationally intense to create. Monolithic signatures were also found to be overly detailed and specific so that benign noise or ad hoc changes (such as, slightly varying resource assignments or activity execution orders) are typically reported as anomalies [4]. This could hinder an organization as benign process executions could be unnecessarily terminated.

To address these limitations this work proposes a novel unsupervised anomaly detection heuristic. Instead of monolithic signatures it applies small sets of independent association rules. Hereby, individual rules can easily be replaced if a process changes. As association rules only represent direct relations between variables (e.g., activity A and C occur during the same execution) they are easy to understand but lack in expressiveness compared to other formalisms, such as, Liner Temporal Logic (LTL) – which we see as an *advantage*. We assume that the more expressive rules become the more likely it is that they are misunderstood and the more computational intense it is to mine them, cf. [10,14], hence, simple formalisms (e.g., association rules) foster mining and understandability.

Further, the proposed approach prevents false positives as it supports noise and ad hoc changes. This is because process executions, differently to monolithic signatures, no longer must be completely represented by the signatures but only by a fraction of the rules which a signature is composed of, cf. [4]. This also enables to provide more details about the individual anomalies, as it can be reported which rules a process execution (trace resp.) supports and which not, but also the anomaly severity. The latter is composed of the aggregated automatically calculated rule significance of each non-supported rule.

Let P be a process that should be monitored for anomalies and let L hold all execution traces t of P. The key idea is to represent the given behavior in L as a set of association rules R. To analyze if a novel execution trace $t' \notin L$ is anomalous it is determined which rules in R are supported by t' (ex post). Supported means that a trace complies to the conditions specified by the rule. If it is found that t' has a lower rule support than the trace $t \in L$ that is most similar to t' then t' is identified as anomalous and an alarm is triggered.

This work follows the design science research methodology, cf. [21]. For this, design requirements were derived from existing work on anomaly detection in the security domain, in general, and the process domain. As a result an existing rule formalism, i.e., association rules, lays the foundations for a novel anomaly detection approach, cf. Sects. 2 and 3. This artifact is evaluated in two ways, cf. Sect. 4. First, its feasibility is shown by performing a cross validation with real life process executions and injected anomalies. Secondly, the findings are compared with related work (cf. Sect. 5) and the achieved root cause analysis

capabilities are discussed, cf. Sects. 3 and 6. Stakeholders of the proposed approach are process managers and security experts.

2   Prerequisites and General Approach

This paper proposes an anomaly detection heuristic to classify process execution
traces as anomalous or not. For this association rules are mined from a set of
recorded execution traces L (i.e., a log). This is beneficiary as L: (a) represents
real process execution behavior; (b) incorporates manual adaptions, noise, and
ad hoc changes; (c) is automatically generated during process executions; and (d)
is independent from abstracted/outdated documentation, cf. [11]. The proposed
approach is *unsupervised* as the traces in L are not labeled, formally:

Definition 1 (Execution Log). *Let L be a set of execution traces $t \in L$;*
$t := \langle e_1, \cdots, e_n \rangle$ holds an ordered list of execution events $e_i := (ea, er, es, ec)$;
e_i represents the execution of activity $e_i.ea$, by resource $e_i.er$, which started at
timestamp $e_i.es \in \mathbb{R}_{>0}$ and completed at $e_i.ec \in \mathbb{R}_{>0}$; t's order is given by $e_i.es$.

This notion represents information provided by process execution log formats
such as, the eXtensible Event Stream, and enables the analysis of the control,
resource, and temporal perspective. Accordingly, the first event in the running
example, cf. Table 1, is $e_1 = (\text{A}, \text{Mike}, 1, 5)$. *Auxiliary* functions: $\{\cdots\}^0$ returns a
random element from a set/bag. $c := a \oplus b$ appends b to a copy c of the collection
given by a; $\langle \cdot \rangle^l$ retains the last element of a list; $\langle \cdot \rangle_i$ retains the list item with
index $i \in \mathbb{N}_{>0}$ while $\langle \cdot \rangle_i^+$ retains all list items with an index $> i$.

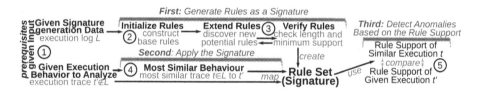

Fig. 1. Proposed rule based process anomaly detection approach – overview

Figure 1 provides an overview of the proposed anomaly detection heuristic;
algorithms are presented in Sect. 3. Firstly, a signature R is created for a process
P based on the associated execution log L (rule mining), i.e., L is assumed as
given input ①. Here, a signature R is a set of rules that represent behavior
mined from L. A rule represents, e.g., that activity C should be a successor of A.

Here, rule mining is inspired from the Apriori algorithm [2] which mines value
relationships in large unordered datasets as association rules. However, process
execution traces are temporally ordered, e.g., based on the start timestamp of
each event. To take this aspect into account association rules are extended into
Anomaly Detection Association Rules (ADAR) – (rule for short) – to detect
control, resource, and temporal anomalies in process execution traces, formally:

Definition 2 (ADAR). $r := \langle rp_1, \cdots, rp_m \rangle$ *is a* `rule` *with* `conditions` $rp_j :=$ (ra, rd). *Let* $t \in L$ *be a trace. All conditions in* r *must be matched by* t *to conclude that* t *supports* r, *cf. Definition 4. The condition indices* j *represents their expected order in* r, *i.e.,* rp_j *must be matched by a trace before* rp_{j+1} *can be. Each rule condition* rp_j *represents an expected activity* $rp_j.ra$ *and an optional execution duration represented as classes, i.e.,* $rp_j.rd \subseteq \{$`low, avg, high`$\}$ *based on* L.

The following projections were defined for rules (r) and conditions (rp): $rt(r) \mapsto \{$`control, temporal, SoD, BoD`$\}$ represents that a rule can specify control, temporal, or resource behavior. The latter focuses on the assignment of resources to activities which is analyzed in the form of Separation of Duty (SOD) or Binding of Duty (BoD), cf. [7]. Further, as rules (conditions) are mined based on given traces (events); we define projections for rules (conditions) on their trace (event) so that $rtr(r) := t$ $(re(rp) := e)$.

Imagine the rule $r_1 := \langle rp_1, rp_2 \rangle$ where $rt(r_1) = $ `control`, $rp_1 = (\mathtt{A}, \cdot)$, and $rp_2 = (\mathtt{B}, \cdot)$ is matched with the running example in Table 1. As r_1 is a control flow rule, execution duration classes are not relevant/defined. While trace t_1 supports r_1 the second does not. This is because an execution of activity A succeeded by an execution of activity B is only given in trace t_1. If $rp_1.ra = \mathtt{A}$ and $rp_2.ra = \mathtt{C}$ then the rule would be supported by both traces t_1 and t_2.

Table 1. Exemplary running example log L containing the exemplary traces t_1 and t_2

Event e	Process P	Trace t	Activity ea	Resource er	Start timestamp es	End timestamp ec
e_1	P_1	t_1	A	Mike	1	5
e_2	P_1	t_1	B	Mike	6	9
e_3	P_1	t_1	C	Sue	12	16
e_4	P_1	t_2	B	Mike	18	21
e_5	P_1	t_2	A	Tom	22	29
e_6	P_1	t_2	C	Sue	32	38

The applied rule mining consists of *three* stages. Initially, see ②, the basis for the mining is laid by converting each event e, given by L's traces, into an individual rule. Hence, at this stage each rule only holds a single condition, so that $\forall r \in R; |r| = 1$. In the following these initial set of rules (the individual rules in R, resp.) is repeatedly extended and verified to create the final anomaly detection rule set. When assuming that the running example log only consists of t_1 then the initial rule set is $R := \{r_1, r_2, r_3\}$ where each rule consists of a single rule condition, e.g., $r_1 := \langle rp_1 \rangle$ where $rp_1.ra = \mathtt{A}$ given that $rt(r_1) = $ `control`.

Subsequently, rule *extension* and *verification* approaches are iteratively applied. The rule extension, see ③, extends each rule in R by one additional rule condition in each possible way to identify new potential rules. For example, to extend r_1 all successors of activity A (i.e., the last rule condition, $\mapsto r_1^l$, in r_1, cf. Definition 3) are determined, i.e., activity B and C. Secondly, activity B and

C are utilized to extend the ADAR r_1 into $r_1' := \langle rp_1, rp_2' \rangle$ and $r_1'' := \langle rp_1, rp_2'' \rangle$ where $rp_2'.ra = $ B and $rp_2''.ra = $ C. Formally, this is defined as:

Definition 3 (Extending individual ADARs). *Let $\mathcal{E}$ be the set of all events in a log L and $\mathcal{RP}$ be the set of all rule conditions. Let $r := \langle rp_1, \cdots, rp_m \rangle$ be a rule and $t := \langle e_1, \cdots, e_n \rangle \in L$ be an execution trace with $rtr(r) = t$. Rule extension function $ext : R \times L \mapsto R$ extends r by:*

$$ext(r, t) := \{r \oplus torp(e, rt(r)) | e \in \{e' \in t | e'.es > e''.es; e'' = re(r^l)\}\} \qquad (1)$$

where $torp : \mathcal{E} \times \{control, temporal, SOD, BOD\} \mapsto \mathcal{RP}$ converts an event e into a rule condition rp, cf. Definition 5; Sect. 3.1 defines how this is performed for each of the four rule types, i.e., $rt(r) \in \{control, temporal, SOD, BOD\}$.

Finally, rule verification is applied ④. Each rule is verified by analyzing its support, i.e., $sup(r, L) := |\{t \in L | mp(r, t) = \mathtt{true}\}| / |L|$ (function mp is defined in Definition 4). Here the support of a rule represents the percentage of traces in L a rule could be successfully mapped on (match the rule conditions, resp.). If the support (i.e., the percentage of traces $t \in L$ a rule supports) of a rule is below user configurable $mins \in [0, 1]$ then the rule is removed from R. Subsequently, the rule extension and verification steps are applied repeatedly till the rules in R are extended to a user configurable maximum length of $rl \in \mathbb{N}_{\geq 1}$ rule conditions.

The verification variables $mins$ and rl enable to fine tune rules for specific use cases and process behavior. For example, we found that the mining of longer rules resulted in stricter signatures than the mining of short rules. In comparison choosing a low $mins$ value could result in overfitting the signatures and a high amount of rules. Further discussions on the variables are given in Sect. 4.

Definition 4 (ADAR mapping). *Let $r := \langle rp_1, \cdots, rp_m \rangle$ be a rule (cf. Definition 2) and $t := \langle e_1, \cdots, e_n \rangle$ an execution trace, cf. Definition 1. Mapping function $mp : R \times L \mapsto \{\mathtt{true}, \mathtt{false}\}$ determines if r is supported by (matching to, resp.) t. Rule type $rt(r)$ determines the matching strategy to be applied, see Sect. 3.1.*

The rule r_1', as given previously, achieves a support of 0.5 for the running example, cf. Table 1. This is because it expects activity A is succeeded by activity B. Accordingly, it can only be mapped onto trace t_1 but not on t_2. Imagine, that $mins$ was defined as 0.9, then r_1' would be removed during the verification phase from R as $0.5 < 0.9$. In comparison rule r_1'' would not be removed as it is supported by both traces t_1 and t_2 (i.e., $sup(r_1'', L) = 1$ so that $1 \not< 0.9$). Rule r_1'' matches to (is supported by, resp.) traces where A is succeeded by activity C.

Finally, the mined rules R (the signature) are applied to classify a given process execution trace $t' \notin L$ as anomalous or not. For this a trace $t \in L$ is identified that is most similar to t', see ⑤. The similarity between traces is measured based on the occurrence of activities in the compared traces, cf. Definition 6. Then t' and t are mapped onto R's rules to determine the aggregated support of both traces. Finally, if the aggregated support of t' is below the aggregated support of t then the given trace t' is classified as being anomalous, see ⑥.

3 ADAR Based Anomaly Detection

This section presents the algorithms for the approach set out in Fig. 1.

3.1 Anomaly Detection Association Rule Mining

The applied ADAR mining approach, cf. Algorithm 1, combines the main mining steps described in Fig. 1. This is the rule set *initialization*, along with the iteratively applied rule *extension* (cf. Definition 3) and *verification* steps (cf. Definition 4). Depending on the *user chosen* rule type $ty \in$ {control, temporal, SoD, BoD} different algorithms are applied to mine either control, temporal, or resource behavior given in L into rules. While each rule type is mined individually, rules of all types can be combined in a single signature (i.e., the union of all individual rule sets).

Algorithm ruleMining(*log L, min support mins, max rule length rl, rule type ty*)

> **Result:** set of mined rules R (i.e., a signature)
> $R := \varnothing$; // initially the rule set (signature, resp.) is empty
> **foreach** $t \in L$ **do** // initializing the rule set R with base rules
> > **foreach** $e \in t$ **do**
> > > $R := R \cup \{\langle torp(e, ty)\rangle\}$ // initial base rule with one condition, Definition 5
> >
> > **for** $rcsize := 0$ **to** rl **do** // generate rules up to a size of rl conditions per rule
> > > $R := \{ext(r, rtr(r))|r \in R\}$ // extend rules in R, cf. Definitions 2 and 3
> > > **foreach** $r \in R$ **do**
> > > > **if** $sup(r, L) < mins$ // verification, calc. rule support, cf. Definition 4
> > > > **then**
> > > > > $R := R \setminus \{r\}$ // remove r from R because its support is too low
> >
> > **return** R // final set of mined rules R for behavior given by the log L

Algorithm 1. Mines rules for a given execution log L and rule type ty.

Definition 5 (Transforming events to ADAR conditions). *The auxiliary function $torp(e, ty) : rp$ transforms an* **event** *e into a rule* **condition** *rp. Depending on the chosen rule type (given by variable ty) one out of the* **three** *following rule condition mining approaches is applied. For example, if $ty = $* control *then the following control flow rule mining approach is used.*

Mining Control Flow ADARs. Control flow rules represent expected activity orders, e.g., that activity A should be succeeded by activity C during a process execution. Hereby, control flow rules enable to identify process misuse, cf. [3], such as, the execution of a financially critical "bank transfer" activity without the previous execution of a usually mandatory "transfer conformation" activity.

Event to Control Condition. Accordingly, transforming an event e, during rule extension, into a rule condition: $rp = (e.ea, \cdot)$. Hence, when transforming e_1, as given by the running example in Table 1, into a rule condition rp then $rp = (\text{A}, \cdot)$.

Control ADAR Support. Trace t supports a control flow rule r if t holds all activity executions specified by the rule conditions in $rp \in r$, ①. Further the activity executions must occur in accordance to the order of rule conditions in r, ②. This represents that activity executions are mutual dependent on each other. ① and ② are verified by Algorithm 2 to determine if a trace t supports the control flow rule r. Accordingly, a rule $r = \langle rp_1, rp_2 \rangle$ where $rp_1.ra = \mathtt{A}$ ($rp_1 = (\mathtt{A}, \cdot)$, resp.) and $rp_1.ra = \mathtt{B}$ (meaning that activity $\mathtt{A}$ must be succeeded by $\mathtt{B}$) would only be supported by trace t_1 but not by t_2 in the running example, cf. Table 1.

Algorithm ControlSupport(*trace t, control flow rule r*)
> **Result:** if r is supported by $t \mapsto$ true or not $\mapsto$ false
> **for** $j = 1$ **to** $|r|$ // $|\langle \cdot \rangle|$ retains the length of the list $\langle \cdot \rangle$ **do**
> > **for** $i = 1$ **to** $|t|$ **do**
> > > **if** $t_i.ea = r_j.ra$ // verify control flow rule condition matching **then**
> > > > $t := t_i^+$; $r := r_j^+$; **break** // remove successfully matched parts of t, r
> **return** $|r| = 0$?*true:false* // return true if r fully matches to t else false

Algorithm 2. Checks if a trace t supports the control flow rule r.

This work applies a relaxed rule matching. Hence, a rule is assumed as supported by a trace as long as this trace contains at least a single combination of events that match to the rule conditions. This enables to deal with loops and concurrency but also it enables to be flexible enough to not struggle with noise and ad hoc changes. However, as found during the evaluation it is still specific enough to differentiate benign and anomalous process executions, cf. Sect. 4.

Mining Temporal ADARs. Temporal rules focus on activity durations. Those were found to be a significant indicator for fraud and misuse, cf. [5,8]. However, while control flow rules focus on representing distinct values (e.g., explicitly activity $\mathtt{A}$ is expected) this is not possible for temporal rules. This is because distinct durations, e.g., one second or one hour, are so specific that even a minor temporal variation, which we assume as being likely, would render a rule to be no longer supported by a trace. This can, potentially, result in false positives.

To tackle this challenge we propose *fuzzy temporal rules*. These rules are not representing durations with explicit values but with duration classes. These classes represent, for example, that the expected duration of activity $\mathtt{A}$ is roughly comparable or below/above the average execution duration of activity $\mathtt{A}$ – given by the traces in L. In this work three duration classes are in use, i.e., $PDC :=$ $\{\mathtt{low}, \mathtt{avg}, \mathtt{high}\}$. Increasing the number of classes would be possible but it was found that this can result in overfitting the generated temporal rules (signature).

Event to Temporal Condition. A temporal rule condition consists of an expected activity execution along with its expected duration classes. For this the activity execution duration is represented as a subset of the possible duration classes PDC. So, based on an event e a temporal condition rp is constructed

---

Algorithm TempClass(*event e, log L, duration classes PDC, widen* $w \in [0;1]$)

  Result: set of representative duration classes $DC \subseteq PDC$ for e
  // calculate durations D for L, *min* and *max* duration, duration class timespan
  part, duration d of event e, relative class timespan widening *wspan*
  $D := \{e'.ec - e'.es | e' \in t, t \in L : e'.ea = e.ea\}$
  $min := \{d | d \in D, \forall d' \in D; d \leq d'\}^0$; $max := \{d | d \in D, \forall d' \in D; d \geq d'\}^0$
  $part := (max - min)/|PDC|$; $d := e.ec - e.es$; $wspan = part \cdot w$; $i := 0$
  foreach $pdc \in PDC$ // check for each class in PDC if it is representative **do**
    $start := min - wspan + part \cdot i$; $end := start + wspan \cdot 2 + part$; $i := i + 1$
    if $d \geq start \wedge d \leq end$ **then**
      | $DC := DC \cup \{pdc\}$
  return DC // set of representative duration classes for event e

Algorithm 3. Determines for a log L the duration class for an event e.

by defining the expected activity, i.e., $rp.ra := e.ea$ and selecting one or more duration class which are expected to be observed, e.g. $rp.rd := \{\texttt{low}\} \subseteq PDC$.

Algorithm 3 determines the representative duration classes for an event e based on L. Hereby, variable $w \in [0;1]$ "widens" the covered timespan of each duration class so that the rule support calculation becomes less strict to prevent overfitting. Compare Fig. 2. It depicts the three duration classes of PDC and how widening affects them. For example, while the activity duration ① can clearly be represented by class $\texttt{low}$ this is not the case for the duration ②. As this duration is between the $\texttt{avg}$ and the $\texttt{high}$ class the "widening" (w) comes into effect, so that ② is represented by both classes. Accordingly, the two exemplary constraints $rp_1.rd = \{\texttt{avg}\}$ and $rp_2.rd = \{\texttt{high}\}$ would both match to ②.

Hence, when converting e_1, see the example in Table 1, into a temporal rule condition rp_1 it would be defined as $rp_1.ra = \texttt{A}$ while $rp_1.rd = \{\texttt{low}\}$ when using a widening factor of $w := 0.1$. Given this widening factor, the average class for activity $\texttt{A}$ would match durations between 3.9 and 4.1. In comparison event e_5 would convert into a condition rp_2 so that $rp_2.ra = \texttt{A}$ while $rp_2.rd = \{\texttt{high}\}$.

Fig. 2. Duration class representation, motivating example

Temporal ADAR Support. A trace t supports the temporal rule r (i.e., $rt(r) = \texttt{temporal}$) if t holds all activity executions specified by the rule conditions in r with the expected durations. In addition, these activity and duration pairs must occur in the expected order given by r's conditions. To verify this Algorithm 2, is extended by calculating and comparing duration classes, cf. Algorithm 4.

Mining SoD and BoD ADARs. Separation and Binding of Duty rules represent expected relative pairs of activities and resource assignments, cf. [7], i.e., all

Algorithm TempRuleSupport(*trace* t, *temporal rule* r, *duration classes* PDC, $w \in [0;1]$)

  Result: true if r is supported by t; false otherwise
  for $j = 1$ **to** $|r|$ // $|\langle \cdot \rangle|$ retains the length of the list $\langle \cdot \rangle$ **do**
    for $i = 1$ **to** $|t|$ **do**
      if $t_i.ea = r_j.ra \wedge ((TempClass(t_i, L, PDC, w) \cap t_i.rd) \neq \emptyset)$ **then**
        $t := t_i^+$; $r := r_j^+$; **break** // cf. Algorithm 2
  return $|r| = 0$?*true:false* // return true if r fully matches t else false

Algorithm 4. Checks if trace t supports the temporal rule r

activities covered by a SoD rule must be executed by different resources while all activities covered by a BoD rule must be executed by the same resource. Failing to support resource rules can be an indicator for fraudulent behaviour, cf. [3,6].

Event to Resource Condition. Converting an event e into a SoD or BoD rule condition rp is performed by extracting the activity related to e, i.e., $rp.ra := e.ea$. Accordingly, for e_1, in Table 1, $rp.ra = $ A holds.

Resource ADAR Support. To verify if a trace t is supporting a resource rule r, a set is generated that holds all resources $RS := \{e.r|e \in t \wedge e.ea \in \{rp.ra|rp \in r\}\}$ that have executed activities which are specified in r's conditions (cf., $rp.ra$). For a BoD rule it is expected that all executions utilize the same resource, i.e., $|RS| = 1$. For a SoD rule the amount of resources taking part in the activity executions should be equal to the amount of conditions, i.e., $|RS| = |r|$. Accordingly, a rule $r = \langle rp_1, rp_2 \rangle$ where $rp_1.ra = $ A, $rp_2.ra = $ B, and $rt(r) = $ SoD would only be supported by trace t_2 but not by t_1 (cf. Table 1). This is because for trace t_1 the set $RS = \{$Mike$\}$ (i.e., $|RS| = 1$) while $|r| = 2$ so that $|RS| \neq |r|$.

3.2 ADAR Based Anomaly Detection

The mined ADARs (i.e., a signature) are applied to classify a given trace $t' \notin L$ as anomalous or not. For this the artificial *likelihood* of t' is calculated and compared with the likelihood of the trace $t \in L$ that is most similar to t'. If t' is identified as less likely it is assumed as being anomalous, cf. Definition 6. Hereby, the presented approach follows and exploits the common assumption that anomalous behavior is less likely than benign behavior, cf. [4,6,8]. The artificial likelihood of a trace is determined by aggregating the overall support (based on L) of the rules which the trace is supporting, cf. Definition 4. This implies: the more rules a trace supports the more likely it and its occurrence is assumed to be.

Definition 6 (Anomaly detection). *Let L be an execution log and t' be an execution trace with $t' \notin L$. Let further R be a set of rules, i.e., signature, that was mined for L and let $\mathcal{R}$ be the set of all signatures. Anomaly detection function $adec : \mathcal{R} \times \mathcal{L} \mapsto \{true, false\}$ with*

$$adec(R,t) := \begin{cases} true & if \quad \sum_{\substack{r \in R \\ mp(r,t')=true}} sup(r,L) < \sum_{\substack{r \in R \\ mp(r,tsim(t',L))=true}} sup(r,L) \\ false & otherwise. \end{cases}$$

where $tsim(t', L)$ returns the trace $t \in L$ that is most similar to $t' \notin L$.

The proposed anomaly detection approach requires to identify, for a given trace $t' \notin L$, the most similar trace $t \in L$. For this function $tsim(t, L) : t$ is applied. In detail: both traces are first converted into bags of activities (each one holds activities executed by the respective trace). Subsequently, the Jaccard similarity $J(\{\cdots\}, \{\cdots\})$, cf. [19], between both activity bags is calculated. This means: the more equal activities[1] the traces contain (have executed) the more similar they are, i.e., $J(\{\text{A}, \text{C}\}, \{\text{B}, \text{C}\}) = |\{\text{A}, \text{C}\} \cap \{\text{B}, \text{C}\}| \, / \, |\{\text{A}, \text{C}\} \cup \{\text{B}, \text{C}\}| = 0.\overline{3}$. Overall the underlying similarity measure can be user chosen but this approach was found to be simple, fast, and sufficient during the evaluation, cf. Sect. 4.

Dealing with Change and Flexibility. The proposed approach applies the common assumption that benign behavior is *more likely* than anomalous behavior, cf. [4]. Nevertheless benign *noise* and *ad hoc changes* still occur in the signature mining data (i.e., L) but also in the traces that are analyzed for anomalies, cf. [4]. These kinds of behavior can, if the applied signature is too strict or overfitting, be misinterpreted as being anomalous and so result in *false positives*. Hence, the proposed approach applies three strategies to mitigate this risk:

Similarity: given traces are not compared with strict fixed signatures or thresholds. Instead the signature and the expected behavior is individually and automatically adapted for the trace that is analyzed by dynamically selecting a similar trace in L which is utilized as a source for comparable behavior.

Rule significance: the significance and impact of each rule is dynamically calculated during the anomaly detection phase. For this the rule support given by L (i.e., the percentage of traces in L that the rule supports) is utilized. Hence, each rule gets automatically assigned an individual significance.

Signature strictness: a trace must not match to all rules a signature R is composed of. Instead the applied approach aggregates the support of each rule so that a trace can "compensate" an unsupported rule by supporting other rules. Such a relaxed approach, in comparison to stricter existing work, cf., [1], provides a basis to deal with noise and ad hoc changes, cf. [5].

Fostering Root Cause Analysis and Understandability. One of the main driver for this work was the need for and lack of *root course analysis* capabilities in the process anomaly detection domain, cf. [4]. An insufficient support for root cause analysis can, for example, harden it to choose and apply *appropriate* countermeasures for identified anomalies. This aspect is tackled by:

Severity: existing process anomaly detection approaches frequently generate *binary* results, i.e., they mark traces either as anomalous or not, cf. [4]. However, this does not allow to assess the severity of an anomaly. Hence, we propose to utilize the aggregated rule support of a signature/traces (t vs. t') as an indicator for the deviation severity between a trace and a signature.

---

[1] Activity equivalence is considered as label equivalence here.

Granularity: binary detection results are also insufficient to perform a thorough anomaly analysis as they do not indicate which specific parts of a trace did not comply to the utilized signatures. In comparison, the proposed approach comprises the signature from multiple fine granular rules which can be individually reported as supported or not; enabling to fine granularly report which parts of a given trace were affected by an identified anomaly.

Simplicity and clarity: during the evaluation the mined signatures were found to contain a relative low amount of rules (e.g., below 100 temporal and control rules). Given the low amount of short and simple rules those can, likely, easily be grasped and taken into account by experts when analyzing process executions (traces resp.) which were found to be anomalous.

For root cause analysis, first of all, the proposed approach estimates the severity of each anomaly, which enables to quickly rank anomalies as less or more crucial. Secondly, it reports which rules are supported or not. This enables to quickly grasp the differences between expected and observed behaviour. Hereby, it can, for example, become evident that a combined resource and control rule violation origins from an inexperienced employee which can be contacted to explain and sort out the situation (countermeasure selection). Alternatively, it could become evident that a large amount of rules are not supported and that the execution must be terminated, as a countermeasure, to prevent further harm.

4   Evaluation

The evaluation utilizes *real life* process execution logs from multiple domains and artificially injected anomalies in order to assess the anomaly detection performance and feasibility of the proposed approach. It was necessary to inject artificial anomalies as information about real anomalies are not provided by today's process execution log sources. The utilized logs were taken from the BPI Challenge 2015[2]. (BPIC) and Higher Eduction Processes (HEP), cf. [20].

The BPIC logs hold 262,628 execution events, 5,649 instances, and 398 activities. The logs cover the processing of building permit applications at five (BIPC_1 to BIPC_5) Dutch building authorities between 2010 and 2015. The HEP logs contain 28,129 events, 354 instances, and 147 activities – recorded from 2008 to 2011 (i.e., three academic years $\mapsto$ HEP_1 to HEP_3). Each trace holds the interactions of a student with an e-learning platform (e.g., exercise uploads). All logs contain sufficient details to apply the proposed approach (e.g., execution events, activities, timestamps, resource assignments, etc.).

The logs were evenly randomly separated into training (for signature generation) and test data (for the anomaly detection performance evaluation). Subsequently, 50% randomly chosen test data entries were randomly mutated to inject artificial anomalies. By randomly choosing which, how many, and how frequently mutators are applied on a single chosen test data entry (trace, resp.) this

---

[2] http://www.win.tue.nl/bpi/2015/challenge – DOI: https://doi.org/10.4121/uuid:
31a308ef-c844-48da-948c-305d167a0ec1.

work mimics that real life anomalies are diverse and occur in different strengths and forms. Further the application of mutators enables to generate labeled non-anomalous (i.e., non-mutated) and anomalous (i.e., mutated) test data entries. Hereby, it becomes possible to determine if both behavior "types" are correctly differentiated by the proposed approach (cross validation). The applied mutators inject random control flow, temporal, and resource anomalies:

(*a*) *Control Flow* – two mutators which randomly mutate the order and occurrence of activity execution events; and (*b*) *Temporal* – randomly chosen activity executions get assigned new artificial execution durations; and (*c*) *Resource* – activity/resource assignments are mutated to mimic, for example, BoD anomalies.

The applied mutators were adapted from our work in [5,6]. Combining multiple mutators enables to represent the diversity of real life anomalies. In addition, the applied random training and test data separation also evaluates if the proposed approach is capable of dealing with *benign* noise and ad hoc changes by not identifying them as anomalous. This is, because the test data contains benign behavior which is not given by the training data (e.g., bening ad hoc changes). The following results are an average of 100 evaluation runs. This enables to even out random aspects, such as, the random data separation and trace mutation.

Metrics and Evaluation. Here, the feasibility of the presented approach is analyzed. For this, a cross validation is performed to determine if known anomalous (mutated) execution traces are correctly differentiated from known non-anomalous (non-mutated) ones. Through this four performance indicators are collected: True Positive (TP) and True Negative (TN), i.e., that anomalous (TP) and non-anomalous (TN) traces are correctly identified. False Positive (FP) and False Negative (FN), i.e., that traces were incorrectly identified as anomalous (FP) or non-anomalous (FN). Finally, these indicators are aggregated into:

(*a*) *Precision* $P = TP/(TP+FP)$ – if identified anomalous traces were in fact anomalous; and (*b*) *Recall* $R = TP/(TP+FN)$ – if all anomalous traces were identified (e.g., overly generic signatures could result in overlooking anomalies); and (*c*) *Accuracy* $A = (TP+TN)/(TP+TN+FP+FN)$ – a general anomaly detection performance impression; $TP, TN, FP, FN \in \mathbb{N}$; $P, R, A \in [0;1]$.

An optimal result would require that TP and TN are high while FP and FN are low so that the accuracy becomes close to one. Further, the F_β-measure, Eq. 2, provides a configurable harmonic mean between P and R, cf. [9]. Hence, $\beta < 1$ results in a precision oriented result while $\beta = 1$ generates a balanced result.

$$F_\beta = \frac{(\beta^2 + 1) \cdot P \cdot R}{\beta^2 \cdot P + R} \tag{2}$$

Results. The results were generated based on the BPIC 2015 and HEP process execution logs and following proof of concept implementation: https://github.com/KristofGit/ADAR. The implementation was found to be calculating a signature within minutes and required only seconds to classify a trace as anomalous or not. Once generated the signatures can be reused and easily adapted by adding new rules or removing old ones, e.g., to address concept drift. To ensure that for both data sources roughly the same amount of traces is analyzed only traces which did take place during 2015 were used from the BPIC 2015 logs.

Primary tests were applied to identify appropriate configuration values, e.g., the maximum rule length $rl := 3$ (control and temporal) and $rl := 2$ (resource). Longer rules can result in stricter potentially overfitting signatures. This is because longer rules contain more details and thus provide less flexibility than shorter ones. In comparison, the minimum support a rule has to achieve $mins$ during the mining phase, to be accepted as a part of the signature, was set to 0.9/0.8 for control/temporal rules. For this variable it was found that higher values could potentially result in a very small rule set or in finding no rules at all. In comparison, using a lower value could result in finding a very high amount of rules. This does not necessarily result in better anomaly detection results as it increases, as we found, the risk of generating overfitting signatures.

Finally, the fuzzy temporal rule generation can be configured based on the chosen temporal class widening variable w which was set to 0.2. Lowering this value would result in stricter signatures (temporal rules, resp.) that could potentially struggle when dealing with noise and ad hoc changes while a higher value would result in potentially overlooking anomalies as the signatures become less strict. Given the low amount of configuration variables we assume that existing automatic optimization algorithms should, likely, be able automatically find optimal settings for the proposed approach based on given training data.

The average evaluation results are shown in Table 2. Overall, an average accuracy of 81% was achieved along with an average precision of 77% and an average recall of 89%. Given these results we conclude that the proposed approach is feasible to identify the injected anomalies in the analyzed process execution data. Moreover, it becomes visible that the detection of diverting anomalous behavior becomes harder the more diverse and complex the benign behavior is (e.g., because of noise or ad hoc changes). Accordingly the anomaly detection performance of the BPIC 2015 logs are lower than the results for the HEP logs. Nevertheless, an average anomaly detection accuracy of 75% was achieved for the more challenging BPIC 2015 process execution log data.

Table 2. Anomaly detection performance of the presented approach

	HEP_1	HEP_2	HEP_3	BPIC5_1	BPIC5_2	BPIC5_3	BPIC5_4	BPIC5_5
Precision	0.86	0.87	0.85	0.73	0.70	0.78	0.69	0.69
Recall	0.98	0.98	0.97	0.90	0.85	0.75	0.87	0.84
Accuracy	0.91	0.91	0.90	0.78	0.74	0.77	0.73	0.73
$F_{0.5}$-measure	0.88	0.88	0.87	0.76	0.73	0.77	0.72	0.72
F_1-measure	0.92	0.92	0.91	0.80	0.77	0.77	0.77	0.76

5   Related Work

The most comparable work [18] also applies association rules for anomaly detection in processes. However, rules are largely manually generated (e.g., a user defines the expected maximum activity duration) and order dependencies between activities are not verified. Our survey on anomaly detection in business processes [4] reveals several shortcomings in most existing work: (*a*) only single process perspectives are supported; and (*b*) the analysis of anomalies is not fostered/supported; and (*c*) monolithic signatures are frequently utilized – which are hard to grasp, struggle with noise and ad hoc changes, but also cannot be partially updated whenever the underlying process changes. Likely, these limitations hinder the application of process anomaly detection in the real world.

Further, compliance checking approaches are related, cf. [14,15]. These approaches utilize, e.g., rule based, definitions of expected process behavior to analyze its definition and execution for deviations and, partially, deviation root causes, cf. [14,16]. However, such work typically does not take noise and ad hoc changes into account, possibly resulting in false positives. Further the expected behavior definitions are typically assumed as given as their creation can result in extensive (manual) efforts that require in depth domain and process knowledge.

In the security domain, anomaly detection and root cause analysis are major research areas. However, existing approaches are too specialized to be applied to process data, cf. [5,12], because they focus on single unique use cases and data formats, such as, specific network protocols, e.g., the Session Initiation Protocol, cf. [17]. These approaches can hardly be generalized and applied to process execution logs which hold different data, formats, and contextual attributes.

One could argue that instead of applying anomaly detection the process definition could be secured by applying security focused modeling notations, cf. [7]. In real world scenarios, this would require to be aware of all potential sources for security incidents during the design phase and to constantly update the processes to meet novel security challenges. In comparison the proposed anomaly detection approach is self learning and can also deal with process changes automatically.

6   Discussion and Outlook

This paper focuses on two main challenges (*a*) the detection of anomalies in process executions; and (*b*) taking a first step towards supporting the analysis of detected anomalies. We conclude that this paper was able to meet the first challenge as the conducted evaluation showed an average anomaly detection recall of 89%. This goes with a substantial simplification of the generated signatures compared to previous work in [6] (complex likelihood graphs vs. short rules) which fosters the understandability of the signatures and identified anomalies.

As the identified anomalies (and related traces) can be complex and hard to understand we argue that anomaly detection approaches should support experts when analyzing anomalies along with the related alarms. For this experts would, as we assume, require, inter alia, information about (*a*) which part of an execution trace is affected by an anomaly; and (*b*) the anomaly severity, cf. [13].

It is shown how such information can be provided by the proposed rule based anomaly detection approach. To our knowledge this is the first *process anomaly detection* approach that does so. We assume that it is necessary to identify but also to understand anomalies to choose appropriate anomaly countermeasures.

Future work will concentrate on expanding and evaluating the proposed approaches' root cause analysis capabilities. For this, visualization and management tools will be created that enable to handle the provided information (e.g., which rules are supported or not) in an interactive manner. Further, user studies will be performed to assess the benefits of the provided information in detail.

References

1. Van der Aalst, W.M., de Medeiros, A.K.A.: Process mining and security: detecting anomalous process executions and checking process conformance. Theor. Comput. Sci. **121**, 3–21 (2005)
2. Agrawal, R., Srikant, R., et al.: Fast algorithms for mining association rules. In: Very Large Data Bases, vol. 1215, pp. 487–499 (1994)
3. Bezerra, F., et al.: Anomaly detection using process mining. Enterp. Bus. Process Inf. Syst. Model. **29**, 149–161 (2009)
4. Böhmer, K., Rinderle-Ma, S.: Anomaly detection in business process runtime behavior-challenges and limitations. arXiv arXiv:1705.06659 (2017)
5. Böhmer, K., Rinderle-Ma, S.: Multi instance anomaly detection in business process executions. In: Carmona, J., Engels, G., Kumar, A. (eds.) BPM 2017. LNCS, vol. 10445, pp. 77–93. Springer, Cham (2017). https://doi.org/10.1007/978-3-319-65000-5_5
6. Böhmer, K., et al.: Multi-perspective anomaly detection in business process execution events. In: Debruyne, C., et al. (eds.) Cooperative Information Systems, pp. 80–98. Springer, Heidelberg (2016). https://doi.org/10.1007/978-3-319-48472-3_5
7. Brucker, A.D., Hang, I., Lückemeyer, G., Ruparel, R.: SecureBPMN: modeling and enforcing access control requirements in business processes. In: Access Control Models and Technologies, pp. 123–126. ACM (2012)
8. Chandola, V., Banerjee, A., Kumar, V.: Anomaly detection: a survey. Comput. Surv. **41**(3), 15 (2009)
9. Chinchor, N., Sundheim, B.: MUC-5 evaluation metrics. In: Message Understanding, pp. 69–78. Computational Linguistics (1993)
10. Czepa, C., et al.: Plausibility checking of formal business process specifications in linear temporal logic, pp. 1–8 (2016)
11. Greco, G., Guzzo, A., Pontieri, L.: Mining taxonomies of process models. Data Knowl. Eng. **67**(1), 74–102 (2008)
12. Gupta, M., Gao, J., Aggarwal, C.C., Han, J.: Outlier detection for temporal data: a survey. Knowl. Data Eng. **26**(9), 2250–2267 (2014)
13. Julisch, K.: Clustering intrusion detection alarms to support root cause analysis. Inf. Syst. Secur. **6**(4), 443–471 (2003)
14. Ly, L.T., Maggi, F.M., Montali, M., Rinderle-Ma, S., van der Aalst, W.M.: Compliance monitoring in business processes: functionalities, application, and toolsupport. Inf. Syst. **54**, 209–234 (2015)

15. Ly, L.T., Rinderle-Ma, S., Knuplesch, D., Dadam, P.: Monitoring business process compliance using compliance rule graphs. In: Meersman, R., et al. (eds.) OTM 2011. LNCS, vol. 7044, pp. 82–99. Springer, Heidelberg (2011). https://doi.org/10. 1007/978-3-642-25109-2_7
16. Ramezani, E., Fahland, D., van der Aalst, W.M.P.: Where did i misbehave? Diagnostic information in compliance checking. In: Barros, A., Gal, A., Kindler, E. (eds.) BPM 2012. LNCS, vol. 7481, pp. 262–278. Springer, Heidelberg (2012). https://doi.org/10.1007/978-3-642-32885-5_21
17. Rieck, K., Wahl, S., Laskov, P., Domschitz, P., Müller, K.-R.: A self-learning system for detection of anomalous SIP messages. In: Schulzrinne, H., State, R., Niccolini, S. (eds.) IPTComm 2008. LNCS, vol. 5310, pp. 90–106. Springer, Heidelberg (2008). https://doi.org/10.1007/978-3-540-89054-6_5
18. Sarno, R., et al.: Hybrid association rule learning and process mining for fraud detection. Comput. Sci. **42**(2), 59–72 (2015)
19. Tan, P.N., Steinbach, M., Kumar, V.: Introduction to Data Mining, 1st edn. Addison-Wesley, Boston (2005)
20. Vogelgesang, T., et al.: Multidimensional process mining: questions, requirements, and limitations. In: CAISE Forum, pp. 169–176. Springer, Heidelberg (2016)
21. Wieringa, R.: Design Science Methodology for Information Systems and Software Engineering. Springer, Heidelberg (2014). https://doi.org/10.1007/978-3-662-43839-8

AB Testing for Process Versions
with Contextual Multi-armed
Bandit Algorithms

Suhrid Satyal[1,2](✉), Ingo Weber[1,2], Hye-young Paik[1,2], Claudio Di Ciccio[3],
and Jan Mendling[3]

[1] Data61, CSIRO, Sydney, Australia
{suhrid.satyal,ingo.weber}@data61.csiro.au
[2] University of New South Wales, Sydney, Australia
hpaik@cse.unsw.edu.au
[3] Vienna University of Economics and Business, Vienna, Austria
{claudio.di.ciccio,jan.mendling}@wu.ac.at

Abstract. Business process improvement ideas can be validated
through sequential experiment techniques like AB Testing. Such
approaches have the inherent risk of exposing customers to an inferior
process version, which is why the inferior version should be discarded as
quickly as possible. In this paper, we propose a contextual multi-armed
bandit algorithm that can observe the performance of process versions
and dynamically adjust the routing policy so that the customers are
directed to the version that can best serve them. Our algorithm learns
the best routing policy in the presence of complications such as multiple
process performance indicators, delays in indicator observation, incomplete or partial observations, and contextual factors. We also propose a
pluggable architecture that supports such routing algorithms. We evaluate our approach with a case study. Furthermore, we demonstrate that
our approach identifies the best routing policy given the process performance and that it scales horizontally.

Keywords: Multi-armed bandit · Business Process Management
AB Testing · Process Performance Indicators

1 Introduction

Business improvement ideas often do *not* lead to actual improvements [9,10].
Contemporary Business Process Management Systems (BPMSs) enable quick
deployment of new process ideas, but they do not offer support for validating
the improvement assumptions existent in the new version. Support for validating
such assumptions during process redesign is also limited.

The AB testing approach from DevOps can be adopted in Business Processes
Management to provide fair validation support. A new process version can be
deployed alongside the older version on the same process engine such that these

© Springer International Publishing AG, part of Springer Nature 2018
J. Krogstie and H. A. Reijers (Eds.): CAiSE 2018, LNCS 10816, pp. 19–34, 2018.
https://doi.org/10.1007/978-3-319-91563-0_2

versions (A and B) are operational in parallel. User requests can be routed to either of these versions using various instance routing algorithms. Based on the performance of each version, the routing configuration can be dynamically adjusted to ultimately find the best performing version. This general idea has been introduced in AB-BPM [14]. However, the routing algorithm proposed in AB-BPM does not address scenarios where the process performance is measured through multiple Process Performance Indicators (PPIs) which may be available at different times. It also does not provide support for evaluating processes for which some of these PPIs may never be available, and processes that are affected by external factors (e.g. the weather condition).

In this paper, we address these shortcomings by revising the routing mechanism of AB-BPM. We propose a pluggable instance router architecture that allows routing algorithms to asynchronously collect and evaluate PPIs. We also propose a routing algorithm, *ProcessBandit*, that can be plugged into the instance router. ProcessBandit finds a good routing policy in the presence of delays and incompleteness in observing PPIs, and also when true performance depends on contextual external conditions. We show that our approach identifies the best routing policy given the performance, and that it scales horizontally. We also demonstrate the overall approach using a synthetic case study.

The remainder of the paper starts with a discussion on the background, key requirements, and related work in Sect. 2. Section 3 describes the architecture of the instance router and the details of ProcessBandit algorithm. In Sect. 4, we analyze the behaviour of the algorithm, and study a use case. Section 5 discusses our approach and draws conclusions.

2   Background

2.1   Problem Description and Requirements

Business process improvement efforts are often analysed by measuring four performance dimensions: time, cost, flexibility, and quality. Improvement decisions have to reflect trade-offs between these dimensions [7,13]. In many cases, shortcomings in one dimension may not be compensated by improvements in other dimensions. For example, a low user satisfaction cannot be compensated with faster performance. In addition, the relationships between these dimensions may not be intuitive. This is illustrated by an anecdote of a leading European bank. The bank improved their loan approval process by cutting turnaround time down from one week to few hours. However, this resulted in a steep decline in customer satisfaction: customers with a negative notice would complain that their application might have been declined unjustifiably; customers with a positive notice would inquire whether their application had been checked with due diligence.

This anecdote shows that customer preferences are difficult to anticipate before deployment and that there is a need to carefully test improvement hypotheses in practice. Up until now, only the AB-BPM approach [14] supports the idea of using principles of AB testing to address these problems. As an early proposal, AB-BPM has a number of limitations regarding utilization of PPIs.

Table 1. Requirements of AB testing system and our approach

	Requirement	Approach
R1	Performance of a process execution is determined by multiple PPIs	Reward design encapsulating all PPIs
R2	Individual PPIs are available at different times	Asynchronously fetch PPIs and update rewards
R3	Most process instances do not provide all PPIs	Maintain ratio of complete and incomplete rewards
R4	Performance of a process instance is affected by *contextual factors*	Identify and integrate contextual factors in the algorithm

First of all, there may be multiple process performance indicators involved in determining a better version. For instance, both user satisfaction score and process execution time are acceptable PPIs for a process instance. Second, not all of the required Process Performance Indicators (PPIs) may be observable at the same time. A PPI such as the user satisfaction score is obtained at different times with delays of varying length. This means that the evaluation mechanism should support a PPI to be collected and aggregated asynchronously, i.e., at different points in time. Finally, some process instances will not produce all of the PPIs. It is likely, for example, that the number of users who do not respond to requests for providing satisfaction scores will outnumber those who do. Therefore, we should also be able to handle the missing or incomplete PPI observations.

Another aspect to consider is the effect of *contextual factors*. The performance of a process can be influenced by factors such as resource constraints, the environment, and market fluctuation. One example of influence of weather has been observed in "teleclaims" process of an insurance company [1]. The call centers of this company receive an incoming call volume of 9,000 per week. However, during a storm season, the volume can reach up to 20,000 calls per week. In order to manage this influx, the managers manually escalate the cases to maintain quality and meet deadlines. Identifying and acting on such contextual factors is crucial to find the best process version.

From the above analysis, we derive four key requirements and propose approaches outlined in Table 1. To address these requirements, we have implemented a two-pronged solution: a pluggable architecture for instance routing, and a routing algorithm that asynchronously learns about process performance.

2.2 Related Work

AB testing is a commonly used approach for performing randomized sequential experiments. This approach is widely used to test micro changes in web applications [6,10,11]. In applying AB testing to business process versions, performing randomized experiments can inadvertently introduce risks such as loss of revenue. Risks in this context are higher than that for standard web applications, where the changes and the effects are small (e.g. the placement of buttons). Therefore, user requests should be distributed according to performance of process versions.

We introduced the idea of AB testing for business process versions in AB-BPM [14], where we modeled this routing challenge as a *contextual multi-armed bandit problem* [2,4,12]. We proposed LtAvgR, which is based on LinUCB [5,12] – a well-known contextual multi-armed bandit algorithm. LtAvgR dynamically adjusts how user requests are routed to each version by observing numerical rewards derived from process performance. LtAvgR defines an experimentation phase where observing rewards is emphasized over optimal routing, and a post-experimentation phase where the best routing policy is selected based on the observed rewards. LtAvgR updates its learning by averaging historical rewards, which enables it to support long-running processes. However, LtAvgR can only handle scenarios where all PPIs are available at the same time. In this paper, we propose *ProcessBandit*, an algorithm that addresses this limitation.

Multi-armed bandit algorithms have been adopted for various kinds for experiment designs [4]. However, work on the effect of feedback delay and impact of partial rewards is not well studied. Furthermore, the effect of sparseness of some rewards, such as those with user satisfaction scores, have not been considered for multi-armed bandits. Temporal-difference (TD) learning can be used to converge towards the best routing configuration in presence of delayed rewards [17, Chap. 6]. Silver et al. [16] propose an asynchronous concurrent TD learning approach for maximizing metrics such as customer satisfaction by learning from partial customer interactions. This approach can be used in sequential scenarios like marketing campaigns where interactions can affect customer state. We propose a simpler multi-armed bandit algorithm that handles asynchronous learning with partial rewards, adapted for scenarios where only one interaction (processes instantiation) needs to be observed.

Approaches for prediction based on imbalanced data include techniques such as oversampling and undersampling [8, Chap. 2]. Such techniques make assumptions about what balanced data should look like, and do not introduce any new knowledge. In our scenario, imbalance occurs when only a subset of all PPIs are observed. Since AB testing aims to remove implicit assumptions, we avoid sampling techniques. Instead, we ensure that the routing algorithm learns mostly through observations that have all PPIs.

3   Solution

Our solution consists of two parts. First, we propose a pluggable and scalable architecture that facilitates a routing algorithm to learn the best routing policy even when PPIs are missing, delayed, or incomplete, and when the performance is affected by contextual factors. Second, we propose a routing algorithm named *ProcessBandit* that learns routing policies by utilizing this architecture.

3.1   The Instance Router

The instance router is a modular system composed of an Asynchronous Task
Queue, the Controller, the Routing Algorithm, the Context Module, the Tasks
Module, and the Rewards Module. Figure 1 shows the architecture of the system.

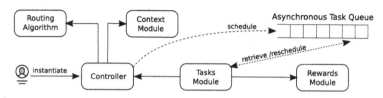

Fig. 1. The architecture of instance router

Algorithm 1. ProcessBandit Instance Routing

Input: $\alpha \in \mathbb{R}_+$, $\lambda \in \mathbb{R}$, $M \in \mathbb{N}$
    // α is the LinUCB's tuning parameter; λ and M are experimentation decay and length
Output: arm id

1   $I \leftarrow$ empty set   // set of contextual factors
2   **for** $t = 1, 2, 3, ..., T$ **do**   // t is the request count
3   $\quad$ Observe features of all arms $a \in A_t : x_{t,a} \in \mathbb{R}^d$
4   $\quad$ constructFeatureVector(I)   // Algorithm 3
5   $\quad$ **for** $a \in A_t$ **do**
6   $\quad\quad$ **if** a *is new* **then**
7   $\quad\quad\quad$ $A_a \leftarrow I_d$, $b_a \leftarrow 0_d$ // identity and zero matrices of dimension $d \times d$, resp.
8   $\quad\quad$ $\hat{\theta}_a \leftarrow A_a^{-1} b_a$
9   $\quad\quad$ $p_{t,a} \leftarrow \hat{\theta}_a^{\mathsf{T}} x_{t,a} + \alpha \sqrt{x_{t,a}^{\mathsf{T}} A_a^{-1} x_{t,a}}$
10 $\quad$ arm $a_{\text{linucb}} = \underset{a \in A_t}{\arg\max}\, p_{t,a}$ with ties broken arbitrarily
11 $\quad$ **if** $t \leqslant M$ **then**   // experimentation phase
12 $\quad\quad$ $pr_{\exp} \leftarrow$ sample y from $Exp(\lambda)$ s.t. $x = t$
13 $\quad\quad$ Choose arm $a_t = a_{\text{linucb}}$ or $a_{\text{alternate}}$ with probability $pr_{\exp}$
14 $\quad\quad$ Schedule update task
15 $\quad$ **else**
16 $\quad\quad$ Choose arm $a_t = a_{\text{linucb}}$

The instance router assigns an instance of the deployed processes, version A
or B, to an incoming request. Upon receiving a request, the Controller invokes
the Context Module to extract contextual information from the request and
construct a feature vector. If required by the Routing Algorithm, the Context
Module captures and stores *hypothesized* contextual factors associated with each
request. These hypothesized contextual factors are set at the start of AB tests.
If a contextual factors is confirmed through the analysis of the stored values, the
Context Module integrates the contextual factor in the feature vector. Using this
feature vector, the Controller invokes the Routing Algorithm, which instantiates
a process and returns an identifier. This process instance identifier is used by
the Controller to schedule an update task on the Asynchronous Task Queue.

An update task asynchronously polls the BPMS for PPIs of the instantiated process, and calculates a numerical reward using the PPIs. When only a subset of the desired PPIs are available, update tasks are re-scheduled by the Tasks Module so that that the missing PPIs can be collected and evaluated at a later point in time. In such scenarios, temporary rewards are calculated using the available PPIs. Reward calculation is delegated to the Rewards Modules. The Routing Algorithm can learn a routing policy through these numerical rewards.

3.2   ProcessBandit Algorithm

We propose *ProcessBandit*, a routing algorithm that can be plugged into the architecture. The algorithm asynchronously observes PPIs associated with a particular request, distributes requests to process versions, observes process performance, and learns the best routing policy given the process performance.

The pseudo code for sampling a process version (or "arm" in multi-armed bandit terminology) to test its performance is shown in Algorithm 1. The algorithm maintains an average of complete, incomplete, and overall rewards for each d-dimensional context in relevant matrices, indicated as b. These values are updated asynchronously according to the performance of each process instance.

The algorithm consists of experimentation (P1) and post-experimentation (P2) phases. When contextual factor detection is enabled, the experimentation phase is further divided into pre-contextual factor (P1A) and post-contextual factor (P1B) phases. The phases are configured using an exponential decay function $exp(\lambda)$, experimentation request threshold M, and pre-contextual factor reward threshold. Request count is incremented on process instantiation. Reward count is incremented when a reward calculated using all PPIs is received.

The algorithm uses an approach similar to LinUCB [12] to select a candidate arm a_{linucb} such that the expected reward is maximized. When the algorithm is in experimentation phase, it either chooses a_{linucb} or the alternate arm based on the probability sampled from the exponential decay function. Asynchronous reward updates are scheduled for all decisions made in the experimentation phase.

Asynchronous Reward Update. We define an ideal PPI vector p_{ideal} as the vector that represents the best possible values for each PPI. We also introduce a reference vector p_{ref}, which defines values that can be used as a substitute for missing PPIs. In AB testing scenarios, historical data of one process version is available. This can be used to inform the choice of p_{ref}. Finally, we define the effective vector p_{eff} as the vector that contains all PPIs used to evaluate a reward. If not all PPIs are available at the time of observing a completed process instance, an effective vector p_{eff} is constructed using components of reference vector p_{ref} instead of the unavailable PPIs. If/when these PPIs are made available, an update is applied by removing the effect of previous p_{eff}, and then using the new effective vector. This helps us address requirements *R1* and *R2*.

Algorithm 2. Asynchronous Update

Input: τ, x_{t,a_t}, A_{a_t}, b_{a_t} // τ: ratio of (in)complete rewards, others as in Algorithm 1

1 $A_{a_t} \leftarrow A_{a_t} + x_{t,a_t}\, x_{t,a_t}^t$
2 Construct p_{eff} and derive reward
3 **begin with** context x_{t,a_t}
4       $r_{\text{incomplete}} \leftarrow$ avg. incomplete reward, $r_{\text{complete}} \leftarrow$ avg. complete reward
5       $r_{\text{avg}} \leftarrow$ average overall reward
6   **if** old reward **then** update $r_{\text{incomplete}}$, and r_{complete}
7   **if** new reward **then**
8   **if** reward ratio $\geqslant \tau$ **or** bootstrap period **then**
9   **if** all PPIs seen **then** update r_{complete}, increment r_{complete} count
10   **else** update $r_{\text{incomplete}}$, increment $r_{\text{incomplete}}$ count
11   **else**
12   **if** all PPIs seen **then** update r_{complete}, increment r_{complete} count
13   **else** update $r_{\text{incomplete}}$ as moving average
14   update r_{avg}
15 Update b_{a_t} such that r_{avg} represents x_{t,a_t}

Using p_{eff}, rewards can be calculated through a *point-based* or *classification based approach*, as illustrated in Fig. 2. In the point-based method, the Rewards Module constructs p_{eff}, applies weights to PPIs (if any), and then normalizes all components of p_{eff} and p_{ideal}. After the normalization, it calculates the effective reward as the euclidean distance between p_{eff} and p_{ideal}. Therefore, the objective of the algorithm is to choose versions that produce shorter distances between the effective vector and the ideal vector.

The point-based approach is intuitive and easy to implement. However, it makes the implicit assumption that a decrease in one PPI can be compensated by an increase in another PPI [3, Chap. 2]. In many real-world scenarios, this may not be the case. For example, while the increase in costs may be compensated with better processing times, lower user satisfaction may not be compensated with any other metric. In addition, granular and insignificant differences in distance can accumulate and produce an effect on routing. In such scenarios, a better approach is to classify performance into categories aligned with business goals.

Therefore, in the classification-based approach domain experts design reward classes and assign weights to each class. Weights represent the relative importance of each class. p_{eff} is constructed as above and a reward is assigned as the weight of the class it falls on. The most important class C_i has the highest weight w_i. As depicted in Fig. 2, the most important class C_1 has the highest weight w_1, C_2 has a lower weight w_2, and so on. The objective of the algorithm is to choose versions that produce the highest average weight.

The algorithm specification is independent of how reward values are derived from
PPIs. Without loss of generality we typically choose rewards on a negative scale (e.g.
$w_1 \mapsto -1, \ldots, w_5 \mapsto -5$). The learning rate
and convergence are, however, dependent on
the quantity of the reward. It is possible for
the algorithm to be misled by a large quantity
of rewards derived from partially observed
metrics. If only a small percentage of process instances provide information about all
PPIs, the effect of rewards derived from these
process instances can be diluted by rewards

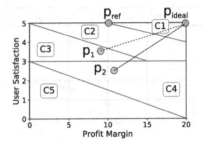

Fig. 2. Reward design approaches.
Rewards can be categorical or distance based.

derived using incomplete PPI observations from other process instances. To
ensure that such dilution does not occur, the algorithm keeps track of the *ratio of
complete and incomplete rewards*, τ, for each version in each context. To accommodate τ, Algorithm 2 starts in bootstrap mode for the first few requests. During
bootstrap, rewards are collected regardless of τ. The algorithm accepts a partial
reward either at the bootstrap period when the number of complete rewards is
below a certain threshold, or when the reward ratio is less than or equal to τ.
The usage of reward ratios in this manner addresses requirement *R3*.

Contextual Factor Detection and Context Integration. Algorithm 3
shows our solution to requirement *R4* – the context integration mechanism.
The algorithm starts in the pre-contextual factor phase (P1A). In this phase,
contextual feature vectors are constructed using information available with the
user requests (e.g., age group). Hypothesized contextual factors are observed
and stored by the controller for future analysis. When the pre-contextual factor
reward threshold is reached, the correlation between the hypothesized contextual factors and process performance is analysed. If the correlation is above
a pre-determined threshold, the algorithm state is reset to accommodate new
contextual information. This marks the beginning of the post-contextual factor
experimentation phase (P1B). From this point onward, contextual feature vectors are constructed using the information from user requests and the observed
values of the contextual factors. Finally, when the experimentation request count
is achieved, the algorithm switches to the post-experimentation phase (P2). In
this phase, the algorithm stops learning from new requests. However, to account
for long delays between process instantiation and reward observation, the algorithm continues learning from the requests made in phase P1B.

In summary, we address requirements R1 and R2 through asynchronous
partial reward updates using effective and ideal vectors. Requirement R3 is
addressed by maintaining a user defined reward ratio τ between complete and
incomplete rewards, and handling updates accordingly. Finally, R4 is addressed
by identifying and integrating contextual features in contextual feature vectors.

Algorithm 3. Contextual Factor Detection

```
1  def constructFeatureVector(I)                    // I: set of contextual factors
2      if reward count < pre-contextual factor reward count then
3          collect data points for hypothesized contextual factors

4      if reward count = pre-contextual factor reward count then
5          test hypothesized contextual factor using pearson correlation
6          if contextual factors are found then
7              update set I
8              reinitialize all variables in Algorithm 1

9      collect data points for set I
10     construct and return feature vector
```

---

4 Evaluation

In this section, we analyse the behaviour of the approach and specifically the
ProcessBandit algorithm in the presence of contextual factors, and in scenarios
where an important PPI is available only for a small number of requests. We
also evaluate the response times of the algorithm under various infrastructure
settings. Finally, we demonstrate the approach using an example process.

The instance router is prototyped using Python and served by Nginx HTTP
server[1] and uWSGI application server[2]. We use Redis[3] as the asynchronous task
queue and data store. Two worker processes operate on the asynchronous queue.
Tasks that require rescheduling are scheduled after 1 s.

4.1 Convergence Characteristics

In the following experiments, we study how ProcessBandit routes requests to
process versions and whether the AB tests converge to the best routing policy given the rewards. We consider two *baselines*: a naïve randomized routing
algorithm with uniform request distribution, *random-udr*, and LTAvgR [14].

Our experiment setup consists of a simulated BPMS which returns two PPIs,
user satisfaction and profit margin, for two process versions. We assume two process versions, A and B, which perform differently based on the context, X and Y,

Table 2. PPI configuration.

Context	Contextual factor $f = 1$				Contextual factor $f = 2$			
	Profit margin		User satisfaction		Profit margin		User satisfaction	
	Version A	Version B	Version A	Version B	Version A	Version B	Version A	Version B
X	9	11	3	2.5	11	9	2.5	3
Y	11	9	2.5	3	9	11	3	2.5

---

[1] https://nginx.org/ Accessed 15-06-2017.
[2] https://uwsgi-docs.readthedocs.io/en/latest/ Accessed 15-06-2017.
[3] https://redis.io/ Accessed 15-06-2017.

and an contextual factor f. Table 2 summarizes the PPIs returned by each version under various conditions. These PPIs are mapped to the reward design models shown in Fig. 2. p_{ideal} represents user satisfaction score of 5 and profit margin of 20%. p_{ref} represents user satisfaction score of 5 and profit margin of 10%. We chose an optimistic reference point with the philosophy that users who do not provide satisfaction scores are happy, and that profit margin is good. Rewards are derived using the classification model in Fig. 2 with weight mapping of $\{C_1 \mapsto w_1, C_2 \mapsto w_2, \ldots, C_5 \mapsto w_5\}$ such that $w_1 = -1, w_2 = -2, \ldots, w_5 = -5$.

We define the following key terms that we use in the experiments below:

t_{ppi1}: the time between request invocation and observation of the first PPI,
t_{obs}: the time between request invocation and observation of the full reward,
d: delay between the first and the second PPI such that $t_{obs} = t_{ppi1} + d$,
ρ: ratio between the average request inter-arrival rate and t_{obs}.

Convergence is shown by evaluating *regret* over time. Regret is defined as the difference between the sum of rewards associated with the optimal solution and the sum of rewards collected by pulling the chosen arm [19]. The objective of our algorithm is to find a configuration where the average regret of future actions tends to zero. Graphically, this is the case when the cumulative regret curve tends to become parallel to the x-axis. Given the initial uncertainty about the performance of the versions, the algorithm needs to start with experimentation, and hence by necessity accumulate some regret at first.

Overall behavior. Figure 3 shows cumulative regret of the algorithms using various probabilistic decay functions such that $M = 500$, $d = 0.2 \cdot t_{ppi1}$ and $f = 1$ for all process instances. In this experiment, we emulate business processes by sampling the completion time of each version from the process execution data of one of the processes from the BPIC 2015 Challenge [18]. The best routing policy can be found only if contextual factor f, context information, and both PPIs are available. To ensure that these algorithms can be compared, regret for LtAvgR and random-udr is calculated using the weights of reward classes in the same manner as ProcessBandit. We observe that ProcessBandit correctly distributes requests to better performing versions, and converges to the best routing policy. However, LtAvgR consistently makes the wrong decision because it never sees the actual value of the second PPI, and is incapable of updating past rewards.

Figure 4 shows the cumulative regret of ProcessBandit with various delays between the observation times of the first and second PPI. In this experiment, we use deterministic completion times for each process instance so that the value of the PPI delay is the same for all observations. We can observe that the regret curves have similar characteristics, and that the algorithm converges to the best routing policy in all cases. There are some small differences in cumulative regret in all scenarios. However, these differences do no support the idea of a conclusive relationship between the delays and overall regret. The magnitude of cumulative regret can be affected by the non-determinism inherent in the algorithm's experimentation phase, and the order in which the PPIs were observed.

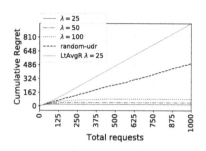

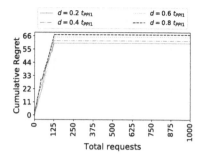

Fig. 3. Cumulative Regret of various algorithms with $d = 0.2 \cdot t_{\mathrm{ppi1}}$.

Fig. 4. Convergence at various reward delays.

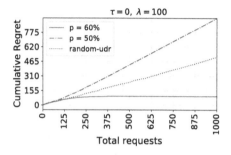

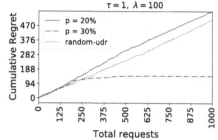

Fig. 5. Convergence with various reward ratio and experiment phase parameters.

Partial rewards and failures. In this section we evaluate convergence characteristics of ProcessBandit when only one PPI can be observed for some instances. We use an experiment setup with $f = 1, \rho = 160$, $d = 0.2 \cdot t_{\mathrm{ppi1}}$, and a constant execution time for all process instances. Each process instance returns the first PPI (profit margin) immediately after execution. The other PPI (user satisfaction) is either never returned, or returned after a delay – which can be expected if, e.g., users are asked to participate in a short survey.

We define p as the percentage of process instances that return both PPIs. We compare regret characteristics of ProcessBandit with *random-udr* because *random-udr* is agnostic to p. Figure 5 shows the convergence characteristics for parameter values of $\lambda = 100$, $M = 500$, and $\tau = 0$, respectively $\tau = 1$. It depicts behavior for values of p that highlight when convergence happens (e.g., 30% for $\tau = 1$) and when not (20% in the same configuration). We observe that by increasing τ from 0 to 1, the algorithm can converge when p is smaller.

The resilience to partial rewards depends on the values of λ, τ and M. There must be enough *complete* observations in every context so that it is possible to reach a state where the current reward ratio is equal or below τ. In some cases where the best routing policy is found, the algorithm can temporarily perform worse than *random-udr*. Increasing λ to 500 (not shown in figures), convergence is achieved with $p = 20\%$. This is further improved to $p = 10\%$ when τ is increased to 1.5, and finally to $p = 2.5\%$ when M is increased to 750.

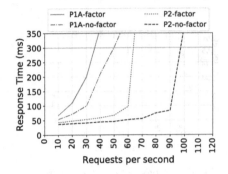
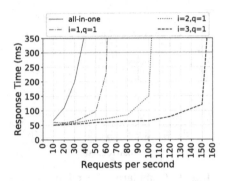

Fig. 6. Response times of various stages and configurations of ProcessBandit.

Fig. 7. Response time in Phase P1A. i and q are number of servers for instance router and task queue respectively.

4.2   Response Time

We measure end-to-end response time and throughput using two servers, one for the instance router and one for the BPMS. We define our SLA metrics in terms of response time and correctness: our system adheres to SLA if it serves 100% of requests under 300 ms. We host these components on Amazon EC2 M4 large instances with 2 vCPUs and 8 GB RAM. We use the reward setup described in Table 2. Figure 6 shows the response times of these configurations, each named with the convention *Phase-Configuration*.

Response times are shown as the average of all requests during a 5 min burst of the corresponding workload. We observe that the throughput is lower and the responses are slower when contextual factor detection is enabled. This is caused by the instance router making additional request to the BPMS to observe the value of f. Performance can be improved by adjusting sampling rates of f. For example, factors like weather condition can be sampled every minute instead.

We observe that the CPU utilization is generally high (above 90%) and proportional to the workload but memory utilization is low (5–7%). The *random-udr* algorithm serves up to 400 requests per second under our SLA criterion. On the same infrastructure, *ProcessBandit* achieves a throughput that is between 10% and 25% of *random-udr*, depending on the configuration.

Based on this observation, we conduct a second experiment to test the horizontal scalability of *ProcessBandit* at its slowest configuration. This configuration, deployed on a single machine, serves a baseline. Then we deploy the instance router and the asynchronous task queue in separate servers, and horizontally scale the number of the instance router servers. We evaluate response-time and throughput for three deployment configurations. The results are shown in Fig. 7. Because instance router was the bottleneck, we observe that increasing the number of instance router servers increases the throughput.

Table 3. User satisfaction model

Outcome	Duration	Version	Satisfaction	
			Ret.	New
Approved	≤5 wks	A	4	5
		B	5	5
	≥5 wks	A	3	4
		B	4	4
Rejected	≤5 wks	A	2	3
		B	3	3
	≥5 wks	A	1	2
		B	2	2

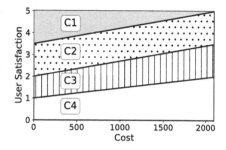

Fig. 8. Performance classification

4.3 AB Test with Synthetic Process

We demonstrate our approach using process versions from the domain of helicopter pilot licensing, as introduced in [14]. The process consists of six activities: Schedule, Eligibility Test, Medical Test, Theory Test, Practical Test, and License Processing. We here add two contextual factors associated with an applicant – age group and applicant type (new or returning). The probability of success in the Medical Test activity is set to be higher for younger age groups. For other activities, success probabilities are the same regardless of age groups.

Activities in Version A of the process are ordered sequentially such that a scheduling activity occurs before each test activity. In Version B, one scheduling activity is performed at the start, which determines the schedules of all the tests, thus reducing the costs of having multiple scheduling activities. We use the activity costs and durations outlined in [14]. Using these process versions, we design an experiment where the process performance is determined by two PPIs: satisfaction score, and cost. Rewards are derived from four categories shown in Fig. 8. Satisfaction scores are derived from the outcome and the duration of the process. Satisfaction score is high if the license is approved and processing is fast, and low otherwise. Satisfaction scores also depend on whether the applicant is new or returning – we assume that returning applicants are harsher on the older version. This is shown in Table 3. While the age group is treated as known context, applicant type is treated as a hypothetical contextual factor.

To simulate a scenario where the satisfaction score is not always available, we assume that satisfaction scores are collected within 60 days after process completion. Applicants are notified four times after

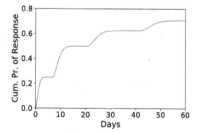

Fig. 9. Probability of receiving satisfaction score over time

Table 4. Performance of versions A and B

Age group	Applicant	Reward class distribution		Samples	
		Version A	Version B	Version A	Version B
18–24	Returning	C1 - 0% C2 - 60.0% C3 - 40.0% Avg Reward = −2.4	C1 - 59.17% C2 - 11.42% C3 - 29.41% Avg Reward = **−1.7**	29	**304**
	New	C1 - 52.67% C2 - 11.03% C3 - 36.3% Avg Reward = **−1.84**	C1 - 47.06% C2 - 5.88% C3 - 47.06% Avg Reward = −2.0	**299**	34
25–40	Returning	C1 - 0% C2 - 28.57% C3 - 71.43% Avg Reward = −2.71	C1 - 56.16% C2 - 11.64% C3 - 32.19% Avg Reward = **−1.76**	27	**307**
	New	C1 - 61.48% C2 - 14.84% C3 - 23.67% Avg Reward = **−1.62**	C1 - 50.0% C2 - 12.5% C3 - 37.5% Avg Reward = −1.88	**299**	34
40+	Returning	C1 - 0% C2 - 57.14% C3 - 42.86% Avg Reward = −2.43	C1 - 56.14% C2 - 14.04% C3 - 29.82% Avg Reward = **−1.74**	24	**310**
	New	C1 - 50.88% C2 - 21.05% C3 - 28.07% Avg Reward = **−1.77**	C1 - 42.86% C2 - 14.29% C3 - 42.86% Avg Reward = −2.0	**300**	33

process completion – on the 7th, 14th, 21st, and 42nd day. The cumulative probability of response is assumed to jump after a notification, a behavior similar to the response rates of web-based career survey [15]. Response probabilities are shown in Fig. 9.

With this setup, the algorithm needs to account for two PPIs $(R1)$, the delay in receiving satisfaction scores $(R2)$, availability of satisfaction scores $(R3)$, and the effect of applicant type on the PPIs $(R4)$. The results of performing AB tests on this setup are shown in Table 4. We observe that in all cases more requests are sent to the version that performs better on average (shown in bold).

5   Discussion and Conclusion

Summary. We introduce *ProcessBandit*, a dynamic process instance routing algorithm that learns a routing policy based on process performance. The algorithm is supported by a modular architecture. ProcessBandit meets all of our requirements and, while not very fast, can be scaled horizontally. It makes sound decisions in scenarios where performance is determined by delayed PPIs which may be fully observed only for some process instances. It also identifies contextual factors at runtime and uses these factors to make routing decisions.

Discussion. ProcessBandit assumes that overall performance can be summarized using the mean. Rewards are averaged per context and version, which helps the algorithm learn its routing policy. As shown in Table 4, the rewards are not always normally distributed. If other statistical properties are important in decision-making, the mechanism for estimating rewards should be changed in the algorithm. Average performance can be ineffective in scenarios where performance deviations from the norm are small but very important. For example, a tiny number of instances may take exceptionally long time. Rewards for such cases are received late and have negligent effect on the mean. Such cases can be handled by an upper bound on the duration: if the process does not complete within an acceptable time, a strong negative reward can be assigned.

There is a threat to external validity by using synthetic datasets. In previous work [14], we demonstrated that AB-BPM can work on real-world datasets, by deriving a single PPI from the available data. Despite our best efforts, we could not find or produce a real-world dataset that combined all features (multiple PPIs, context, etc.) to evaluate ProcessBandit. We aimed to minimize this risk by producing synthetic datasets based on parameters taken from the literature [15], industry (described in [14]), and a BPI Challenge [18] where possible.

Our contextual factor detection approach is based on correlation. In our experiments, we set the correlation thresholds low, so that the context sensitivity could be evaluated. The detection of contextual factors does not need to be based on correlation. Our approach is not tied to how these contextual factors are identified. Contextual factor detection is however a challenge per se that needs further investigation.

Conclusion and Future Work. Unlike prior work on AB testing, our solution provides a risk-managed approach tailored for the requirements of business processes. We demonstrate that this solution meets these requirements by evaluating the behaviour of the routing algorithm, the horizontal scalability of the approach, and its effectiveness in a synthetic business process. Our future plans include extension of the approach to accommodate other statistical properties in reward evaluation and upper bound on duration. We also plan to collaborate with domain experts to conduct field tests in the industry.

Acknowledgements. The work of Claudio Di Ciccio has received funding from the EU H2020 programme under MSCA-RISE agreement 645751 (RISE_BPM).

References

1. van der Aalst, W.M.P., Rosemann, M., Dumas, M.: Deadline-based escalation in process-aware information systems. Decis. Support Syst. **43**(2), 492–511 (2007)
2. Agrawal, S., Goyal, N.: Thompson sampling for contextual bandits with linear payoffs. In: International Conference on Machine Learning, ICML (2013)

3. Branke, J., Deb, K., Miettinen, K., Słowiński, R. (eds.): Multiobjective Optimization. LNCS, vol. 5252. Springer, Heidelberg (2008). https://doi.org/10.1007/978-3-540-88908-3
4. Burtini, G., Loeppky, J., Lawrence, R.: A survey of online experiment design with the stochastic multi-armed bandit. CoRR abs/1510.00757 (2015)
5. Chu, W., Li, L., Reyzin, L., Schapire, R.E.: Contextual bandits with linear payoff functions. In: International Conference on Artificial Intelligence and Statistics, pp. 208–214 (2011)
6. Crook, T., Frasca, B., Kohavi, R., Longbotham, R.: Seven pitfalls to avoid when running controlled experiments on the web. In: ACM SIGKDD, pp. 1105–1114 (2009)
7. Dumas, M., Rosa, M.L., Mendling, J., Reijers, H.A.: Fundamentals of Business Process Management. Springer, Heidelberg (2013). https://doi.org/10.1007/978-3-642-33143-5
8. He, H., Ma, Y.: Imbalanced Learning: Foundations, Algorithms, and Applications. Wiley, Hoboken (2013)
9. Holland, C.W.: Breakthrough Business Results with MVT: A Fast, Cost-Free "Secret Weapon" for Boosting Sales, Cutting Expenses, and Improving Any Business Process. Wiley, Hoboken (2005)
10. Kohavi, R., Longbotham, R., Sommerfield, D., Henne, R.M.: Controlled experiments on the web: survey and practical guide. Data Min. Knowl. Discov. **18**(1), 140–181 (2009)
11. Kohavi, R., Crook, T., Longbotham, R., Frasca, B., Henne, R., Ferres, J.L., Melamed, T.: Online experimentation at Microsoft. In: Workshop on Data Mining Case Studies (2009)
12. Li, L., Chu, W., Langford, J., Schapire, R.E.: A contextual-bandit approach to personalized news article recommendation. In: International Conference on World Wide Web (2010)
13. Reijers, H.A., Mansar, S.L.: Best practices in business process redesign: an overview and qualitative evaluation of successful redesign heuristics. Omega **33**(4), 283–306 (2005)
14. Satyal, S., Weber, I., Paik, H., Di Ciccio, C., Mendling, J.: AB-BPM: performance-driven instance routing for business process improvement. In: Carmona, J., Engels, G., Kumar, A. (eds.) BPM 2017. LNCS, vol. 10445, pp. 113–129. Springer, Cham (2017). https://doi.org/10.1007/978-3-319-65000-5_7
15. Sauermann, H., Roach, M.: Increasing web survey response rates in innovation research: an experimental study of static and dynamic contact design features. Res. Policy **42**(1), 273–286 (2013)
16. Silver, D., Newnham, L., Barker, D., Weller, S., McFall, J.: Concurrent reinforcement learning from customer interactions. In: ICML, pp. 924–932 (2013)
17. Sutton, R.S., Barto, A.G.: Introduction to Reinforcement Learning, 1st edn. MIT Press, Cambridge (1998)
18. Teinemaa, I., Leontjeva, A., Masing, K.O.: BPIC 2015: Diagnostics of building permit application process in dutch municipalities. BPI Challenge Report 72 (2015)
19. Vermorel, J., Mohri, M.: Multi-armed bandit algorithms and empirical evaluation. In: Proceedings of the ECML European Conference on Machine Learning, pp. 437–448 (2005)

Filtering Spurious Events from Event Streams of Business Processes

Sebastiaan J. van Zelst[1]([✉]), Mohammadreza Fani Sani[2], Alireza Ostovar[3],
Raffaele Conforti[4], and Marcello La Rosa[4]

[1] Eindhoven University of Technology, Eindhoven, The Netherlands
`s.j.v.zelst@tue.nl`
[2] RWTH Aachen University, Aachen, Germany
`fanisani@pads.rwth-aachen.de`
[3] Queensland University of Technology, Brisbane, Australia
`alireza.ostovar@qut.edu.au`
[4] University of Melbourne, Melbourne, Australia
`{raffaele.conforti,marcello.larosa}@unimelb.edu.au`

Abstract. Process mining aims at gaining insights into business processes by analysing event data recorded during process execution. The majority of existing process mining techniques works offline, i.e. using static, historical data stored in event logs. Recently, the notion of online process mining has emerged, whereby techniques are applied on live event streams, as process executions unfold. Analysing event streams allows us to gain instant insights into business processes. However, current techniques assume the input stream to be completely free of noise and other anomalous behaviours. Hence, applying these techniques to real data leads to results of inferior quality. In this paper, we propose an event processor that enables us to filter out spurious events from a live event stream. Our experiments show that we are able to effectively filter out spurious events from the input stream and, as such, enhance online process mining results.

Keywords: Process mining · Event stream · Filtering
Anomaly detection

1 Introduction

Nowadays, information systems can accurately record the execution of the business processes they support. Common examples include order-to-cash and procure-to-pay processes, which are tracked by ERP systems. Process mining [1] aims at turning such event data into valuable, actionable knowledge, so that process performance or compliance issues can be identified and rectified. Different process mining techniques are available. These include techniques for automated

R. Conforti and M. La Rosa—Part of the work was done while the authors was at the Queensland University of Technology.

J. Krogstie and H. A. Reijers (Eds.): CAiSE 2018, LNCS 10816, pp. 35–52, 2018.
https://doi.org/10.1007/978-3-319-91563-0_3

process discovery, conformance checking, performance mining and process variant analysis. For example, in process discovery we aim at reconstructing the underlying structure of the business process in the form of a process model, while in conformance checking we assess to what degree the recorded data aligns with a normative process model available in the organisation.

The vast majority of process mining techniques are defined in an *offline* setting, i.e. they work over historical data of completed process executions (e.g. over all orders fulfilled in the past six months). They are typically not adequate to directly work in *online* settings, i.e. from live streams of events rather than historical data. Hence, they cannot be used for operational support, but only for a-posteriori analysis. Online process mining provides a wealth of opportunities. For example, when applying conformance checking techniques, compliance deviations could be detected as soon as they occur, or better, their occurrence could be predicted in advance. In turn, the insights gained could be used to rectify the affected process executions on the fly, avoiding the deviations to occur altogether.

As a result, several process mining techniques have recently been designed to specifically work online. These include, for example, techniques for drift detection [16,19,20], automated discovery [7,15,25], conformance checking [8,26] and predictive process monitoring [18]. Such techniques tap into an *event stream* produced by an information system. However, they typically assume the stream to be free of noise and anomalous behaviour. In reality however, several factors cause this assumption to be wrong, e.g. the supporting system may trigger the execution of an inappropriate activity that does not belong to the process, or the system may be overloaded resulting in logging errors. The existence of these anomalies in event streams easily leads to unreliable results. For example, in drift detection, sporadic stochastic oscillations caused by noise can negatively impact drift detection accuracy [16,20].

In this paper, we propose a general-purpose event stream filter designed to detect and remove *spurious events* from event streams. We define a spurious event as an event emitted onto the stream, whose occurrence is extremely unlikely, given the underlying process and process context. Our approach relies on a time-evolving subset of behaviour of the total event stream, out of which we infer an incrementally-updated model that represents this behaviour. In particular, we build a collection of *probabilistic automata*, which are dynamically updated to filter out spurious events.

We implemented our filter as a stream processor, taking an event stream as an input and returning a filtered stream. Using the implementation, we evaluated accuracy and performance of the filter by means of multiple quantitative experiments. To illustrate the applicability of our approach w.r.t. existing online process mining techniques, we assessed the benefits of our filter when applied prior to drift detection.

The remainder of this paper is structured as follows. In Sect. 2, we discuss related work, while in Sect. 3 we present background concepts introducing (online) process mining concepts. In Sect. 4, we present our approach, which we evaluate in Sect. 5. We conclude the paper and discuss several avenues for future work in Sect. 6.

2   Related Work

Work in the areas of online process mining and noise filtering are of particular relevance to the work presented in this paper. In the area of online process mining, the majority of work concerns automated process discovery algorithms. For example, Burattin et al. [7] propose a basic algorithm that lifts an existing offline process discovery algorithm to an online setting. Additionally, in [6], Burattin et al. propose an online process discovery technique for the purpose of discovering declarative models. Hassani et al. [15] extend [7] by proposing the use of indexed prefix-trees in order to increase memory efficiency. Finally, van Zelst et al. [25] extend [7,15] and generalize it for a large class of existing process discovery algorithms. More recently, event streams have been used for online conformance checking [8,26] and online concept drift detection [19,20]. In the context of online conformance checking, Burattin et al. [8] propose an approach that uses an enriched version of the original process model to detect deviant behaviour. In [26], van Zelst et al. propose to detect deviant behaviour by incrementally computing prefix-alignments. In the context of online concept drift detection, Ostovar et al. [20] detect drifts on event streams by monitoring the distribution of behavioural abstractions (i.e. α^+ relations) of the event stream across adjacent time sliding-windows. In follow-up work, Ostovar et al. [19] extend [20] to allow for concept drift characterization.

With respect to noise filtering in context of event logs, three approaches are described in literature [10,13,23]. The approach proposed by Wang et al. [23] relies on a reference process model to repair a log whose events are affected by labels that do not match the expected behaviour of the reference model. The approach proposed by Conforti et al. [10] removes events that cannot be reproduced by an automaton constructed using frequent process behaviour recorded in the log. Finally, Fani Sani et al. [13] propose an approach that uses conditional probabilities between sequences of activities to remove events that are unlikely to occur in a given sequence.

Existing noise filtering techniques have shown to improve the quality of process mining techniques [10,13], yet they are not directly applicable in an online context. Similarly, online process mining techniques do not address the problem of noise in event streams. Our approach bridges the gap between these techniques providing, to the best of our knowledge, the first noise filter for business process event streams.

Finally, the problem of detecting spurious events from event stream of business processes shares similarities with the problem of outlier detection in temporal data, e.g. reading sensor data. In this context, we observe three types of techniques: (i) techniques detecting if entire sequences of events are anomalous; (ii) techniques detecting if a single data point within a sequence is an outlier; and (iii) techniques detecting anomalous patterns within a sequences. For a detailed discussion about techniques for outlier detection in temporal data we refer to the works by Gupta et al. [14] for events with continuous values, and by Chandola et al. [9] for events with discrete values.

3   Background

Here we introduce our notation and basic concepts such as event logs and event streams.

3.1   Mathematical Preliminaries and Notation

Let X denote an arbitrary set and let $\mathcal{P}(X)$ denote the power set of X. We let $\mathbb{N}$ denote the set of natural numbers including 0. $\mathbb{B} = \{0, 1\}$ represents the boolean domain. A multiset M over X generalizes the notion of a set and allows for multiple instances of its elements, i.e. $M \colon X \to \mathbb{N}$. We let $\mathcal{M}(X)$ denote the set of all possible multisets over X. We write a multiset M as $[x_1^{i_1}, ..., x_n^{i_n}]$ where $M(x_j) = i_j$ for $1 \leq j \leq n$. If for $x \in X$, $M(x) = 0$ we omit it form multiset notation, and, if $M(x) = 1$ we omit x's superscript. A sequence σ of length n is a function $\sigma \colon \{1, ..., n\} \to X$. We write $\sigma = \langle x_1, ..., x_n \rangle$, where for $1 \leq i \leq n$ we have $\sigma(i) = x_i$. The set of all sequences over set X is denoted X^*. Given an n-ary Cartesian product $X_1 \times X_2 \times \cdots \times X_n$ and corresponding element $e = (x_1, x_2, ..., x_n)$, for $1 \leq i \leq n$, we write $\pi_i(e) = x_i$. We overload notation to define projection of sequences, i.e. let $\sigma = \langle e_1, e_2, ..., e_m \rangle \in (X_1 \times X_2 \times \cdots X_n)^*$, we have $\pi_i(\sigma) = \langle \pi_i(e_1), \pi_i(e_2), ..., \pi_i(e_m) \rangle$ for $1 \leq i \leq n$. A pair $(X, \nleq)$ is a partial order if $\nleq$ is a *reflexive*, *anti-symmetric* and *transitive* binary relation on X.

Our approach builds on the notion of a *probabilistic automaton* (PA). Such automaton is an extension of a conventional non-deterministic automaton, where each transition has an associated probability of occurrence.

Definition 1 (Probabilistic Automaton). *A probabilistic automaton* (PA) *is a 6-tuple* $(Q, \Sigma, \delta, q_0, F, \gamma)$, *where Q is a finite set of states, Σ is a finite set of symbols, $\delta \colon Q \times \Sigma \to \mathcal{P}(Q)$ is a transition relation, $q^0 \in Q$ is the initial state, $F \subseteq Q$ is the set of final states, $\gamma \colon Q \times \Sigma \times Q \to [0, 1]$ is the transition probability function.*

Additionally we require:

1. $\forall q, q' \in Q, a \in \Sigma (q' \in \delta(q, a) \Leftrightarrow \gamma(q, a, q') > 0)$: *if an arc labelled a connects q to q', then the corresponding probability is non-zero.*
2. $\forall q \in Q \setminus F (\exists q' \in Q, a \in \Sigma (q' \in \delta(q, a)))$: *non-final states have outgoing arc(s).*
3. $\forall q \in Q (\exists a \in \Sigma, q' \in Q(q' \in \delta(q, a)) \Rightarrow \displaystyle\sum_{\{(a, q') \in \Sigma \times Q \mid q' \in \delta(q, a)\}} \gamma(q, a, q') = 1)$: *the sum of probabilities of outgoing arcs of a state equals one.*

For given $q, q' \in Q$ and $a \in \Sigma$ s.t. $\delta(q, a) = q'$, $\gamma(q, a, q')$ represents the probability of reaching state q' from state q by means of label a. We write such probability as $P(a \mid q \to q')$ and we define $P(a \mid q) = \sum_{q' \in Q} P(a \mid q \to q')$.

3.2   Event Logs

Modern information systems track, often in great detail, what specific activity is performed for a running instance of the process, i.e. a *case*, at a certain point in time. Traditional process mining techniques aim to analyse such data, i.e. *event logs*, in a static/a-posteriori setting. Consider Table 1, depicting an example of an event log. Each line refers to the execution of an activity, i.e. an *event*, in context of a process instance, which is identified by means of a *case-id*. In this example the case-id equals the id of the ticket for which a compensation request is filed. In general the case-id depends on the process under study, e.g. a customer type or product-id are often used as a case-id.

Table 1. Example event log fragment.

Event-id	Case-id	Activity	Resource	Time-stamp
...	...	...	...	...
571	412	*decide (e)*	*Ali*	*2017-11-10 13:47*
572	417	*register request (a)*	*Marcello*	*2017-11-10 14:14*
573	412	*reject request (g)*	*Mohammad*	*2017-11-10 14:18*
574	417	*examine causally (b)*	*Mohammad*	*2017-11-10 14:33*
575	417	*check ticket (d)*	*Marlon*	*2017-11-10 14:06*
576	417	*decide (e)*	*Mohammad*	*2017-11-10 14:51*
577	417	*pay compensation (f)*	*Mohammad*	*2017-11-10 15:03*
578	504	*register request (a)*	*Ali*	*2017-11-10 15:05*
...	...	...	...	...

Consider the events related to case-id 417. The first event, i.e. with id 572, describes that *Marcello* executed a *register request* activity. Subsequently, *Mohammad* performed a *causal examination* of the request (event 574). In-between event 572 and 574, event 573 is executed that relates to a case with id 412, which reflects that several process instances run in parallel. An event e, i.e. the execution of an activity, is defined as a tuple $(\iota, c, a) \in \mathcal{E}$, where $\mathcal{E} = \mathcal{I} \times \mathcal{C} \times \mathcal{A}$, $\mathcal{I}$ denotes the universe of event identifiers, $\mathcal{C}$ denotes the universe of case identifiers, and $\mathcal{A}$ denotes the universe of activities. Typically, more event attributes are available, e.g. the resource(s) executing the activity and/or the time-stamp of the activity. However, here we only consider the ordering of activities in context of a process instance, i.e. the *control-flow perspective*.

Definition 2 (Event Log, Trace). *Given a collection of events $E \subseteq \mathcal{E}$, an event log L is a partially ordered set of events, i.e. $L = (E, \preceq)$ s.t. $\forall e = (\iota, c, a), e' = (\iota', c', a') \in E(\iota = \iota' \Rightarrow (c = c' \wedge a = a'))$.*
 A trace related to case $c \in \mathcal{C}$ is a sequence $\sigma \in E^*$ for which:

1. $\forall 1 \leq i \leq |\sigma|(\pi_2(\sigma(i)) = c)$; Events in σ relate to case c.
2. $\forall e \in E(\pi_2(e) = c \Rightarrow \exists 1 \leq i \leq |\sigma|(\sigma(i) = e))$; Each event related to c is in σ.
3. $\forall 1 \leq i < j \leq |\sigma|(\sigma(i) \neq \sigma(j))$; All events in σ are unique.
4. $\forall 1 \leq i < j \leq |\sigma|(\sigma(j) \npreceq \sigma(i))$; Events in σ respect their order.

The partial order of the events of an event log is usually imposed by means of recorded times-stamps. A log is partially ordered due to, for example, inherent parallelism of and/or mixed time-stamp granularity. A *trace* is a sequence of events related to the same case identifier that respect the partial order. Consider the trace related to case 417 of Table 1, which we write as $\langle (572, 417, \text{register request}), (574, 417, \text{examine causally}), (575, 417, \text{check ticket}),$

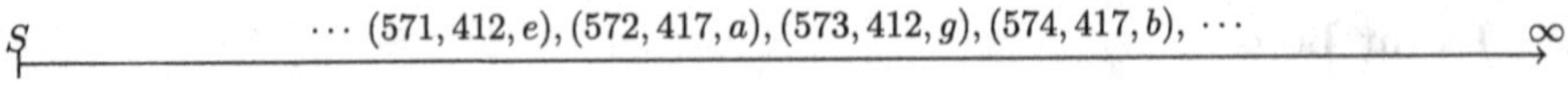

$$\cdots (571, 412, e), (572, 417, a), (573, 412, g), (574, 417, b), \cdots$$

Fig. 1. Example event stream S.

$(576, 417, \text{decide}), (577, 417, \text{pay compensation}))$,   or   simply $\langle (572, 417, a),$ $(574, 417, b), (575, 417, d), (576, 417, e), (577, 417, f) \rangle$ using short-hand activity names. Most process mining techniques ignore the event- and case-identifiers stored within events and simply distil a sequence of activities from the given trace. For example, by projecting the example trace, we obtain $\langle a, b, d, e, f \rangle$. When adopting such view on traces, a multitude of cases exist which project onto the same sequence of activities.

3.3   Event Streams

We adopt the notion of online/real-time *event stream*-based process mining, in which the data is assumed to be an infinite sequence of events. Since in practice, several instances of a process run in parallel, we have no guarantees w.r.t. the arrival of events related to the same case. Thus, new events related to a case are likely to be emitted onto the stream in a dispersed manner, which implies that our knowledge of the activities related to cases changes over time.

Definition 3 (Event Stream). *An event stream S is a (possibly infinite) sequence of unique events, i.e. $S \in \mathcal{E}^*$ s.t. $\forall 1 \leq i < j \leq |S|(S(i) \neq S(j))$.*

Consider Fig. 1, in which we depict a few of the (short-hand) events that we also presented in Table 1. The first event depicted is $(572, 417, \text{register request})$, the second event is $(574, 417, \text{examine causally})$, and so on. We assume that one event arrives per unit of time, i.e. we do not assume the existence of a multi-channel stream. Moreover, we assume that the order of event arrival corresponds to the order of execution.

4   Approach

In this section, we present our approach. We aim to build and maintain a collection of probabilistic automata which we use to filter out spurious events. Each automaton represents a different view on the behaviour of the underlying process, as described by the event stream. The main idea of the approach is that dominant behaviour attains higher occurrence probabilities within the automata compared to spurious behaviour.

4.1   General Architecture

The proposed filter uses a subset of all behaviour emitted onto the stream and is intended to be updated incrementally when new events arrive on the stream.

Since we need to maintain the possibly infinite event stream in finite memory, we need to "forget" behaviour observed in the past. Hence, we account for removal of events as well.

In Fig. 2, we depict the main architecture of the proposed filter. We assume an input event stream S that contains spurious events. As indicated, events related to different cases are typically dispersed over an event stream. Hence, we need means to track, given case c, what behaviour was received in the past for case c. The exact nature of such data structure is outside the scope of this paper. We assume the existence of a finite event window $w \colon \mathcal{C} \times \mathbb{N} \to \mathcal{E}^*$, where $w(c,t)$ represents the sequence of events stored in the event window at time t. As such, the event window maintains a set of relevant recently received events, grouped by case-identifier.

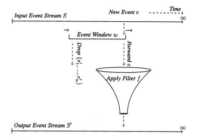

Fig. 2. Schematic overview of the proposed filtering architecture.

In order to determine what events need to be removed from the event window (i.e. we need to maintain a finite view of the stream), we are able to use a multitude of existing stream-based approaches, e.g. we are able to us techniques such as (adaptive) sliding windows [4,5], reservoir sampling [3,22] or (forward) decay methods [11]. The only strict assumption we pose on w, is that event removal respects the order of arrival w.r.t. the corresponding case. Thus, whenever we have a stream of the form $\langle ...,(\iota, c, a),(\iota', c', a'),(\iota'', c, a''), ...\rangle$, we assume event (ι, c, a) to be removed prior to event (ι'', c, a''). A new event e is, after storage within w, forwarded to event filter f. From an architectural point of view we do not pose any strict requirements on the dynamics of the filter. We do however aim to let filter f reflect the behaviour captured within window w. Hence, the filter typically needs to process the event within its internal representation, prior to the actual filtering. For the newly received event, the filter f either decides to emit the event onto output stream S', or, to discard it.

4.2 Automaton Based Filtering

Given the general architecture, in this section, we propose an instantiation of filter f. We first present the conceptual idea of the use of probabilistic automata for the purpose of spurious event filtering, after which we describe the main approach.

Prefix-Based Automata. In our approach, a collection of probabilistic automata represents recent behaviour observed on the event stream. These automata are subsequently used to determine whether new events are, according to the probability distributions described by the automata, likely to be spurious or not. Each state within an automaton refers to the recent history of cases as described by recently received events on the event stream. The probabilities of

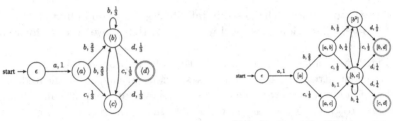

(a) Prefix Size 1/Identity Abstraction   (b) Prefix Size 2/Parikh Abstraction

Fig. 3. Two examples of prefix-based automata, based on traces $\langle a, b, b, c, d \rangle$, $\langle a, b, c, b, d \rangle$ and $\langle a, c, b, b, d \rangle$.

the outgoing arcs of a state are based on cases that have been in that state before, and subsequently moved on to a new state by means of a new event. Upon receiving a new event, we assess the state of the corresponding case and check, based on the distribution as defined by that state's outgoing arcs, whether the new event is likely to be spurious or not.

We construct probabilistic automata in which states represent recent behaviour for a newly related event based on its case-identifier, i.e. *prefix-based automata*. In prefix-based automata, a state q represents a possible *prefix of executed activities*, whereas outgoing arcs represent those activities $a \in \mathcal{A}$ that are likely to follow the prefix represented by q, and their associated probability of occurrence. We define two types of parameters, that allow us to deduce states in the corresponding prefix automaton based on a prefix, i.e.:

1. *Maximal Prefix Size*; Represents the size of the prefix to take into account when constructing states in the automaton.
2. *Abstraction*; Represents an abstraction that we apply on the prefix in order to define a state. We identify the following abstractions:
   - *Identity*; Given $\sigma \in \mathcal{A}^*$, the identity abstraction $\boldsymbol{id}$ yields the prefix as a state, i.e. $\boldsymbol{id} \colon \mathcal{A}^* \to \mathcal{A}^*$, where $\boldsymbol{id}(\sigma) = \sigma$
   - *Parikh*; Given $\sigma \in \mathcal{A}^*$, the Parikh abstraction $\boldsymbol{p}$ yields a multiset with the number of occurrences of $a \in \mathcal{A}$ in σ, i.e. $\boldsymbol{p} \colon \mathcal{A}^* \to \mathcal{M}(\mathcal{A})$, where:

$$\boldsymbol{p}(\sigma) = \left[a^n \mid a \in \mathcal{A} \land n = \sum_{i=1}^{|\sigma|} \left(\begin{cases} 1 & \text{if } \sigma(i) = a \\ 0 & \text{otherwise} \end{cases} \right) \right]$$

   - *Set*; Given $\sigma \in \mathcal{A}^*$ the set abstraction $\boldsymbol{s}$ indicates the presence of $a \in \mathcal{A}$ in σ, i.e. $\boldsymbol{s} \colon \mathcal{A}^* \to \mathcal{P}(\mathcal{A})$, where $\boldsymbol{s}(\sigma) = \{a \in \mathcal{A} \mid \exists 1 \leq i \leq |\sigma|(\sigma(i) = a)\}$

In Fig. 3, we depict two different automata based on the traces $\langle a, b, b, c, d \rangle$, $\langle a, b, c, b, d \rangle$ and $\langle a, c, b, b, d \rangle$, which we assume to occur equally often. In Fig. 3a, we limit the prefix size to 1 and use the identity abstraction. Note that any of the possible abstractions in combination with prefix size 1 always yields the same automaton. Consider the state related to abstraction $\langle b \rangle$ which states that

$P(b \mid \langle b \rangle) = P(c \mid \langle b \rangle) = P(d \mid \langle b \rangle) = \frac{1}{3}$, i.e. we are equally likely to observe activity b, c or d after $\langle b \rangle$. In Fig. 3b, we limit the prefix size to 2 and use the Parikh abstraction. In this case, since we use a larger prefix size, we have more fine-grained knowledge regarding the input data. For example in Fig. 3a, the automaton describes that sequence $\langle a, b, d \rangle$ is likely, whereas in Fig. 3b we have $P(d \mid [a, b]) = 0$.

Incrementally Maintaining Collections of Automata. As new events are emitted on the stream, we aim to keep the automata up-to-date in such a way that they reflect the behaviour present in event window w. Let $k > 0$ represent the maximal prefix length we want to take into account when building automata. We maintain k prefix-automata, where for $1 \leq i \leq k$, automaton $PA_i = (Q_i, \Sigma_i, \delta_i, q_i^0, F_i, \gamma_i)$ uses prefix-length i to define its state set Q_i. As exemplified by the two automata in Fig. 3, the prefix length influences the degree of generalization of the corresponding automaton. Moreover, increasing the maximal prefix length considered is likely to generate automata of larger size, and thus it is more memory intensive.

Upon receiving a new event, we incrementally update the k maintained automata. Consider new event $e = (\iota, c, a)$ arriving at time t and let $\sigma = \sigma' \cdot \langle a \rangle = w(c, t)$. To update automaton PA_i we apply the abstraction of choice on the prefix of length i of the newly received event in σ', i.e. $\langle \sigma'(|\sigma'| - i + 1), ..., \sigma'(|\sigma'| - i + i) \rangle$ to deduce corresponding state $q_{\sigma'} \in Q_i$. The newly received event influences the probability distribution as defined by the outgoing arcs of $q_{\sigma'}$, i.e. it describes that $q_{\sigma'}$ can be followed by activity a. Therefore, instead of storing the probabilities of each γ_i, we store the weighted outdegree of each state $q_i \in Q_i$, i.e. $\deg_i^+(q_i)$. Moreover, we store the individual contribution of each $a \in \mathcal{A}$ to the outdegree of q_i, i.e. $\deg_i^+(q_i, a)$ with $\deg_i^+(q_i, a) = 0 \Leftrightarrow \delta(q_i, a) = \emptyset$. Observe that $\deg_i^+(q_i) = \sum_{a \in \mathcal{A}} \deg_i^+(q_i, a)$, and, that deducing the probability of activity a in state q_i is trivial, i.e. $P(a \mid q_i) = \frac{\deg_i^+(q_i, a)}{\deg_i^+(q_i)}$.

Updating the automata based on events that are removed from event window w is performed as follows. Assume that we receive a new event e at time $t > 0$. For each $c \in \mathcal{C}$, let $\sigma_c' = w(c, t - 1)$, $\sigma_c = w(c, t)$ and let $\Delta_c(t) = |\sigma_c'| - |\sigma_c|$. Observe that for any case c that does not relate to the newly received event, we have $\Delta_c(t) \geq 0$, i.e. some events may have been dropped for that case, yet no new events are received, hence $|\sigma_c| \leq |\sigma_c'|$. In a similar fashion, for the case c that relates to the new event e, we have $\Delta_c(t) \geq -1$, i.e. either $|\sigma_c| = |\sigma_c'| + 1$, or, $|\sigma_c| \leq |\sigma_c'|$. Thus, to keep the automata in line with the events stored in the event window, in the former case we need to update the automata if $\Delta_c(t) > 0$, i.e. at least one event is removed for the corresponding case-id, whereas in the latter case we need to update the automata if $\Delta_c(t) \geq 0$. Therefore, we define $\Delta_c'(t) = \Delta_c(t)$ for the former case and $\Delta_c'(t) = \Delta_c(t) + 1$ in the latter case. Henceforth, if for any $c \in \mathcal{C}$, we have $\Delta_c'(t) > 0$, we need to update the maintained automata to account for removed events. To update the collection of k maintained automata, for each $1 \leq i \leq \Delta_c'(t)$ we generate sequences $\langle \sigma'(i) \rangle$,

$\langle \sigma'(i), \sigma'(i)+1 \rangle, ..., \langle \sigma'(i), ..., \sigma'(i+k) \rangle$ (subject to $|\sigma'| > i+k$). For each generated sequence we apply the abstraction of choice to determine corresponding state q, and subsequently reduce the value of $\deg^+(q)$ by 1. Moreover, assume the state q corresponds to sequence $\langle \sigma'(i), \sigma'(i+1), ..., \sigma'(i+j) \rangle$ with $1 \le i \le \Delta'_c(t)$ and $1 \le j < k$, we additionally reduce $\deg^+(q, a)$ by 1, where $a = \sigma'(i+j+1)$. As an example, consider that we use maximal prefix length 2, i.e. $k = 2$, a identity abstraction, and assume that for some $c \in C$ we have $\sigma' = \langle a, b, c, d, e \rangle$ and $\sigma = \langle b, c, d, e \rangle$, i.e. the event related to activity a is removed. We have $\Delta'_c(t) = 1$, thus we generate sequences $\langle \sigma'(1) \rangle = \langle a \rangle$ and $\langle \sigma'(1), \sigma'(1+1) \rangle = \langle a, b \rangle$. Since we use identity abstraction these two sequence correspond to a state in their associated automaton, and we reduce $\deg^+(\langle a \rangle)$, $\deg^+(\langle a \rangle, b)$, $\deg^+(\langle a, b \rangle)$ and $\deg^+(\langle a, b \rangle, c)$ by one.

Filtering Events. After receiving an event and subsequently updating the collection of automata, we determine whether the new event is spurious or not. To assess whether the newly arrived event is spurious we assess to what degree the probability of occurrence of the activity described by the new event is an outlier w.r.t. the probabilities of other outgoing activities of the current state. Given the set of k automata, for automaton $PA_i = (Q_i, \Sigma_i, \delta_i, q_i^0, F_i, \gamma_i)$ with prefix-length i $(1 \le i \le k)$, we characterize an automaton specific filter as $f_i: Q_i \times \Sigma_i \to \mathbb{B}$. Note that an instantiation of a filter f_i often needs additional input, e.g. a threshold value or range. The exact characterization of f_i is a parameter of the approach, however, we propose the following instantiations:

- *Fractional*; Considers whether the probability obtained is higher than a given threshold, i.e. $f_i^F: Q_i \times \Sigma_i \times [0,1] \to \mathbb{B}$, where, $f_i^F(q_i, a, \kappa) = 1$ if $P(a \mid q_i) < \kappa$.
- *Heavy Hitter*; Considers whether the probability obtained is higher than a fraction of the maximum outgoing probability, i.e. $f_i^H: Q_i \times \Sigma_i \times [0,1] \to \mathbb{B}$, where, $f_i^H(q_i, a, \kappa) = 1$ if $P(a \mid q_i) < \kappa \cdot \max_{a' \in \mathcal{A}} P(a' \mid q_i)$.
- *Smoothened Heavy Hitter*; Considers whether the probability obtained is higher than a fraction of the maximum outgoing probability subtracted with the non-zero average probability. Let $NZ = \{a \in \Sigma_i \mid P(a \mid q_i) > 0\}$, we define $f_i^{SH}: Q_i \times \Sigma_i \times [0,1] \to \mathbb{B}$, where, $f_i^{SH}(q_i, a, \kappa) = 1$ if $P(a \mid q_i) < $
$$\kappa \cdot \left(\max_{a' \in \mathcal{A}} P(a' \mid q_i) - \frac{\sum\limits_{a' \in NZ} P(a' \mid q_i)}{|NZ|} \right).$$

For a newly received event, each automaton, combined with a filter of choice yields a boolean result indicating whether or not the new event is spurious. In context of this paper we assume that we apply the same filter on each automaton. Moreover, we assume that when any of the k maintained automata signals an event to be spurious, the event itself is spurious. Note that maintaining/filtering the automata can be parallelized, i.e. we maintain an automaton on each node within a cluster.

5   Evaluation

We implemented our filter as an open-source plugin for both ProM [21] and RapidProM [2]. The filter source code is available at https://github.com/s-j-v-zelst/prom-StreamBasedEventFilter. All raw results, including process models, associated event data, scientific workflows and charts are available at https://github.com/s-j-v-zelst/research/releases/tag/2018_caise.

Using the RapidProM plugin, we conducted a two-pronged evaluation. First, we assessed filtering accuracy and time performance on randomly generated event streams, based on *synthetic* process models, i.e. a collection of process models that resemble business processes often present in organizations. Second, we assessed the applicability of our filter in combination with an existing class of online process mining techniques, namely concept drift detection. In the latter experiment we used both synthetic and real-life datasets.

5.1   Filtering Accuracy and Time Performance

For this first set of experiments, we generated several event streams using 21 variations of the loan application process model presented in [12]. These variations are inspired by the change patterns of [24]. Out of 21 stable models, we generated 5 different random event logs, each containing 5000 cases, with a varying amount of events. For each generated log we randomly inserted spurious events with probabilities ranging from 0.025 to 0.15 in steps of 0.025. In these experiments we use a simple sliding window with fixed size as an implementation for w. Given a sliding window of size $|w|$, the first $|w|$ events are used for training and are ignored. Each event arriving after the first $|w|$ events that relates to a case that was received within the first $|w|$ events is ignored.

Accuracy. We assess the impact of a wide variety of parameters on filtering accuracy. These are the prefix size, the particular abstraction used, the filtering technique and the filter threshold. The values of these parameters are presented in Table 2. Here, we mainly focus on the degree in which prefix size, abstraction, filtering method and window size influence the filtering quality. The results for each of these parameters are presented in Fig. 4. Note that, to reduce the amount of data points, we show results for noise levels 0.025, 0.05, 0.1 and 0.15, and threshold levels 0.05–0.25.

For the maximal prefix size (see Fig. 4a), we observe that a prefix-size of 1 tends to outperform prefix-sizes of 3 and 5. This is interesting as it shows that, for this collection of models and associated streams, ignoring history improves the results. Note that, for maximal prefix length k, we use k automata, and signal an event to be spurious whenever one of these signals that this is the case. Using a larger maximal prefix-length potentially identifies more spurious events, yielding higher recall values. However, a side effect is potentially lower precision values. Upon inspection, this indeed turns out to be the case, i.e. the differences in F1-score are explained by higher recall values for increased maximal prefix lengths, however, at the cost of lower precision.

Table 2. Parameters of *Data generation* and *experiments* with *synthetic data*

Data generation	
Artefact/parameter	*Value*
Number of models	21
Number of event logs, generated per model	5
Probability of spurious event injection, per event log	$\{0.025, 0.05, ..., 0.15\}$
Experiments	
Window size	$\{2500, 5000\}$
Prefix size	$\{1, 3, 5\}$
Abstraction	$\{\texttt{Identity } (id), \texttt{ Parikh } (p), \texttt{ Set } (s)\}$[a]
Filter	$\{\texttt{Fractional } (f^F), \texttt{ Heavy Hitter } (f^H),$ $\texttt{Smoothened Heavy Hitter } (f^{SH})\}$
Filter threshold (κ)	$\{0.05, 0.1, ..., 0.5\}$

[a]Only Sequence is used i.c.m. with size 1 since all abstractions yield the same result.

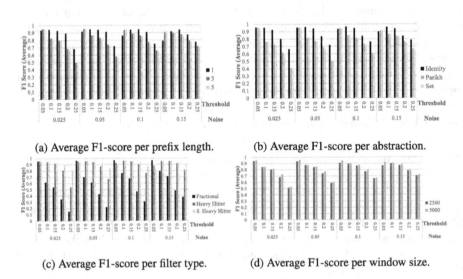

(a) Average F1-score per prefix length.

(b) Average F1-score per abstraction.

(c) Average F1-score per filter type.

(d) Average F1-score per window size.

Fig. 4. Average F1-score for different prefix sizes, abstractions, filtering methods and window sizes, per threshold/noise combination.

As for the abstraction used (see Fig. 4b), we observe that the *Identity*- outperforms both the *Parikh*- and the *Set* abstraction (for these results prefix length 1 is ignored.). The results are explained by the fact that within the collection of models used, the amount of parallelism is rather limited, which does not allow us to make full use of the generalizing power of both the *Parikh* and *Set* abstraction.

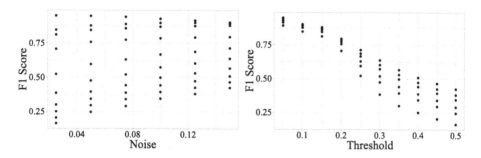

Fig. 5. Average F1-score per noise (a)/threshold level (b).

At the same time, loops of short length exist in which order indeed plays an important role, which is ignored by the two aforementioned abstractions. Upon inspection, the recall values of all three abstractions is relatively equal, however, precision is significantly lower for both the *Parikh-* and *Set* abstraction. This can be explained by the aforementioned generalizing power of these abstractions, and, in turn, explains the difference in F1-score.

For the filter method used (see Fig. 4c), we observe that the *Smoothened Heavy Hitter* and *Heavy Hitter* outperform the *Fractional* filter for increasing threshold values. This is explained by the fact that the fractional filter poses a rigorous requirement on events to be considered non-spurious, e.g. threshold $\frac{1}{4}$ requires an activity to occur at least in 25% of the observed cases. The other two filters solve this by using the maximal observed value, i.e. if a lot of behaviour is possible, the maximum value is lower and hence the requirement to be labelled non-spurious is lower.

Finally, we observe that an increased sliding window size does not affect the filter results significantly (see Fig. 4d). Since the process is stable, i.e. there is no concept-drift within the generated streams, this indicates that both window sizes used are large enough to deduce automata that allow us to accurately filter the event stream.

Figure 5 shows how the average F1-score varies based on percentage of noise and threshold level. We observe that the F1-score slightly converges for the different threshold levels as noise increases (cf. Fig. 5a). Interestingly, in Fig. 5b, we observe that for relatively low threshold values, the range of F1-score values for various noise levels is very narrow, i.e. the filtering accuracy is less sensitive to changes in the noise level. This effect diminishes as the threshold increases, leading more scattered yet lower F-score values. Observe that, these observations coincide with the *Kendall rank correlation coefficient values* of 0, 1792 (Fig. 5a) and −0, 8492 (Fig. 5b) respectively. We conclude that, for the dataset used, the threshold level seems to be the most dominant factor in terms of the F1-score.

Time Performance. Window w maintains a finite representation of the stream, thus, memory consumption of the proposed filter is finite as well. Hence we focus on time performance, which we measured in RapidProM, using one stream per base model with 15% noise, and several different parameter values. The experiments were performed on an Intel Xeon CPU (6 cores) 3.47 GHz system with 24 GB memory. Average event handling time was ~0.017 ms, leading to handling ~58.8 events per ms. These results show that our filter is suitable to work in real-time settings.

5.2   Drift Detection Accuracy

In a second set of experiments, we evaluate the impact of our filter on the accuracy of process drift detection. For this, we chose a state-of-the-art technique for drift detection that works on event streams [20]. We apply our filter to the event streams generated from a variety of synthetic and real-life logs, with different levels of noise, and compare drift detection accuracy with and without the use of our filter.

Experimental Setup. For these experiments, we used the 18 event logs proposed in [20]. These event logs were generated by simulating a model featuring 28 different activities (combined with different intertwined structural patterns). Additionally, each event log contains nine drifts obtained by injecting control-flow changes into the model. Each event log features one of the twelve *simple change patterns* [24] or a combination of them. Simple change patterns may be combined through the insertion ("I"), resequentialization ("R") and optionalization ("O") of a pattern. This produces a total of six possible *nested change patterns*, i.e. "IOR", "IRO", "OIR", "ORI", "RIO", and "ROI". For a description of each change pattern we refer to [20].

Starting from these 18 event logs, we generated 36 additional event logs (two for each original event log) containing 2.5% and 5% of noise (generated inserting random events into traces of each log). This led to a data set of 54 event logs (12 simple patterns and 6 composite patterns with 0%, 2.5%, and 5% noise), each containing 9 drifts and approximately 250,000 events.

Results on Synthetic Data. In this experiment, we evaluated the impact that our approach has on the accuracy of the drift detection technique proposed in [20]. Figure 6 illustrates F1-score and mean delay of the drift detection before and after the application of our filter over each change pattern.

The filter successfully removed on average 95% of the injected noise, maintaining and even improving the accuracy of the drift detection (with F1-score of above 0.9 in all but two change patterns). This was achieved while delaying the detection of a drift by less than 720 events on average (approximately 28 traces).

When considering noise-free event streams (cf. Fig. 6a), our filter preserved the accuracy of the drift detection. For some change patterns ("rp", "cd", "IOR", and "OIR"), our filter improved the accuracy of the detection by increasing its precision. This is due to the removal of sporadic event relations, that cause stochastic oscillations in the statistical test used for drift detection. Figure 6b and 6c show that noise negatively affects drift detection, causing the F1-score to drop, on average, to 0.61 and 0.55 for event streams with 2.5% and 5% of noise, respectively. This is not the case when our filter is applied, where an F1-score of 0.9 on average is achieved.

Finally, in terms of detection delay, the filter on average increased the delay by 370, 695, and 1087 events (15, 28, and 43 traces) for the logs with 0%, 2.5%, and 5% noise, respectively. This is the case since changes in process behaviour immediately following a drift are treated as noise.

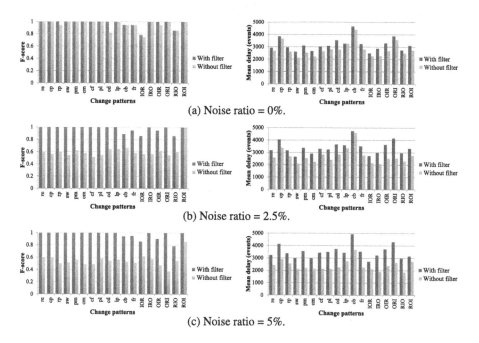

(a) Noise ratio = 0%.

(b) Noise ratio = 2.5%.

(c) Noise ratio = 5%.

Fig. 6. Drift detection F1-score and mean delay per change pattern, obtained from the drift detection technique in [20] over filtered vs. unfiltered event streams.

Results on Real-Life Data. In this experiment, we checked if the positive effects of our filter on drift detection, observed on synthetic data, translate to real-life data. For this, we used an event log containing cases of Sepsis (a life-threatening complication of an infection) from the ERP system of a hospital [17]. Overall, the event log contains 1,050 cases with a total of 15,214 events belonging to 16 different activities.

For this experiment, we attempted the detection of drift over the last 5,214 events, as the first 10,000 events are used to train the filter. Figure 7 plots the *P-value* curves of the statistical tests used for drift detection, both without (left figure) and with (right figure) the use of our filter. When comparing these two curves, what appears evident is that drifts detected after the 2067th event and after the 4373th event are no longer there after the application of our filter. In the

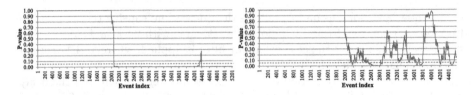

Fig. 7. P-value without filtering (left) and with our filtering (right) for the Sepsis log.

experiments with synthetic logs, we observed that our filter reduced the number of false positives (drift detected when it did actually not occur). To verify if this was also the case for the real-life event log, we profiled the direct-follows dependencies occurring before and after the drifts.

The profiling showed that while direct-follows dependencies "IV Antibiotics $\longrightarrow$ Admission NC" and "ER Sepsis Triage $\longrightarrow$ IV Liquid" could be observed several times across the entire event stream, the infrequent direct-follows dependencies "Admission NC $\longrightarrow$ IV Antibiotics" and "IV Liquid $\longrightarrow$ ER Sepsis Triage" appeared only in the proximity of the two drifts. These two infrequent dependencies cause a change in the $\alpha+$ relations between the activities (changing from causal to concurrent), which then results in the detection of the drifts. These infrequent dependencies are removed by our filter. In light of these insights, we can argue that the two drifts detected over the unfiltered event stream are indeed false positives, confirming what we already observed on the experiments with synthetic logs, i.e. that our filter has a positive effect on drift detection accuracy.

5.3 Threats to Validity

The collection of models used within the synthetic experiments related to filtering accuracy, i.e. as presented in Sect. 5.1, represent a set of closely related process models. As such these results are only representative of models that exhibit similar types and relative amounts of control-flow constructs compared to the process models used. Similarly within these experiments, the events are streamed trace by trace, rather than using event-level time stamps. Note that, since the process is stable we expect the automata to be based on a sufficient amount of behaviour, similar to streaming parallel cases. Finally note that, we do observe that our filter can be applied on real-life data, yet whether the results obtained are valid is hard to determine due to the absence of a ground-truth.

6   Conclusion

We proposed an event stream filter for online process mining, based on probabilistic automata which are updated dynamically as the event stream evolves. A state in these automata represents a potentially abstract view on the recent history of cases emitted onto the stream. The probability distribution defined by the outgoing arcs of a state is used to classify new behaviour as spurious or not.

The time measurements on our implementation indicate that our filter is suitable to work in real-time settings. Moreover, our experiments on accuracy show that, on a set of stable event streams, we achieve high filtering accuracy for different instantiations of the filter. Finally, we show that our filter significantly increases the accuracy of state-of-the-art online drift detection techniques.

As a next step, we plan to use our filter in combination with other classes of online process mining techniques, such as techniques for predictive process monitoring and automated process discovery.

Currently, filtering is immediately applied when an event arrives, taking into account only the recent history for that event. To increase filtering accuracy, we plan to experiment with different buffering strategies for incoming events, to keep track both of the recent history as well as the immediate future for each event. We also plan to test different strategies for adapting the length of the sliding window used to build our automata. For example, in our experiments we often observed windows with a large number of events with low relative frequency, due to a high number of parallel cases and due to case inactivity. The hypothesis here is that in these cases a larger window leads to less false positives.

Acknowledgments. This research is funded by the Australian Research Council (grant DP150103356), and the DELIBIDA research program supported by NWO.

References

1. van der Aalst, W.M.P.: Process Mining - Data Science in Action. Springer, Heidelberg (2016). https://doi.org/10.1007/978-3-662-49851-4
2. van der Aalst, W.M.P., Bolt, A., van Zelst, S.J.: RapidProM: Mine Your Processes and Not Just Your Data. CoRR abs/1703.03740 (2017)
3. Aggarwal, C.C.: On biased reservoir sampling in the presence of stream evolution. In: Proceedings of the VLDB 2006, pp. 607–618. VLDB Endowment (2006)
4. Babcock, B., Datar, M., Motwani, R.: Sampling from a moving window over streaming data. In: Proceedings of the ACM SODA 2002, pp. 633–634. SIAM (2002)
5. Bifet, A., Gavaldà, R.: Learning from time-changing data with adaptive windowing. In: Proceedings of the SDM 2007, pp. 443–448. SIAM (2007)
6. Burattin, A., Cimitile, M., Maggi, F.M., Sperduti, A.: Online discovery of declarative process models from event streams. IEEE TSC **8**(6), 833–846 (2015)
7. Burattin, A., Sperduti, A., van der Aalst, W.M.P.: Control-flow discovery from event streams. In: Proceedings of the CEC 2014, pp. 2420–2427. IEEE (2014)
8. Burattin, A., Carmona, J.: A framework for online conformance checking. In: Teniente, E., Weidlich, M. (eds.) BPM 2017. LNBIP, vol. 308, pp. 165–177. Springer, Cham (2018). https://doi.org/10.1007/978-3-319-74030-0_12
9. Chandola, V., Banerjee, A., Kumar, V.: Anomaly detection for discrete sequences: a survey. IEEE Trans. Knowl. Data Eng. **24**(5), 823–839 (2012)
10. Conforti, R., La Rosa, M., ter Hofstede, A.H.M.: Filtering out infrequent behavior from business process event logs. IEEE TKDE **29**(2), 300–314 (2017)
11. Cormode, G., Shkapenyuk, V., Srivastava, D., Xu, B.: Forward decay: a practical time decay model for streaming systems. In: Proceedings of the ICDE 2009, pp. 138–149. IEEE (2009)

12. Dumas, M., La Rosa, M., Mendling, J., Reijers, H.A.: Fundamentals of Business Process Management. Springer, Heidelberg (2013). https://doi.org/10.1007/978-3-642-33143-5

13. Fani Sani, M., van Zelst, S.J., van der Aalst, W.M.P.: Improving process discovery results by filtering outliers using conditional behavioural probabilities. In: Teniente, E., Weidlich, M. (eds.) BPM 2017. LNBIP, vol. 308, pp. 216–229. Springer, Cham (2018). https://doi.org/10.1007/978-3-319-74030-0_16

14. Gupta, M., Gao, J., Aggarwal, C.C., Han, J.: Outlier detection for temporal data: a survey. IEEE Trans. Knowl. Data Eng. **26**(9), 2250–2267 (2014)

15. Hassani, M., Siccha, S., Richter, F., Seidl, T.: Efficient process discovery from event streams using sequential pattern mining. In: Proceedings of the SSCI 2015, pp. 1366–1373. IEEE (2015)

16. Maaradji, A., Dumas, M., La Rosa, M., Ostovar, A.: Detecting sudden and gradual drifts in business processes from execution traces. IEEE TKDE **29**(10), 2140–2154 (2017)

17. Mannhardt, F.: Sepsis Cases - Event Log. Eindhoven University of Technology (2016). https://doi.org/10.4121/uuid:915d2bfb-7e84-49ad-a286-dc35f063a460

18. Marquez-Chamorro, A., Resinas, M., Ruiz-Cortes, A.: Predictive monitoring of business processes: a survey. IEEE Trans. Serv. Comput. (2017). https://doi.org/10.1109/TSC.2017.2772256

19. Ostovar, A., Maaradji, A., La Rosa, M., ter Hofstede, A.H.M.: Characterizing drift from event streams of business processes. In: Dubois, E., Pohl, K. (eds.) CAiSE 2017. LNCS, vol. 10253, pp. 210–228. Springer, Cham (2017). https://doi.org/10.1007/978-3-319-59536-8_14

20. Ostovar, A., Maaradji, A., La Rosa, M., ter Hofstede, A.H.M., van Dongen, B.F.V.: Detecting drift from event streams of unpredictable business processes. In: Comyn-Wattiau, I., Tanaka, K., Song, I.-Y., Yamamoto, S., Saeki, M. (eds.) ER 2016. LNCS, vol. 9974, pp. 330–346. Springer, Cham (2016). https://doi.org/10.1007/978-3-319-46397-1_26

21. Verbeek, H.M.W., Buijs, J.C.A.M., van Dongen, B.F., van der Aalst, W.M.P.: XES, XESame, and ProM 6. In: Soffer, P., Proper, E. (eds.) CAiSE Forum 2010. LNBIP, vol. 72, pp. 60–75. Springer, Heidelberg (2011). https://doi.org/10.1007/978-3-642-17722-4_5

22. Vitter, J.S.: Random sampling with a reservoir. ACM TOMS **11**(1), 37–57 (1985)

23. Wang, J., Song, S., Lin, X., Zhu, X., Pei, J.: Cleaning Structured event logs: a graph repair approach. In: Proceedings of the ICDE 2015, pp. 30–41. IEEE (2015)

24. Weber, B., Reichert, M., Rinderle-Ma, S.: Change patterns and change support features - enhancing flexibility in process-aware information systems. DKE **66**(3), 438–466 (2008)

25. van Zelst, S.J., van Dongen, B.F., van der Aalst, W.M.P.: Event stream-based process discovery using abstract representations. KAIS **54**, 407–435 (2017)

26. van Zelst, S.J., Bolt, A., Hassani, M., van Dongen, B.F., van der Aalst, W.M.P.: Online Conformance Checking: Relating Event Streams to Process Models using Prefix-Alignments. IJDSA (2017). https://doi.org/10.1007/s41060-017-0078-6

The Relational Process Structure

Sebastian Steinau$^{(\boxtimes)}$, Kevin Andrews, and Manfred Reichert

Institute of Databases and Information Systems, Ulm University, Ulm, Germany
{sebastian.steinau,kevin.andrews,manfred.reichert}@uni-ulm.de

Abstract. Using data-centric process paradigms, small processes such as artifacts, object lifecycles, or Proclets have become an alternative to large, monolithic models. In these paradigms, a business process arises from the interactions between small processes. However, many-to-many relationships may exist between different process types, requiring careful consideration to ensure that the interactions between processes can be purposefully coordinated. Although several concepts exist for modeling interrelated processes, a concept that considers both many-to-many relationships and cardinality constraints is missing. Furthermore, existing concepts focus on design-time, neglecting the complexity introduced by many-to-many relationships when enacting extensive process structures at run-time. The knowledge which process instances are related to which other process instances is essential. This paper proposes the relational process structure, a concept providing full support for many-to-many-relationships and cardinality constraints at both design- and run-time. The relational process structure represents a cornerstone to the proper coordination of interrelated processes.

Keywords: Process interactions · Data-centric processes
Many-to-many relationships · Relational process structure

1 Introduction

With the emergence of data-centric process support paradigms as, for example, the object-aware [4] or artifact-centric [8] approaches, the focus has shifted away from large, monolithic process models. Instead, the use of small processes showing limited complexity is preferred. An example of such a small process may be the lifecycle process of an artifact or object. With the advent of microservices, which may be used to implement processes, and the vision that devices in the Internet of Things become capable of running their own processes, small process models are an enticing way of specifying business processes. In general, each of these small processes does not constitute a business process by itself. Instead, to reach a specific business goal, these processes need to interact and collaborate, i.e., small processes are not executed in isolation, but are interdependent. Therefore, a coordination mechanism is needed to properly manage these process interactions. It is paramount for a coordination mechanism to be aware of every process relationship at design-time and, especially, at run-time.

© Springer International Publishing AG, part of Springer Nature 2018
J. Krogstie and H. A. Reijers (Eds.): CAiSE 2018, LNCS 10816, pp. 53–67, 2018.
https://doi.org/10.1007/978-3-319-91563-0_4

Existing approaches for the interaction-focused modeling of business processes, e.g., Proclets [11], have investigated the multiplicity of process relationships in the context of *one-to-many relationships*. These approaches use cardinality constraints to manage the relations between different processes. However, it is not possible to have the necessary awareness of process relationships by solely using cardinality constraints. Furthermore, processes may exhibit *many-to-many relationships*, which previously have not been considered. While the challenges regarding many-to-many relationships in an artifact-centric business process setting have been investigated, a solution that enables a coordination mechanism to have complete awareness over all process relationships is still missing. Specifically, [2] states an open challenge: *The need for a coordination mechanism to describe which processes interact with which other processes in settings that use many-to-many-relationships.* This challenge reveals its complexity when considering the processes at run-time, i.e., when considering process instances. In general, multiple instances of a process may be related to multiple instances of another process. Figure 1 shows a schematic example of interrelated process instances and their dependencies. Identifying the relations constitutes a prerequisite to adequately resolve and consider these dependencies at run-time, i.e., for coordinating processes. This challenge is central to every coordination mechanism that considers one-to-many and many-to-many relationships.

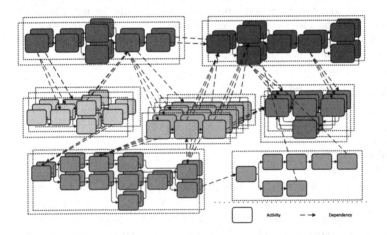

Fig. 1. Processes and their dependencies at run-time

This paper proposes a concept that solves this challenge and enables *process relation awareness* at both design- and run-time: the *relational process structure*. The complexities of managing and monitoring processes and their relations are covered by the relational process structure, which may be used in a generic fashion, i.e., the relational process structure is agnostic to specific coordination mechanisms or approaches. At design-time, the relational process structure allows identifying processes and capturing the relations between them.

It inherently supports many-to-many relationships between processes and considers cardinality constraints. At run-time, the approach automatically keeps track of instantiated processes and relations created between different process instances. This enables a coordination mechanisms to apply queries to the relational process structure, e.g., to determine which processes are related to a particular object instance. It also enforces the cardinality constraints of the different process relationships with the process instances. As the number of instances and, therefore, the relational process structure may become large, different measures are employed to reduce query times. The run-time considerations set the relational process structure apart form conventional approaches, e.g., ER diagrams.

The remainder of the paper is organized as follows. The problem statement is elaborated in Sect. 2. Section 3 discusses the design-time aspects of the relational process structure. Section 4 deliberates on the dynamic aspects and functions of the relational process structure during run-time. Considerations concerning the optimizations of a relational process structure at run-time are examined in Sect. 5. Section 6 presents the application of the concept in the PHILharmonicFlows prototype. Section 7 discusses related work before concluding the paper with a summary and an outlook in Sect. 8.

2   Problem Statement

The primary challenge of achieving *process relation awareness* is keeping track of process instances and their relations at run-time. In other words, it should be possible to identify the number and specific identifiers of related process instances at any point in time and for all process instances. The logistics process described in Example 1 exhibits many of the characteristics of interrelated processes. Moreover, the problem of interrelated processes may also be observed in domains such as Human Resources or Healthcare [5], indicating the need for keeping track of processes and their relations.

Example 1 (Simplified logistics process). An online retailer has various products on offer. A customer may place an order at the website of the retailer, requesting one or more products. The retailer handles the order by creating a bill for the customer. Once the bill has been paid, the retailer gets the ordered products from the storage and packages them for delivery. An order may be split into multiple packages, which may be distributed among multiple deliveries. A delivery may further comprise packages from different orders. If packages cannot be delivered, they are assigned to a subsequent delivery. The customer order is completed once all packages of the order have been delivered.

Order, product, bill, package, and delivery represent the processes in this example, i.e., they can be viewed as *entities* with a *lifecycle process*. The processes are interdependent. For example, a product cannot be delivered without an existing order, a product must be packaged in order to be delivered, and an order must have all its deliveries completed before it may be completed itself.

Furthermore, these processes exhibit one-to-many and many-to-many relationships. For example, an order may contain one or more products, as does a package. A delivery may contain multiple packages. An order may be distributed across deliveries, and a delivery may be associated with one or more orders, i.e., it contains products from different orders, constituting a many-to-many relationship. Therefore, relations also indicate a dependency between processes.

At run-time, an instance of an order is connected to specific product instances. Other order instances may be connected to specific, but different product instances. Later, product instances will establish relations to a specific package instance. A delivery will obtain relations to specific package instances and, consequently, will establish a relation to the order instances the packages belong to. Furthermore, a delivery is not directly connected to instances of product, but transitively via a *path* of relations. In particular, this means that dependencies exist between processes having no direct relation, e.g., the execution of a delivery instance depends on the execution status of its transitively related product instances. It is crucial that a coordination mechanism is aware of these relations.

Given the description of Example 1, it may be perceived as fairly static, with little changes in the number of related process instances over time. In principle, however, the nature of the run-time is highly dynamic. Process instances may be created or deleted at any point in time. During this time, relations may be established between process instances and, consequently, may be removed later. Due to the existence of transitive relations, the creation of new relations may not only make a connection to one process instance, but to an entire substructure of related process instances. For example, with the creation of a relation between a package and a delivery and assuming the package contains products, the delivery is now related transitively to a specific number of product instances. In general, this might have significant consequences for the coordination of these processes.

The concept of the *relational process structure* aims to provide a complete map to the relations of different process instances at run-time. Further, it must keep track of the dynamic changes that occur during run-time, delivering accurate and up-to-date information to the coordination mechanism that manages these interdependent processes. The relational process structure is intended as a generic, but capable solution to this challenge. Any possible coordination approach is required to have process relation awareness. The relational process structure serves as a foundation on which specialized approaches for coordinating interdependent process instances may build on.

To be capable of monitoring process instances and their relations, a design-time model must first identify which types of processes may exist in a given context and what relations may be established between them. This achieves process relation awareness at design-time, and coordination mechanisms are able to use the explicitly known process and relation types to define the coordination needed between these processes.

3   Design-Time Specification of the Relational Process Structure

At design-time, a relational process structure serves to capture the types of processes and relations that exist in the context of the overall business process. A relational process structure distinguishes between design-time entities, denoted as *types*, and run-time entities, denoted as *instances*. At run-time, several instances may be created by instantiating a type. The relational process structure does not assume that processes have a specific structure or use a pre-defined modeling notation; it is agnostic to the modeling paradigm and notation of the process, as well as to the specifics of the coordination mechanism used.

In principle, this enables a relational process structure to be used with any approach that deals with multiple processes and their relations. At design-time, the relational process structure is denoted as a *relational process type structure*. A formal definition of a relational process type structure and the basic process type definition are given in Definitions 1 and 2, respectively. The definitions use shorthand notations to identify types and instances. Superscript notation ^{T} denotes a design-time entity, i.e., a type, whereas superscript ^{I} denotes a run-time entity, i.e., an instance. The dot (.) represents the access operator.

Definition 1. *A relational process type structure d^T has the form (n, Ω^T, Π^T) where*

- *n is a unique identifier (name) of the relational process type structure.*
- *Ω^T is the set of object types ω^T (cf. Definition 2).*
- *Π^T is the set of relation types π^T (cf. Definition 3).*

The relational process type structure defines the context in which process types and relation types exist. Through Ω^T and Π^T, it provides an entry point for external clients, e.g., a coordination mechanism. It is represented as a graph where process types are the vertices and relation types are represented as edges.

Definition 2. *A process type ω^T has the form (d^T, n, θ^T) where*

- *d^T is the relational process type structure to which this process type belongs (cf. Definition 1).*
- *n is the unique identifier (name) of the process type.*
- *θ^T is a process specification.*

A process type requires an identifier, which has to be unique in the given context d^T, i.e., the relational process type structure. The identifier n may be indicated as ω_n^T. The details of the process specification θ^T is unimportant for the relational process structure, i.e., the relational process structure is paradigm-agnostic. Regarding Example 1, the process types include ω_{Order}^T, ω_{Bill}^T, $\omega_{Delivery}^T$, $\omega_{Package}^T$, and $\omega_{Product}^T$. However, their relations have not been identified yet. A process type stores two sets of *relation types*, identifying its incoming and outgoing relations. A relation type is represented as a directed edge between process types. A formal definition of a relation type is given in Definition 3.

Definition 3. *A relation type π^T represents an m:n relation and has the form* $(\omega^T_{source}, \omega^T_{target}, m_{upper}, m_{lower}, n_{upper}, n_{lower})$ *where*

- ω^T_{source} *is the source process type, i.e., π^T is directed.*
- ω^T_{target} *is the target process type.*
- m_{upper} *is an upper bound on the number of process instances ω^I_{target} with which ω^I_{source} may be related. Default: $m_{upper} = \infty$.*
- m_{lower} *is a lower bound on the number of process instances ω^I_{target} with which ω^I_{source} may be related. Default: $m_{lower} = 0$.*
- n_{upper} *is an upper bound on the number of process instances ω^I_{source} with which ω^I_{target} may be related. Default: $n_{upper} = \infty$.*
- n_{lower} *is a lower bound on the number of process instances ω^I_{source} with which ω^I_{target} may be related. Default: $n_{lower} = 0$.*

A relation type possesses a source and a target process type, i.e., the relation type is *directed*. It represents a many-to-many $(m : n)$ relationship between its source and target. The directed edges are used to indicate a semantic hierarchy among processes, which can often be observed (cf. Fig. 2). Each of the cardinalities m and n may be restricted by an upper and lower bound. Using the cardinality restrictions, a relation type may also be reduced to a one-to-many (1:n, m:1) or one-to-one (1:1) relationship. The actual number of process instances related to each other at run-time is restricted by the lower and upper bounds. For example, the online shop from Example 1 may establish that a delivery must contain at least 3 packages before it will be sent out, in order to reduce shipping costs. A relational process type structure uses a cardinality annotation for relation types of the following form:

$$m_{lower}..m_{upper} : n_{lower}..n_{upper}$$

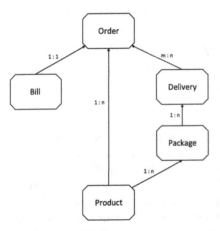

Fig. 2. Relational process model of the logistics process example

If the lower and upper bound of a cardinality are equal, or each bound possesses its default value, the notation may be shortened accordingly or replaced by the arbitrary cardinality, i.e., m or n. Figure 2 shows the relational process type structure for the processes from Example 1. Relation types have been identified and added between the process types ω^T_{Order}, ω^T_{Bill}, $\omega^T_{Delivery}$, $\omega^T_{Package}$, and $\omega^T_{Product}$. From Definitions 1–3, it can be seen that the relational process structure supports many-to-many relationships and cardinality constraints.

Two process types are said to be *related*, either *transitively* or *directly*, if there exists any *path* of relation types between them. The existence of a path may be determined with function $path^T$ as set out in Definition 4. Note that a path itself, as an entity, is defined as an ordered set of relation types, whereas function $path^T$ determines whether a path exists between two process types.

Definition 4. *Let* $d^T = (n, \Omega^T, \Pi^T)$ *be a relational process type structure. Then: function* $path^T : \Omega^T \times \Omega^T \to \mathbb{B}$ *determines whether a directed path of relations* π^T *from* $\omega_i^T \in \Omega^T$ *to* $\omega_j^T \in \Omega^T$ *exists, i.e., if* ω_i^T, ω_j^T *are related.*

$$path^T(\omega_i^T, \omega_j^T) := \begin{cases} true & \exists \pi_{out}^T \in \omega_i^T.\Pi_{out}^T : \pi_{out}^T.\omega_{target}^T = \omega_j^T \\ path^T(\omega_k^T, \omega_j^T) & \exists \pi_{out}^T \in \omega_i^T.\Pi_{out}^T : \pi_{out}^T.\omega_{target}^T = \omega_k^T, \\ & \omega_i^T \neq \omega_k^T \neq \omega_j^T \\ false & otherwise \end{cases}$$

With function $path^T$, it becomes possible to define two sets $L_{\omega^T}^T$ and $H_{\omega^T}^T$ (cf. Definition 5) that describe other processes in relation to a particular process type ω^T. The terms *lower-level* and *higher-level* hereby refer to the direction of the relations of the respective paths. There terms may describe the kind of a process relation directly, i.e., a process type may be a lower-level process of ω^T.

Definition 5. *Let* $\omega^T = (d^T, n)$ *be a process type.*

(a) The set of lower-level process types $L_{\omega^T}^T$ *is defined as*
$$L_{\omega^T}^T = \{ \omega_k^T \mid \omega_k^T \in \omega^T.d^T.\Omega^T, \ path(\omega_k^T, \omega^T) \} \cup \{\omega^T\}$$
(b) The set of higher-level process types $H_{\omega^T}^T$ *is defined as*
$$H_{\omega^T}^T = \{ \omega_k^T \mid \omega_k^T \in \omega^T.d^T.\Omega^T, \ path(\omega^T, \omega_k^T) \} \cup \{\omega^T\}$$

Function $path^T$ and sets $L_{\omega^T}^T$ and $H_{\omega^T}^T$ can be defined for process instances in the same way. For the sake of clarity, process type ω^T in the subscript of sets L and H may be replaced by the identifier $\omega^T.n$. The sets facilitate the definition of several concepts in respect to process relations, and can be used for a run-time optimization as proposed in Sect. 5.

A relational process structure is represented as a directed, acyclic graph. Using directed edges to represent relations provides several benefits. First, the direction of a relation corresponds directly to its cardinalities. The target process type always possesses cardinality m, whereas the source process type always has cardinality n, allowing for easy assignment of cardinalities in the relational process structure. Second, directed edges allow two processes to be related to one common process without the relational process structure containing a cycle. In Fig. 2, process types ω_{Order}^T and $\omega_{Delivery}^T$ are (transitively) related to process type $\omega_{Product}^T$, i.e., $\omega_{Order}^T, \omega_{Delivery}^T \in L_{Product}^T$. With directed edges, the relations do not form a cycle. If undirected edges had been used, the same relational process structure would contain a cycle. Acyclic graphs with undirected edges correspond to trees, which are too restrictive to represent a relational process structure, as they prohibit commonly found relations between processes, such as the one involving ω_{Order}^T, $\omega_{Delivery}^T$, and $\omega_{Product}^T$.

A relational process type structure interdicts the existence of cycles in its graph, i.e., it is represented as an acyclic graph. The reason for the acyclicity of the relational process structure is that cycles circumvent the cardinality restrictions of a relation type. This is caused by the fact that a relational process type structure explicitly considers transitive relations. Assume that the graph of a relational process type structure contains a cycle and a specific relation π^T with $\pi^T.n_{upper} = 3$ and $\pi^T.\omega^T_{target} = \omega^T_a$ is part of this cycle, as shown exemplarily in Fig. 3a. Then, a process instance ω^I_{a1} of type ω^T_a could be related to more than three process instances of type $\pi^T.\omega^T_{source} = \omega^T_b$. Each relation type of the cycle may be instantiated multiple times, first connecting ω^I_{a1} to an instance ω^I_{b1}. Subsequently, ω^I_a is transitively connected to another instance ω^I_{a2} of the same process type. The cycle can be repeated arbitrarily often, as shown in Fig. 3b. Ultimately, ω^I_{a1} is transitively related to more than three instances of ω^T_b using relation type π^T, which renders the cardinality restriction on π^T moot. With the acyclicity of the graph, the relational process structure ensures that cardinality constraints must not be circumvented, as would be the case when using ER-diagrams having undirected edges and cycles.

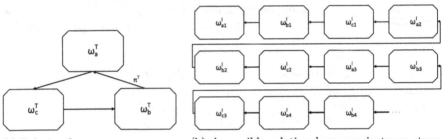

(a) Relational process type structure with cycle

(b) A possible relational process instance structure resulting from the cyclic type structure

Fig. 3. Transitive relations at run-time with cyclic type structure

Note that a process instance may still be related to more instances of a specific type than allowed by the cardinality restrictions on the direct relation between them. There may be other relations that transitively connect to the same process type, using a different path through the relational process type structure graph. For example, ω^T_{Order} in Fig. 2 is connected to $\omega^T_{Product}$ directly via $\pi^T_{Order-Product}$ and transitively via $\pi^T_{Order-Delivery}$. As a consequence, an order instance may be related to products of other orders sharing the same delivery. It is therefore crucial to take specific relation types into account when determining related instances of type $\omega^T_{Product}$ at run-time. The ability to define such sophisticated queries is only enabled by the relational process type structure. In turn, this also highlights the complexity that arises when considering transitive one-to-many and, especially, many-to-many relationships.

Transitive relations also possess a cardinality, determined by the cardinality of the relations on the path representing the transitive relation. It is possible to determine the cardinality of all transitive relations by calculating the transitive closure of the relational process type graph, e.g., by using a modified

Floyd-Warshall Algorithm. Since the relational process type structure employs directed edges to represent many-to-many relationships, the modified Floyd-Warshall Algorithm needs to be employed twice: One time going in the direction of the edges and the second time going against. Transitive cardinalities help to discover modeling errors of the relational process structure at design-time.

The relational process structure shows similarity to diagrams at design-time. This may help a domain expert or process modeling expert to gain an entry to the relational process structure and to use it effectively. A relational process structure captures process types and allows displaying the relations between them. This enables process relation awareness at design-time. A coordination mechanism may use this information to specify the necessary coordination restrictions and enforce them at run-time. However, the relational process structure extends beyond the design-time and also demonstrates benefits at run-time. The specifics of these benefits will be discussed in Sect. 4.

4   Run-Time Support of the Relational Process Structure

At run-time, the process and relation types may be instantiated. The created process instances and the interconnecting relation instances form a *relational process instance structure* (cf. Definition 6). At run-time, the continuous instantiation of processes and relations creates a highly dynamic environment, in which the relational process structure evolves dynamically as well.

Definition 6. *A relational process instance structure d^I has the form (d^T, Ω^I, Π^I) where*

- d^T *is the relational process type structure from which d^I has been instantiated.*
- Ω^I *is the set of process instances ω^I (cf. Definition 7).*
- Π^I *is the set of relation instances π^I (cf. Definition 8).*

Analogous to the relational process type structure, a relational process instance structure provides context in which instantiated processes and relations exist. In general, a multitude of different process instance structures, i.e., contexts, may exist in parallel during run-time. Furthermore, the graph of the relational process instance structure consists of process instances (cf. Definition 7) as vertices and relation instances (cf. Definition 8) as edges.

Definition 7. *A process instance ω^I has the form $(\omega^T, d^I, l, \theta^I)$ where*

- ω^T *is the process type from which ω^I has been instantiated.*
- d^I *is the relational process instance structure to which this object instance belongs (cf. Definition 6).*
- l *is the identifier (label) of the process instance. Default is $\omega^T.n$.*
- θ^I *is a process instance specification derived from $\omega^T.\theta^T$.*

Definition 8. *A relation instance π^I represents an m:n relation and has the form $(\pi^T, \omega^I_{source}, \omega^I_{target})$ where*

- π^T is the relation type from which π^I has been instantiated.
- ω^I_{source} in the source process instance, i.e., π^I is directed.
- ω^I_{target} is the target process instance.

As process and relation types may not just be instantiated once, but many times, additional complexity ensues in comparison to the relational process type structure. The process instances create a large and complex network, with possibly several independent sub-structures. However, in its basic structure, the process instances and relations evolve according to the specification of the process type structure. Thereby, a relational process instance structure resembles the corresponding relational process type structure. Regarding Example 1 and the relational process type structure depicted in Fig. 2, one possible relational process instance structure is depicted in Fig. 4.

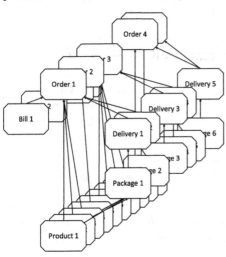

Fig. 4. Relational process instance structure for the logistics example

As instances of processes and relations may only be created if a corresponding type has been specified, tracking instances at run-time is greatly facilitated. When creating a relation π^I between two process instances ω^I_i and ω^I_j, it is first checked with the type structure whether the cardinality constraints of the relation type permit creating more relation instances between these specific process instances. The corresponding check can be performed efficiently, as process instances store their incoming and outgoing relation instances. By counting the number of relations instances π^I in $d^I.\Pi^I$ where $\pi^I.\omega^I_{target} = \omega^I_i$ that are instantiated from type π^T. This number can then simply be checked against the cardinality restrictions of π^T; the relation type may then be instantiated accordingly. By checking the minimum cardinality, the relational process type structure is capable of determining whether additional instances are needed.

The relational process instance structure keeps track of all instances by interconnecting them through $\pi^I.\omega^I_{target}$ and $\pi^I.\omega^I_{source}$ and storing them in $d^I.\Pi^I$ By keeping track which process types have been instantiated and what relation instances have subsequently been created between them, the relational process instance structure has full process relation awareness.

The relational process instance structure evolves over time and alters shape. When a coordination mechanism makes a query, e.g., about which products are contained in a specific package, the relational process instance structure is able to provide a reliable result. Subsequent additions to the package, i.e., new products are instantiated and new relations are created between them and the package, alter the result of the query. Should the same query be made at a later point in time, the relational process instance structure returns the updated result. Regarding the relational process structure, the term "query" is used to indicate the extraction of data from the relational process structure by an external agent. It is assumed that the agent corresponds to the coordination mechanism that requires the knowledge of process relationships to properly coordinate the process instances involved. Regarding the formal specification of queries, this paper remains intentionally vague to avoid unnecessary limitations, also regarding future extensions of the concept. However, examples are given throughout this paper, and the query "Which process instances are related to process instance ω^I?" serves as a reference.

The main run-time benefit of the relational process structure is the provision of detailed information on the relations of any given process. The creation of the graph representing the relational process instance structure comes at very little cost in terms of computation time. However, querying the graph in order to obtain desired information is, in general, computationally expensive. Some kinds of queries can be answered efficiently by the relational process instance structure, e.g., obtaining all directly related process instances of specific process instance ω_a^I requires the aggregation of all source processes $\{\omega^I \mid \omega^I \in d^I.\Omega^I, \exists \pi^I : \pi^I.\omega_{target}^I = \omega_a^I \wedge \pi^I.\omega_{source}^I = \omega^I\}$. However, as transitive relations constitute integral parts of the structure, most queries require a *depth-first search* or *breadth-first search* of the relational process structure. These possess time complexity of $O(|\Omega| + |\Pi|)$, in terms of processes Ω and relations Π of the relational process structure. Obviously, the time needed for queries increases when the relational process structure instance grows. For the remainder of the paper, it is assumed that queries use a depth-first search.

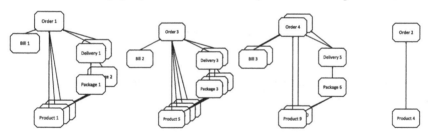

(a) Sub-structure A   (b) Sub-structure B   (c) Sub–structure C   (d) Sub-structure D

Fig. 5. Sub-structures of the overall process instance structure

The main problem, however, is a different one. The continuously evolving process instance structure requires that a coordination mechanism also queries the relational process instance structure continuously in order to keep updated. Figures 5a–d show different sub-structures of the relational process instance structure from Fig. 4. The sub-structures show all interrelated process instances. When any of these sub-structures are altered, e.g., by adding a new relation to another process instance, another query must be made so that a coordination mechanism can discover the change. The sub-structures may even be combined, e.g., by connecting Delivery 5 in Fig. 5c with Order 2 in Fig. 5d. A significant change in (transitive) process relations may be observed, requiring even more queries to discover the specific changes in regard to each process instance involved.

While the relational process instance structure is responsible for increasing the query count due to dynamic changes and therefore prolonging the individual query execution time, another factor has not been considered yet. A coordination mechanism is not only affected by dynamic changes in process relations and process instance count, but also by changes regarding the progress of process instance execution. In principle, the progress of process instances is independent from the evolution of the relational process instance structure, i.e., process instances may change their execution status while the relational process instance structure remains unchanged. However, to discover changes in process status, additional queries must be issued to determine execution status changes of process instances. In general, not only process status, but any metadata might be queried. An optimization to address this problem is presented in Sect. 5.

5   Query Performance Optimization

For alleviating the problem of continuously querying the relational process instance structure and the resulting degraded performance due to the many depth-first searches, the caching of query results is not feasible. The continuous changes to the relational process instance structure and the progress of the process instances frequently require the invalidation of the cached query results. For this reason, a practical use of *query result caching* would have negligible performance benefits. As the query count lies outside the control of the relational process structure, reducing this number to improve overall performance is impossible.

However, as many queries are likely to include the determination of related process instances, reducing individual query time becomes necessary. Therefore, caching the related process instances for each process instance would reduce the number of depth-first searches that have to be performed. In effect, for each process instance $\omega_i^I \in d^I.\Omega^I$, sets L_i^I and H_i^I (cf. Definition 5) cache the result of depth-first searches and, thereby, maintain references to the current lower- and higher-level process instances. We denote this as *related process caching*. Then, a query may use these sets directly while the relational process structure remains unchanged. If the cache can be used, the query complexity is $O(|L^I| + |H^I|) \subset O(|\Omega| + |\Pi|)$. If only one of the sets is used for the query, the query time is reduced accordingly. This allows speeding up query execution time significantly by reducing the number of depth-first searches.

In fact, it is feasible to eliminate depth-first searches entirely by maintaining L_i^I and H_i^I for each process instance i along the construction of the relational process instance structure. The basic idea is as follows: when a relation is newly instantiated, the sets of the target and source objects are updated with the newly related process instances. If this is done beginning with the first instantiated relation, the overall state of the relational process instance structure is always kept consistent. The construction of the relational process instance structure with related process caching and its correctness will be shown exemplarily for set L^I. The construction works analogously for set H^I. In the beginning, suppose two process instances ω_a^I and ω_b^I have been instantiated and no relation exists. Then $L_a^I = \{\omega_a^I\}$ and $L_b^I = \{\omega_b^I\}$. Without loss of generality, a relation is created with ω_b^I as source and ω_a^I as target. Accordingly set $L_a^I = L_a^I \cup L_b^I = \{\omega_a^I, \omega_b^I\}$ is obtained. This is correct, as ω_b^I now has a path to ω_a^I. Set L_b^I remains unchanged, as ω_b^I has gained to new lower-level instances. In the general case, if a new relation is created with ω_j^I as source and ω_i^I as target, and with the postulation that both sets L_i^I and L_j^I contain the correct process instances, set $L_i^I = L_i^I \cup L_j^I$. Every $\omega_k^I \in L_i^I$ now fulfills $path^I(\omega_k^I, \omega_i^I)$ due to the newly created relation.

While L_i^I is correct, the overall process instance structure is inconsistent, as higher-level process instances of ω_i^I, i.e., set H_i^I, are related to the new lower-level process instances in L_j^I. Therefore, L_j^I must be propagated to the process instances in H_i^I, i.e., $\forall \omega_h^I \in H_i^I : L_h^I \cup L_j^I$. As can be seen, the overall process instance structure is consistent, i.e., every process instance has the correct lower-level process instances cached. Consequently, querying may be performed faster. As a drawback, the related process instance caching increases the time for creating a new relation between process instances. However, as the number of newly instantiated relations is expected to be significantly lower than the number of queries, a significant overall performance increase can be achieved.

6   Application to the Object-Aware Approach

To provide a first evaluation of the relational process structure, it is applied to the object-aware process support paradigm [4]. The relational process structure is used to organize object types and their corresponding lifecycle processes, extending the previous data model with many-to-many relationships. The lifecycle processes of different objects are coordinated using *semantic relationships* [9], a concept that explicitly requires the identification and monitoring of relations between objects. Thereby, the object-aware approach takes full advantage of the capabilities of the relational process structure. The specification of semantic relationships requires the identification of relation types at design-time, i.e., it uses the relational process type model. At run-time, the exact relations between objects are crucial for a proper coordination. The necessary information is provided by the relational process instance structure.

The relational process structure, as presented in this paper, has been fully realized in the implementation of the object-aware approach, named PHILharmonicFlows. More precisely, the rudimentary implementation of the data model has been replaced with the relational process structure, offering many-to-many relation support. This replacement also improved the run-time capabilities of PHILharmonicFlows significantly, offering full-fledged run-time support. Various preliminary test results of the performance of the PHILharmonicFlows system show that the optimizations presented in Sect. 5 reduces query time by orders of magnitude, while increasing the time for creating a relation instance only moderately. A thorough evaluation with reliable numbers of PHILharmonicFlows performance will be presented in a future publication. Currently, PHILharmonicFlows, and with it the relational process structure, is moving towards a highly scalable architecture using microservices [1].

7   Related Work

The coordination of large process structures with focus on the engineering domain is considered in [6,7]. The COREPRO approach explicitly considers process relations with one-to-many cardinality. However, it does not consider many-to-many relationships and transitive relations. Furthermore, relationships cannot be restricted by cardinality constraints.

Artifacts consist of a lifecycle model and an information model [8]. The lifecycle model is described suing the Guard-Stage-Milestone (GSM) metamodel [3]. The information model may store any information required for the operation of the artifact. While relations of artifacts play a significant role in the specification of a business process, a concept similar to the relational process structure is missing. Nonetheless, process modelers may use the information model and the GSM lifecycle to replicate the same functionality. However, it creates high efforts for the process modeler and is error-prone. Leveraging the functionality of the relational process structure is therefore beneficial and easily realizable.

Proclets [10,11] are lightweight processes with a focus on process interactions. They interact via messages called *performatives*. Proclets allow specifying the cardinality for a message multicast, i.e., the number of Proclets receiving a performative. However, this number is fixed at design-time. The specification of Proclets does not include many-to-many-relationships between Proclets. Each Proclet requires a direct channel to exchange a performative. For this reason, transitive relations are not considered in the Proclet approach. While Proclets specify cardinality constraints, the exact recipients of a performative at run-time are unknown. Common to all these approaches is that their focus is almost exclusively on design-time issues. The many challenges arising from providing run-time support are not considered.

8   Summary and Outlook

The concept of the relational process structure allows modeling processes and their relations at design-time, accounting for many-to-many relationships, transitive relations, and cardinality constraints between processes. At run-time, the design-time information is used to automatically track process instances and their relations to other process instances. Thereby, the cardinality constraints may be enforced. At any point in time it is possible to obtain accurate information about process instances and their relations. Optimizations for increasing performance during run-time have been proposed. The whole relational process structure has been implemented in the PHILharmonicFlows system.

The relational process structure, as presented in this paper, allows tracking many-to-many relationships and enforcing cardinality constraints. However, several extensions to enhance the functionality and performance may be added in the future, e.g., restricting the overall number of process instances of a certain type, which also poses new challenges. While the implementation in PHILharmonicsFlows shows the applicability and general viability of the relational process structure concept, an evaluation with user studies and performance assessments of the system will be conducted to ascertain and solidify the benefits of this concept.

Acknowledgments. This work is part of the project ZAFH Intralogistik, funded by the European Regional Development Fund and the Ministry of Science, Research and the Arts of Baden-Württemberg, Germany (F. No. 32-7545.24-17/3/1)

References

1. Andrews, K., Steinau, S., Reichert, M.: Towards hyperscale process management. In: 8th International Workshop on Enterprise Modeling and Information Systems Architectures (EMISA), pp. 148–152. CEUR-WS.org (2017)
2. Fahland, D., de Leoni, M., van Dongen, B.F., van der Aalst, W.M.P.: Many-to-many: some observations on interactions in artifact choreographies. In: 3rd Central-European Workshop on Services and their Composition (ZEUS), vol. 705, pp. 9–15. CEUR-WS.org (2011)
3. Hull, R., et al.: Introducing the guard-stage-milestone approach for specifying business entity lifecycles. In: Bravetti, M., Bultan, T. (eds.) WS-FM 2010. LNCS, vol. 6551, pp. 1–24. Springer, Heidelberg (2011). https://doi.org/10.1007/978-3-642-19589-1_1
4. Künzle, V., Reichert, M.: PHILharmonicFlows: towards a framework for object-aware process management. J. Softw. Maint. Evol. Res. Pract. **23**(4), 205–244 (2011)
5. Lenz, R., Reichert, M.: IT support for healthcare processes - premises, challenges, perspectives. Data Knowl. Eng. **61**(1), 39–58 (2007)
6. Müller, D., Reichert, M., Herbst, J.: Data-driven modeling and coordination of large process structures. In: Meersman, R., Tari, Z. (eds.) OTM 2007. LNCS, vol. 4803, pp. 131–149. Springer, Heidelberg (2007). https://doi.org/10.1007/978-3-540-76848-7_10
7. Müller, D., Reichert, M., Herbst, J.: A new paradigm for the enactment and dynamic adaptation of data-driven process structures. In: Bellahsène, Z., Léonard, M. (eds.) CAiSE 2008. LNCS, vol. 5074, pp. 48–63. Springer, Heidelberg (2008). https://doi.org/10.1007/978-3-540-69534-9_4
8. Nigam, A., Caswell, N.S.: Business artifacts: an approach to operational specification. IBM Syst. J. **42**(3), 428–445 (2003)
9. Steinau, S., Künzle, V., Andrews, K., Reichert, M.: Coordinating business processes using semantic relationships. In: 19th IEEE Conference on Business Informatics (CBI), pp. 33–43. IEEE Computer Society Press (2017)
10. van der Aalst, W.M.P., Barthelmess, P., Ellis, C.A., Wainer, J.: Workflow modeling using proclets. In: Scheuermann, P., Etzion, O. (eds.) CoopIS 2000. LNCS, vol. 1901, pp. 198–209. Springer, Heidelberg (2000). https://doi.org/10.1007/10722620_20
11. van der Aalst, W.M.P., Barthelmess, P., Ellis, C.A., Wainer, J.: Proclets: a framework for lightweight interacting workflow processes. In. J. Coop. Inf. Syst. **10**(04), 443–481 (2001)

User-Oriented IS Development

Support of Justification Elicitation:
Two Industrial Reports

Clément Duffau[1,2(✉)], Thomas Polacsek[3], and Mireille Blay-Fornarino[2]

[1] AXONIC, Sophia Antipolis, France
duffau@i3s.unice.fr
[2] Université Côte d'Azur, I3S, CNRS UMR 7271, Sophia Antipolis, France
[3] ONERA, Toulouse, France

Abstract. The result of productive processes is commonly accompanied by a set of justifications which can be, depending on the product, process-related qualities, traceability documents, product-related experiments, tests or expert reports, etc. In critical contexts, it is mandatory to substantiate that a product's development has been carried out appropriately which results in an inflation of the quantity of justification documents. This mass of document and information is difficult to manage and difficult to assess (in terms of soundness). In this paper, we report on the experience gained on two industrial case studies, in which we applied a justification elicitation approach based on justification diagrams and justification pattern diagrams in order to identify necessary and sufficient justification documentation.

Keywords: Requirements elicitation · Justification · Certification
Requirements engineering · Quality requirements

1 Introduction

In critical contexts, it is usual to provide documentation to explain why product development is trustworthy. Here, we use the term critical in a very general sense: it qualifies an activity that may have very negative consequences for a product development or for a project. In this context, the purpose of this documentation is to convince that development process has been managed correctly and/or the design process properly followed a standard. It is not the matter to convince the customer about the features of the final product, but to convince the *accreditation client* that he cans be confident in the final product. An accreditation client can be a project manager, certification authorities or the client of the product. Unlike a usual final customer who will focus on product requirements regardless of the chosen development methodology, the accreditation client is mainly concerned by the achievement of quality requirements.

We can draw a parallel between the accreditation client's activities and activities in the field of simulation. Thus, Verification, Validation and Accreditation (VV&A) [2] defines an accreditation activity that involves an authority to certify

© Springer International Publishing AG, part of Springer Nature 2018
J. Krogstie and H. A. Reijers (Eds.): CAiSE 2018, LNCS 10816, pp. 71–86, 2018.
https://doi.org/10.1007/978-3-319-91563-0_5

that a model or simulation can be used for a specific usage. To this end, it is necessary to have comprehensive documentation, a set of justifications, explaining not only the results but also input data, hypotheses, applied techniques, etc. The accreditation activities, or the certification activities, consist to collect and evaluate this documentation.

Historically, accreditation activities are strongly linked to critical contexts such as aeronautics, health, railway and automotive. However, the need to produce a set of justifications to convince of the validity of an activity is now extending in areas such as risk management and strategic decision.

Pointed out by Knauss [15], in the context of projects related to safety and security, quality requirements are difficult to obtain. To cope with this need of justifications, in order to respond to standards and to be sure of the completeness of the justifications, we observe a widespread practice of recording, tracing and motivating everything. However, some documents are *useless* according to justification purposes because they do not provide justifications (e.g. logs), are redundant (e.g. same information in different formats or repeated in several documents).

In addition, sometimes the development of a project is difficult to predict. The activities to be carried out during a development cycle can lead to changes depending on, for instance, hardware evolutions or experimental tests feedback. So, justification activities must be adapted to this evolution to ensure that all the required justifications are provided at the end of the development stage. In these conditions, it is essential to capitalize on the strategies that have been followed to obtain these justifications.

In [24], Polacsek introduced a new kind of diagram, namely the *Justification Diagram* (JD)[1], to support accreditation and certification activities. The JD allows to organize in diagram form the various elements, formal and informal, that contribute to the justification of a result. It shows the rationale of the documentation and presents this information in a comprehensive way.

In this article, starting from the definition of JD given in [24], we will introduce a new concept: *Justification Pattern Diagram* (JPD), which is an abstraction of JD (cf. Sect. 2). Through two industrial reports, whose domains, objectives and life cycles are very different, we explain how we used these diagrams to elicit justifications. A medical technologies use case, in Sect. 3, focuses on certification of software in medical devices in the context of agile development and conformance to several tangled standards. In Sect. 4, an aeronautic use case focuses on a strategic decision totally independent of certification purpose but requiring confidence. In Sect. 5, we compare our approach to related work and Sect. 6 is dedicated to lessons learned. Section 7 concludes and gives some perspectives.

---

[1] In the first version, these diagrams are called the *Argumentation Diagram*, but to avoid any ambiguity with dialogical argumentation, we chose to use the term *Justification Diagram*.

2 Justification Diagrams and Justification Pattern Diagrams

2.1 Justification Diagrams (JD)

Justification Diagrams come from argumentation theory, specifically from the Toulmin argumentation schema [27]. The aim of JD is to define a comprehensive notation to explain why a result is trustable. It captures the rationale logical structure of all evidence that leads to the acceptance of a high-level property. On the top, there is a *conclusion* and, on the bottom, the leaves are the *evidences* that lead to the conclusion. Actually, a JD is the juxtaposition of reasoning steps, where a reasoning step is the transition from *supports* to a *conclusion*, as presented in [24].

The cornerstone of this model is the notion of *strategy*. The strategy is the inference relation, it explains clearly how, from supports, it is possible to infer a conclusion. Examples of strategies are: an expert committee review, the application of a standard, the use of software or the validation by an expert. To understand what underlies the strategy and why the strategy is acceptable, two concepts are added: (a) *rationale*, the justification for why guarantees are acceptable, and (b) *usage domain*, contexts in which the strategy can be applied.

Note that, in a project, JD is built alongside of a development process that accompanies the various stages of Verification and Validation activities: it is constructed by aggregating artefacts (e.g. documents, spreadsheets, simulation). Experiments on the use of JDs have already been conducted in various aeronautical case studies [3, 24] and integrated in a software in the context of medical experimental studies [8].

2.2 Justification Pattern Diagrams (JPD)

In this article, we extend the concept of JD presented in [24] with *Justification Pattern Diagrams*. JPD is an abstraction of JD, it designs the expected JDs. Indeed, before starting a project, identifying what is needed in terms of justifications is crucial.

Designing JPDs is a complex task. In fact, it requires a global vision of the process, a good knowledge of the applicable standards, the definition of the expected usages of the product and it relies on previous means of justification. By means of JPD, experts reason on the type of artefacts (e.g. tests coverage, requirements) more than on the documents themselves (e.g. Jacoco document results, Business Requirements Document). When a JD results of a day-to-day justification process, a pattern-diagram is a canvas more or less strict that gives a guide to conduct a project in terms of justifications.

An example of JPD and JD is given in Fig. 1. On the left side of the Figure, the JPD gives, in an abstract way, the justification for *software integration tests* in a medical context. To validate this integration, the *traces of the integration tests* is required and tests for each module must be validated. This validation is carried out in compliance with the standard IEC 62304 which describes acceptable software life cycle processes and activities in medical. The JD, on the right side of the Figure, conforms to this JPD. All required elements by the standard are present in the JPD. For instance, the *unit tests execution traces* (respectively *integration tests execution traces*) correspond to *JUnit logs*, (respectively *SoapUI logs*). To assess on the *system integration tests* validity, without the approval of the accreditation client, the conclusion must be limited to internal purposes. This restriction can only be lifted when an external audit is conducted.

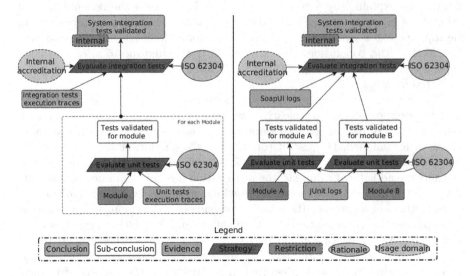

Fig. 1. Justification Pattern-Diagram (JPD) on right and the associated Justification Diagram (JD) on left in the context of automated tests execution.

The next two sections show how we used these diagrams in two industrial reports. As Franch et al. highlighted in [11], we try to fill the gap between academic research, practitioner and regulatory adoption by driving our research thanks to the following two industrial use cases. The first case study is part of an agile development process and the use of JDs for justification requirement elicitation relies on case reasoning [21]. The second case study is part of a V-Model development lifecycle and relies on a priori design.

3   Application in a Healthcare Company to Software Certification Purpose

3.1   Context

AXONIC is a young company including a dozen people, founded in 2014. It develops neurostimulation devices to address different pathologies bounded to the nervous system. According to the destination of the devices, the AXONIC's development process has to be compliant with different standards. For instance, the standard IEC 62366 is a general canvas that takes care about summative evaluation, user training and user documentation in a general medical context. Additionally to this standard, for a medical product used at the patient's home, the collateral standard IEC 62366-1 must also be fulfilled too: it introduces especially error-prone aspect, *e.g.*, physical security to prevent a child to be able to use the device. AXONIC has also to apply standards relating to hardware and software development, e.g. ISO 13485 that describes regulatory requirements on quality management system and IEC 62304 that gives a canvas of development for software in medical equipment.

In this case study, we focus on attesting conformity for clinical studies delivery dedicated to software. To this end, a set of documents has to be produced in order to ensure confidence in the product and in the development process. Guidelines can be found to help development processes to be compliant with applicable European standards [1]. These guidelines compile practices and provide tools to support the applicants in producing the appropriate evidence of compliance, i.e. the justifications. Thus, standards (e.g. ISO 14971 refined in IEC 62304 and IEC 62366) lead to complex tangled justification activities. Moreover, application contexts of medical standards can be so diversified and so complex that AXONIC wants to reuse justification patterns and capitalize on them in other projects, so that to avoid repeating the recurrent tasks several times.

In addition to technical standards, development cycle standards has to be applied. For instance, in the context of this study, we apply Agile development cycle, this development process is described in the AAMI guidance [1].

The elicitation of justification requirements is carried out at the kick-off of the project and takes into account the applicable standards and guidelines. We show, in the next subsections, how we use JPDs and JDs to elicitate the justification artefacts and their dependencies throughout the development of a new device.

3.2   Justification Requirements Elicitation

In this case study, the stakeholders involved for the justification parts are: one researcher/practitioner (PhD student), a quality management team (two people) and technical leaders (three people). The quality team is responsible for standard's watchfulness, design process, audit preparation, etc. The technical leaders ensure that justifications planned in the process are properly produced and defend these justifications in audits. To conduct the study, the researcher is included in day-to-day activities from meetings (e.g. project monitoring, risks analysis assessment, technical meetings and audits) to system production itself.

After one year of involvement in AXONIC, the researcher designed a preliminary JPD for a new project, based on company's practices (see Fig. 2). Then, during the next year, the researcher, the quality management team and the technical leaders iterated to produce justifications, improve this JPD and take into account new applicable standards. To build the JPDs at each stage of the development project, the team has iterated on the basis of the following process:

1. Before starting the development of a new stage, the researcher designs a JPD according to quality management team and technical leader requirements. This new JPD can be a refinement of the JPD of the previous step, for instance, in order to include new applicable standards. For example, The JPD in Fig. 3 is a

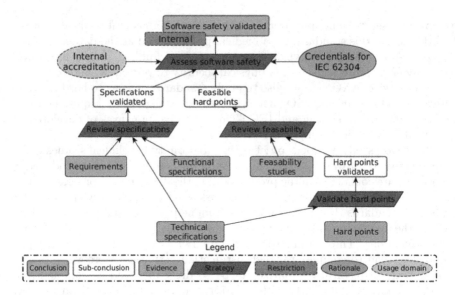

Fig. 2. JPD illustrates the result of kick-off of a new project. Requirements must lead to functional and technical specifications that must be reviewed to be validated. Due to this refinement, technical hard points are found and feasibility studies need to prove that they can be overcome.

revision of the JPD described in Fig. 2. At this step, to *assess software safety*, it is mandatory to have the architecture of the device validated. As shown on the diagram, this validation is carried out by a review which corresponds to the addition of the strategy: *Review architecture*. This architecture review requires previous justifications about feasibility review. Moreover, to conform to standard ISO 14971 new justification steps have been added to asses risk management. IEC 62304 and ISO 14971 are tangled in terms of software risk analysis. Thanks to JPD, we identify the previous artefacts that can be shifted to answer to these new requirements but also identify missing justifications.

2. Technical leaders (i) identify justification items that must be produced and (ii) develop tools to produce them (*e.g.*, extracting documents from the production toolchain, adding a plug-in to existing tools) [9]. Based on the JPD, a JD is created. If a JD from the previous iteration exists, common elements are automatically populated in the new JD by copying the elements present in the previous one. If there is no diagram of the previous iteration, all justifications artefacts will have to be produced during the stage.

3. During the development stage, technical leaders eventually define new activities, such as expert validation, committee review or external tests and the development team produces the necessary justifications of the JD. Indeed, while the development process evolves, new justifications or ways to justify appear. So, they are added to the current JD. Note that, at this point, we have a shift between the expectations designed in the JPD and the reality captured by the JD.

4. At the end of the development stage, Quality team and researcher analyze the difference between the JPD and the JD. The comparison helps to identify (i) missing justifications that must be produced and (ii) new artefacts that must be introduced as justification requirements in the JPD. After a deep analysis on the reason of occurrence of these differences, we can align the JD on the JPD: refute real-world practices, or align the JPD on the JD: accept real-world evolution. For example, during the project, AXONIC introduced the standard IEC 62366 in their practices. While producing justifications, we noted that documents produced for ISO 14971 and for IEC 62366 was actually much easier to produce together than in a separate way. This fact, different from what expected in the JPD, leads us to integrate this practice in the JPD.

During this study, we designed five JPDs, for the five main stages of the project, and we used these JPDs to produce an associated JD at each project iteration. We also versioned JPDs and JPs in the same way that the source code. In fact, it is useful to attest of good practices during all the development process and to retrieve state of justifications at each stage of the project. In the future, these JPDs will be reused as an initial input for new projects. Moreover, other JPDs will be designed for other domains (e.g., hardware, mechanics) and finally to justify all quality requirements met a such system.

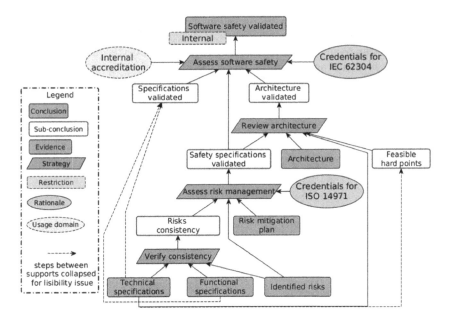

Fig. 3. Project stages further, this JPD keeps supports defined by the previous stages and adapts it to match new requirements of the current stage.

4   Application in an Aircraft Manufacturing to Workload Assessment Purpose

4.1   Context

To ensure that production costs are not taken into account too late in the aircraft design processes, it is crucial to perform an assessment of these costs at the early stages of the design. With this assessment, production costs could become one of the selection criteria among different possible aircraft designs as well as the mass, the noise, the thermal efficiency, etc. To do so, it is necessary to establish a strong interaction between the design engineering and the production and, therefore, to be able to perform a reliable assessment of the workload for a given design [25].

The current workload assessment is based on a very detailed design and requires several complex and time-consuming computations. To get a faster estimation, at the cost of a less reliable assessment, a new process has been defined. To remain in line with the aircraft manufacturer's practice, this process has to be part of a V cycle. It starts at very preliminary design and finishes with advanced design, and reuses previous results from one stage of a cycle to the next as much as possible. Confidence is gained at each step of the process. With this process, it is possible to assess the workload very quickly, in the preliminary phase, with little information and a lot of uncertainty. This workload assessment process is based on steps and a new assessment is made at each step.

The computation of workload assessment is very critical. A lot of decisions could be taken with this information, such as design choices, but also manpower allocation, cost assessment, etc. We used JPDs to have a better understanding of evidence on which this assessment is based on and clearly identify the necessary validation operations to have confidence in the final result.

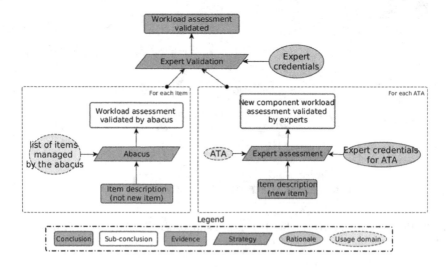

Fig. 4. Justification Pattern-Diagram at stage 1

4.2   Justification Requirements Elicitation

JPD are used to define the documentation and the list of justifications, in order to have a reliable process and to perform justification requirements elicitation. In this context, JPD is useful: firstly, to define the list of necessary evidence for this new workload assessment process; secondly, to identify activities required for justification (activities that will be added to the process afterward).

Therefore, we use the JPD to understand the rationale on workload assessment computation and to identify the evidence that must be produced at each step of the process.

The diagrams we introduce are a very simplified version of our final diagrams. We chose, for legibility and confidentiality reasons, to skip some minor details which do not contribute to the understanding. Moreover, in this case study, we worked on the aircraft modifications (design changes), so on the additions as well as the removals of components, but we chose to focus here only on the additions.

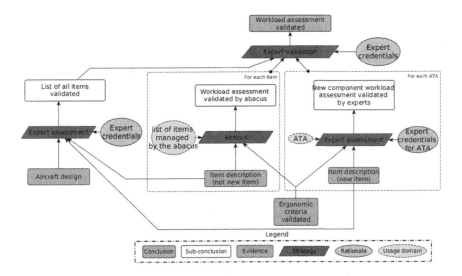

Fig. 5. Justification Pattern-Diagram at stage 2

At the beginning of this study, we focused on understanding the process of how manufacturing departments perform workload assessment. For this end, we interviewed aircraft architects, workload experts and people who are involved in the workload assessment process. We also spent two days on the assembly line to gain first-hand knowledge of their approaches.

From there, we made guileless JDs to represent what could be a good justification for a concrete example. Then, we refined these JDs with the different actors and we converged on three diagrams. Having three diagrams means that the process is in three steps. Indeed, each JD represents an accreditation step, a step where the workload assessment value must be justified. This process should describe the production activities of justification artefacts present in JDs.

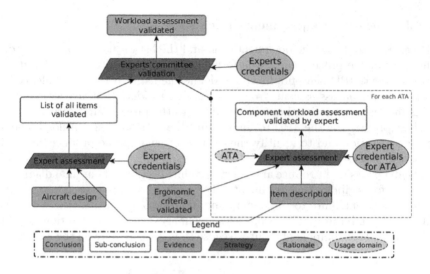

Fig. 6. Justification Pattern-Diagram at stage 3

Finally, on the basis of these JDs, we developed the JPDs, which were also confronted against reality and approved regarding practice feedbacks. In the first stage, when we have a very preliminary aircraft design, an abacus software is used to compute workload assessment (see Fig. 4). The usage domain of this software is the list of aircraft parts known by the software. The confidence in this assessment is based on the confidence we have in the software: we thus have a strategy based on the abacus software. If an aircraft's part is not in the abacus (right part of the Fig. 4), for instance it is a new component, then the workload assessment is performed by a workload expert. This assessment is done by an expert and there is an expert for each ATA[2]. Doing this JPD allowed to highlight, in a first version, the need to have an expert of the assembly line that validates the result of the software and, in a second time, that this expert also needs to validate the calculation carried out by the workload expert. In our diagram, this validation is the strategy to pass from the workload assessments to the final conclusion: the workload assessment is validated. This strategy introduces a new activity, an expert validation, which must be added to the workload assessment process.

In the second stage (see Fig. 5), all previous justifications are kept but strategies are enriched to enforce confidence. In this stage, the aircraft design is more detailed, we must therefore have the complete list of elements, nothing should be missing. This verification introduces a new element, the list of all items is validated, which is guaranteed by an expert. So, the confidence in the accuracy of the list is based on the trust on the component expert. Finally, to increase

---

[2] ATA chapters are defined by the Air Transport Association of America. It is a common referencing standard for all commercial aircraft documentation. An ATA chapter represents a aircraft domain like Air Conditioning & Pressurization (ATA 21), Electrical Power (ATA 24) or Pneumatics (ATA 36).

confidence in the final result a new document is used: the workload impact of ergonomic criteria.

In the last stage (see Fig. 6), the two most important points are: the introduction of an experts committee and the removal of the abacus software. Here, we have an advanced aircraft design and this is the last step before crucial decisions are made. Thus, all workload assessment needs to be performed by an expert and not by a machine. In the first version of this diagram, we had not changed the top validation strategy (*"Expert validation"* see Fig. 5), but it appeared that, given the critical nature of the final result, all the calculations had to be validated by an experts' committee. Note that, the committee valid not only the workload assessment, but also the list of parts. The characterization of usage domain and rationale for the strategy *"experts' committee validation"* was an opportunity to clearly define the members, the roles and the credentials of this committee.

5   Related Work

Close to our work, some methods use a goal-oriented approach to support requirement elicitation and organization. Language and methodology like KAOS [17], *i** [7] and globally all the *Goal-oriented Requirements Languages* (GRL) tackle challenges like evolution of requirements but also internal practices with a goal-oriented approach [20] and also are focused on accreditation client requirements [19]. All these goal modeling languages focus on *who* and *what*, but they do not try to capture the *why*. To fill this gap, Van Zee et al. introduce a complementary approach between GRL and argument diagrams to capture the rationale of goal models inside mapped argument diagrams [28]. They add a representation of arguments pros and cons a system modeled with *i**. Contrary to us, they focus on the justification of a choice versus alternatives and not on the justification that this choice is compliant with some standards.

In the field of knowledge organization, a numerous of methodologies are argumentation based. They all have the same root: *Issue-Based Information Systems* (IBIS) [16]. *IBIS* has been designed to support and to document decision processes, but it is now used to organise knowledge and justification ([14] presents a set of case studies and commentaries on how *IBIS* is used in practice). All the notations coming from *IBIS* have in common to try to capture the rationales behind the decisions taken during a design process, as well as the different alternatives that have finally failed. For example, the *Questions, Options and Criteria* (QOC) [18] approach identifies *design problems* with questions and *alternatives* with possible answers. In addition, QOC has an evaluation criteria based on requirements and desired properties. With these evaluation criteria, it is possible to rank different options.

However, all this approaches are designed for the early phases of development, for the design and for design choices. On our side, we address a somewhat different problem, while we are interested in the acceptance of a product and not in alternatives. It is therefore no longer a question of keeping track of the alternatives, but of trying, as finely and formally as possible, to explain the reasons and the context why a product is trustable.

In safety, two notations have been imposed to present justifications and valid arguments to convince that a system satisfies safety properties: the *Goal Structuring Notation* (GSN) [13] and *Claim-Argument-Evidence* (CAE) [10]. Even if historically the GSN and CAE were based on the Toulmin schema, nowadays it is no longer true [5]. For Toulmin, the strategy is the cornerstone of reasoning, it gives the reasons why it is possible to pass from supports to a conclusion. In GSN and CAE, strategy is optional. It is possible to jump directly from supports to a conclusion, without usage domain and justification, without explanation. By doing this a large part of the rationale is lost. In addition, both notations are very safety oriented and are not appropriate for a generic purpose.

Finally, we can cite the OMG initiative which aims to establish a meta-model for assurance case called *SACM* for *"Structured Assurance Case Metamodel"* [23]. In this standard, the OMG captures the key concepts of assurance case to structure them in a conceptual model. Therefore, this representation can be applied as the baseline of automation, checks, integration with other tools and also evolutions [12]. For us, the purpose is the same but we introduce in this article a higher level of abstraction. In contrary to *SACM* that reasons at the assurance case instance level (for us JD), we introduce JPD to design template that can be seen as assurance case templates. Even if the community is adopting the envision of templates, there are still domain centric [6] and not presented in a global approach that can be introduced in *SACM*.

Regarding the current practices in highly-regulated industries, compliance with standards is mostly ensured by traceability [26]. In practice, industries define processes that follow recognized guidelines. Compliance with these processes therefore ensures, de facto, compliance with standards. So, the means of compliance are based on proof that the process has been followed, what has been done is consistent with what was to be done, which is ensured by traceability. In our approach, the use of JPD and JD is properties oriented, not process oriented. Our aim is, therefore, to organize the justifications of means of compliance to a property or to a process.

6   Lessons Learned

JPDs Elicitate the Justification Requirements

JPDs synthesize information required to justify a property. In fact, they are an extension of requirement formalisms and methodologies.

JPD Makes the Link Between Verification and Validation (V&V) and Justification Requirements. The relationship between quality requirements and JPDs relies currently on the expertise of quality managers. In the case of the Aircraft Manufacturing application, establishing this relationship was the essential part of the work. As we have seen in this case study, the JPD captures the results of the V&V activities, and therefore captures V&V requirements. However, a systematic check of the alignment between model of quality requirements, like i^* [7], remains a perspective to this work.

JPD Design Requires an Iterative Process. The construction of JPDs is strongly dependent on: the maturity of teams, the aims of the projects (e.g. prospective, production) and the development cycle. In our two experiments, teams must refine the JPDs according to the stages of the project. For cost and progress management purposes, all justifications cannot be required at the beginning of a project. Some of them need to be added or refined along the lifecycle and, sometimes, justification activities must be added during development for instance when an artefact requires justifications that were not be anticipated. Thus, the construction of JPDs is an iterative process in which not only quality experts but also all stakeholders must be involved.

The Two Reports Focus on Long-term Projects in Different Fields. Both reports focus on different contexts that involve different standards and usages. However, they both correspond to long-term projects. As a result, teams could not yet reuse JPDs in the production of other products. However, the iterative nature of the projects treated showed the interest of reusing the JPDs from stage to another. The contexts of the studies also differ in one case by integrating the researcher into the team and, in the other case, by the use of an external consultant. In both, JPDs have proven to be very useful for discussions with all stakeholders.

JDs Help to Manage Justifications

As we have seen, the JD elements are connected to artefacts of justification, like for instance a report or a minutes. Therefore, a JD can be considered as a means to organize the justification documents. Note that, in the aeronautic experiment, JDs are built informally using the JPDs as a guide and they are built with tools in the medical experiment.

JDs Provide a Global Justification Based on Justification Artefacts. In the medical application, JDs refer to the artefacts stored in a data base, the *Electronic Document Management*. From the JD, we were able to generate automatically a textual document of justification: the *Master File*. The *Master File* is the document that quality auditors are used to use. In this project, the quality auditor was very enthusiastic the use a navigable JD instead of the *Master File*. However, for the highly regulated activities, it does not seem possible today to work only with the JD, without *Master File*, which represents an excessively important breakthrough. Nevertheless, according to the team and to the quality auditor, we get a significant benefit by automatically having artefacts updated and referenced in JD, and the ability to generate the expected auditing document is quite promising.

JDs Support the Confidence in Product Justifications. Because JPDs may be seen as a structured guideline, a detailed check list of all needed justification artefacts, they relieve development teams from missing something necessary for the product compliance. For instance, the validation by the quality team that a JPD captures all the standards requirements allows the development team to focus on the development, not on the standards. In addition, JPDs could support the production of justification artefacts. For instance, in the medical experiment, automating the production of several artefacts using tools for continuous integration is one of the key elements of this confidence [9]. Once the tools are configured, they produce always consistent artefacts contrary to humans that are error-prone.

Experiment Context Could be a Threat to Validity. In both experiments, the construction of the JD was done by small teams. We have not tested our approach on large teams, for instance with people specialized in specific quality aspects (e.g. prototyping, clinical studies, CE marking). We don't know what the results would have been on larger teams.

Moreover, while AXONIC constructs an evolving JPD during the project development, aircraft manufacturer constructs different JPDs based on old projects. This is mainly due to a difference of maturity level between the two industries. AXONIC tries to adopt Agile practices and needs to redefine their justification process, while aeronautic industry can monitor previous projects to predict the next one. Thus, it is difficult to establish a standardized methodology for JPD. However, with these two complete different use cases, we can assert that, regardless of the maturity level of the company and the life cycle model, for small teams JPDs are useful.

7 Conclusion and Perspectives

In this article, we have shown, through two industrial case studies, that it is possible to use JPDs and JDs to perform justification elicitation. Projects in aeronautics and medical are long-term developments, so, these projects are still in development and we continue to monitor usage of these notations. Still, feedbacks from industrial partners are good enough to say that we are confident for the validation in long-term usage.

The use of JPDs in the Aeronautic case study has made explicit practices that were previously unstated. These were formulated iteratively on the basis of a study of the practices and a comparison of these with the requirements stated by the experts. In the case of AXONIC, the JPDs made it possible to express, in a concise format, practices that were previously diffused in several standards and guidelines. As JPD and JD are easily accessible to team members, not only experts, they have proved to be a powerful communication support.

Today, requirement engineering efforts tend to either be too shallow in terms of justification, like in most agile approaches, or too heavyweight, especially in certification systems where compliance is the key. With the support of diagrams, it is easier to identify the necessary and sufficient documentation. So, even if we only apply our approach to two use cases carried by critical industries, we think about applying it in more general information system where justifications is valuable (e.g. system with a high quality of service, system where the goal is to convince someone). In this way, we begin to study how we can apply JD and JPD in RockFlows [4]. This meta-learning platform proposes the best data-mining workflow for a given goal, like performance or accuracy, and for a data-set.

Here, the key point for the adoption of this platform is to convince data mining experts that: there is no bias, no gap in the experiments or that the initial experiments protocol has been followed.

Finally, in our experiments, the use of JD and JPD highlight the links between justifications and activities to produce justification. It might be interesting to make the connections between our diagram and existing notations like BPMN [22] or $i*$ [7]. We think that integration of Justification Diagrams into common industrial technologies, like Continuous Integration platforms and Document Management Software, is a key point to industry interest. To address these technologies, automation around Justification Diagrams must be extended.

References

1. AAMI AT: Guidance on the use of agile practices in the development of medical device software. Association for the Advancement of Medical Instrumentation (2012)
2. Balci, O.: Verification, validation, and accreditation. In: Proceedings of the 30th Conference on Winter Simulation, pp. 41–44. IEEE Computer Society Press (1998)
3. Bieber, P., Boniol, F., Durrieu, G., Poitou, O., Polacsek, T., Wiels, V., Martinez, G.: MIMOSA: Towards a model driven certification process. In: Proceedings of the 8th International Congress on Embedded Real Time Software and Systems (ERTS 2016) (2016)
4. Camillieri, C., Parisi, L., Blay-Fornarino, M., Precioso, F., Riveill, M., Cancela Vaz, J.: Towards a software product line for machine learning workflows: focus on supporting evolution. In: Proceedings of the 10th Workshop on Models and Evolution Co-located with ACM/IEEE 19th International Conference on Model Driven Engineering Languages and Systems (MODELS), October 2016
5. Cassano, V., Maibaum, T.S.E.: The definition and assessment of a safety argument. In: 25th IEEE International Symposium on Software Reliability Engineering Workshops, ISSRE Workshops, pp. 180–185. IEEE Computer Society (2014)
6. Chowdhury, T., Lin, C.w., Kim, B., Lawford, M., Shiraishi, S., Wassyng, A., Division, S., View, M.: Principles for systematic development of an assurance case template from ISO 26262. In: ISSREW 2017 (2017)
7. Dalpiaz, F., Franch, X., Horkoff, J.: istar 2.0 language guide. CoRR abs/1605.07767 (2016)
8. Duffau, C., Camillieri, C., Blay-Fornarino, M.: Improving confidence in experimental systems through automated construction of argumentation diagrams. In: ICEIS 2017 (2017)
9. Duffau, C., Grabiec, B., Blay-Fornarino, M.: Towards embedded system agile development challenging verification, validation and accreditation: application in a healthcare company. In: ISSREW 2017 (2017)
10. Emmet, L., Cleland, G.: Graphical notations, narratives and persuasion: a pliant systems approach to hypertext tool design. In: Blustein, J., Allen, R.B., Anderson, K.M., Moulthrop, S. (eds.) Proceedings of the 13th ACM Conference on Hypertext and Hypermedia, HYPERTEXT 2002, pp. 55–64. ACM (2002)
11. Franch, X., Fernández, D.M., Oriol, M., Vogelsang, A., Heldal, R., Knauss, E., Travassos, G.H., Carver, J.C., Dieste, O., Zimmermann, T.: How do practitioners perceive the relevance of requirements engineering research? an ongoing study (2017). arXiv:1705.06013

12. González, V., Génova Fuster, G., Álvarez Rodríguez, J.M., Llorens Morillo, J.B., et al.: An analysis of safety evidence management with the structured assurance case metamodel (2017)
13. Kelly, T., Weaver, R.: The goal structuring notation - a safety argument notation. In: Proceedings of Dependable Systems and Networks 2004 Workshop on Assurance Cases (2004)
14. Kirschner, P.A., Buckingham-Shum, S., Carr, C.S.: Visualizing Argumentation: Software Tools for Collaborative and Educational Sense-Making. Computer Supported Cooperative Work. Springer, London (2003). https://doi.org/10.1007/978-1-4471-0037-9
15. Knauss, E., Liebel, G., Schneider, K., Horkoff, J., Kasauli, R.: Quality requirements in agile as a knowledge management problem: more than just-in-time. In: 2017 IEEE 25th International Requirements Engineering Conference Workshops (REW), pp. 427–430 (2017)
16. Kunz, W., Rittel, H.: Issues as elements of information systems. Working Paper 131, Institute of Urban and Regional Development, University of California, Berkeley, California (1970)
17. van Lamsweerde, A.: Requirements Engineering - From System Goals to UML Models to Software Specifications. Wiley, New York (2009)
18. MacLean, A., Young, R.M., Bellotti, V.M.E., Moran, T.P.: Questions, options, and criteria: elements of design space analysis. Hum. Comput. Interact. **6**(3), 201–250 (1991)
19. Massey, A.K., Holtgrefe, E., Ghanavati, S.: Modeling regulatory ambiguities for requirements analysis. In: Mayr, H.C., Guizzardi, G., Ma, H., Pastor, O. (eds.) ER 2017. LNCS, vol. 10650, pp. 231–238. Springer, Cham (2017). https://doi.org/10.1007/978-3-319-69904-2_19
20. Nguyen, C.M., Sebastiani, R., Giorgini, P., Mylopoulos, J.: Requirements evolution and evolution requirements with constrained goal models. In: Comyn-Wattiau, I., Tanaka, K., Song, I.-Y., Yamamoto, S., Saeki, M. (eds.) ER 2016. LNCS, vol. 9974, pp. 544–552. Springer, Cham (2016). https://doi.org/10.1007/978-3-319-46397-1_42
21. Nisbett, R.E.: Rules for Reasoning. Psychology Press, London (1993)
22. OMG: Business Process Model and Notation (BPMN), Version 2.0, January 2011
23. OMG: Structured assurance case meta-model (SACM) (2013)
24. Polacsek, T.: Validation, accreditation or certification: a new kind of diagram to provide confidence. In: 10th IEEE International Conference on Research Challenges in Information Science, RCIS, pp. 59–466 (2016)
25. Polacsek, T., Roussel, S., Bouissiere, F., Cuiller, C., Dereux, P.-E., Kersuzan, S.: Towards thinking manufacturing and design together: an aeronautical case study. In: Mayr, H.C., Guizzardi, G., Ma, H., Pastor, O. (eds.) ER 2017. LNCS, vol. 10650, pp. 340–353. Springer, Cham (2017). https://doi.org/10.1007/978-3-319-69904-2_27
26. Rempel, P., Mäder, P., Kuschke, T., Cleland-Huang, J.: Mind the gap: Assessing the conformance of software traceability to relevant guidelines. In: Proceedings of the 36th International Conference on Software Engineering, ICSE 2014, pp. 943–954. ACM (2014)
27. Toulmin, S.E.: The Uses of Argument. Cambridge University Press, Cambridge (2003). updated Edition, first published in (1958)
28. van Zee, M., Marosin, D., Bex, F., Ghanavati, S.: RationalGRL: a framework for rationalizing goal models using argument diagrams. In: Comyn-Wattiau, I., Tanaka, K., Song, I.-Y., Yamamoto, S., Saeki, M. (eds.) ER 2016. LNCS, vol. 9974, pp. 553–560. Springer, Cham (2016). https://doi.org/10.1007/978-3-319-46397-1_43

Configurations of User Involvement and Participation in Relation to Information System Project Success

Phillip Haake[1(✉)], Johanna Kaufmann[2], Marco Baumer[3],
Michael Burgmaier[3], Kay Eichhorn[3], Benjamin Mueller[4],
and Alexander Maedche[1]

[1] Karlsruhe Institute of Technology (KIT), Karlsruhe, Germany
phillip.haake@partner.kit.edu
[2] University of Passau, Passau, Germany
[3] University of Mannheim, Mannheim, Germany
[4] University of Groningen, Groningen, The Netherlands

Abstract. Information system (IS) project success is crucial given the importance of these projects for many organizations. We examine the role of user involvement and participation (UIP) for IS project success in terms of perceived usability in 16 cases, where an IS has been implemented in an organization. Qualitative Comparative Analysis (QCA) enables us to research multiple IS project configurations. We identify the participation of the appropriate users in the requirements analysis phase as the key condition for IS project success. Our research corroborates anecdotal evidence on key factors and informs practitioners about the most effective way to conduct UIP.

Keywords: Configurational theory · Qualitative Comparative Analysis (QCA)
IS project success · User involvement · User participation

1 Introduction

A large number of information system (IS) implementation projects still fail to reach their objectives [1]. Many attempts have been made to develop an understanding of the critical success factors for IS projects. However, the project success rate has not improved significantly in the past [1], despite the comprehensive knowledge gathered by practitioners and researchers. Researchers argue that project success is a multidimensional construct with different dimensions of success [2]. Each dimension can be of different importance in different projects. This indicates that there is a multitude of possible configurations for successful or unsuccessful projects and therefore an incongruence of IS project success as the phenomenon of interest and the analysis methods commonly used for its evaluation. Our interpretation of this issue in research is in line with previous efforts in general project success research [3]. This enables us to go beyond the identification of factors for project survival [4] to the evaluation of constellations of relative project success. It is the basic premise of our configurational approach that a configuration consists of a constellation of characteristics, which are

© Springer International Publishing AG, part of Springer Nature 2018
J. Krogstie and H. A. Reijers (Eds.): CAiSE 2018, LNCS 10816, pp. 87–102, 2018.
https://doi.org/10.1007/978-3-319-91563-0_6

conceptually distinct and commonly occur together [5]. Generally, we view the different characteristics of an IS project as parts of a project's configuration. Thus, we propose a remedy for the incongruence of research method and project success as the phenomenon of interest. We use the set-theoretic method of Qualitative Comparative Analysis (QCA) [6], specifically, the fsQCA method, which is generally established in other fields such as political science [7].

Many different characteristics of IS projects' can have a substantial effect on IS project success. Thus, a multitude of different configurations or characteristics can be related with a successful IS project. However, user involvement and participation (UIP) have been identified as some of the most important factors to ensure overall IS project success [1]. In particular, Bano and Zowghi [8] suggested to analyze the level and degree of UIP required to achieve project success in different project phases, as they did not find any such guideline in the literature. Furthermore, there is still a lack of knowledge about the appropriate timing for UIP, even though it has mostly been suggested that UIP is important in requirements analysis or during testing to positively influence project success [8]. Inspired by this, we focus our research on factors of UIP related to project success. We measure project success in terms of end-users' perceived usability, as UIP normally serves the purpose of improving the user interface and work processes of a system [1]. The analysis of the appropriate point in time, the kind of UIP, users' involvement and motivation, and the types of involved users are our main research interests. Based on these insights, we propose the following research question: *How are different forms of user involvement and participation in IS implementation projects related with IS project success?*

We answer this research question with an analysis of multiple-case studies of the relationship of project success and different forms of UIP in 16 different IS implementation projects. The cases are completed IS projects with a user interface for commercial users. We employ a mixed methods research approach combining qualitative and quantitative research methods to answer the research question, as such an approach is required to study the complex and multifaceted relationship between UIP and project success [8, 9]. We contribute to the IS project research by presenting additional evidence for the most effective point in time for end users' involvement and participation in an IS project. In our analysis, we are further highlighting that successful UIP is not only about when to best involve users, but also about showing what kind of involvement works best when and what pattern of relative configurations of such involvement can contribute to further improve IT project success. Therefore, our study advances previous IS project success research by moving beyond the often anecdotal evidence and experience reports by practitioners about the best point in time for user participation in IS implementation projects. We are able to show the applicability of a configurational approach and specifically fsQCA for IS project analysis. Our research also benefits practitioners because it gives them an indication to focus their resources on aspects of a project for which intensive UIP is crucial instead of using it in all phases of the project.

2   Theoretical Background

Generally, IS project success has been measured based on the dimension of IS product success [10]. It is possible to measure IS product success based on indirect indicators such as, for example, user satisfaction [11]. Usability is a very important software product characteristic [12] and can help to achieve user satisfaction. It is a higher design objective and an attribute of software quality [13]. The definitions of usability are often bound to the measurement method [13]. Hence, numerous definitions for usability exist [14]. We will adopt the process-oriented ISO 9241 standard as the definition for the purpose of this research [15]. We chose the ISO standard because it is one of the most prominent and widely used definitions. It does not focus on the characteristics required for the user interface but rather on the procedures of a system. Hence, we decided to use the values for the System Usability Scale (SUS) [16] as an indication for project success. Especially, because it is very established and allows for benchmarking based on values gathered in other projects [17, 18].

User participation and involvement (UIP) have been identified as positively related to IS project success in previous research [9], although earlier literature reviews have produced conflicting results [19]. Harris and Weistroffer [20] name several advantages of UIP including preventing the adoption of unneeded, costly features and an improved quality of IS due to requirements that are more precise. The role of users in the provision of the tacit process and work context knowledge, which is necessary to evaluate requirements, is also highlighted by other researchers [8]. If users participate in a project, they are also more likely to claim ownership of a system [21]. Based on such an understanding, user participation can be seen as an antecedent of user involvement [22]. McGill and Kobas [23] have shown that such participating users perceive a new system as more useful and will have a more positive attitude towards a project. In part, this can be attributed to the fact that users develop a more realistic expectation regarding the features and connected capabilities of the IS [24].

The terms *user involvement* and *user participation* have often been used synonymously by researchers [8]. This has happened despite of early efforts to develop distinctive definitions for these two aspects of project management. For instance, Barki and Hartwick [25] introduced the following definition: User involvement is a "subjective psychological state of the individual, defined as the importance and personal relevance of a system to a user", while user participation is a "set of behaviors and activities [that] users perform" [26, p. 53]. We are going to follow this definition in this paper. Users can therefore be involved in an IS project without participating and performing any activities on their part [8]. Discussing UIP in more detail also requires a definition of the actual users of an IS. Broadly defined, users are all non-technical employees of an organization who are affected by the IS [27]. Thus, we define a user as someone with direct interaction with the system or as someone who is going to have it in the future.

With regard to the ideal point in time for UIP, Bano and Zowghi [8] state that it is widely believed that user participation in early project phases is most effective. However, it is not enough to just involve users in any project stage, instead this has to be done in an appropriate manner. Bano and Zowghi [8] also argue that the different project phases require different types and levels of UIP for an ideal contribution to project success. Considering different phases in more detail, in requirements analysis UIP helps to better understand the requirements of the users [8]. In development and customization for an IS implementation project it helps that the user requirements are purposefully transformed into technical solutions [27, 28]. Moreover, user participation in the testing phase can ensure that the user requirements are actually fulfilled by the developed system. End-user training helps users to learn how to use the system and therefore contributes to project success [29]. Based on the aforementioned insights, we classify the phases for user participation (see Fig. 1) [8]. Nonetheless, in all phases there can also be "token" user participation that does not really influence overall system success, but is rather a half-hearted measure to gather input from users [28]. The point in time, the degree and level of UIP can also influence project success [8].

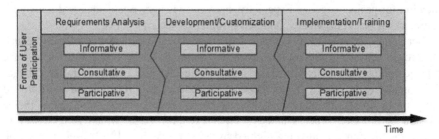

Fig. 1. Research framework [9, 31]

Damodaran [30] (see Fig. 1) developed an approach, which we adapted for the assessment of user participation over the course of a project. There are three levels of user participation of which the latter is the most extensive form: informative, consultative, and participative. The informative form of user participation means that users provide information to and receive information from the project team. That implies that users affect the project indirectly, but do not actively participate. If users have a consultative role, they comment on predefined services or a range of facilities. For instance, they comment different types of artifacts developed during the project [8]. In a participative role users influence decisions that are related to the whole system [30]. In such a setup, at least some users can be understood as part of the project team [8]. The level of user participation and the types of participating users is an additional characteristic of the particular configuration of a case.

3   Research Methodology

In this chapter, we present our approach of two analysis steps. First, we analyzed the individual cases. This ensured a thorough understanding of the cases as individual configurations, which is necessary for a successful QCA approach [31]. The data for the individual cases was gathered individually by some of the co-authors. Second, we present our cross-case analysis approach based on configurational theory. This second step is an integrated analysis of the data gathered by some co-authors in the first round.

3.1   Data Collection

While we gather data on the independent variables using semi-structured interviews with project members, we use an online survey among software users to measure the perceived product success based on usability.

Selection of the Cases. We selected cases with substantial differences in the aspects under investigation [32]. For instance, cases differed in type and degree of UIP as well as in the complexity of the tasks represented in the software concerned in the project. This ensures some generalizability of our research results. Furthermore, we needed to ensure the comparability of the projects. In general, we chose implementation projects of standard software and projects in which software was developed and implemented. Thereby, we considered implementation projects of completely new IS as well as projects of substitutions of an old IS. Moreover, it was a prerequisite that the considered software has a user interface and that business users work with it. Considering the context of the cases, we selected projects taking place in companies in Germany as well as in public organizations. Furthermore, we got the opportunity to conduct interviews in Columbia because of the contacts of one of the authors. We gained access to the different projects via personal connections of individual researchers and cold calling efforts. Lastly, as different effects of the project configuration on project success are subject to our analysis, we made sure to gather users' evaluation in relation to the circumstances of the project and not to other factors that occurred afterwards. Therefore, we only selected IS which still were in the shakedown phase [33]. This phase "refers to the period from 'going live' until 'routine use' has been achieved and can typically last anywhere from 6 months to a year" [34].

Interviews. For each case, some co-authors conducted two interviews: one with the IT project manager and one with a project member, ideally a user of the software. We chose this approach to obtain a certain breadth of opinions [35]. We conducted semi-structured interviews [35]. As recommended by Myers [35] we allowed interviewees to tell their own stories but created an interview guide before the interviews series. This allowed us to structure the interviews and to ensure that we asked all necessary questions and collected all the information required for the multi-case analysis. We asked interviewees questions about project characteristics, such as the introduced software and the conditions of the project, and the UIP characteristics of the case. On-site visits allowed us to conduct face-to-face interviews. We interviewed the project manager separately from the other project member to ensure that the two interviewees did not influence each other. If it was not possible to have a face-to-face interview, we conducted them either via phone or via video-call. We recorded and transcribed the interviews, if interviewees agreed. Otherwise, we took notes during the interviews to enable coding in a later stage.

Online-Survey. We used an online-survey among users of the particular software for each of the cases to measure the level of perceived usability. We chose a web-based survey form as it requires less effort of the respondent than other survey forms [36]. We used the System Usability Scale (SUS) developed by Brooke (1996) with a 5-Point-Likert-Scale to measure the usability of the software implemented in the particular projects [17, 18]. We chose it for three reasons. First, it is flexible enough to evaluate a wide range of different interface technologies [18] which is important to ensure that the survey works for all considered projects. Second, it is quick and easy to use for both participants and administrators [18]. Third, the result of the SUS is a single score that can be easily understood by many people and which is comparable to the results of other published studies and examined software systems [18, 37]. One researcher translated all statements into German, a second person translated them back into English, and a third person confirmed parity with the original statements. If this was not case, we revised the translations. The same process was executed for the Spanish survey.

3.2   Data Analysis

As suggested by Eisenhardt [38], we employed a two-stage approach comprising a within-case analysis and a cross-case analysis to analyze the case study data. First, the within-case analysis is used to develop a case study write-up for each of the cases [38]. Second, we conducted a cross-case analysis with the fsQCA approach [6, 26] to find patterns across cases and thus to identify generalizable findings. We chose fsQCA as a data analysis technique because it allows to determine the logical conclusions that are supported by a given data set.

Within-Case Analysis. We analyzed the considered cases separately to gather detailed information about the projects. We used a coding strategy to reduce, organize, and classify the data [35]. ATLAS.TI was our software for the coding and analysis process of qualitative data because it is widely used and well-established for qualitative analysis [39]. We followed three not consecutive coding steps: open coding, axial coding, and selective coding [40]. When we analyzed the survey data, we compared the results of the online survey for the SUS with published averages. For instance, Sauro [41] compared the results of 446 studies and respectively more than 5,000 observations and states that the average SUS score is 68 (similarly 67.6 for business to business applications) whereby the bottom third has a score up to 60 and the upper third a score higher than 72. You find a brief overview of the cases in Table 1.

Cross-Case Analysis. After analyzing each case separately, we conducted a cross-case analysis using the fsQCA method. We use the work of Schneider and Wagemann [7, 31] as well as of Thiem and Duşa [42] as guidelines for the application of fsQCA. The argument for choosing fsQCA is especially based on the categorization of the outcome [31] that it allows. We use the software called fs/QCA, Version 3.0 [43] for the initial analysis and replicated our analysis with a QCA-package for R [42, 44]. Furthermore, we also used the QCA-package for R in the data calibration process, which is necessary

Table 1. Overview of the selected cases.

Case name	Description	End-users	SUS	# of Responses
UniPortal (E)	Campus management system: All end user groups except "students"	~1,350 employees	40	29
UniPortal (S)	A campus management system with a focus on the user group "students"	~12,000 students	51	80
Construction ERP	ERP system of a Colombian company active in concrete formworks	~40 end-users	73	33
Residential Soft	An administration tool for residential complexes	~1,500 end-users	78	48
SkillSoft	A testing and managing platform for IT skills of users	~5,000 end-users	76	38
UniAlerts	Business intelligence software	~15,000 students	80	43
LabSales	A sales force tool pharmaceutical companies	~300 sales agents	72	34
MGIS	Geographic information system for the purpose of land consolidation	~750 users	65	46
FGIS	FGIS is a geographic information system for forestal planning	~25 users	68	12
MIS	IS for project managers in the area of land consolidation	~900 users	57	108
CAD-WS	A computer-aided design (CAD) system for construction purposes	~750 users	66	23
DMS	DMS is a data management system add-on to an ERP-system used by the organization	~400 users	54	11
Money	Software for calculating prepayment penalties	~100 users	84	17
Ticket-Reporting	Standard reporting system with some customizations	~100 users	74	14
CorporateWiki	Corporate groupware system	~180 users	66	18
ChemLawTool	ChemLawTool is a task management tool for managing compliance	~100 users	63	21

when using fsQCA. We used the framework for the assessment of user participation, that we developed based on Damodaran's [30] forms of user participation, for the assessment of user participation across the different projects. Thus, it served as a qualitative anchor during set calibration by direct assignment [31]. We identified the three main phases for user participation from our model in all projects. In light of the available data, fsQCA confines us to a reasonable number of conditions [31].

Requirements analysis and design are closely related and connected in most projects and thus are not clearly separable. User participation in the **development and customization phase** did take place when developers used agile methods. For instance, users participated by providing feedback on prototypes in intermittent user feedback cycles. Furthermore, we assessed the participation in the **implementation and testing phase**. This was user participation during training and adjustment efforts. We also judge if the user group participating in the project was **representative for all users** of the implemented IS. We distinguished between groups that were 'not representative' and 'largely non-representative', meaning that for instance several departments were affected by the new software, but only a small number of users of one department participated, and groups being representative for the user group, meaning that a higher number of users of affected departments was included. 'Representative, few users' was assigned if the participating users indeed represented the whole user group from a functional point of view but were only a very small share in respect of the whole user group.

We also assessed the **user involvement** in the different projects based on users' perception of attributes of the IS. For instance, we evaluated whether users saw the software as a burden or as important and personally relevant (see Table 2). The assessment of the representativeness of the users was based on the analysis of interviews for each case. As a second calibration method, we used transformational assignment. In this approach, we made use of continuous functions to map base variable values to fuzzy values. Thus, we had to define three thresholds, one for full exclusion, the crossover threshold, and one for full inclusion [42]. While the full exclusion value defines the threshold of a condition for not being a member of a set (=0), the full inclusion value defines the threshold for a full member (=1).

Moreover, the crossover threshold defines the boundary between a condition being a set member and not being a set member. We based our transformational assignment on a positive end-point concept, which implies that the set membership scores increase with increasing values of the base variable. We used transformational assignment for the calibration of the outcome of usability from the survey data in our study based on the calculation in R with the QCA package [42]. Specifically, we used the piecewise logistic function, which is the standard function in the QCA package in R. We set the full exclusion threshold to 50, the crossover threshold to 62, and the full inclusion threshold to 73 [45]. The raw data matrix containing the assessed conditions and outcome for all projects (Table 2) shows the results of the within-case analysis, whereas Table 3 contains all fuzzy-values on which we base the fsQCA.

Table 2. Conditions and outcome for the fsQCA.

Name	Assignment type	Fuzzy-set calibration				Source
		0	0.33	0.66	1	
Condition						
User Participation in Requirements Analysis (UPR)	Direct	None	Informative	Consultative	Participative	[30]
User Participation in Development/Customization (UPD)	Direct	None	Informative	Consultative	Participative	[30]
User Participation in Implementation (UPI)	Direct	None	Informative	Consultative	Participative	[30]
Degree of User Representation (DUR)	Direct	Non-representative	Largely non-representative	Representative, few users	Representative	[8]
User Involvement (UI)	Direct	Software (SW) is seen as a burden	SW rather not important and personally relevant	SW rather important and personally relevant	SW important and personally relevant	[25]
Outcome						
Usability (SUS)	Transform	Full-exclusion threshold: 50	Crossover threshold: 62	Full-inclusion threshold: 73		[45]

Table 3. Overview of data for truth table minimization.

Case	UPR	UPD	UPI	DUR	UI	SUS
UniPortal (E)	0.33	0.33	0.66	0.66	0.33	0
UniPortal (S)	0	0	0.66	0.66	0.33	0.04
Construction ERP	1	0	0.66	0.66	1	1
Residential Soft	0.66	0	0.66	0.66	0.66	1
SkillSoft	1	1	0.66	1	0.66	1
UniAlerts	1	1	0.66	0.66	1	1
LabSales	0.66	0	0.66	1	0.66	0.95
MGIS	0.33	0.66	0.66	1	0.66	0.63
FGIS	1	0	0.33	1	0.66	0.77
MIS	0	0.33	0.33	0.33	0	0.29
CAD-WS	0.33	0	0.33	0.33	0.66	0.68
DMS	0.33	0.33	0.66	0.66	0.33	0.17
MONEY	1	1	1	0.66	1	1
TicketRep	0.66	1	1	1	1	1
CorporateWiki	0.66	0.33	0.33	0.66	0.33	0.68
ChemLawTool	0.66	0	0.66	0.33	1	0.55

4 Findings

When conducting a cross-case analysis using fsQCA, it is the primary goal to identify necessary and sufficient conditions for the examined outcome [7]. A condition is necessary if it is present in all cases in which the outcome is present. A consistency value of 1 [7], or at least higher than 0.9 [46] can indicate a necessary condition. We did not identify a necessary condition in this step of the analysis (see Table 4).

Table 4. Truth table for high system usability.

UPR	UDP	UPI	DUR	UI	SUS	Consistency	PRI	#	Case name
1	0	1	1	1	1	0.885	0.807	3	LabSales, Construction ERP; ResidentialSoft
1	1	1	1	1	1	0.876	0.843	4	SkillSoft, Uni-Alerts; Money; TicketRe
1	0	0	1	1	1	0.866	0.743	1	FGIS
1	0	0	1	0	1	0.836	0.689	1	CorporateWiki
1	0	1	0	1	0	0.774	0.571	1	ChemLawTool
0	1	1	1	1	0	0.738	0.539	1	MGIS
0	0	0	0	1	0	0.706	0.469	1	CAD-WS
0	0	0	0	0	0	0.560	0.242	1	MIS
0	0	1	1	0	0	0.546	0.264	3	DMS, Uni-Portal (S, E)

This test is essential, since a condition which is necessary for an outcome as well as for the negated outcome is a trivial condition [7]. We also conducted an analysis for the sufficient conditions. A condition is sufficient if it is part of the configuration of the considered outcome in any case, implying that there is no case among the considered ones where the condition is present, but not the outcome. The sufficient conditions for the outcome are identified by the creation of a truth table (see Table 4) followed by using the enhanced Quine-McCluskey algorithm to minimize the Boolean output function [47]. We also conducted this analysis for the negative outcome to make sure that there are not contradictory paths in comparison to the positive outcome [7]. The required levels of consistency and subsequently coverage determine the inclusion of a configuration. We used a minimum consistency value of 0.8 for analyzing sufficiency, which is above the suggested minimal threshold of 0.75 [7, 47].

The application of the Quine-McCluskey algorithm in the aforementioned software solutions generates complex, parsimonious and intermediate solutions (see Table 5). We focus on the intermediate solution for our analysis because of the theoretical underpinnings of the minimization process [47]. In our particular case, the results for the intermediate and the parsimonious solution are identical, which indicates that logical remainders did not have a particular influence on the results of the minimization process. We focus our description of the analysis on the identical parsimonious and intermediate solution *UPR*DUR*. They suggest that only the combination of the conditions of early user participation during the requirements phase conducted with the appropriate users is relevant for project success in terms of usability. Besides its high

Table 5. Truth table minimization results.

Type of solution	Minimized configuration	Consistency	Cov.r	Cov.u	Case name
Complex	UPR* ~ UPD* ~ UPI*DUR	0.88	0.32	0.06	FGIS (0.67, 0.77), CorporateWiki (0.66, 0.68)
	UPR*UPI*DUR*UI	0.93	0.60	0.33	ConstructionERP (0.66, 1), ResidentialSoft (0.66, 1), SkillSoft (0.66, 1), UniAlerts (0.66, 1), LabSales (0.66, 0.95), Money (0.66, 1), TicketRep (0.66, 1)
Parsimonious/ Intermediate	UPR*DUR	0.91	0.70	0.70	SkillSoft (1, 1), FGIS (1, 0.77), ConstructionERP (0.66, 1), ResidentialSoft (0.66, 1), UniAlerts (0.66, 1), LabSales (0.66, 0.95), Money (0.66, 1), TicketRep (0.66, 1), CorporateWiki (0.66, 0.68)

level of consistency, this solution covers many different cases and therefore provides a good explanation for a large share of the outcome. Furthermore, this result indicates that user participation as a condition of IS project success is inseparable from the assessment of the appropriate users for the participatory practice. The projects with a membership in this solution all have some form of consultative or participative user participation in the requirements analysis phase.

The results of the fsQCA underline that UIP is more effective when the participants represent a large share of the affected users, a finding that links well to previous research results [30]. In sum, project success is very likely for the majority of the cases when the appropriate users participate, and this takes place in the phase of requirements analysis. This is our key finding which reinforces findings based on anecdotal evidence that claimed that user participation of representative end-users is related to IS implementation projects success. Thus, the research in this paper has two major contributions. First, it provides a richer perspective on UIP configurations related with IS project success. Second, it reaffirms prior research results in a more comprehensive way with an analysis of the configurations of different forms and timing of UIP in relation to IS implementation project success. In addition, our focus on perceived usability for end-users as a measure for success is a new approach that veers of the traditional time, budget, and quality approach towards a more end-users focused approach in the evaluation of IS implementation project success.

5 Discussion

Our analysis of multiple cases provides a detailed look at comprehensive patterns that link different forms of UIP across different project phases to IS project success. Through our novel configurational perspective, we contribute to research with an improvement in the level of detail of the understanding and empirical backing for the notion that user participation and involvement is especially beneficial in the requirements analysis phase [8]. Thus, our fsQCA provides additional empirical evidence on this finding. Furthermore, a higher level of user participation without participation of the appropriate user group (mainly end users) is unlikely to be associated with a significant improvement in the outcome. This finding can also be linked to previous research, which indicated that ineffective management of the participation and involving possibly the wrong users could also have adverse effects on project success [8]. This indicates the need to involve actual end-users and not just their managers or respective power users, who are not representative for the majority of the user group. The results of our analysis also cast doubt on the net effect of user participation in development as well as testing. However, a qualification of this assessment is necessary, as the lack of participation in the development phase can be due to the lack of prominence of truly agile development/customization approaches in most projects. Nonetheless, our results reinforce the notion that the participation of users in the development phase is much more prone to complications, therefore costly, and less effective than user participation in the requirements analysis phase. The low coverage of the complex solution, which encompasses user participation during development, indicates this. We also add to the research domain by providing an analysis of the intertwined conditions of user participation in the requirements phase and the appropriate degree of user representation. Our research also improves the understanding when and under which conditions well-established heuristics actually help to explain or even predict project success. It can be the objective of future research to evolve the theoretical understanding of the observed relationships.

However, there are also limitations for the interpretation of the results of our multiple case study. The number of cases that we were able to analyze, and their particular nature influenced the analysis. Furthermore, we used a limited number of conditions for the analysis. Only a larger number of cases can help to reduce these limitations. A larger number of cases would allow introducing more conditions in the QCA, or a different focus of a reanalysis of the available data set with a different theoretical motivation. A different theoretical perspective could, as aforementioned, include the condition of project complexity as it has already been established that the degree of user participation and involvement can be related to the complexity of the project [20]. Furthermore, we need to address the perception that there are possible threats to validity for the empirical analysis presented in this paper. First, the fsQCA is a deterministic technique for logical analysis, which means that the analysis results show that a particular condition, in our case user participation in the requirements analysis phase, most commonly occurs together with the outcome of a high level of usability as perceived by end-users. The observation of this particular configuration as parsimonious solution for the minimization process of the truth table does not mean that the user participation in the requirements analysis phase caused the high level of perceived usability. This can only be deducted from the observation based on the qualitative information from the cases and prior anecdotal evidence that suggests such an interpretation. Thus, there is no direct causal link, but the observation that a particular condition is almost always part of the configuration with a high level of usability. Second, the variation in the level of perceived usability by end-users cannot only be attributed to the conditions that we covered in the analysis presented in the paper, but also other factors. These other factors might include as aforementioned the project complexity, the specificity of the software solution and the user group. However, as fsQCA is designed to determine logical relationships it is not a probabilistic approach that causally determines an influence on a relationship. Thus, there is no direct threat to the validity of our results, as the results are only an indication based on logical analysis of empirical data.

Our analysis of the cases also has several practical implications. Project managers should make sure that they focus their attention on user participation in the phase of user requirements analysis. In particular, they should make sure that actual end-users of the IS participate. This should help to increase the perceived usability of the software and thereby increase the productivity of end users.

6   Discussion

We conducted a multiple-case study and highlighted the effect of UIP in the requirements phase on IS project success. We are able to show that the participation of the appropriate users in the requirements analysis phase is a key condition for IS project success. In addition, we are able to show that a higher level of user participation will not result in a significant improvement in the outcome, if users participate that cannot contribute as much as true end-users can. While these findings are a first empirical basis for the analysis of the relationship of different forms of end-user participation in IS implementation projects, further research on more different cases is still needed to broaden the theoretical basis for a preference for user participation in the requirements analysis phases. Specifically, the different forms of user participation in this particular project phase and their relationship with project success as experienced by end-users can be the subject of future research.

References

1. The Standish Group International: Chaos Manifesto. The Standish Group International, Boston (2013)
2. Jetu, F., Riedl, R.: Determinants of information systems and information technology project team success: a literature review and a conceptual model. Commun. Assoc. Inf. Syst. **30**(27), 455–482 (2012)
3. Verweij, S.: Producing satisfactory outcomes in the implementation phase of PPP infrastructure projects: a fuzzy set qualitative comparative analysis of 27 road constructions in the Netherlands. Int. J. Project Manag. **33**(8), 1877–1887 (2015)
4. Wagner, E.L., Newell, S., Piccoli, G.: Understanding project survival in an ES environment: a sociomaterial practice perspective. J. Assoc. Inf. Syst. **11**(5), 276–297 (2010)
5. Meyer, A.D., Tsui, A.S., Hinings, C.R.: Configurational approaches to organizational analysis. Acad. Manag. J. **36**(6), 1175–1195 (1993)
6. Ragin, C.C.: Fuzzy-Set Social Science. University of Chicago Press, Chicago (2000)
7. Schneider, C.Q., Wagemann, C.: Set-Theoretic Methods for the Social Sciences: A Guide to Qualitative Comparative Analysis. Cambridge University Press, Cambridge (2012)
8. Bano, M., Zowghi, D.: A systematic review on the relationship between user involvement and system success. Inf. Softw. Technol. **58**, 148–169 (2015)
9. Fiss, P.C.: Case studies and the configurational analysis of organizational phenomena. In: Byrne, D., Ragin, C.C. (eds.) The SAGE Handbook of Case-Based Methods, pp. 415–431. SAGE, Thousand Oaks (2009)
10. Byrd, T.A., Thrasher, E.H., Lang, T., Davidson, N.W.: A process-oriented perspective of IS success: examining the impact of is on operational cost. Omega **34**(5), 448–460 (2006)
11. Saleh, Y., Alshawi, M.: An alternative model for measuring the success of IS projects: the GPIS model. J. Enterp. Inf. Manag. **18**(1), 47–63 (2005)
12. Grudin, J.: Interactive systems: bridging the gaps between developers and users. In: Baecker, R.M., Grudin, J., Buxton, W.A.S., Greenberg, S. (eds.) Readings in Human-Computer Interaction - Toward the Year 2000, pp. 293–303. Morgan Kaufmann, San Francisco (1995)
13. Folmer, E., Bosch, J.: Architecting for usability: a survey. J. Syst. Softw. **70**(1–2), 61–78 (2004)
14. Madan, A., Dubey, S.K.: Usability evaluation methods: a literature review. Int. J. Eng. Sci. Technol. **4**(2), 590–599 (2012)
15. International Organization for Standardization: ISO 9241-11: Ergonomic Requirements for Office Work with Visual Display Terminals (VDTs): Part 11: Guidance on Usability (1998)

16. Schaefer, M., Xu, B., Flor, H., Cohen, L.G.: Effects of different viewing perspectives on somatosensory activations during observation of touch. Hum. Brain Mapp. **30**(9), 2722–2730 (2009)

17. Tullis, T., Albert, W.: Measuring the User Experience – Collecting, Analyzing, and Presenting Usability Metrics, 2nd edn. Morgan Kaufmann, Waltham (2013)

18. Bangor, A., Kortum, P.T., Miller, J.T.: An empirical evaluation of the system usability scale. Int. J. Hum. Comput. Interact. **24**(6), 574–594 (2008)

19. Ives, B., Olson, M.H.: User involvement and MIS success: a review of research. Manag. Sci. **30**(5), 586–603 (1984)

20. Harris, M., Weistroffer, H.: A new look at the relationship between user involvement in systems development and system success. Commun. Assoc. Inf. Syst. **24**(42), 739–756 (2009)

21. Hope, K.L., Amdahl, E.: Configuring designers? Using one agile project management methodology to achieve user participation. New Technol. Work Employ. **26**(1), 54–67 (2011)

22. Barki, H., Hartwick, J.: Measuring user participation, user involvement, and user attitude. MIS Q. **18**(1), 59–82 (1994)

23. McGill, T., Klobas, J.: User developed application success: sources and effects of involvement. Behav. Inf. Technol. **27**(5), 407–422 (2008)

24. Baronas, A., Louis, M.: Restoring a sense of control during implementation: how user involvement leads to system acceptance. MIS Q. **12**(1), 111–124 (1988)

25. Barki, H., Hartwick, J.: Rethinking the concept of user involvement. MIS Q. **13**(1), 53–63 (1989)

26. Ragin, C.C.: Redesigning Social Inquiry: Fuzzy Sets and Beyond. University of Chicago Press, Chicago (2008)

27. Carmel, E., Whitaker, R.D., George, J.F.: PD and joint application design: a transatlantic comparison. Commun. ACM **36**(6), 40–48 (1993)

28. Lynch, T., Gregor, S.: User participation in decision support systems development: influencing system outcomes. Eur. J. Inf. Syst. **13**(4), 286–301 (2004)

29. Sabherwal, R., Jeyaraj, A., Chowa, C.: Information system success: individual and organizational determinants. Manag. Sci. **52**(12), 1849–1864 (2006)

30. Damodaran, L.: User involvement in the systems design process-a practical guide for users. Behav. Inf. Technol. **15**(6), 363–377 (1996)

31. Schneider, C.Q., Wagemann, C.: Standards of good practice in Qualitative Comparative Analysis (QCA) and fuzzy-sets. Comp. Sociol. **9**(3), 397–418 (2010)

32. Pettigrew, A.M.: Longitudinal field research on change: theory and practice. Organ. Sci. **1**(3), 267–292 (1990)

33. Markus, M.L., Tanis, C.: The enterprise system experience — from adoption to success. In: Markus, M.L., Tanis, C., Zmud, R.W. (eds.) Framing the Domains of IT Management: Projecting the Future Through the Past, pp. 173–207. Pinnaflex Educational Resources, Cincinnati (2000)

34. Sykes, T.A., Venkatesh, V., Rai, A.: Explaining physicians' use of EMR systems and performance in the shakedown phase. J. Am. Med. Inform. Assoc. **18**(2), 125–130 (2011)

35. Myers, M.D.: Qualitative Research in Business & Management. Sage, London (2009)

36. Dillman, D.A., Smyth, J.D., Christian, L.M.: Internet, Phone, Mail, and Mixed-Mode Surveys: The Tailored Design Method. Wiley, Hoboken (2014)

37. Brooke, J.: SUS: a retrospective. J. Usability Stud. **8**(2), 29–40 (2013)

38. Eisenhardt, K.M.: Building theories from case study research. Acad. Manag. Rev. **14**(4), 532–550 (1989)

39. Scientific Software Development GmbH: ATLAS.ti (2015)

40. Strauss, A.L., Corbin, J.M.: Basics of Qualitative Research: Techniques and Procedures for Developing Grounded Theory. Sage Publications, Thousand Oaks (1998)

41. Sauro, J.: A Practical Guide to the System Usability Scale: Background, Benchmarks & Best Practices. Measuring Usability LLC, Denver (2011)

42. Thiem, A., Duşa, A.: Qualitative Comparative Analysis with R - A User's Guide. Springer, Heidelberg (2013). https://doi.org/10.1007/978-1-4614-4584-5

43. Ragin, C.C., Davey, S.: Fuzzy-Set/Qualitative Comparative Analysis 3.0 (2016)

44. Duşa, A.: User manual for the QCA (GUI) package in R. J. Bus. Res. **60**(5), 576–586 (2007)

45. Bangor, A., Kortum, P., Miller, J.: Determining what individual SUS scores mean: adding an adjective rating scale. J. Usability Stud. **4**(3), 114–123 (2009)

46. Ragin, C.C.: Set relations in social research: evaluating their consistency and coverage. Polit. Anal. **14**(3), 291–310 (2006)

47. Ragin, C.C., Rihoux, B.: Qualitative comparative analysis using fuzzy sets (fsQCA). In: Rihoux, B., Ragin, C.C. (eds.) Configurational Comparative Methods: Qualitative Comparative Analysis (QCA) and Related Techniques, pp. 87–121. Sage, Thousand Oaks (2009)

Human and Value Sensitive Aspects of Mobile App Design: A Foucauldian Perspective

Balbir S. Barn[1](✉)(iD) and Ravinder Barn[2](iD)

[1] Middlesex University, London, UK
b.barn@mdx.ac.uk
[2] Royal Holloway University of London, Egham, UK
r.barn@rhul.ac.uk

Abstract. Value sensitive concerns remain relatively neglected by software design processes leading to potential failure of technology acceptance. By drawing upon an inter-disciplinary study that employed participatory design methods to develop mobile apps in the domain of youth justice, this paper examines a critical example of an unintended consequence that created user concerns around Focauldian concepts including power, authority, surveillance and governmentality. The primary aim of this study was to design, deploy and evaluate social technology that may help to promote better engagement between case workers and young people to help reduce recidivism, and support young people's transition towards social inclusion in society. A total of 140 participants including practitioners (n = 79), and young people (n = 61) contributed to the data collection via surveys, focus groups and one-one interviews. The paper contributes an important theoretically located discussion around both how co-design is helpful in giving 'voice' to key stakeholders in the research process and observing the risk that competing voices may lead to tensions and unintended outcomes. In doing so, software developers are exposed to theories from social science that have significant impact on their products.

Keywords: Governmentality · Value sensitive design · Co-design

1 Introduction

The software engineering community does not traditionally evaluate research artefacts from theoretical positions located in social sciences. Given software pervades our daily lives and our social transactions, this is a potentially serious deficit. We argue that the design of information systems in the context of widespread ubiquity, mobile device based deployment and hidden data interchange places new challenges on designers and implementers of systems. In particular the need to preserve key human (moral) values such as privacy, security

© Springer International Publishing AG, part of Springer Nature 2018
J. Krogstie and H. A. Reijers (Eds.): CAiSE 2018, LNCS 10816, pp. 103–118, 2018.
https://doi.org/10.1007/978-3-319-91563-0_7

and autonomy within the design process is paramount. We propose that identification of values and tracing their subsequent governance through software
design process remains relatively neglected, and potentially detrimental to final
acceptance of software if not done. Scholars such as Van den Hoven, writing on
the role of value sensitive design for ICT, made a similar and earlier case:

"...these values will have to be expressed in the design, architecture and
specifications of systems. If we want our information technology - and the
use that is made of it - to be just, fair and safe, we must see to it that it
inherits our good intentions [25].

Van dan Hoven however was making the case from a philosophical perspective
and the *engineering* of values into the design process remains elusive.

The role of information systems especially in their new guise of *apps* delivered
through sensor rich smartphones is particularly pertinent. Significantly, organisations responsible for these apps should recognise that their corporate actions,
with respect to design and deployment of such systems, have a profound impact
on all aspects of societal welfare including concerns around invasion of privacy
and security concerns around the sharing of data.

This paper draws upon an empirical inter-disciplinary study involving social
and computer scientists engaged in building mobile app based social technology
to promote positive engagement between case workers and young people in youth
offending teams in England. Such intended positive use is in direct contrast to
prior use of technology in this domain, which has largely been for two reasons
[18]. Firstly, in its attempt to manage risk, private firms such as G4S and SERCO
are contracted to electronically monitor the movements of young offenders in the
community. Secondly, technology is employed as a tool for data management and
this signals its own tactics of surveillance and discipline. Both efforts are part of
the general move towards neoliberalism in public services and the so-called new
public management models that emerged in the early 2000s [19].

The thrust of this paper explores (moral) universal values and their incorporation into the design of a mobile app. The paper presents an analysis of
the French social theorist Michel Foucault's writings on 'how the human subject is governed and fashioned by disciplinary power' [6, :221]. Examination of
software practice and the resultant artefacts from such a lens became significant
because the methodological approaches (co-design) used in the mobile app study
elicited a class of non functional requirement we refer to as a *value*. In the context of technology artefacts, values are what Friedman refers to as: ownership
and property; privacy, freedom from bias, universal usability, trust, autonomy,
informed consent and identity. She defines values as: *what a person or group of
people consider important in life* [11]. Further, reinforcement, or erosion of such
values occurs through the use of software either through deliberate design or
through accident.

This exploration is used to appraise the development of our own social technology as a tool for positive engagement in the youth justice sector and to critique Foucault's idea of surveillance in the context of the findings from our study.
In this exploration, we contribute a critical discussion of the potential risks of

designing technologies that have unplanned side effects around surveillance and propose that software engineering practice has to find ways to account for these human impacts arising from technology. While recognising the case study limitations of our work, the risks and concerns identified in this paper have relevance to designers of software for widespread consumer use. Our intention is to invoke discussion amongst engineers who would not normally consider such concerns from a social science perspective.

The remainder of the paper is structured in structured in four segments. In Sect. 2, key Foucauldian theoretical positions on governmentality are sketched out together with an outline of the context within which youth justice is currently located [13, 14, 20]. In Sect. 3, we provide the reader with a background to our study aims and methods, together with an understanding of co-design and value sensitive approaches as tools for inclusion and empowerment. Section 4 presents a qualitative discussion on our findings related to the concerns generated from adopting a Foucauldian perspective and the impact on values in the software design process [10]. In Sect. 5, concluding remarks are made.

2   Theoretical Background

2.1   Governmentality, Discipline and Knowledge

The French social theorist, Michel Foucault, has left an important legacy in his writings on the genealogy of the modern state. His studies on power, knowledge, surveillance, and governmentality have a broad cross-disciplinary appeal. Although, Foucault died before the advent of the Internet in public spaces, his theories lend themselves to an understanding of information technology. Indeed, Foucault's reach is such that he is beginning to wield some influence beyond the Social Sciences to the disciplines of Information Systems and Computer Science [1, 5, 28].

Several writers have suggested the increasing importance of Foucault's theoretical ideas to our understanding of the ways in which the state seeks to manage crime and criminal justice processes in modern society [13, 15]. One of the central planks of Foucault's work, generally discussed, is the notion of governmentality. Governmentality as a concept arose originally from Foucault's lectures at the College de France as part of a broader concept of what he called the 'art of government'. The governmentality thesis holds that the modern state wields tremendous power in the government of its populations through an ensemble of institutions, procedures, analyses and a series of social practices including codification. Notions of power and episteme are central to this. In line with Foucault's own later work where he attempted to respond to criticism of his work for its deterministic and narrowly defined approach to state power, we adopt a broad understanding of governmental power and focus on human agency and resistance to the processes of subjectification [4, 13].

Within the framework of neoliberal governmentality [12] we seek to advance Foucault's idea of surveillance in the context of the findings from our study. Foucault's application of, 19th century English political philosopher, Jeremy

Bentham's Panopticon, to understand the 'art of government' is of key relevance for our purposes. Bentham's architectural design of a Panopticon prison sought to ensure that discipline and subjectification were to be achieved through this structure in which a guard in a central tower could see into the cells and maintain power and surveillance at all times [2]. Foucault [9, :203] outlined that the major effect of the Panopticon was to 'induce in the inmate an illusion of conscious and permanent visibility that assures the automatic functionality of power', and in this process, the inmate 'becomes the principle of his own subjection'. It is Foucault's analysis of discipline and the use of the Panopticon as an analytical tool for discussing institutions and society that the Panopticon in the form of *panopticism* has become the mostly widely used metaphor and explanatory theory for surveillance today.

The idea of the Panoptic society in our current information technology age is extremely apt. Willcocks [28] reminds us that although the word technology appears in Foucault's work, he rarely defines it for the reader. It is terms such as 'technologies of power', 'political technology of the body', 'disciplinary technologies', and 'technologies of the self' that have a fascinating appeal for contemporary scholars as they lend themselves to be employed as useful explanatory tools.

Gane, writing in 2012, proposes a heuristic typology for panoptic governmentality [12]:

- surveillance and discipline: where the state watches over the market and over its citizens, where watching is sufficient and intervention only happens when necessary;
- surveillance and control: where subjects are not limited to physical space and non-state actors such as commercial organisations also do the watching;
- surveillance to promote competition: the state or its proxy actors strive to create conditions for the freedom of markets, and through it, achieve legitimacy.
- interactivity: an inversion of the panopticon architecture so that the many watch the few.

As we will observe in later sections, these typologies are apparent in the technology described in this paper.

It is important to provide the youth justice context within which our social technology is located. We draw on the work of other writers who have paved the way to advance an argument of governmentality and youth justice.

2.2   Governmentality and Youth Justice

The discipline of criminology has a long tradition of studying youth crime to identify risk factors that pre-dispose young people to become involved in criminal activity. Such factors are generally located within a socio-economic context, and psycho-social behaviours and practices. Some writers have classified such knowledge to make actuarial predictions of would be young offenders [7].

The influence of risk predictive studies is evident in the technologies of government that are operational within youth justice. Arguably, these include the use of the ASSET risk assessment form[1] in youth offending teams in England and Wales, and the strategies of responsibilisation. Data collected from such instruments and further coupled with data from other information systems deployed within the sector allows comparisons and consolidation across time and space. Ultimately this can support an 'economic' rationality – the increasing reliance upon an "analytical language" of risks and rewards of objectives/targets [13].

Although Foucault did not consider the notion of risk in his work, we can see that the risk paradigm in contemporary neoliberal society fits in very well as technology of governmentality. Through their regular risk assessments (using the ASSET form instrument), the young offenders become marked/visible, and are served with the tools of self-discipline to become good neoliberal subjects. It is the management of risk and responsibility, during their involvement with the youth offending team, that will lead to subjectification. Here, the 'technology of responsibilisation' serves as a tool of governmentality. A recent study of 29 young offenders, distils the youth justice policy and practices of responsibilisation in three ways – 'reconfiguring the field of governance, extending the reach of governance and the ethical construction of the subject' [20, :433]. The key focus here is on governmentality of the new liberal state and the ways in which it operates to exercise power through key mechanism and processes (government/non-government agencies, civil society, bio-power, risk-based reasoning) to achieve 'governance at a distance' through 'mobile mechanisms' and thereby construct 'non-deviant, neoliberal citizens' [13,20]. Such governmentality relies on a powerful discourse of 'evidence-based practice', efficiency, and effectiveness in the fight to prevent crime. Muncie reminds us that such a discourse 'of 'what works' is deceptively benign, pragmatic and non-ideological' [17, :778].

3   Study Aims and Methods

The socio-technical context for this research concerns young people in the UK Youth Justice system. Research suggests that engagement with young offenders to help promote social inclusion and prevent re-offending remain key challenges for public policy and youth justice service providers [24].

This study aimed to explore how social technology could be developed and adopted for the purposes of promoting better engagement between young offenders and their case workers. Our MAYOT (Mobile Applications for Youth Offending Teams) project developed a personalised mobile app for use by young people and their case workers in youth offending teams. The app provides relevant, timely information to a young person as well as features such as ease of access to their case history, relevant contacts such as professional networks, peer networks and their family networks. Given that, currently, digital tools that could engender closer engagement and encourage co-creation between case workers and young offenders are not available, we set out to address this gap.

---

[1] https://www.gov.uk/government/publications/asset-documents.

The study adopted a mixed-methods approach to determine the current and intended/desired use of technology. A quantitative questionnaire was employed to establish the patterns of communication between young people and case workers. A total of 33 young people and 43 case workers contributed to this self-completion survey, from three youth offending services, representing inner-city (*Site 1*), urban (*Site 3*), and rural locations (*Site 2*) in England. The questionnaire sought to gain insight from case workers on their existing use of technologies.

The core of the method was the requirements elicitation process approach adopted. A combination of co-design and value-sensitive design approaches (VSD) [11] were used in the collection of data and the building of the mobile app. Co-design (and its earlier form of participatory design [3]) is a well established design approach for working with end-users.

Co-design involves potential (un-trained) end users working with researchers and designers using tools provided to jointly create artefacts that lead directly to the end product [22]. Yoo et al. (2013) state that co-design has become a dominant user study methodology in the fields of product design, service design, interaction design and Human Computer Interaction (HCI) [29].

In our study, we advanced the use of co-design methodology in software engineering by embedding a VSD approach within it. Here, values include privacy, trust, freedom from bias, universal usability, autonomy, informed consent, identity and others.

VSD emerged to integrate moral values (and more broadly ethics) with the design of systems to address the issue raised by Wiener (1985) when he argued that we should be the masters of technology, not worshippers [27]. A key premise of VSD is that it seeks to design technology that accounts for human values throughout the design process (over and beyond the identification of functionality and visual appearance) of systems. Leading advocates of VSD have included those focused on technology such as Terry Winograd, Friedman [11], and Nissenbaum [8] whose work identified issues of freedom from bias in systems. That is, computer systems should not systematically and unfairly discriminate against certain individuals or groups of individuals in favour of others [11]. Others such as Van der Hoven have explored value sensitive design through a philosophical lens, such as 'just' design.

Our study used co-design through a series of participative co-design workshops in a mix of inner-city, urban and rural settings. 17 case workers and 10 young people participated across two co-design workshops to contribute ideas to the design and development of the mobile app. Following the first co-design workshop in our inner-city location, mock-ups were created and represented in screen captures as co-designed requirements. These were presented to a new set of case workers and young people in the second workshop in our rural location, to capture their perspectives on the planned design. A software prototype, that we called the MAYOT app was developed. The requirements leading to the design of prototype were independently evaluated in our third urban research site. Here, a total of 11 respondents (7 case workers, and 4 young people) participated in a co-design workshop to provide us with their perspectives on the requirements. Self-completion questionnaires were also completed by these respondents to give us a sense of their everyday use of technology and techniques of communication between case workers and young people.

Following the data collection processes outlined above, the software comprising a web-based application (for use by the case worker) and a mobile app (for the young person) was developed and then deployed in our case study sites. A further admin-web interface provided administrative functions for use by the researchers. Interviews were conducted with 26 respondents (14 young people and 12 case workers) who had made use of the software. These interviews reflected the views and experiences of users in all three of our research sites.

Participation was voluntary for all respondents and the use of the software was subject to ethical guidelines from the British Sociological Association. Participants were reassured that the technology was supplementary and data arising from its use would not be used to adversely affect participants.

Data analysis involved descriptive analysis of the survey demographic data and Internet use, and a thematic analysis of the co-design workshops and interviews. The qualitative data analysis software, NVIVO, was used to assist with the thematic analysis and to code the key terms and analyse data with greater ease [21].

4   Qualitative Findings and Discussion

We now present a discussion and analysis of our qualitative findings drawn from transcripts from interviews and workshops within the context of Foucauldian notions of governmentality.

4.1   Youth Justice and Information Technology Infrastructures

Our findings suggest that although there is an appetite for the use of technology to assist communication with young people, youth offending services had not yet 'entered the 21st century'. With the exception of a few individual case workers who, at times, sent SMS text messages via their mobile phones to the young people with whom they were working, traditional methods of communication were in existence such as letters, phone calls, appointment cards, and so on. Such methods were rationalised as providing clear evidence in cases where young people were being breached for non-compliance.

In each of the three case study sites, no organisational wifi network was available, and workshops had to be conducted by the research team setting up their own wifi network in the working area.

Case workers had access to personal computers at work to assist them with ASSET data input, and general record keeping; however they did not have access to other devices to promote better communication:

S1-CW1 (Site 1, CaseWorker 1): " . . . the mobile phones we have are very out-dated, um there aren't enough laptops so if you want to work remotely there often isn't a laptop and when you do get one the bloody thing won't log on most of the time. So, I think you know, I think we're slipping behind in terms of where technology is. We should all be having ipads or you know tablets to, with a good connection, 3G or 4G connection, cos there's all the stuff about security - this is always the argument - 'we can't make them secure' but you know the police use them in, in patrol, children's services use them."

The young people in the study also showed alertness to concerns around data. The notion of data security was expressed vociferously by the young people themselves who indicated fear and anxiety about the potential theft and loss of their mobile phone, and thereby their personal data. Young people were reported to change their phone numbers frequently as a consequence of loss of their phones through theft, or confiscation of their phone by the police. Such circumstances coupled with the youth offending team concerns about security of personal data relating to young people led us to ensure that we built appropriate safeguards in our planned social technology [23]. Both the young people and case workers were cognisant of how electronica data from multiple sources and over periods of time could be used for the purposes of control:

S3-YP1 (Site 1, Young Person 1): "yeah, but its just like they can go into more detail probably, and its just a lot easier for them to..."

S3-CW1 (Site 3, Case Worker 1): "...I mean its, its kind of wise to be careful in these, in this day and age isn't it? About anything digital or electronic... it is almost healthy to be a bit paranoid about . . . personal details."

4.2   Governmentality, Human Agency and Resistance

Working within the framework of co-design and a value-sensitive approach, we held separate workshops with case workers and young people to ascertain their qualitative experiences of using apps, including likes/dislikes, usefulness, cost, and ease of use [26]. Case workers and young people were invited to tell us how communication between them could be enhanced using digital technology. Many of these discussions began from exploring current methods of communication, and areas of need. The ideas put forward by the first set of case workers were later presented to other case workers for their reflection and input. These were subsequently showcased to young people for their reactions to the building of the app. Young people were not only asked to critique these initial ideas but also suggest other possible need scenarios. This participatory research design proved to be extremely useful in helping to embrace ethical and practical challenges raised in this process.

Values such as confidentiality, privacy, security, trust, and autonomy began to emerge as we continued with our co-design and value-sensitive approach in our three research sites. Features became associated with specific values and it is evident that some features were perceived to be straightforward and raised little concern or conflict. For example, text messaging, group messaging (including sending messages to family members and a list of useful contacts were not regarded to be problematic by the respondents. It was generally believed that such methods of communication would a useful means of reaching young people to provide them with timely and appropriate information. Many of the ideas emerging from the case workers appear to have a tone of welfare/benevolence to help provide young people with appropriate and timely information. Interestingly, although there was recognition of low levels of literacy among this population, many case workers conceptualized the empowering role of technology:

S1-CW2: "...they're highly phone literate. Many of them won't write on a piece of paper but are happy to text. Also, the mobile phone auto spells which helps them"

Fig. 1. Screenshots of MAYOT App

Garland notes that the governmentality literature presents a paradox whereby 'governing' takes place through our 'freedom' [13]. The paradox exists because of a conflation of two concepts: agency and freedom. Agency refers to the capacity of an agent to act on some decision based on perceived pertinent information. Foucault would consider this ability as a necessary condition for rules at a distance in the social sphere. Freedom, however is a capacity for an agent to choose an action without external constraint. Do we choose to buy a particular product? Or is it because of a marketing campaign that, perhaps, through a process of subliminal invasion, we choose to buy that product? Thus a central element of governmentality is that of creation and simulation of agency while simultaneously reconfiguring the constraints upon the freedom of choice of the agent. Arguably, the app features were designed to create agency within the young people. Simulateneously, these features are a re-configuration of constraints as these features were designed to be a vehicle of interaction between young people and case workers albeit with potential reference to concerns of surveillance and control by the case workers. Features such as access to contact details of close friends and family who can support the young person with respect to reminders for appointments creates a form of governance that rests upon the "willingness of individuals to exercise a 'responsibilised' autonomy" [13].

Thus, young people's narratives attest to the belief that to receive information directly to their phones in various forms including appointment reminders, progress charts, intervention plans, useful contacts, 'stop and search', 'drugs awareness', and health could be beneficial to them. This was particularly so, and as they reported, they invariably misplaced/lost paper information that had been given to them. Crucially however, young people were all too familiar with the governmentality processes at work and were conscious of the re-configuration of constraints:

> S2-YP2: "Yeah, because if you forget when your appointment is, and you don't bother ringing, then you can end up back in court".
> S2-YP3: "Yeah, you get a warning and that, and you're just causing more problems for yourself".
> S1-YP3: "... if you had a reminder, then you wouldn't miss it and then..."

Having said this, case workers and young people both reported that their digital communication was largely one-way, i.e., case workers pushed out information via texts but young people did not reciprocate. Moreover, young people were said to 'delete texts,' 'block calls', 'block calls but send/receive texts'. Our findings show that such techniques were employed, possibly, as a reaction to governmentality. Such practices manifested themselves where young people found themselves being 'sanctioned or breached'. Moreover, it was not uncommon for young people to describe their weekly contact with the youth offending services in terms which suggested little value for them, that is, interaction was said to constitute simply 'turning up to your appointments', or just as 'a load of bollocks' [20].

In addition to the more information related app features, case workers also believed that some of their young people who had particular conditions attached to their orders, such as curfews and exclusion zones would benefit from alerts if these young people wittingly or unwittingly stayed out beyond the curfew or ventured into prohibited areas.

> S1-CW3: "... maybe bespoke... some young people are prohibited from going into certain areas so maybe their phone could vibrate if they are getting close to that area".

In their accounts of young people's experiences of private sector providers, contracted to run this operation for the youth justice sector, case workers expressed their disquiet about the fact that they invariably came to know about these incidents when it was too late:

S2-CW3: "Curfews, in my view, are not at all supportive. They' re punitive. I mean what's happening at the moment, it's a bit technical from a business point of view, because of the way the contract is set up with Serco, who are the enforcement, um, who actually run it, sorry, in this area, the curfews. The contract is set up in such a way that in fact the, um, the company don't let us know until the young person's accrued 2 h of absence from their curfew. So, kids being kids, you know they test it. And they, the sort of test each night and of course each time they're, sort of like, staying away from the house or whatever for 20 min, uh and then stretching it out maybe for 40 min. Well then, you know, they've accrued an hour and they think 'oh its not working' so they don't bother the next night. Stop out for two hours and then we get a notice saying, you know, they're in breach".

The overarching theme from our interviews with the case workers was that their relationship with the young people was one of make or break. In this regard, they wanted to do everything within their power to ensure that they were able to build a relationship of trust to ensure positive engagement.

S2-CW4: "I think the important thing is that you, you make them feel safe... So you spend the first part of your order building rapport... so that they can start to engage with you and trust you and feel safe with you..."

When the above ideas were presented to the young people in a subsequent workshop by means of screen capture images, to help illustrate these additional app features such as curfew alerts and exclusion zone alerts, we received a mixed response. The initial reaction was one of acceptance, however it soon transpired that young people resented the marker/visibility and hence the governmentality of these techniques:

Re: Both exclusion zone and curfew alerts:

S2-YP1: "So your phone's gonna vibrate when you cross?"
S2-YP2: "Yeah, that would be alright."
S2-YP3: "yeah, that would be quite useful."

The Exclusion zone feature had a both a stronger reaction and a change of position from the young people:

S2-YP4: "...so it actually tells the YOT workers and that, that I'm in that area?"
S2-YP2: "You're basically just trying to get a tracker onto our phones"
S3-YP1: "I wouldn't download it at all".

The marker/visibility concern became more apparent as young people expressed a desire to re-balance the overall power-relation between themselves and their case workers. Recognising the potential of the exclusion zone feature as a tracking concern, they asked incisively:

S3-YP1: "yeah, but how do we know, like, on their side of the app, they haven't got something they can click on to find out where you are, like. . ."
S3-YP2: ". . . yeah, you should show us their side of the app"

Concerns about being tracked were described within the context of an infringement of their sense of privacy and autonomy, and young people reported strategies to evade detection by law enforcement agencies. Understandably, young people voiced their apprehension about the possibility of being 'tagged' by default.

S3-YP2: ". . . they could just be watching what road you're walking up, where you're going to"

Despite the view that app had potential tracking facilities, there was also recognition, on the part of young people, that the app could be a source of power for them in cases where the police data/perception was inaccurate or out of date:

S3-YP2: "it would be good if like police try to stop you or something and they've got the details and stuff and they put it through the system or you're not allowed in this area and you pull out your phone and be like yeah well I'm not in that area. You know what I mean? To prove them wrong."

It seemed that the power of aesthetics was also of importance to young people. It was suggested that the MAYOT app logo should not attract attention in a way that young people were left to explain to their peers why they had this app on their phone. Preference for a design that was less conspicuous was expressed.

In the final prototype for the MAYOT app, the features went through a number of design changes representing the perspectives of case workers, young people and the designers. For example, young people's concerns that the exclusion order alert feature violated their privacy was considered seriously by the research team to ensure that a balance was struck between relative individual privacy and security of information. Given that perceived autonomy and control have been identified as key components of empowerment, we were keen to build social technology, which afforded such capability. Thus, young people, in being able to exercise choice and autonomy about whether they wished to receive alerts and have access to a map of the exclusion zone, or simply have access to the exclusion zone without the alerts. They were thus assured that they were not being regulated but empowered through timely and appropriate information. The latter overcame the concern about GPS location and the fear of being tracked.

The unintended consequences of the MAYOT app loomed large as we persevered with the deployment phase of this study [16]. It was evident that although case workers recognised the value of empowering young people through appropriate social and personal information sharing, their role demanded that the app be used in a way that could evidence misdemeanours. Thus, whether it was proof that a text was received but ignored, or whether a young person ventured into the excluded zone, the practitioners wanted to know to be able to present this as evidence to a breach panel and/or youth court.

4.3   The Researcher and the Co-design Process

A key contribution of this research is the experimental evidence use of novel co-design methods for eliciting requirements when working within a challenging environment. When offered the chance to become first order participants in the design of potential new technologies for their use, our young people demonstrated the necessary engagement for designers to benefit from their knowledge but raised critical challenges of how dilution of values such as privacy and autonomy can affect acceptance of technology. Such challenges only become apparent when participatory design approaches are augmented with value sensitive concerns as a central objective of the design.

The act of conducting and then reflecting on our co-design approach raises several lessons for those working in areas of social need. Researchers and design teams need to ensure that their interactions clearly demonstrate that they are not part of the governmentality infrastructure. How to do that persuasively without risk to the eventual deployment of software applications is potentially difficult. Our research demonstrates that taking a co-design approach augmented with value sensitive concerns, at least provides a vehicle for discussion and exposure of these concerns. Tracing how value concerns evolve over the lifetime of the design and deployment of technology is currently relatively neglected area of research in software engineering practice.

Co-design methods need to ensure that future issues of technology can also be evaluated effectively. Unintended consequences of technology can work in multiple directions. For example, the exclusion zone feature offered potential for un-intended use (by this we mean: not the intentional design purpose) for both case workers and young people. Some case workers and magistrates wanted to use location data collected from the exclusion zone to support breaching. Young people suggested that they would like to see features that empower them to deal more competently with the police such as knowledge of stop and search legal rights.

> S3-YP2: "yeah. You could read it on your phone and be like 'you can't do this under this Act'"

5   Conclusions

At the time of publication, the technology, the web portal and the app represent the current state of an accepted solution and further deployments are planned. Our findings suggest that practices of new public management in the use of technology have become embedded in the field of youth justice. Increasingly, young people in the youth justice system, whilst they are significant users of smart phone technologies, are alert to the pervasive nature of governmentality. In particular, they recognise that any sense of autonomy offered by technology only occurs within a constrained sense of freedom. Consequently they offer resistance and actively seek ways of working that can address the power relations that are afforded through the introduction of technology.

Gane's four categories of neoliberal governmentality introduced in Sect. 2, when revisited reveal the following. Surveillance and discipline manifest themselves in the aspects of the app features such as the exclusion zone and curfew alert. By installing the app and allowing these features to be active is sufficient for discipline to be present. Although the app collects data for when a young person may have violated an exclusion zone area, the data is not reported to the case worker. Hence, the intervention when necessary is not operationalised. Surveillance and control, on other hand, require that governmentality agencies or commercial agencies to which there has been devolved power, actively monitor mobile entities and carry out actions arising from the monitoring. Our findings recognised how this could be reified through the MAYOT app but the strict ethical guidelines and our involvement in the deployment of the technology and control of what data was available to case workers ensured that risk was mitigated.

The governmentality to promote competition category is interesting should software such as MAYOT app or similar become commercialised and supported through existing commercial software providers. In such an event the state creates conditions for the freedom of markets in the provision of mobile apps for youth offending teams and through audit processes (surveillance) evaluates the success of such apps through some defined measures. Further, within the confines of the research presented in this paper, Youth Offending Team managers could engender competition between case workers and measure the extent to which case workers are using the app.

In this paper, we have explored the relevance of Michel Foucault's writings in order to appraise the development of our social technology intended as a vehicle of engagement in the youth justice sector. The research has demonstrated that analysis of software systems from this perspective yields important observations that could ultimately affect the acceptance of new technology. The young people in our study exhibited an awareness of the governmentality agenda and actively sought ways to overcome it. Critically, we have presented a participatory design methodology that has been augmented with value sensitive concerns. This approach has provided an important lens by which we have been able to identify those values that have potential to affect the final use of any proposed technology. The research findings used in this paper suggest that appropriate management of such value sensitive concerns can go some way towards addressing the issues of governmentality.

While we have sought to demonstrate the role of governmentality in the design of technologies that are intended to encourage engagement of young people with their case workers, we acknowledge that our findings are illustrative and exploratory. More research is required to develop techniques and methods that can help software technologists, social policy experts and practitioners collaborate to provide effective tools to work with young people to prevent further offending. Most importantly, these tools should avoid the overt criticisms of governmentality.

References

1. Avgerou, C., McGrath, K.: Power, rationality, and the art of living through socio-technical change. MIS Q. **31**, 295–315 (2007)
2. Bentham, J.: Panopticon letters. In: The Panopticon Writings, pp. 29–96 (1995)
3. Bjerknes, G., Ehn, P., Kyng, M., Nygaard, K.: Computers and Democracy: A Scandinavian Challenge. Gower Pub Co, Brookfield (1987)
4. Donzelot, J.: The mobilization of society. In: The Foucault Effect: Studies in Governmentality, pp. 169–179 (1991)
5. Dorrestijn, S.: Design and ethics of product impact on user behavior and use practices. In: Intelligent Environments (Workshops), pp. 253–260 (2009)
6. Dorrestijn, S.: Technical mediation and subjectivation: tracing and extending Foucault's philosophy of technology. Philos. Technol. **25**(2), 221–241 (2012)
7. Farrington, D., Piquero, A.R., Jennings, W.G.: Offending from childhood to late middle age. Recent Results from the Cambridge Study in Delinquent Development. Springer, New York (2013). https://doi.org/10.1007/978-1-4614-6105-0
8. Flanagan, M., Howe, D.C., Nissenbaum, H.: Values at play: design tradeoffs in socially-oriented game design. In: Proceedings of the SIGCHI Conference on Human Factors in Computing Systems, pp. 751–760. ACM (2005)
9. Foucault, M.: Discipline and Punish: The Birth of the Prison. Vintage, New York (1977)
10. Foucault, M.: Space, Power and Knowledge. Penguin, London (1993)
11. Friedman, B.: Value-sensitive design. Interactions **3**(6), 16–23 (1996)
12. Gane, N.: The governmentalities of neoliberalism: panopticism, post-panopticism and beyond. Sociol. Rev. **60**(4), 611–634 (2012)
13. Garland, D.: Governmentality and the problem of crime: foucault, criminology, sociology. Theor. Criminol. **1**(2), 173–214 (1997)
14. Goldson, B., Muncie, J.: Youth Crime and Justice. Sage, London (2015)
15. Lyon, D.: An electronic panopticon? a sociological critique of surveillance theory. Sociol. Rev. **41**(4), 653–678 (1993)
16. Merton, R.K.: The unanticipated consequences of purposive social action. Am. Sociol. Rev. **1**(6), 894–904 (1936)
17. Muncie, J.: The construction and deconstruction of crime. In: The Problem of Crime, vol. 1, p. 5 (1996)
18. Nellis, M.: The 'tracking' controversy: the roots of mentoring and electronic monitoring. Youth Justice **4**(2), 77–99 (2004)
19. Newman, J.: Beyond the new public management? modernizing public services. In: New Managerialism, New Welfare, pp. 45–61 (2000)
20. Phoenix, J., Kelly, L.: 'You have to do it for yourself' Responsibilization in Youth Justice and young people's situated knowledge of youth justice practice. Br. J. Criminol., azs078 (2013)
21. Ritchie, J., Spencer, L., Bryman, A., Burgess, R.G.: Analysing Qualitative Data (1994)
22. Sanders, E.B.-N.: Generative tools for co-designing. In: Scrivener, S.A.R., Ball, L.J., Woodcock, A. (eds.) Collaborative Design, pp. 3–12. Springer, London (2000). https://doi.org/10.1007/978-1-4471-0779-8_1
23. Shay, L.: Deconstructing the relationship between privacy and security. IEEE Technol. Soc. Mag. **33**, 29 (2014)
24. Smith, D., Goldson, B., Muncie, J.: Youth crime and justice: research, evaluation and evidence. In: Youth, Crime and Justice: Critical Issues, p. 78 (2006)

25. Hoven, J.: ICT and value sensitive design. In: Goujon, P., Lavelle, S., Duquenoy, P., Kimppa, K., Laurent, V. (eds.) The Information Society: Innovation, Legitimacy, Ethics and Democracy In Honor of Professor Jacques Berleur s.j. IIFIP, vol. 233, pp. 67–72. Springer, Boston, MA (2007). https://doi.org/10.1007/978-0-387-72381-5_8
26. Venkatesh, V., Morris, M.G., Davis, G.B., Davis, F.D.: User acceptance of information technology: toward a unified view. MIS Q. **27**, 425–478 (2003)
27. Wiener, N.: The machine as threat and promise. Norbert Wien. Collected Works Commentaries **4**, 673–678 (1985)
28. Willcocks, L.P.: Michel Foucault in the social study of ICTs critique and reappraisal. Soc. Sci. Comput. Rev. **24**(3), 274–295 (2006)
29. Yoo, D., Huldtgren, A., Woelfer, J.P., Hendry, D.G., Friedman, B.: A value sensitive action-reflection model: evolving a co-design space with stakeholder and designer prompts. In: Proceedings of the SIGCHI Conference on Human Factors in Computing Systems, pp. 419–428. ACM (2013)

Context-Aware Access to Heterogeneous Resources Through On-the-Fly Mashups

Florian Daniel[(✉)], Maristella Matera, Elisa Quintarelli, Letizia Tanca,
and Vittorio Zaccaria

Dipartimento di Elettronica, Informazione e Bioingegneria,
Politecnico di Milano, Milan, Italy
{florian.daniel,maristella.matera,elisa.quintarelli,letizia.tanca,
vittorio.zaccaria}@polimi.it

Abstract. Current scenarios for app development are characterized by rich resources that often overwhelm the final users, especially in mobile app usage situations. It is therefore important to define design methods that enable dynamic filtering of the pertinent resources and appropriate *tailoring* of the retrieved content. This paper presents a design framework based on the specification of the *possible contexts* deemed relevant to a given application domain and on their mapping onto an *integrated schema* of the resources underlying the app. The context and the integrated schema enable the instantiation at runtime of templates of app pages in function of the context characterizing the user's current situation of use.

Keywords: Context-aware data access · Service selection · Mashups
CAMUS

1 Introduction

In the last decades, the pervasive introduction of ICT technologies in our society has changed the way people access information. Traditional data management systems have left the place to sophisticated data integration systems that combine and expose rich information extracted from all kinds of sources and make it available through different media devices. Also, users have changed their attitudes and behavior and are now "digital" and "social", independently of their current usage situation and device. Yet, this flexibility does not come at a low price, as finding information that is most suitable to the users' current context may require a significant time and effort, especially if the used software does not leverage on the users' context [15].

As highlighted in [13], in order to facilitate the development of software (for any kind of device) that is able to take into account the user's context, it is

This research is supported by the IT2Rail project (EU H2020 program, grant agreement no: 636078) and the Italian project SHELL (CTN01 00128 111357).

J. Krogstie and H. A. Reijers (Eds.): CAiSE 2018, LNCS 10816, pp. 119–134, 2018.
https://doi.org/10.1007/978-3-319-91563-0_8

important to define design methods that natively support the dynamic selection and filtering of pertinent resources and the consequent tailoring of retrieved data. Treating the dimensions characterizing the *context of use* as first-class design artifacts can enable context-awareness [15] in the data access layer and can guide the definition of context-aware queries over available heterogeneous resources.

To respond to the need for a development method with native support for context-awareness, this paper presents a design framework for the fast development of apps that revolves around (i) the explicit specification of the *context dimensions* deemed relevant in a given application domain and (ii) their mapping onto an *integrated schema of the available resources*. The scenarios we support have all the ingredients of data mashups or service compositions. However, differently from conventional mashup approaches, our mashups do not produce stand-alone applications, but serve as small, on-the-fly data integrations to be embedded into generic applications that require the fetching of data from heterogeneous resources. These "mini mashups" are formulated as context-agnostic queries, automatically turned into context-aware queries by the framework.

Running Example. To illustrate our method, we make use of a tourism scenario where an app personalizes the provided contents on the basis of the traveler contexts (e.g., current location and time, possible disabilities, user's preferences about topics and means of transportation). The app gathers contents about restaurants, hotels and itineraries from different resources, i.e., Web APIs and datasets that may be public or made available by the service provider who offers the app. In the scenario, local, proprietary data are, for instance, user profiles, buying histories, or similar core assets of the application to be developed. We in particular assume that data about affiliated hotels and discounts are stored in a local database table (HOTEL). As for remote sources, we assume the app leverages on two external services for the calculation of itineraries (ITINSVC1 and ITINSVC2) and on one service to search for restaurants (RESSVC). As there may be multiple providers offering similar services, a service selection at runtime may be needed – again, taking into account the user's context.

Paper Structure. We next introduce the concepts and artifacts of the proposed method; then, in Sect. 3, we go into the details of the Resource Schema, the Context Dimension Tree, and context-agnostic and -aware queries. Next we show how to interpret and execute queries and discuss an implementation in GraphQL (Sect. 4). Before closing we discuss related works.

2   Approach

We model context dimensions using the Context Dimension Model (CDM) [3], a specific abstract representation that, on the side of the model of the non-contextual features, provides an intuitive way to visually depict context information and the conceptual relationships that exist among the properties of context in a given scenario. The approach is in line with the idea of using two separate feature models for contextual and non-contextual requirements discussed in [13]. We propose the use of the CDM in the development of context-aware applications to achieve two goals: (i) the concise and human- and machine-readable *representation* of all context dimensions and properties relevant to a given development scenario and (ii) the *simplification* of the development of context-aware application features. The approach turns the specification of all the possible contexts of the considered scenario into a first-class development artifact that not only serves a documentation purpose but also an operational one. The context dimensions themselves can be aggregated into categories: *information on the user* (knowledge of habits, emotional state, physiological conditions), *user's social environment* (co-location of others, social interaction, current tasks), *physical environment* (location, time, and physical conditions like noise, light, pressure, air quality). The dimensions in the latter category can be automatically derived by means of appropriate sensors, while the previous ones may require direct user input or suitable default settings.

Figure 1 illustrates the resulting method and development steps. The process starts from the identification and description of the available resources, which can be both *local data sources* ①, for example relational data bases, or third-party *Web services* represented by their *service descriptors* ②. An integrated schema of the available data is then defined, so as to present the developer with one abstract model of the data only. We call this schema the *resource schema* ③, and we express it as a *relational model*. The resource schema is generally defined

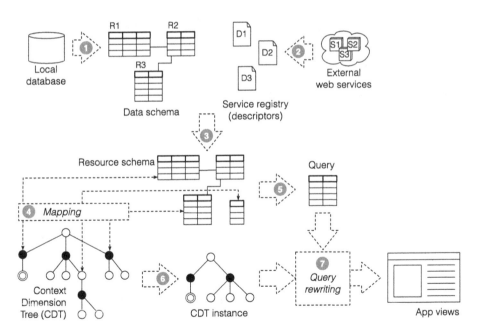

Fig. 1. Overview of approach to the development of context-aware, mobile apps starting from internal data sources, external services and a Context Dimension Tree.

manually, but it could be generated automatically depending on the regularity of the schema of the selected resources. The resource schema is accompanied by a CDT (defined in the next section) that captures all the execution contexts the application may run in ④. Each node of the CDT is mapped to the resource schema (e.g., by means of views expressed in relational algebra), in order to support automatic query rewriting at runtime.

The next step is the design of the queries that will feed the pages of the final application, e.g., the pages in a mobile app that allow the user to access and interact with the content ⑤. The resource schema focuses on the integration of data, independently of their use, and is thus context-agnostic. It is the CDT that defines which of the elements in the resource schema are related to context in the given application domain. The queries expressed over the resource schema are thus *context-agnostic* too. Neglecting context properties in this phase allows the developer to focus on the core functionality of the application, deferring adaptation concerns. The queries are typically written manually; without loss of generality, we express them in relational algebra.

At runtime, the CDT can be instantiated with concrete values coming from the *context sensors* (sensing devices, user inputs, external sources) ⑥. That is, the runtime environment of the application automatically updates the tree with context information to characterize the usage scenario the user is currently involved in. The availability of a CDT instance enables the derivation of *context-aware* queries from the context-agnostic queries, by suitably enriching them with context information ⑦. This step can be performed fully automatically, e.g., based on conventional, view-based query rewriting techniques [11].

The simplification of the development process proposed for context-aware applications therefore consists in (i) the use of a resource schema that hides technological details and data provenance issues and (ii) the automatic rewriting of context-agnostic queries defined on the resource schema into context-aware queries. Both features alleviate the developer from tasks that are typically tedious and time consuming. In the following sections we describe the core ingredients of the method. We will also show how the conceptual approach can be naturally mapped into state-of-the-art implementations making use of GraphQL.

3   Resource Schema and Context-Aware Queries

The initial activities in the development process are the *selection* of the resources of interest, which can be both local and remote data sources, and their *technical description*, specifying the details that are needed to access them. For local data sources such details refer to (i) the *endpoint* of the data source (e.g., its IP address), (ii) the *port* though which the source can be accessed, and (iii) the *username* and *password* identifying the user that represents the application. To programmatically access external Web services (e.g., SOAP/WSDL or RESTful services), it is necessary to specify: (i) the service *endpoints* (one or more URIs), (ii) the *operations* offered by the service, (iii) the respective *input parameters* and *output data schemas* and serializations (e.g., JSON or XML), and (iv) possible *authentication* details (e.g., usernames/passwords or developer keys). Data sources may differ in the communication protocols they use (plain HTTP vs. SQL connectors), their implementation technologies (as long as they expose an HTTP or SQL interface), the data formats and schemas they use (as long as data can be correctly extracted). All these properties are specified in the respective registry entries.

For the sake of brevity, we do not further detail all technicalities here: they represent state-of-the-art development practice. What is relevant in our framework is the specification of the *access patterns* that can be used to access the services, as most services support different mandatory or optional input parameters to access data [5]. The specification of access patterns will be illustrated in Sect. 3.2.

3.1   Resource Schema and Context-Agnostic Queries

Once the different resources are registered in the system, it is possible to derive the resource schema of the available data so as to present the developer with a unique schema of the data. The schema aims to represent the data provided by the resources and their relationships in a way that accommodates the requirements posed by the specific application domain. For example, we can think of the tables introduced in our running example as the result of a modeling activity that produces a data schema representing all the resources selected for the given application domain.

Considering the resources identified for our running example, a possible resource schema could be the following one, where RESTAURANT represents the data accessible through the service RESSVC, HOTEL represents the data stored in the local database HOTELTAB, and ITINERARY represents the itineraries computed by the two services ITINSVC1 and ITINSVC2:

RESTAURANT(name, address, phone, type, cuisine_type, playground)
HOTEL(name, address, category, childcare)
ITINERARY(from, to, type, directions, price)

The identification of the previous relations depends on opportunistic choices in the app design. For example, the attribute ITINERARY.`directions` is not further specified, as it is not necessary to further split and query direction descriptions. `directions` could however contain a map image, a list of instructions, e.g., about how to go from a hotel to a restaurant, or similar. The resource schema also makes only use of logical **addresses** (city, street, number), instead of physical GPS coordinates.

In addition to these application-specific design choices, the resource schema keeps track of the provenance of each attribute in the defined relations. This is achieved using an additional table: SOURCE(id, attribute, registry_entry) that tracks the necessary, minimal meta-data: for each attribute in the resource schema, it contains a link to the registry entry that may provide data for the attribute. For the attribute ITINERARY.`directions` the table will thus contain two entries, i.e., ITINSVC1 and ITINSVC2. These meta-data will be used only at query execution time and are not accessible to the developer, who instead is now able to write her context-agnostic queries. For example, the following query

$$Q_1 = \Pi_{\substack{name,address,\\phone,type,\\directions}} \sigma_{cuisine\_type=\$VAL} \text{RESTAURANT} \bowtie_{address=to} \text{INTINERARY}$$

is used to instantiate the page of the application that, at interaction time, allows the user to choose a type of cuisine: the user choice will replace the parameter $\$VAL$, the `from` attribute for the calculation of itineraries will be taken from the context. The app thus shows the restaurants matching the cuisine choice along with the itineraries to reach them. In Sect. 3.4, we show how to inject context into Q_1 to obtain its context-aware version.

3.2 Resource Mapping

Given the above resource schema, each access pattern to a resource can now be expressed as a view over it. This equips the pure technical registry entry, that tells how to interact with the service, with a semantical mapping of the service to the resource schema that enables the context-aware service selection at runtime. For instance, the chosen restaurant search service can be expressed as follows:

$$\text{RESSVC} \equiv \Pi_{\substack{name,address,phone,\\type,cuisine,playground}} \sigma_{\substack{address=\$optional\wedge\\type=\$optional\wedge\\playground=\$optional\wedge\\cuisine\_type=\$optional}} \text{RESTAURANT}$$

In bold we highlight two keywords that are needed to express a service's access pattern: the values **$mandatory** and **$optional** tell, respectively, if an attribute is a *mandatory* or *optional* input of the service. Both keywords are automatically replaced at runtime by their respective values. Without proper values for mandatory inputs, the service cannot be invoked; optional inputs may be used to restrict the output data produced by the service.

In addition to mandatory and optional inputs, it is also possible to specify *constant* values for some input parameters. Doing so binds the view representing the access pattern to the given value, expressing that the service is able to provide only data that complies with this restriction. If we take, for instance, two services that provide itinerary information (e.g., ITINSVC1 about a city's local transport network and ITINSVC2 about a national railway network), we may obtain the following two access patterns:

$$\text{ITINSVC1} \equiv \Pi_{\substack{from,to,type,\\itinerary,price}} \sigma_{\substack{from\$mandatory\wedge\\to=\$mandatory\\\wedge type=\$optional}} \text{ITINERARY}$$

$$\text{ITINSVC2} \equiv \Pi_{\substack{from,to,type,\\itinerary,price}} \sigma_{\substack{from=\$mandatory\wedge to=\$mandatory\\\wedge type=\text{``train''}}} \text{ITINERARY}$$

The second access pattern explicitly binds the attribute `type` to the value "train" as ITINSVC2 is able to provide only data about train connections, while ITINSVC1 may provide data about all among trains, undergrounds, busses and trams within its geographical area of competence (for simplicity, we do not represent this limitation here).

Analogously, the HOTELTAB table of the local database can be expressed as follows (observe that, for consistency with the mapping of Web services, *all* attributes are *optional* as relations do not have access patterns):

$$\text{HOTELTAB} \equiv \varPi_{\substack{name,address,\\ category,childcare}} \sigma_{\substack{name=\$optional \wedge address=\$optional\\ category=\$optional \wedge childcare=\$optional}} \text{HOTEL}$$

3.3   Context-Aware Queries

The Context Dimension Model [3] allows one to represent Context Dimension Trees (CDTs). An example of CDT for the touristic scenario of the running example is shown in Fig. 2. *Dimension nodes*, depicted in black, represent the different perspectives describing context (e.g., user type and transportation), while *concepts*, depicted as white nodes, are the admissible values of each dimension (e.g., the concepts adult, young adult and family with children are values for the user type dimension). *Attributes*, represented by double circles, are parameters whose values are dynamically derived from the environment or provided by the users themselves at execution time, and used to replace a high number of concepts when it is impractical to list them all: e.g., the current position dimension has as child the attribute curr_pos.

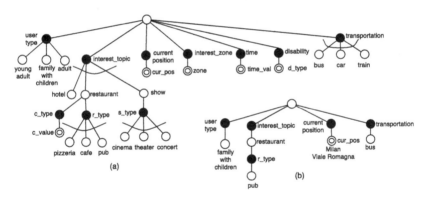

Fig. 2. The CDT of our example (a) and a context instance (b)

A context instance is a subtree of a CDT (also represented as a set of <dimension=value> pairs) where the parameter nodes are replaced with concrete values. Figure 2(b) shows graphically the instance C = {user type = family with children, r_type = pub, current position = Milan Viale Romagna, transportation = bus}: a family with children, currently located in Viale Romagna in Milan, moves around by bus and prefers to eat at pubs.

3.4   Query Rewriting

Following an approach similar to [3], we propose that the designer associates each *context element* (i.e., `<dimension=value>`) of the CDT with one or more relations of the resource schema, filtered on the basis of the `value` of the context element itself. For instance, Table 1 shows the expressions associated with the context elements in C.

Table 1. Relational algebra expressions associated with the context instance C.

Context Element	Relational Algebra Expressions
`user type = family with children`	$\sigma_{playground="yes"}$RESTAURANT
	$\sigma_{childcare="yes"}$HOTEL
`r_type= pub`	$\sigma_{type="pub"}$RESTAURANT
`current position = Milan Viale Romagna`	$\sigma_{from="Milan\ Viale\ Romagna" \wedge to=\$Value}$ITINERARY
	$\sigma_{address="Milan\ Viale\ Romagna"}$RESTAURANT
	$\sigma_{address="Milan\ Viale\ Romagna"}$HOTEL
`transportation = bus`	$\sigma_{type="bus"}$ITINERARY

Suppose now that at run time, while in context C, the app prompts the user with the context-agnostic query Q_1 from Sect. 3.1, where restaurants can be selected on the basis of a cuisine type chosen by the user. Q_1 will be expanded with the user's choice and automatically rewritten using the context-aware expressions in Table 1. The result is the context-aware query Q_2 represented in Fig. 3 as a standard syntax tree for relational algebra.

Fig. 3. Query Q_2 where P_1 : *cuisineType* = *"indian"* $\wedge$ *r_type* = *"pub"* $\wedge$ *playground* = *"yes"* $\wedge$ *address* = *"Milan Viale Romagna"* and P_2 : *type* = *"bus"* $\wedge$ *from* = *"Milan Viale Romagna"*.

Note how the agnostic query is extended by adding conditions related to the context dimensions (e.g., the user's current position or the type of user) to the selection and join operations: in Table 1, the expression $\sigma_{from="Milan\ Viale\ Romagna" \wedge to=\$Value}$ITINERARY, associated with the context element `current position = Milan Viale Romagna`, contains a parameter referring to the destination of the itinerary, that takes now the value specified by the join condition of the agnostic query Q_1.

4   Query Interpretation and Execution

In this section we describe the synthesis of a query execution plan from the context-aware query and the consequent selection of the related data and services; in the last subsection we then give an account of the real execution flow of our framework, as it has been made concrete in a research prototype [8].

4.1   Query Interpretation and Service Selection

Conceptually, the generation of the query execution plan proceeds as follows:

1. We produce a tree-like representation of the relational algebra query that is extended with an explicit representation of the corresponding predicates.
2. We map a service to one or more query primitives in the original tree by exploiting the structure of their access pattern represented as a *tile pattern*[1].
3. We visit the mapped tree to produce a schedule of web services invocations and database assesses.

For example, let us consider query Q_2 of Fig. 3. To see how this query can be mapped to an actual query execution plan involving service requests, we consider the services described in Sect. 3.2. Every service exposes a potential set of filtering predicates to be used when accessing it. To capture this information, the tile pattern covers not only the abstract relational algebra operation associated with the service, but also the syntactic structure of valid predicates. Here and in the rest of this section we extend with a double-line arrow the representation of the associated logical predicate. For example, Fig. 4(a) shows the subtree pattern in the original query the ITINSVC1 service can answer to (Sect. 3.2). The pattern specifies that such a service can be selected when part of the original query aims to select itineraries where:

- the **type** attribute is optional (as represented by the parentheses),
- the **from** attribute is mandatory,
- the **to** attribute is mandatory.

Figure 4(b) shows, similarly, the pattern associated with RESSVC. ITINSVC2 pattern is not shown, as its restriction `"type=train"` is not compatible with the query's request for a bus (`"type=bus"`).

We now can expand query Q_2 by exposing the logical structure of its predicates. We then note that the only registered access patterns that allow valid matches are RESSVC and ITINSVC1 because ITINSVC2 cannot pattern match. Figure 5 shows the result of the match operation; since we deal with a logically correlated sub-query, we mark with a small circle each tile pattern leaf that corresponds to an attribute involved in a join condition appearing above it in the tree.

---

[1] A tile pattern is a tree template with one or more wildcards that can match any subtree of the original query. Note that, in this view, data coming from lower nodes is an "input" to a service, while the root of the node is the "output" of the service.

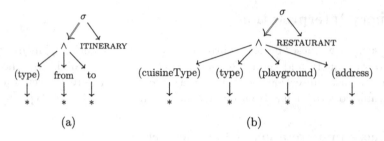

Fig. 4. Itinerary service tile (ITINSVC1) and restaurant service tile (RESSVC).

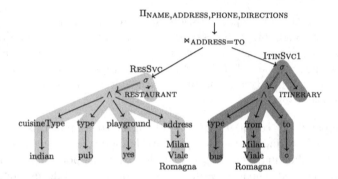

Fig. 5. Example context aware query where predicates have been expanded and services have been verified to match (see annotation).

After having identified the services by pattern matching, we schedule the service invocation by traversing the tree in post-order and executing joins with a nested join strategy. In particular, we use the convention that the left relation is always the outer relation and exploit the commutativity of the join operator to reorder the nodes in the tree and optimize the query execution. In particular, we order nodes from left to right by ascending number of correlated attributes. In this case, RESSVC is invoked first (using as parameters the values of the leafs that allowed the match), while ITINSVC1 is invoked for each tuple returned by RESSVC by using the "address" attribute value as the "to" parameter value.

More refined methods to enlarge the space of solutions considered for tile pattern matching can be also adopted [5]. For example, to address the case when multiple valid services can serve a specific query, we already experimented techniques (i) to invoke them all and then fuse their outputs and also (ii) to apply some ranking strategy, for example based on service quality criteria, and then select the best service [6,8].

4.2 Prototype Implementation

As a concrete implementation of the conceptual approach discussed in the previous sections, we developed a prototype based on node.js. The prototype

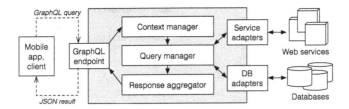

Fig. 6. Functional architecture of the prototype implementation.

implements a GraphQL server-side runtime [1] that supports the execution of queries over a GraphQL schema. The schema consists of simple type declarations and describes the data to be exposed to the front-end applications, independently of any specific database, storage engine or service access logic. This feature natively supports the implementation of our (virtual) resource schema inside a GraphQL API and the proxying of incoming queries to the respective resources.

The architecture of the prototype is shown in Fig. 6. To describe its main functions, let us consider a mobile app whose task is to render the example query Q_1 as defined in Sect. 3.1. The process for query execution starts with the mobile app sending a GraphQL query with both the query parameters and the context data. Then the following actions are taken:

- A *Context Manager*, given an instance of the CDT and possible user inputs, decides if/which user inputs overwrite which context parameter.
- A *Query Manager* selects and invokes the corresponding data sources according to the techniques presented above (leveraging on suitable service/DB adapters for the communication with the sources).
- A *Response Aggregator* composes the integrated result set into a JSON structure and sends it back to the mobile app.

Given the internal logic of the GraphQL API and the nature of the CDT, no complex, structural query rewriting is needed: it suffices to ship the set of context properties as parameters along with the original query from the client to the API. Depending on the presence or not of context properties in the CDT, the API can then internally apply suitable selection conditions on retrieved results. As we show next, this means that the effect of adding context essentially translates into enriched selection conditions.

To exemplify the internal logic of the API, let's consider the GraphQL schema in Fig. 7. `Restaurant` and `Itinerary` are the two relations of the resource schema, `CamusContext` is the CDT structure (limited to the properties of the example) and `restaurant` is the type of query supported by the API. Note how the attribute `reachThrough` of `Restaurant` supports the calculation of the join of Q2. Also note how this join must be specified at schema level (and then supported by the internal API implementation), as the schema describes the structure of the output data, rather than that of the underlying data.

```
type Restaurant {                    input CamusContext {
    name:         String                 userType:       String
    address:      String                 rType:          String
    phone:        String                 currentPosition: String
    type:         String                 transportation: String
    cuisine_type: String             }
    playground:   Boolean
    reachThrough: [Itinerary]
}
                                     type Query {
type Itinerary {
    from:         String                 restaurant (cuisine_type: String,
    to:           String                             from: String,
    type:         String                             cdt: CamusContext!):
    directions:   String                             [Restaurant]
    price:        Float              }
}
```

Fig. 7. Resource schema of the example scenario in GraphQL schema language.

```
{
    restaurant (cuisine_type : "indian",
        cdt: {
            userType: "family with children",
            rType: "Pub",
            currentPosition: "Milan Viale Romagna",
            transportation: "bus"
        })
        {
            name
            address
            phone
            reachThrough {
                type
                directions
}}}
```

Fig. 8. Example GraphQL query (Q2) to retrieve restaurants and respective itineraries.

Figure 8 now shows the GraphQL query corresponding to Q_2 as sent from the client to the API. The query asks for restaurants, provides the selection conditions between parentheses, and then lists the properties (projection) it wants to extract about the restaurants. The presence of the reachThrough property among these properties corresponds to the join to be calculated between Restaurant and Itinerary. Instead of flatting out the list of attributes, the query keeps all context properties grouped as one cdt element (for separation of concerns); conceptually, they all correspond to possible selection conditions.

In response to this query, the GraphQL API enacts the resolver illustrated in Fig. 9. First, the existence of possible user inputs that would overwrite context properties is checked (in the example, if an explicit from information is provided, this would overwrite the context's currentPosition property). Then, the Query Manager is invoked with the entities to be retrieved (ordered list), the selection conditions, and the context of the query. Internally, the Query Manager proceeds as described in the previous section in order to produce the set of requested restaurants (including the respective itineraries) as output. Finally, for each retrieved restaurant, the resolver leverages on a so-called data class (Restaurant), which implements GraphQL's projection logic.

```
var root = {

  restaurant: function ({cuisine_type, from, cdt}) {

    if (from) cdt.currentPosition = from; // simplified Context Manager

    var restaurants = QueryManager.get({ entities: ["restaurants", "itineraries"],
                                         cuisine_type: cuisine_type,
                                         context: cdt });

    return restaurants.map (function(restaurant) { // simplified Response Aggregator
      return new Restaurant(restaurant, cdt);
    });
}}
```

Fig. 9. Example implementation of the GraphQL resolver answering Q2.

5 Comparison with Other Work

Different works on mobile app design describe ad-hoc solutions for the development of context-aware applications [16] in which it is difficult to identify reusable abstractions. In [9] the authors follow a more systematic approach, showing how context-aware mobile apps can be built by mashing-up components managing the app logic with reusable *context components* dedicated to capturing context events and activating related operations in the app. The approach does not provide abstractions for context modeling: the designer is in charge of configuring the context components (which basically manage user location and time) by means of parameter settings. However, the work shows how to achieve context-aware applications by means of a lightweight integration of heterogeneous and reusable components. Our approach also exploits mashup techniques. Our context-aware queries can in fact be considered "mini data mashups" integrating on the fly selected data sources. The goal, however, is to promote the adoption of a conceptual layer (i.e., the combined use of the resource schema and the context model), which enables app developers to reason at a high level of abstraction. The adopted conceptual models then drive the automatic selection of services and their dynamic, context-aware querying at runtime.

Some other approaches offer systematic methodologies and design environments. *MoWA* [4] introduce *augmentation*, which consists in adding some scripts on top of context-agnostic pages so that at runtime context can be gathered and processed to trigger page adaptations. In line with our approach, *MoWA* promotes separation of concerns and gives context a first-class role; however, it forces the designer to add a number of scripts for each page to be adapted dynamically at runtime. The advantage of our approach is that context-awareness is achieved at the only cost of defining, during the initial design phases, an adequate conceptual model capturing the most salient context dimensions.

Further works focus on the retrieval of content from heterogeneous services. *MyService* [12] provides expert designers with the possibility to select pre-defined context-based rules on top of a service directory. Based on the chosen rules, proper services are selected at runtime depending on the gathered context, and the code of the final app invoking these services is dynamically generated. This work is in line with our idea to filter services at runtime on the basis of the identified context. However the adopted notion of context is limited to the user location, while CDM is generic enough to cover several other dimensions that, in each given scenario, might characterize the contexts of use. Moreover, in our approach the designers are not required to care about which services have to be invoked at runtime; the system selects those services that best match the identified context instance.

Some other works are characterized by the adoption of context models to guide the access to heterogeneous resources. In [7], the authors use CDM to model the possible contexts and build a platform serving the execution of a context-aware tourism app. The app flexibly collects non-structured data from varying heterogeneous sources, and provides contextual recommendations to the user. This work shows the feasibility of adopting CDM to drive the context-aware selection of services to be invoked at runtime. In addition to this, in this paper we clarify how to select services in an automatic manner and how to build related context-aware queries on top of the selected data sources.

In [2], the authors then present a Model-Driven Engineering (MDE) approach where context meta-models and model-to-code transformations guide the automatic generation of code for the final context-aware apps. The proposed techniques for model-to-code transformations are interesting and are also in line with the goal of our research. However, once meta-models are in place, context modeling for the generation of a specific app requires the designer to define rules (i.e., OCL expressions) specifying the context-aware behaviors to be shown at runtime. Our approach, instead, does not need additional specifications on top of the context model; strategies for service and content filtering are shaped up at runtime, depending on the way the captured context guides the rewriting of context-agnostic queries.

From an application perspective, the proposed method focuses on the context-aware filtering of data integrated from multiple sources. It does not provide for personalized recommendations of data items, a problem that is typically addressed in scenarios similar to our tourism example [10]. Recommending suitable items, once contextual data are fetched, is an orthogonal design issue that we already addressed in our previous work [14].

6   Conclusions and Future Work

The contribution of this paper is a principled definition of the design method underlying CAMUS (Context-Aware Mobile mashUpS) [8], a research project that aims at the conception of high-level abstractions for efficient data and service integration in context-aware mobile applications. The method is general in nature and can be applied in the development of any kind of information system that requires on-the-fly, context-aware data access capabilities (e.g., the method can be used to provide context-aware access to external resources if it is wrapped by a suitable API called from within an application, or it can be used to provide context-aware access to internal data sources in parallel to existing data access channels). The paper specifically focuses on the data preparation and retrieval tasks and shows how to enable the automatic rewriting of context-agnostic queries into context-aware queries by explicitly modeling what is considered context in a given application scenario. The method also provides the conceptual and technological foundation for principled context management, effectively assisting the work of the developer. The prototype implementation shows how GraphQL naturally lends itself as candidate technology for the seamless integration of the (virtual) resource schema with concrete data access logics.

We preliminarily measured the performance of service selection and invocation; results are encouraging and time log-normally distributed [8]. Next, we will generalize our prototype implementation, parameterize it, and devise suitable transformation logics able to transform the resource schema (already expressed in GraphQL schema language), the CDT and the set of context-agnostic queries into full-fledged GraphQL APIs. We also would like to look into contextual data display, context-driven discovery of services from large repositories, and visual design environments for modeling context and designing resource schemas.

References

1. GraphQL. Draft RFC Specification, Facebook (2015). https://facebook.github.io/graphql
2. Achilleos, A., Yang, K., Georgalas, N.: Context modelling and a context-aware framework for pervasive service creation: a model-driven approach. Pervasive Mob. Comput. **6**(2), 281–296 (2010)
3. Bolchini, C., Quintarelli, E., Tanca, L.: CARVE: context-aware automatic view definition over relational databases. Inf. Syst. **38**(1), 45–67 (2013)
4. Bosetti, G.A., Firmenich, S., Gordillo, S.E., Rossi, G.: An approach for building mobile web applications through web augmentation. J. Web Eng. **16**(1&2), 75–102 (2017)
5. Braga, D., Ceri, S., Daniel, F., Martinenghi, D.: Optimization of multi-domain queries on the web. PVLDB **1**(1), 562–573 (2008)
6. Cappiello, C., Matera, M., Picozzi, M.: A UI-centric approach for the end-user development of multi-device mashups. TWEB **9**(3), 11 (2015)
7. Casillo, M., Colace, F., Santo, M.D., Lemma, S., Lombardi, M.: A context-aware mobile solution for assisting tourists in a smart environment. In: HICSS 2017. AIS Electronic Library (AISeL) (2017)
8. Cassani, V., Gianelli, S., Matera, M., Medana, R., Quintarelli, E., Tanca, L., Zaccaria, V.: On the role of context in the design of mobile mashups. In: Daniel, F., Gaedke, M. (eds.) RMC 2016. CCIS, vol. 696, pp. 108–128. Springer, Cham (2017). https://doi.org/10.1007/978-3-319-53174-8_7
9. Daniel, F., Matera, M.: Mashing up context-aware web applications: a component-based development approach. In: Bailey, J., Maier, D., Schewe, K.-D., Thalheim, B., Wang, X.S. (eds.) WISE 2008. LNCS, vol. 5175, pp. 250–263. Springer, Heidelberg (2008). https://doi.org/10.1007/978-3-540-85481-4_20
10. Daramola, O., Adigun, M., Ayo, C.: Building an ontology-based framework for tourism recommendation services. In: Höpken, W., Gretzel, U., Law, R. (eds.) Information and Communication Technologies in Tourism 2009, pp. 135–147. Springer, Vienna (2009). https://doi.org/10.1007/978-3-211-93971-0_12
11. Halevy, A.Y.: Answering queries using views: a survey. VLDB J. **10**(4), 270–294 (2001)

12. Lee, E., Joo, H.-J.: Developing lightweight context-aware service mashup applications. In: ICACT 2013, pp. 1060–1064 (2013)
13. Mens, K., Capilla, R., Hartmann, H., Kropf, T.: Modeling and managing context-aware systems variability. IEEE Softw. **34**(6), 58–63 (2017)
14. Miele, A., Quintarelli, E., Rabosio, E., Tanca, L.: Adapt: automatic data personalization based on contextual preferences. In: IEEE 30th International Conference on Data Engineering, Chicago, ICDE 2014, IL, USA, 31 March–4 April 2014, pp. 1234–1237 (2014)
15. Salber, D., Dey, A.K., Abowd, G.D.: The context toolkit: aiding the development of context-enabled applications. In: CHI 1999, pp. 434–441 (1999)
16. Schaller, R.: Mobile tourist guides: bridging the gap between automation and users retaining control of their itineraries. In: Proceedings of the 5th Information Interaction in Context Symposium, IIiX 2014, pp. 320–323 (2014)

Social Computing and Personalization

CrowdSheet: An Easy-To-Use One-Stop Tool for Writing and Executing Complex Crowdsourcing

Rikuya Suzuki, Tetsuo Sakaguchi$^{(\boxtimes)}$, Masaki Matsubara, Hiroyuki Kitagawa, and Atsuyuki Morishima

University of Tsukuba, 1-1-1 Tennodai, Tsukuba, Ibaraki 305-8577, Japan
rikuya.suzuki.2015b@mlab.info, {saka,masaki,mori}@slis.tsukuba.ac.jp, kitagawa@cs.tsukuba.ac.jp

Abstract. Developing crowdsourcing applications with dataflows among tasks requires requesters to submit tasks to crowdsourcing services, obtain results, write programs to process the results, and often repeat this process. This paper proposes CrowdSheet, an application that provides a spreadsheet interface to easily write and execute such complex crowdsourcing applications. We prove that a natural extension to existing spreadsheets, with only two types of new spreadsheet functions, allows us to write a fairly wide range of real-world applications. Our experimental results indicate that many spreadsheet users can easily write complex crowdsourcing applications with CrowdSheet.

Keywords: Rapid development · Complex crowdsourcing
Expressive power analysis

1 Introduction

Crowdsourcing involving dataflow among tasks (i.e., the results of some tasks affect other tasks) is called *complex crowdsourcing*, and is a promising approach for a wide range of applications [14] such as writing articles and filling in tables. Complex crowdsourcing is ubiquitous even in real-world applications. For example, asking people to provide photos that contain suspected "red imported fire ants" and a location and to filter out obviously different ones, and then asking experts to check whether the remainder really are red imported fire ants, is complex crowdsourcing.

As crowdsourcing allows us to realize things previously impossible with computers only, enabling many people to easily *develop* complex crowdsourcing will have an impact on people's problem-solving abilities. However, the development is not an easy task for many people at present.

As a running example, suppose that we are holding an academic conference with parallel programme sessions (e.g., session 1 A and 2A in room A; session 1B and 2B in room B; session 1 A and 1B at the same time), and want to estimate

© Springer International Publishing AG, part of Springer Nature 2018
J. Krogstie and H. A. Reijers (Eds.): CAiSE 2018, LNCS 10816, pp. 137–153, 2018.
https://doi.org/10.1007/978-3-319-91563-0_9

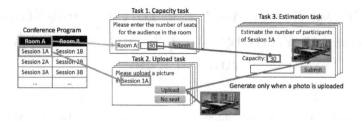

Fig. 1. Flow for complex crowdsourcing with running example: (1) Capacity task; enter the number of audience seats in the room. (2) Upload task; upload a photo taken in the session or report "no seat". (3) Estimation task; estimate the number of attendees in the session.

the number of attendees in each session except demos and posters. To this end, we solve our problem using three types of microtasks. The underlying idea is that we may be able to estimate the number of attendees in each session, if we know the room capacity (i.e., the number of audience seats in each room) and have photos taken in the session. Figure 1 illustrates the flow. First, for each room, we make a "capacity task" to ask workers to enter the number of audience seats in the room. Second, for each session of the program, we create an "upload task" to ask workers to upload a photo taken in the session if the audience seats exist (e.g., in general presentation sessions), or to report that there is no audience seats if the seats are removed in some sessions such as demos and posters. If a worker uploads a photo with audience seats, we submit the third task, called the "estimation task" to ask workers to analyze the picture and the room capacity and to estimate the number of attendees in the session. If it is reported that there is no audience seats (e.g., in demo & poster sessions), we do not issue the estimation task.

The flow can be implemented with current crowdsourcing services as follows: (1) Submit the capacity and upload tasks to the crowdsourcing service: (2) wait for the workers to complete the tasks, and then download the results in some manner: (3) check whether each uploaded picture was taken in a room with audience seats and, if so, submit an estimation task for the session and download the results. Using this procedure, we need to submit tasks to crowdsourcing services, obtain results, write programs, and use tools to process the results. This process is often repeated. If we want to improve data quality, we have to implement our own code for it (e.g., duplicated tasks), which is not separated from the essential logic of the application.

This paper proposes CrowdSheet, an easy-to-use, one-stop tool for implementing complex crowdsourcing (Fig. 2). CrowdSheet provides a spreadsheet interface for easily writing complex dataflows with a variety of microtasks, with crowdsourcing services such as Amazon Mechanical Turk (MTurk) in its backend. Whereas many other solutions focus on how to optimize the execution plans, our focus is on making it easy for a wide range of people to exploit the power of complex crowdsourcing. Our design principle separates the concerns on aspects

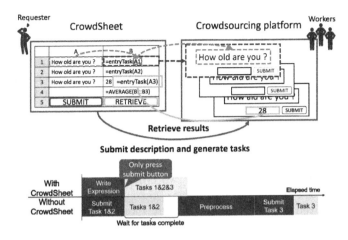

Fig. 2. (Top) CrowdSheet Overview: Tasks are submitted to crowdsourcing platforms according to the dataflow description in spreadsheets. (Bottom) Execution Process: Without CrowdSheet, the user has to wait for the first set of tasks to be completed, download the result, often filter it, and upload the second set of tasks based on the results. The process becomes more complicated if the user has more than three sets of tasks. With CrowdSheet, the user is released immediately after writing expressions and pressing the submit button and the spreadsheet cells will be filled with the results of completed tasks.

such as quality and costs from the essential logic of applications; CrowdSheet accepts independent modules implementing techniques for generating alternative execution plans [24] (omitted in this paper).

With CrowdSheet, the user is released after writing simple expressions and pressing the submit button. Tasks are *automatically* submitted and the spreadsheet cells are *gradually filled* with the results of completed tasks.

Because of its simplicity, the spreadsheet paradigm has been widely accepted by people who are not IT experts. Each cell contains either a value (numerical or string) or a function to compute a value whose parameters are often taken from other cells. CrowdSheet builds on this paradigm and provides two predefined functions to define and invoke tasks, whose parameters are often taken from other cells.

Although the idea is simple, coming up with a *good design* was non-trivial. We carefully designed CrowdSheet as a natural extension of existing spreadsheets with only two additional spreadsheet functions, while guaranteeing reasonable expressive power with our theoretical analysis results. As shown in Sect. 6, more than 60% of 33 participants who usually had experience using spreadsheets could implement complex crowdsourcing using CrowdSheet.

The contributions of this paper are as follows.

Spreadsheet Paradigm for Crowdsourcing. This paper shows a set of functions allows us to implement complex crowdsourcing with the spreadsheet paradigm. The design conforms to a common spreadsheet framework and extends it naturally. A running example is used to demonstrate how CrowdSheet easily facilitates the implementation.

Theoretical Analysis. Theoretical analysis is conducted to identify the expressive power of CrowdSheet. More specifically, a class of programs in another language that is equivalent to CrowdSheet expressions is identified and used to demonstrate its limitation. In addition, we explain that some of the crowdsourcing applications surveyed in [26] can be implemented as long as certain conditions are satisfied.

Implementation. We give constructive proof of our idea by implementing CrowdSheet with Microsoft Excel and two crowdsourcing platforms, although our concept is not restricted to particular spreadsheets and crowdsourcing platforms. Through the explanation, we not only show that the concept is feasible, but also identify the detailed semantics of the extended functions for Crowd-Sheet.

Usability Evaluation. We evaluate whether ordinary spreadsheet users can use CrowdSheet to write complex crowdsourcing applications by recruiting users via a crowdsourcing service and asking them to write applications with CrowdSheet.

2   Related Work

Complex crowdsourcing beyond simply submitting a set of tasks leads to many attractive applications [5,7]. Although many tools have been proposed for *optimizing* complex crowdsourcing [10,18,21], which try to find execution plans considering execution times, monetary costs, and data quality, only a few abstractions that support the *development* of complex crowdsourcing applications have been proposed. For example, CrowdForge is MapReduce abstraction [14], Crowd-Lang uses control flow diagrams [19], CyLog uses a logic-based abstraction [11,20], and Lukyanenko and others discuss conceptual modeling principles for crowdsourcing [16]. Our spreadsheet paradigm is unique in that it not only is different from the others but can also foster *end-user development* of complex crowdsourcing applications, which we believe can make a meaningful difference in the real world. CrowdSheet has the same spirit as the system presented by Kongdenfha et al. [15], although the domain and problems are different. In addition, this paper discusses the appropriateness of the proposed design through theoretical analysis (that shows a natural and small extension of existing spreadsheets leads to wide ranging of expressive power) and usability studies (that show that many ordinary spreadsheet users can implement complex crowdsourcing with CrowdSheet). It is worth noting that the design of CrowdSheet is independent of the optimization techniques, in the sense that it can be combined with existing techniques as long as the description can be mapped to the abstractions used in the techniques. Such mappings can be automatically generated because CrowdSheet has formal semantics.

Various proposals have been made as regards office applications with crowdsourcing platforms in their backends [8,23]. In contrast to them, CrowdSheet

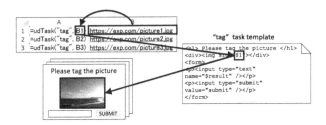

Fig. 3. udTask functions and the generated tasks by user-defined task template

is a *general-purpose* engine for implementing complex crowdsourcing. Google Spreadsheet and CrowdFill [22] actualize another approach related to spreadsheets and crowdsourcing. Whereas in this approach, the spreadsheet interface helps *workers* to enter data efficiently, CrowdSheet helps *requesters* to implement their complex crowdsourcing.

Tyszkiewicz [25] showed that a relational database can be implemented using only spreadsheet functions. Our theoretical result is compatible with it in the sense that the expressive power of CrowdSheet does not allow transitive closure.

Data quality is always an important issue in crowdsourcing [6,12,17]. CrowdSheet adopts a declarative feature that separates the application logic and mechanisms for improving data quality [24], where independent modules rewrite the code generated by CrowdSheet to propose alternative execution plans for improving data quality. Consequently, CrowdSheet can be combined with a variety of techniques (e.g., worker selection and majority consensus) for improving the data quality in crowdsourcing. Incorporation of related novel notions, such as social computing units [9], into our framework is an interesting future work.

3   CrowdSheet

CrowdSheet is a spreadsheet with a crowdsourcing platform in its backend (Fig. 2). It allows requesters to use functions to invoke microtasks and to control dataflows that can be used in the same manner as other spreadsheet functions (e.g., sum and average)[1].

TI-Functions. Task Invocation (TI) Functions invoke tasks. Each TI-function generates a task with specified parameters and submits it to the crowdsourcing platform. Subsequently, it receives the task result when the task is completed. Figure 2 shows an example of this. entryTask (String *question*) generates a task with a specified *question* and a text field, and submits it to the crowdsourcing platform. Another example is choiceTask (String *question*, Range *choices*), which generates and submits a task with a specified *question* and a list of *choices*, and returns the result.

---

[1] In our current implementation, TI-functions cannot be combined with other functions in the same cell. However, this is not an essential limitation because expressions in the cell can refer to the results of TI-functions in other cells.

CrowdSheet also allows us to invoke microtasks with user-defined task templates (Fig. 3). udTask (String *templateID*, Range *parameters*) is a TI-function that submits a task with a specified task template and returns the result. A requester can easily register their user-defined task templates to CrowdSheet. First, the requester writes a task template in HTML containing place holders where the values of *parameters* will be embedded. Next, they upload the file to our system with an associated *templateID* to be used when requesters call the function[2].

C-Functions. CrowdSheet has one C-function, called evaluateIf, which controls task invocations. evaluateIf (String *condition*, String *then-value*, String *else-value*) returns the *then-value* if the specified *condition* holds and returns the *else-value* otherwise. The two values can be specified by cell references. For example, the *condition* could be "A3 < 5," which means cell "A3" is smaller than 5, and the *then-value* refers to a cell containing a TI-function, and the *else-value* could be a value "None". A C-function is different from the usual if-function of spreadsheets in that it allows for lazy evaluation of tasks if it involves references to task invocations. Therefore, the TI-function referred to by the C-function is evaluated only when the C-function finds that the condition is true. C-functions are useful if a requester wants to conditionally invoke tasks, as in the estimation tasks in the running example.

How to Use CrowdSheet. The running example can be easily implemented as shown in Fig. 4. We assume that the sheet already has columns that store rooms

Fig. 4. CrowdSheet description in running example

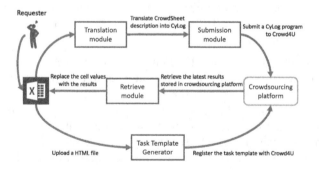

Fig. 5. CrowdSheet implementation

---

[2] Each task template is also associated with the payment information for workers.

(Column H) and sessions (Column A). First, a requester writes and uploads task templates for the three types of tasks. Then, they encode the logic in CrowdSheet as follows: (1) Put TI-functions into Column I of each room (row) for invocation of the capacity task, (2) put references to the results of the capacity tasks in Column C, (3) put TI-functions into Column D of each session (row) for invocation of the upload tasks, and (4) put TI- and C-functions into Columns E and F for invocation of the estimation tasks if the room for the session has audience seats. Note that a user can write TI- and C-functions for only one session and one room, and then drag the mouse to copy them for others.

Then, the user clicks on the "submit" button to submit tasks. When the user clicks on the "retrieve" button, the cells containing TI- or C-functions change to the result value if the task is complete at that time. Naturally, the user can click on the "retrieve" button at any point to obtain the intermediate results.

4   Implementation

Figure 5 shows the main components of our CrowdSheet implementation. We used Microsoft Excel and Crowd4U [13], but can adopt other services, such as Google Sheets and MTurk (We have already implemented modules for MTurk [24]).

In order to use CrowdSheet, the user has to download an Excel file with EXCEL add-ins for CrowdSheet. The sheet contains "submit" and "retrieve" buttons in it.

The submission module (implemented as an EXCEL add-in) passes the CrowdSheet description (i.e., all cells in the spreadsheet) to the translation module, which then translates it into an intermediate expression written in CyLog, a programming language for human-machine computations [11,20]. The logic encoded in the expression is executed by Crowd4U, and the tasks are submitted to crowdsourcing services such as MTurk and the Crowd4U native task pool. Workers then perform tasks on crowdsourcing services. If specified in the expression, Crowd4U dynamically generates tasks using the results of other tasks.

```
Pre1: Session(sid:1, rid:1, sname:"Session 1A");
Pre2: Room(rid:1, rname:"Room A");
Pre3: Session(sid:2, rid:2, sname:"Session 1B");
Pre4: Room(rid:2, rname:"Room B");
Pre5: Session(sid:3, rid:3, sname:"Session 1C");
Pre6: Room(rid:3, rname:"Room C");
Pre7: SRoom(sid:s, rid, sname, rname) <- Session(sid:s, rid, sname), Room(rid, rname);

R1: Capacity(rid, capacity)/open <- Room(rid, rname);
R2: Picture(sid, rid, sname, rname, URI)/open <- SRoom(sid, rid, sname, rname);
R3: !CapacityTask(rid, rname) <- ?Capacity(rid, capacity), Room(rid, rname);
R4: !UploadTask(sid, rid, sname, rname) <- ?Picture(sid, rid, sname, rname);
R5: Capacity(rid, capacity), Picture(sid, rid, URI){
        URI == "no_seat" {
                    Estimation(sid, rid, result:0);
            } else {
                    Estimation(sid, rid, result)/open;
            }
    }
R6: !EstimationTask(sid, rid, sname, rname, capacity, picture)
<- ?Estimation(sid, rid, result), Capacity(rid, rname, capacity), Picture(sid, rid, sname, URI);
```

Fig. 6. Fragment of a CyLog program

When the CrowdSheet user clicks on the "retrieve" button, the retrieve module (implemented as an EXCEL add-in) retrieves the latest results stored in crowdsourcing services at that time and inserts the retrieve results into the cells containing TI- or C-functions.

The user can also write and upload HTML files for user-defined task templates. The task template translator is an independent module that takes each HTML file as input, assigns a template id to it so that the template can be used in TI-functions.

4.1 CyLog Overview

As a means to define the formal semantics and discuss its expressive power, we use CyLog [20], a rule-based language for crowdsourcing. The basic data structure in CyLog is a *relation*, which is a table to deal with a set of *tuples* that conform to the *schema* of the relation. A program written in CyLog consists of three sections: `schema`, `rules` and `views`. The `schema` section describes the schema of the relations. The `rules` section has a set of rules, each of which *fires* (is executed) if its condition is satisfied. The `views` section describes HTML templates to be used as the interface with workers. In the following discussions, we explain only the `rules` section. The `schema` section is straightforward and we assume that they are appropriately given. The HTML templates in the `views` section are supplied by the task template translator in Fig. 5.

Facts and Rules. The main component of a CyLog program is the set of *statements*. Figure 6 shows a set of statements, each of which is preceded by a label (such as `Pre1`) for explanation purposes. A statement is either a *fact* or a *rule*. A rule has the form *head* ← *body*. In the figure, `Pre1` to `Pre6` are facts. `Pre7` and `R1` to `R6` are rules. Each *fact* or *head* is given in the form of an *atom*, while each atom consists of a predicate name (e.g., `Session`) followed by a set of *attributes* (such as `sid`, `rid`, and `sname`). Optionally, each attribute can be followed by a colon with a value (e.g., :"Session 1A") or an alias (e.g., :s). Each *body* consists of a sequence of atoms.

A fact describes that the specified tuple is inserted into a relation. For example, `Pre1` is a fact that inserts a tuple whose values for attributes `sid`, `rid`, and `sname` are 1, 1, and "Session 1A" respectively into relation `Session`[3].

A rule specifies that, for each combination of tuples satisfying the condition specified in the *body*, the tuple described in *head* be inserted into a relation. Atoms in the body are evaluated from left to right and variables are bound to values that are stored in the relation specified by each atom. For example, `Pre7` is a rule that inserts a tuple having `sid`, `rid`, `sname`, and `rname` into relation `SRoom` if `sid`, `rid`, and `sname` are in `Session` and the `rid` and `rname` in `Room`. In other words, for each combination of a tuple in `Session` and a tuple in `Room`

---

[3] CyLog adopts the *named perspective* [3], which means that the variables and values in each atom are associated with attributes by explicit *attribute names*, not by their positions.

whose `rid` attributes match each other, it inserts a tuple having `sid`, `rid`, `sname`, and `rname` values into relation `SRoom`.

Open Predicates. CyLog allows predicates to be `open`, which means that the decision as to whether a tuple exists in the relation is performed by humans when the data cannot be derived from the data in the database. For example, the head of `R1` is followed by `/open` and is an open predicate. If a head is an open predicate, CyLog asks humans to give values to the variables that are not bound to any values in the body (e.g., `capacity` in `R1` and `URI` in `R2`).

Task Predicates. A task predicate (preceded by `!`) represents a relation implementing a task pool, in which each tuple corresponds to a task instance. For example, `R2` defines a task predicate `!CapacityTask` in which each tuple corresponds to one capacity task. In general, each tuple in a task predicate is created for a combination of tuples including (1) ones to supply data to be presented to workers in the task screen and (2) open tuples to store the task results. For example, `R3` creates a capacity task instance for a combination of (1) a tuple in `Room(rid, rname)`, whose `rname` is used to show workers the name of the room in the task screen, and (2) a tuple in `Capacity(rid, capacity)`, where `capacity` contains an open value. In the rule body, `?Capacity(rid, capacity)` is a predicate that holds when it contains an open tuple. Note that open tuples for `Capacity` are supplied by `R1`.

Block Style Rules. Each rule $P \leftarrow P_1, P_2, \ldots, P_n$ can be written in the *block style* $P_1\{P_2\{\ldots\{P_n\{P;\}\ldots\}$. For example, `Pre7` in Fig. 6 can be written as follows:

```
Session(sid:s, rid, sname) {
    Room(rid, rname) {
        SRoom(sid:s, rid, sname, rname);
    }
}
```

where `SRoom(sid, rid, sname, rname)` is the head of the rule. The block style provides a concise expression when we have many rules that have the same body

SR-1: $Cell(loc, value)/open \leftarrow Ready(loc);$
SR-2: $!Task_k(loc, d_1, \ldots, d_{n_k}) \leftarrow ?Cell(loc, value), Param_k(loc, d_1, \ldots, d_{n_k});$

TP-1: $Param_k(loc : (x, y), d_1, \ldots, d_{n_k}) \leftarrow Cell(loc : (x_1, y_1), value : d_1), \ldots,$
$Cell(loc : (x_{n_k}, y_{n_k}), value : d_{n_k});$
TP-2: $Ready(loc) \leftarrow Param_k(loc : (x, y), d_1, \ldots, d_{n_k});$

CP-1: $Cell(loc : loc_1, value : d_1), \ldots, Cell(loc : loc_m, value : d_m), cond\{$
CP-2: $Ready(loc : (x_b, y_b)) \leftarrow Param_k(loc : (x_b, y_b), d_1, \ldots, d_{n_k});$
CP-3: $\} else \{$
CP-4: $Cell(loc : (x_b, y_b), value) \leftarrow Cell(loc : (x_c, y_c), value);$
CP-5: $\}$

Fig. 7. Patterns for CyLog code generation: SR (SR-*is*), TP (TP-*is*), and CP (CP-*is*)

atoms (e.g., $P_1\{P_2; P_3;\}$ for $P_2 \leftarrow P_1; P_3 \leftarrow P_1;$). In addition, the block style allows us to use the `else` clause. For example, R5 inserts a tuple having the result "0" into `Estimation(sid, rid, result)` if the URI value is "no_seat".

4.2   Semantics of CrowdSheet Descriptions

CyLog Program for the Running Example. R1 to R6 in Fig. 6 constitute a set of rules that is equivalent to the CrowdSheet description shown in Fig. 4. (Rules Pre1 to Pre7 are rules we intend to use to explain CyLog and do not have expressions in the figure.)

R1 to R4 generates two types of tasks for asking workers to count the number of seats in the room and take a picture of the session. R5 states that if the results of the capacity task is not "0", the estimation result of the number of audiences is open; otherwise, we take the value "0" as the estimation result. R6 generates a task for estimation of the number of people in the audience in each session, if the estimation result is open.

Mapping to CyLog Rules. The semantics of CrowdSheet descriptions is defined by a method that maps a CrowdSheet description to CyLog rules. A CrowdSheet description is mapped to CyLog rules consisting of the following three components (Fig. 7).

Shared Rules (SR) a set of rules to give common functionalities to generate tasks.
TI-function Pattern (TP) a set of rules generated for each TI-function appearing in the sheet.
C-function Pattern (CP) a set of rules generated for each C-function in the sheet.

The TP or CP for a function specified in the cell at (x, y) is fired only once when their parameter values are ready. Note that (x, y) is a pair of constant values, not variables.

Shared Rules and TI-function Pattern. Assume that we have a TI-function with Template k in a cell at (x, y), and its parameters are stored in cells at $(x_1, y_1), \ldots, (x_{n_k}, y_{n_k})$. Then, the translator generates a one SR (consisting of SR-1 and SR-2) and one TP (consisting of two rules: TP-1 and TP-2) for the function.

SR works as follows. First, SR-1 makes the cell value open if the task is ready. Then, SR-2 generates an instance of the Template-k task if the value is open.

TP works as follows. First, TP-1 constructs a tuple to store parameters for the TI-function at (x, y). Here, $Param_k$ is a predicate to store parameters for TI-functions with Template k, where the key loc stores the locations of the cells at which TI-functions are placed on the spreadsheet. As a result, TP-1 inserts a tuple containing parameters for the TI-function at (x, y) into $Param_k$. The predicate $Cell(loc : (x, y), value)$ stores the value in the cell at (x, y).

Next, TP-2 states that the task is ready to be evaluated if we have all parameters to invoke the task. Here, the predicate $Ready(loc)$ stores the locations of those TI-functions that are ready to be evaluated.

If we had more than one TI-function for the same Template K, the translator would only generate TPs for the TI-functions. The shared rules do not depend on the locations of cells and are commonly used by every TP for Template k.

Rules for C-functions. Assume that we have a C-function in the cell at (x, y), and the condition specified in the C-function requires values in cells at $(x_1, y_1), \ldots, (x_m, y_m)$. Note that the "then" and "else" clauses can refer to values in other cells. Here, we assume that the *then-value* parameter of the C-function refers to a TI-function located at (x_b, y_b) and the *else-value* parameter refers to a value v at (x_c, y_c).

Then, the translator generates rules in the following two steps. First, it generates the CP for the C-function, as shown in Fig. 7. CP-1 to CP-2 make the task at (x_b, y_b) ready only when the condition holds and we have all of its parameters. CP-4 copies the value at (x_c, y_c) to the cell at (x, y) if the condition is not satisfied. Second, for each reference to a TI-function contained in the *then-value* or *else-value* parameters, the original rule for creating Ready tuples (i.e., TP-2) is removed, because the tuple is created by CP-2 when the condition is satisfied.

5   Expressive Power

In this section, we prove that there is a class $\mathcal{P}$ of CyLog programs such that every program in $\mathcal{P}$ has a CrowdSheet description equivalent to it, and vice versa.

5.1   CrowdSheet to CyLog

First, we identify the conditions satisfied by any CyLog program that defines the semantics of a CrowdSheet expression.

Theorem 1. *Let p be a CyLog program converted from a CrowdSheet description. Then, p satisfies all of the following conditions.*

C1. Every task generated by p returns only one value[4].
C2. There exists a natural number N such that p generates at most N tasks.
C3. p generates tasks of predefined task templates only.

Proof. The conditions are derived from the limitations of spreadsheets and the CrowdSheet design. The first condition is derived from the fact that the output of each spreadsheet function is inserted into one spreadsheet cell. The second states that an infinite number of functions can not be written in a sheet because TI-functions have to be explicitly written in cells. The third is derived from the design policy stating that the task-design functionalities are out of

---

[4] Tasks can return more than one value with composite values (omitted owing to lack of space).

the spreadsheet paradigm, i.e., new templates of tasks can not be defined with spreadsheet functions. Every program mapped from a CrowdSheet description (Sect. 4.2) satisfies the three conditions.   □.

Definition 1. *We define $\mathcal{P}$ as a class of CyLog programs that satisfies all conditions C1, C2, C3 identified by Theorem 1.*

5.2   CyLog to CrowdSheet

Second, we prove that every CyLog program in class $\mathcal{P}$ can be converted into a set of CyLog fragments of shared rules, TPs, and CPs. The fundamental idea of the proof is to map each task invocation to a virtual spreadsheet space, such that every task result is associated with a unique location (corresponding to a cell in the virtual spreadsheet). In this manner, every TI-function or C-function can refer to the results of other tasks in the mapped spreadsheet.

First, we consider a simple case; we prove that every program in $\mathcal{P}'$ can be converted into a set of CyLog fragment patterns.

Definition 2. *$\mathcal{P}'$ is a class of CyLog programs in which each program in $\mathcal{P}'$ satisfies all of C1, C2', and C3, where,*

- *C2' p generates exactly N tasks and no function uses the results of other functions as its parameters.*

Theorem 2. *Every program p in $\mathcal{P}'$ can be converted into a set of TPs and shared rules.*

Proof. Let $pred\_seq_{q,k}$ be a sequence of CyLog predicates (filled with constant values) that computes a set of parameters $d_1, \ldots, d_{n_k}$ to be used for invoking tasks with Template k. Here, q is a unique number associated to such a sequence, independent of k (i.e., q is unique in all $pred\_seq_{q,k}$ with any ks). Because the total number of task invocations is fixed by N, the number of task invocations generated by each $pred\_seq_{q,k}$ is also fixed. Let $N_{q,k}$ be that number. In addition, as no task uses the results of other tasks as its parameters, all parameters are constant values. Therefore, we can construct a set of cells in a virtual spreadsheet to store those parameters generated by $pred\_seq_{q,k}$. For that purpose, we "expand" $pred\_seq_{q,k}$ by generating the following CyLog fragments $N_{q,k}$ times.

```
pred_seq_q,k ,  cond_i {
     Cell(loc : l_new1,  value : d1);
       ...
     Cell(loc : l_newk,  value : dk);
}
```

Here, $cond_i (1 \le i \le N_{q,k})$ is a condition to choose a tuple for one task, and l_{new_i} is a constant value to represent a new cell location that has not been used yet. In this way, we obtain all "virtual" cells that can be referred to by task invocations.

Then, we generate the following CyLog fragment for each sequence of parameters for invoking a task with Template k.

TP-1: $Param_{k,q}(loc:l_{new}, d_1, \ldots, d_{n_k})$
   $\leftarrow Cell(id:l_1, value:d_1), \ldots, Cell(id:l_k, value:d_k),$
TP-2: $Ready(loc) \leftarrow Param_{k,q}(loc:l_{new}, d_1, \ldots, d_{n_k});$

Where l_i is the location of the parameter used for the ith parameter to invoke the task, and l_{new} is the new location to store the task result. The combination of the rules and shared rules is equivalent to CyLog rules generated from a CrowdSheet description. Therefore, there is a CrowdSheet description that is equivalent to any p in $\mathcal{P}'$. $\square$

Note that each loc has a unique value such that each task result can be referred to by other tasks as a parameter. loc does not have to be in the form (x, y), as long as the values serve as unique identifiers for spreadsheet cells.

This way, any value to be used as a parameter is assigned to a unique location. The following lemma is important.

Lemma 1. *For every task t in any program in the class $\mathcal{P}$, there exists a number $N_t < N$ s.t. t is ready after N_t tasks are completed.*

Proof. If this does not hold, then N is not a fixed number, which is a contradiction. $\square$

Note that for every task t in programs in the class $\mathcal{P}'$, $N_t = 0$. We use Lemma 1 and gradually increment N_t to prove the following theorem (proof omitted).

Theorem 3. *Every p in $\mathcal{P}$ can be converted into a set of TPs, CPs, and SRs.* $\square$

5.3 Expressive Power in Terms of Applications

An important question that entails from this is the question of what the expressive power means for real-world applications. To answer this question, we reviewed a survey [26] to ascertain the expressive power of CrowdSheet in terms of real-world applications. Note that being able to write an application does not equate to being able to implement it as is. We are interested in whether we can use CrowdSheet to implement applications that are functionally equivalent to the applications in surveyed.

There are two major limitations in CrowdSheet. The first is that each task is explicitly distinguished with others and must be an instance of a task template. This hinders CrowdSheet from implementing certain kinds of games where the interactive user interface is important. Secondly, it does not allow recursive task invocations so that any CrowdSheet expression cannot issue an infinite number of microtasks.

The survey [26] grouped crowdsourcing applications into four groups: voting systems, information sharing systems, games, and creative systems. First, most of the voting systems operated on MTurk and, in most cases, it is implemented with microtask interfaces. Therefore, as the number of tasks is fixed,

CrowdSheet can implement voting systems. Second, information sharing systems, including Wikipedia and YouTube. In this case, the ability to issue an infinite number of tasks is essential. Therefore, we surmise that CrowdSheet is not appropriate to implement such systems. Third, games are crowdsourcing applications that have been developed in line with ESP games [4]. For many games, an interactive interface is essential and a microtask interface only is not enough to implement effective game experiences. CrowdSheet is not appropriate to implement such games. Finally, creative systems, such as The Sheep Market [1] can be implemented if the number of tasks is limited. There are systems that can be implemented with the microtask interface, and we believe that certain kinds of creative systems can be implemented with CrowdSheet.

Our conclusion is that although CrowdSheet is not a perfect tool, it has a reasonable expressive power to be able to implement some of real-world applications.

6   Usability Evaluation

We recruited workers with prior experience using Microsoft Excel and spreadsheet functions and asked them to write CrowdSheet expressions to implement an application that is isomorphic to the running example presented in Sect. 1.

6.1   Settings

Worker Recruitment. We recruited workers as follows. First, we submitted 300 tasks to Yahoo! Crowdsourcing [2] to ask workers to answer questions on how to use Microsoft Excel, in order to find workers who can use spreadsheet functions, i.e., workers who are potentially able to write CrowdSheet expressions. The questions included how to use Excel's if function to solve a given problem in Excel. Three hundred (300) workers participated in the task, with each worker receiving 2 JPY (about 0.02 USD). 110 out of the 300 participants gave correct answers to the questions. Next, we submitted the "CrowdSheet" tasks (which we will explain next) to Yahoo!Crowdsourcing, with the condition that each worker could accept to perform only one of the submitted CrowdSheet tasks and would receive 10 JPY (about 0.09 USD) for the task. The tasks were visible only to the 110 workers who gave correct answers to the first task.

CrowdSheet Tasks. The CrowdSheet task consisted of two parts. In the first part, the workers were asked to write CrowdSheet expressions with only TI-functions for issuing entryTasks and choiceTasks having data flow among them. In the second part, the workers were asked to write CrowdSheet expressions with the combination of TI- and C-function to implement the running example in Fig. 1. Each part began with a tutorial to explain TI-functions entryTask and choiceTask (in the first part) and C-funcion evaluateIf (in the second part), and then asks workers to write CrowdSheet descriptions in the matrix of text forms simulating the CrowdSheet interface.

6.2   Results

Table 1 shows the results obtained. 33 out of the 110 workers accepted to perform the CrowdSheet tasks. 19 workers did not accept to perform the tasks. 58 workers did not access to the task at all. If we look at the numbers for Part 2, which asked workers to write expressions for a more difficult application, 15 workers out of the 33 workers gave correct answers and 6 workers gave correct answers with minor mistakes (such as typo and lack of the closing parenthesis). The remaining 12 workers made essential mistakes. An example of a mistake is a wrong choice of TI-function. For example, a worker issued `choiceTask` task instead of an estimation task, because `choiceTask` was used in the tutorial part. Another example is missing parameters of TI-tasks, which could be corrected if the spreadsheet supported interactive syntax error check. The sum of workers in the first two categories was 21, which is 63.6% of the workers who performed the CrowdSheet tasks. The results suggest that many workers with experience using Excel and its spreadsheet functions were able to use CrowdSheet, even though they were using it for the first time and had received only a simple CrowdSheet functions tutorial.

Table 1. Evaluation Result

	Correct (minor error)	Incorrect	Did not accepted	Ignored	Total
Part 1 (only TI-functions)	31 (3)	2	19	58	110
Part 2 (TI- and C-functions)	21 (6)	12			

The average of the time spent by participants for performing a CrowdSheet task (reading the tutorials and writing CrowdSheet descriptions for two scenarios) was about 10 min (598.2 s). For reference, a skilled programmer who use MTurk needed 28 min to implement tasks with the same data flow as the part 2 (submitting tasks and processing data) with MTurk, excluding the time required for him to wait for workers to complete the first step tasks. This suggests that we can expect many spreadsheet users can quickly learn and easily use CrowdSheet.

7   Summary

This paper proposed CrowdSheet, a spreadsheet for writing and executing complex crowdsourcing applications. We explained the design and a possible implementation, gave a theoretical analysis of its expressive power, and discussed the power and limitations for implementing real-world applications. Further, we presented experimental results showing that many users with spreadsheet experience are able to write complex crowdsourcing applications with CrowdSheet. In future work, we plan to make it generate codes in the form closer to that described by humans. This will make it easier to use CrowdSheet as a tool for implementing the basic functions of applications and rewriting the generated code to add functionalities beyond the expressive power of CrowdSheet.

Acknowledgement. This work was partially supported by JST CREST GrantNumber JPMJCR16E3, Japan.

References

1. The sheep market. http://www.thesheepmarket.com/
2. Yahoo! crowdsourcing. http://crowdsourcing.yahoo.co.jp
3. Abiteboul, S., et al.: Foundations of Databases: The Logical Level, 1st edn. Addison-Wesley Longman Publishing Co., Inc., Boston (1995)
4. von Ahn, L., et al.: Labeling images with a computer game. In: Proceedings of the SIGCHI Conference on Human Factors in Computing Systems, pp. 319–326. ACM (2004)
5. Akiki, P., et al.: Crowdsourcing user interface adaptations for minimizing the bloat in enterprise applications. In: Proceedings of ACM SIGCHI 2013, pp. 121–126 (2013)
6. Allahbakhsh, M., et al.: Quality control in crowdsourcing systems: issues and directions. IEEE Internet Comput. **17**(2), 76–81 (2013)
7. Artikis, A., et al.: Heterogeneous stream processing and crowdsourcing for urban traffic management. In: Proceedings of EDBT 2014, vol. 14, pp. 712–723 (2014)
8. Bernstein, M.S., et al.: Soylent: a word processor with a crowd inside. Commun. ACM **58**(8), 85–94 (2015)
9. Candra, M.Z.C., Truong, H.-L., Dustdar, S.: Provisioning quality-aware social compute units in the cloud. In: Basu, S., Pautasso, C., Zhang, L., Fu, X. (eds.) ICSOC 2013. LNCS, vol. 8274, pp. 313–327. Springer, Heidelberg (2013). https://doi.org/10.1007/978-3-642-45005-1_22
10. Franklin, M., et al.: Crowddb: answering queries with crowdsourcing. In: Proceedings of ACM SIGMOD 2011, pp. 61–72. ACM (2011)
11. Fukusumi, S., Morishima, A., Kitagawa, H.: Game aspect: an approach to separation of concerns in crowdsourced data management. In: Zdravkovic, J., Kirikova, M., Johannesson, P. (eds.) CAiSE 2015. LNCS, vol. 9097, pp. 3–19. Springer, Cham (2015). https://doi.org/10.1007/978-3-319-19069-3_1
12. Hung, N.Q.V., et al.: Erica: expert guidance in validating crowd answers. In: Proceedings of ACM SIGIR 2015, pp. 1037–1038. ACM (2015)
13. Ikeda, K., et al.: Collaborative crowdsourcing with crowd4u. PVLDB **9**(13), 1497–1500 (2016)
14. Kittur, A., et al.: Crowdforge: crowdsourcing complex work. In: Proceedings of the 24th Annual ACM UIST, pp. 43–52. ACM (2011)
15. Kongdenfha, W., et al.: Rapid development of spreadsheet-based web mashups. In: Proceedings of WWW 2009, pp. 851–860. ACM (2009)
16. Lukyanenko, R., et al.: Conceptual modeling principles for crowdsourcing. In: Proceedings of the 1st International Workshop on Multimodal Crowd Sensing, pp. 3–6. ACM (2012)
17. Lukyanenko, R., et al.: The IQ of the crowd: understanding and improving information quality in structured user-generated content. Inf. Syst. Res. **25**(4), 669–689 (2014)
18. Marcus, A., et al.: Crowdsourced databases: query processing with people. In: CIDR (2011). http://hdl.handle.net/1721.1/62827

19. Minder, P., Bernstein, A.: *CrowdLang*: a programming language for the systematic exploration of human computation systems. In: Aberer, K., Flache, A., Jager, W., Liu, L., Tang, J., Guéret, C. (eds.) SocInfo 2012. LNCS, vol. 7710, pp. 124–137. Springer, Heidelberg (2012). https://doi.org/10.1007/978-3-642-35386-4_10

20. Morishima, A., et al.: Cylog/Game aspect: an approach to separation of concerns in crowdsourced data management. Inf. Syst. **62**, 170–184 (2016)

21. Park, H., et al.: Deco: a system for declarative crowdsourcing. Proc. VLDB Endow. **5**(12), 1990–1993 (2012)

22. Park, H., et al.: Crowdfill: collecting structured data from the crowd. In: Proceedings of the 2014 ACM SIGMOD International Conference on Management of Data, pp. 577–588. ACM (2014)

23. Quinn, A.J., et al.: AskSheet: efficient human computation for decision making with spreadsheets. In: Proceedings of CSCW 2014, pp. 1456–1466. ACM (2014)

24. Suzuki, R., et al.: Crowdsheet: instant implementation and out-of-hand execution of complex crowdsourcing. In: Proceedings of ICDE 2018 (2018)

25. Tyszkiewicz, J.: Spreadsheet as a relational database engine. In: Proceedings of ACM SIGMOD 2010, pp. 195–206. ACM, New York (2010)

26. Yuen, M.C., et al.: A survey of crowdsourcing systems. In: 2011 IEEE Third International Conference on Privacy, Security, Risk and Trust (PASSAT) and 2011 IEEE Third Inernational Conference on Social Computing (SocialCom), pp. 766–773. IEEE (2011)

Educating Users to Formulate Questions in Q&A Platforms: A Scaffold for Google Sheets

Oscar Díaz and Jeremías P. Contell[✉]

ONEKIN Web Engineering Group,
University of the Basque Country (UPV/EHU), San Sebastián, Spain
{oscar.diaz,jeremias.perez}@ehu.eus

Abstract. Different studies point out that spreadsheets are easy to use but difficult to master. When difficulties arise, home users and small-and-medium organizations might not always resort to a help desk. Alternatively, Question&Answer platforms (e.g. Stack Overflow) come in handy. Unfortunately, we can not always expect home users to properly set good questions. However, examples can be a substitute for long explanations. This is particularly important for our target audience who might lack the skills to describe their needs in abstract terms, and hence, examples might be the easiest way to get the idea through. This sustains the effort to leverage existing spreadsheet tools with example-centric inline question posting. This paper describes such an effort for Google Sheets. The extension assists users in posing their example-based questions without leaving Google Sheets. These questions are next transparently channeled to Stack Overflow.

Keywords: Spreadsheets · Q&A platforms · Social computing
StackOverflow

1 Introduction

Home users tend to learn "just enough" to keep going [9]. Learning as you go, ideally without consulting the manual, becomes a common practice. It comes as no surprise that these users are frequently faced with doubts about how to proceed. These doubts might be resolved through trial-and-error cycles or resorting to help desks. The former might be cumbersome and unsuccessful, while support for end users tends to be rather limited in most organizations. Indeed, different studies point to the lack of assistance as a major stumbling block for these users' self-reliance [14,24]. In this scenario, Question&Answer platforms (Q&A platforms) come in handy [10]. In Q&A sites, users (askers) post questions, and rely on other community members (answerers) to provide a suitable solution to their information needs [23]. Unfortunately, home users are not always aware of Q&A platforms or, more commonly, they are discouraged

© Springer International Publishing AG, part of Springer Nature 2018
J. Krogstie and H. A. Reijers (Eds.): CAiSE 2018, LNCS 10816, pp. 154–169, 2018.
https://doi.org/10.1007/978-3-319-91563-0_10

by the upfront-cost of posing a question, and the risk of not getting a satisfying answer [11]. Findings confirm that (re)editing likelihood was larger for users with limited familiarity with either *Stack Overflow* or the topic at hand [25], both aspects characterizing our target audience.

The irony is that those who can benefit most from Q&A platforms might lack the skills (and support) to pose "good questions". Studies found how successful inline posting correlates with affect, presentation quality or time [5]. This leads to our research question: *how can home users be assisted in successful question posting in Q&A platforms?*

Our main premise is that examples (and doubts) emerge while conducting tasks. Unfortunately, askers need to move away from the working environment (where questions arise) to the Q&A platform (where questions are posed). This requires askers to re-create the task scenario into the question. As an example, consider *Google Sheets* as the working environment, and *Stack Overflow* as the Q&A platform. When struggling with spreadsheet formulas, end users might resort to the *Stack Overflow* community. This involves moving to this Q&A platform and posting the question. This can be achieved in a rambling textual way, or conveniently formatted using *markdown* (i.e. a markup language for code snippets in *Stack Overflow*). More to the point, in accordance with *StackOverflow*'s own recommendations [18], screenshots or separated shared spreadsheet should be provided for respondents to more reliably reproduce the task scenario. This requires more involvement (and skills) on the asker's side but, in return, it permits potential answerers not only to provide but also to validate their answers with the sample data. Indeed, the presence of code snippets in questions highly correlates with getting prompt and appropriate answers [5]. Hence, users need to balance the upfront investment (i.e. time/effort spent on writing the question) *versus* the risk of obtaining low-quality answers or no answer at all. The issue is that the upfront "attention capital" available for home users might be too limited.

To reduce this upfront investment, we resort to **inline question posting** through a question scaffold. Here, users can seamlessly channel questions to Q&A platforms without leaving their working environment. The aim: reducing the cost of posting questions while increasing the payoff, i.e. obtaining more accurate answers by making questions clearer through examples.

QuestionSheet is a research prototype we built to demonstrate the feasibility of this approach on top of *Google Sheets*. To use *QuestionSheet*, home users starts with an ordinary spreadsheet. Then, in cells where a formula is needed but unknown, the user enters a special *=QUESTION()* formula. This triggers an assisted process that helps users to pose their questions in *Stack Overflow* resorting to examples from the current spreadsheet to illustrate the desired output, all without leaving *Google Sheets*. Through this work, we pursue three main contributions:

– a scaffold design aimed at guiding users towards successful inline question posting (Sect. 3),

– a *Google Sheets* extension for inline question posting in *Stack Overflow* (Sect. 4),
– five use cases that provide insights about the benefits brought by inline question posting (Sect. 5).

We start by framing this work within the related bibliography.

2 Social Computing in Programming

While Personal Computing describes the behavior of isolated users, Social Computing highlights the social aspect when achieved through computers. Q&A platforms, microblogging or wikis can be framed within this term. The social aspect does not stop at collaboratively editing a wiki article or tweeting about a movie. Programming has also been the subject of socialization. A proliferation of approaches to provide this kind of capabilities *within* programming environments soon began. The vision is for users to seamlessly channel questions to Q&A platforms without leaving their working environment. By *channelling* is meant the interplay between *the Q&A platform* and *the working platform* during the question lifecycle. Questions are written from *the working platform*, and next, transparently posted into *the Q&A platform*. And the other way around: *the Q&A platform* is periodically pulled for answers that will eventually be available at *the working platform*. Differences among approaches stem from the working environment and the target social platform to tap into. Next, we cluster references based on two main working environments: *Eclipse* and spreadsheet programs (see Table 1).

Table 1. Approaches to social computing in programming.

Tool	Hosting platform	Social platform	Added value	Asker	Answerer
Fishtail	*Eclipse*	*Stack Overflow*	Discovering SO code examples	Programmer	Altruistic
Seahawk	*Eclipse*	*Stack Overflow*	Indexing into SO threads	Programmer	Altruistic
Smartsheet	Prop. spreadsheet framew	*Mechanical Turk*	Launching crowdsourcing tasks	Data analyst	Rewarded
AskSheet	Prop. spreadsheet framew	*Mechanical Turk*	Launching crowdsourcing tasks	Decision Maker	Rewarded
QuestionSheet	*Google App Sheet*	*Stack Overflow*	Launching SO threads	Home Users	Altruistic

Eclipse is an IDE for professional applications. The success of Q&A services soon led to devise mechanisms "for employees to appropriate status message Q&A as one possible source of stable peer support" [20]. For *Stack Overflow*, *Fishtail* is an *Eclipse* plugin that assists programmers in discovering code examples and documentation on the Web relevant to their current task [17]. Another plugin, *Seahawk*, turns Eclipse-hosted code snippets into hyperlinks which point to relevant *Stack Overflow* question threads [1]. *Seahawk* transparently channels all the communication between *Eclipse* and *Stack Overflow*.

Spreadsheet frameworks also attempt to capitalize on social networks. *Smartsheet* [6] adds crowdsourcing capabilities to spreadsheets. Specifically, it permits to create tasks associated with spreadsheet cells so that cell values are obtained as a result of crowdsourcing tasks. *Smartsheet* transparently handles all the back-end process through *Amazon's Mechanical Turk*. However, *Smartsheet* is not targeted to end users struggling with formulas but to professional programmers looking for data. In a similar vein, *AskSheet* [16] helps users to gather data for decision making in the form of spreadsheets. Decision making spreadsheets are data intensive, and include functions to aggregate data to help users to make a decision. To this end, *AskSheet* provides the *ASK()* function. *ASK()* takes as parameters a range of expected values (e.g. "1 to 10"), the item and attribute labels (e.g. "Galaxy S3", "screen size"), the name of the task ("check basic specs") and the target crowdsourcing platform (*Amazon's Mechanical Turk*). The side-effect is the creation of a crowdsourcing task (a.k.a. HIT in the *Mechanical Turk* parlance). Main contribution rests on the optimization of the number of human tasks required to resolve all the cell dependencies in order to reduce the economic cost use as incentive to the human workers.

Our work is akin to these efforts insofar as inlining Q&A capabilities within the working environment. In so doing, questions are posted in context, facilitating focus and avoiding moving to the Q&A platform. However, previous approaches consider different target audiences. As for *askers*, previous works target professional programmers where coming up with good questions is not an issue but rather the smooth integration with the working environment. On the other hand, *answerers* might vary from rewarded ones *versus* altruistic ones. This introduces a remarkable difference: on money-rewarded systems (e.g. *Amazon's Mechanical Turk*), the success is driven mainly by the amount of the incentive whereas in altruistic sites (e.g. *Stack Overflow*), success very much depends on how interesting the problem is and how well it is described.

Table 1 highlights how our work differs from previous approaches as for the combination (asker, answerer) being considered: (home users, altruistic respondents). If answerers are altruistic, then we should delve into what factors impact respondent engagement (Sect. 3). If askers are home users, then we should strive to provide some kind of scaffold that acts upon the engaging factors. If they are left on their own, studies confirm that newcomers tend to have lower chances of having their questions properly answered [4].

3 Scaffold Design for Successful Question Posting

This section gathers studies about how askers can increase the chances of eliciting a successful answer. Specifically, our approach pivots around the model for successful question posting on *Stack Overflow* presented in [5]. Calefato et al. focus on three main factors askers can act upon: affect (i.e. the positive or negative sentiment conveyed by text), presentation quality, and time (i.e. the moment at which the question is posted). Figure 1 depicts the main metrics introduced by Calefato et al. together with pairs *(coefficient estimate, odds ratio)* resulting from their experiment[1]. Next, we elaborate on means to assist users to increase these metrics, i.e. the scaffold.

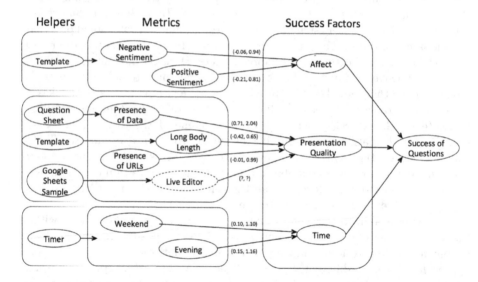

Fig. 1. A model for successful question posting (adapted from [5]). Arrows are labeled by pairs *(coefficient estimate, odds ratio)* taken from [5]. Live editors are mentioned in this study but not investigated.

Affect. Successful questions adopt a neutral emotional style. This insight is aligned with previous investigations where expressing sentiment, either positive or negative, might decrease the chances of getting help [2].

Scaffold insight: Parse user text for sentiment analysis. Alternatively, a boilerplate template can be used.

---

[1] The sign of the coefficient estimate indicates the positive/negative association of the predictor with the success of a question. The odds ratio weighs the magnitude of this impact: the closer its value to 1, the smaller the impact of the parameter on the chance of success. A value lower than 1 corresponds to a negative association of the predictor with success (negative sign of the coefficient estimate), and the opposite for a value higher than 1.

Presentation Quality. Successful questions are "short, contain code snippets, and do not abuse with uppercase characters" [5]. The importance of attaching code snippets to questions was highlighted in a separate questionnaire where 95% of respondents strongly agreed with this view. An interesting twist was put forward by one of the respondents about the importance of using live editors, which would permit answerers to fiddle directly with the code samples provided by askers.

The literature backs the use of examples when posing questions [19]. Good examples can be a substitute for long explanations. This is particularly important for home users who might lack the skills to describe their needs in abstract terms, and hence, examples might be the easiest way to get the idea through. Good-practice manuals exist about how to write good questions [8,10,18]. Nasehi et al. advice for questions to have enough details (but not too many), enough depth (without drifting from the core subject), examples (if applicable) as well as including the avenues already investigated by the asker [15]. This effort pays off in terms of getting more high-quality answers, and hence, reducing the risks of obtaining inappropriate answers, or even no answer at all [13,22].

Scaffold insight: Provide means for examples to be easily extracted from the working environment.

Time Slice. Experiments found that questions posted during the weekend are more likely to be answered than questions posted during the week [3]. In addition, the most successful time slices correspond to 3:00–6:00 PM of West Coast US time, that is, when most experts were available.

Scaffold insight: Provide a timer that decouples question specification from question posting. Users with a different time zone can work out their questions at the most appropriate time, and let the timer post them at the USA working hours or leave it till the weekend.

4   A scaffold for inline question posting for GoogleSheet

This Section tackles inline question posting using *Google Sheets* as the working environment, and *Stack Overflow (SO)* as the Q&A platform. The outcome is *QuestionSheet*, an extension for *Google Sheets* available for download at https:// goo.gl/RMD3p3. Readers are encouraged to watch this video to see *QuestionSheet* at work: https://goo.gl/wSy767.

Functional requirements are those that derive from handling the question lifecycle. This includes: question description (i.e. title, message, answers, votes, respondent data, etc); question sharing (i.e. seamless propagation of the query from the working environment to the Q&A platform); answer awaiting (i.e. once the question is posted, mechanisms should be in place for the working environment to monitor the Q&A platform); answer resolution (i.e. once answers show up, mechanisms are required to handle answer resolution and integration into the working environment).

As for the non-functional requirements, we prevail "compatibility", i.e. "the degree to which an innovation is perceived as being consistent with the existing

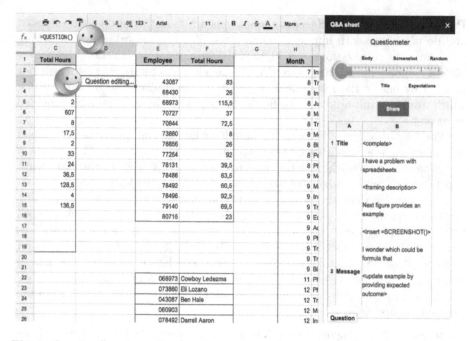

Fig. 2. *Question()* makes a *QuestionSheet* to be displayed as a side panel: the question and the working sheet are shown side by side. Smileys are overlaid to highlight points of interest.

values, needs, and past experiences of potential adopters" [21]. In our setting, we depart from a working platform to which the user is familiarized with (i.e. Google Sheet). Compatibility calls for the new functionality (i.e. inline question posting) to mimic as much as possible *the modus operandi* of Google Sheet. Targeting home users, consistency lowers the entry bar and eases user adoption [12].

The rest of this Section is structured along the stages of the question lifecycle. As the running example, consider the spreadsheet in Fig. 2. Employee data is scattered among two tables. One table holds the employee's ID and hours worked. The table below keeps the ID and the employee's name. The user wonders how the employee's name could be replicated besides the ID column in the first table. Rather than re-typing this information, he looks for a function that achieves this.

4.1   Question Description

For the sake of compatibility, *QuestionSheet* is realized through two spreadsheet functions: *QUESTION()*, and *SCREENSHOT(Range)*.

Initialization. To use *QuestionSheet*, users start with an ordinary spreadsheet. Then, in cells where a formula is needed but unknown, the user enters a special *=QUESTION()* formula. Figure 2 shows the case for our running example (notice the expression *=QUESTION()* at the formula bar) From the perspective of *Google Sheets*, *QUESTION()* is just another formula. Hence, *QUESTION()* can appear any place a formula is expected. The difference stems from the output. Traditional formulae output data. However, *QUESTION()* outputs a formula, and it has a side effect: poping up a *QuestionSheet* at the right-side. A *QuestionSheet* holds a cell for each of the question elements (e.g. Title, Message, etc). Function *QUESTION()* then obtains its arguments from this sheet[2]. As any other formula, *QUESTION()* output is re-evaluated as its companion *QuestionSheet* changes its content.

Description. At the onset, the *Message* cell contains a template that serves as a basic guideline for users. A hallmark of our approach is to enrich these messages with examples, i.e. providing input values and their expected outputs. For our running case, this entails adding the employee names by the IDs (red shaded in Fig. 3). The user is prompted to introduce the expected outputs that the sought formula would return. This not only facilitates understanding by potential respondents (and hence, the chances of getting a more accurate answer) but it also offers a way to validate the solution. Indeed, the blog of *Stack Overflow* itself advises to provide shared spreadsheet where potential respondents can more reliably reproduce the task scenario and validate their solutions [18]. Including expected outcomes in the working sheets might come in handy, but it also pollutes spreadsheets with spurious data. To avoid smudging the spreadsheet, the data introduced for illustration purposes is transient, i.e. they are kept from *QUESTION()* first enactment till the question is posted. Once the question is posted, no trace is left in the working sheet. Transient data is highlighted with a red shaded.

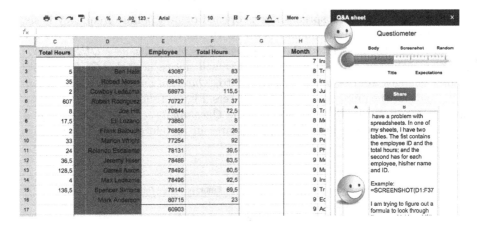

Fig. 3. Function *Screenshoot()* inlays sheet snippets into the message. (Color figure online)

---

[2] A spreadsheet can hold different *QUESTION()* formulae, each with its own *QuestionSheet*.

Enriching Questions with Sample Data. Sample data from the working sheet can be inlaid into the question's message through *SCREENSHOT(Range)*. Figure 3 illustrates the use of this function. *SCREENSHOT()* obtains (1) a screenshot, and (2), a detached sample spreadsheet for the selected *Range*[3]. The screenshot helps message understanding while the detached spreadsheet allows respondents to easily try out their formulae. The actual thread can be found at https://goo.gl/c9vbCF.

4.2   Question Sharing

QuestionSheets can be posted immediately or handed to the timer. In this way, question specification is decoupled from question posting, letting users decide the most appropriate time slide for posting their questions. So far, the timer admits three values: *immediate, USA_evening* and *USA_weekend* based on the studies of [5]. When posted, *QuestionSheets* are transparently turned into their question-thread counterparts. Therefore, questions have a double representation: one in *SO* (as question threads) and one in *Google Sheets* (as *QuestionSheets*). These two representations need to be kept in sync, i.e. answers in *SO* are propagated to the *QuestionSheet* counterpart while changes in the *Q&A* spreadsheet might lead to re-edits of the question in *SO* (see later). Appropriate drivers map the *QuestionSheet* representation to the data structure expected by the Q&A platform at hand.

4.3   Answer Awaiting

Once the question is posted, drivers periodically polls *SO* for answers. This happens when either the spreadsheet is opened or in a polling frequency basis. By default, this polling frequency is set to 15'. In this way, answers in *SO* are propagated to the associated *QuestionSheet*. This in turn, causes *QUESTION()* to be re-evaluated. At this point, the user will notice how the *QUESTION()* display changes from *"Question posted..."* to *"Question answered (n) ..."* where *"n"* stands for the number of available answers. Figure 4 shows the case when two answers are available. It is not need for the *QuestionSheet* to be visible. Similar to the warnings about available app updates in mobile phones, the existence

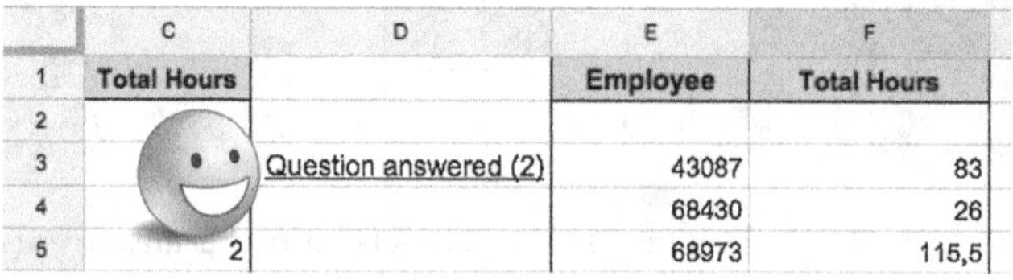

Fig. 4. Adding answers makes function *QUESTION()* be re-evaluated showing the number of SO answers so far.

---

[3] This shared table is attached to the question's message as a URL.

of answers do not force users to move to resolution right away. Based on the urgency, users might select the first answer that shows up or rather wait to see if additional answers come up.

4.4   Answer Resolution

SO threads are mapped back into *QuestionSheets*. Information collected from *SO* includes the answer itself, the contributor's reputation, and the number of votes (if any) (see Fig. 5). Users can move between answers by clicking on the tabs. This causes *QUESTION()* to be re-evaluated. Specifically, *QUESTION()* accesses the cell containing the answer, and attempts to extract the formula from the text. In most case, *QUESTION()* is successful since most contributors follow *SO's* guidelines of using markdown for code. If so, *QUESTION()* returns the formula. This in turn, causes *Google Sheets* to evaluate the obtained formula, and return a value (e.g. *Ben Hale*). This value is compared with the expected outcome (should outcomes be provided at question time), shading the answer tab with green, yellow or red, depending on the number of expect outcomes the answer has matched. In this way, a single click on the question tab promptly permits users to see whether the select answer fits their expectations or not. No need to understand what the formula is about. Formulas are judged by their outputs. This highlights the importance of providing expected outcomes with proper coverage.

Users can wait till an answer produces the desired outputs. Once provided, they can make an answer final. To this effect, users click on the *stop-sharing* button (see top of Fig. 5). This ends the question lifecycle: the *QuestionSheet* is deleted together with the transient data. As for the SO counterpart, the query thread is finished along with *SO* good practices[4]. The only cue about a formula being obtained through SO, is by a comment left attached to the corresponding

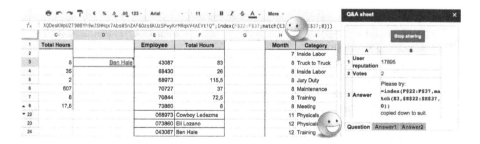

Fig. 5. Answer Resolution. Answers are accessible via tabs (see Smiley at the bottom). Clicking on a tab has a three-fold effect: (1) the sought formula is obtained from the SO thread using pattern matching; this formula is displayed in the formula bar; and (3) the associated *QUESTION()* is re-evaluated to show what would be the formula output (e.g. *Ben Hale*). (Color figure online)

---

[4] http://stackoverflow.com/help/someone-answers.

cell. The comment holds the URL to the *SO* thread counterpart. This is the only trace left back about how the formula was obtained.

5   Evaluation

This section reports on a formative evaluation conducted through five case studies.For *QuestionSheet*, the experiment does not stop at posing the question but expand along the question lifecycle, including waiting for the SO community to answer. This potentially long lifespan is what sustains the use of use cases for formative evaluation.

Table 2. Task description. The *QuestionSheet*-generated SO thread is included for reference.

Case Study	Statement	Sample table	Sought Formula / Thread URL
T1	Work out award nominee students, i.e. those whose marks are among the 25% top marks on his/her class		*if(B2>=percentile(B$2:B$51;0,75); "YES";"NO")* https://goo.gl/KDNxms
T2	Count borrowed books by type (e.g. REF, AA) where type is embedded as part of the call-number ID		*ArrayFormula(query({A2: A\regexextract(B2:B; "^[A-Z]*")}; "Select Col2, count(Col1) where Col2 <>" group by Col2 label count(Col1) ""))* https://goo.gl/CHy9b3
T3	Place employees' names by the Employee columns		*index(F$22:F$37,match(E3, $E$22:$E$37,0))* https://goo.gl/c9vbCF
T4	Calculate working time subtracting 30' lunch break, if applicable		*mod(B2-A2;1)-if(mod(B2-A2;1) >time(4;0;0); time(0;30;0); 0)* https://goo.gl/heqhr3
T5	Calculate average calories for "moderate" food, with the formula being resilient to additional intakes being inlaid		*averageif(A2:A; "moderate"; B2:B)* https://goo.gl/WuR6me

5.1   The Experiment

Subjects. Participants were recruited locally. For participants to qualify as "home users", they should have at least one year exposure to *Google Sheets* but not having created more than six sheets in this period. Five students qualified. Participants were given a brief introduction to *QuestionSheet* where a running example was developed.

Case Studies. Scenarios were carefully selected so that questions should not be neither too trivial nor too complicated. For the validity of the experiment, it is most important to find the right balance. Too easy cases will not justify the effort of *QuestionSheet* as most users will know the answers without resorting to SO. In addition, easy questions will get prompt answers no matter whether the description of the query is enhanced with examples or not (one of the aspects to be tested out). On the other hand, too complicated cases will be also difficult for home users to stumble upon in their daily activities. Table 2 displays the selected cases inspired by real doubts risen in forums. The table also includes the sought formula for readers to ponder the scenarios' complexity. In addition, a link to the *QuestionSheet*-generated question in Stack Overflow is also included.

Table 3. Constructs of diffusion theory adapted for *QuestionSheet*.

Construct	Definition: The degree to which *QuestionSheet* innovation is perceived as
Relative advantage	...being better that typing the question directly on StackOverflow
Compatibility	...consistent with the existing values, needs, and past experiences of GoogleSheet users
Complexity	...being difficult to use

Methodology. Each subject addresses one of previous scenarios. To prevent scenario understanding from polluting the performance measures on the usage of *QuestionSheet*, the task was split into parts. First, we made sure subjects understand the scenario, i.e. they were asked to complete the sample spreadsheet with the values the expected formula should return (red shaded in Table 2). This helped to ensure scenarios were properly understood. Next, subjects were asked to post their questions using *QuestionSheet*. Once the testing session was over, participants were asked to fill in a questionnaire that rates different aspects of *QuestionSheet* through Likert scales (see Fig. 7). In addition, open comments were also welcome. The questionnaire builds upon a reduction of Roger's model of Diffusion of Innovations that includes only those constructs consistently related to technology adoption behavior: relative advantage, complexity and compatibility [21]. Table 3 elaborates these constructs for the *QuestionSheet* case.

5.2 Results

Elapsed Time. That is, the time it took for the SO community to answer the *QuestionSheet*-generated questions. We frame this outcome w.r.t. the distribution of time-to-answer in SO, more specifically, we obtained the elapsed-time first-answer distribution for *"google-spreadsheet"*-tagged questions in SO

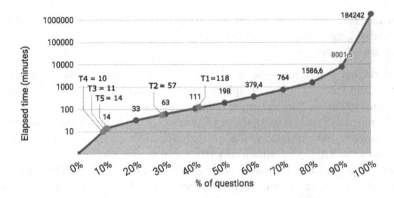

Fig. 6. Elapsed-time first-answer distribution for *"google-spreadsheet"*-tagged questions at *Stack Overflow* as for September, 2017. Questions without answer were removed.

(see Fig. 6)[5]. We use this figure to frame our results: T3, T4 and T5 are among the top 10%, T2 in the 30% and T1 in the 50%. We believe these results to be rather good. More to the point, if we consider that questions were posted by SO newcomers with no reputation, where reputation is being indicated as a main correlator of response speed [5]. This anecdotal evidence supports the findings of Calefato et al. as for the use of examples as main answer attractors.

Questionnaire. In general, users perceive *QuestionSheet* as beneficial w.r.t. directly typing the question in SO (questions 1–4 in Fig. 7). Important enough, users ranked high *QuestionSheet* compatibility with *Google Sheets modus operandi* (question 5). This is a main incentive to user adoption.

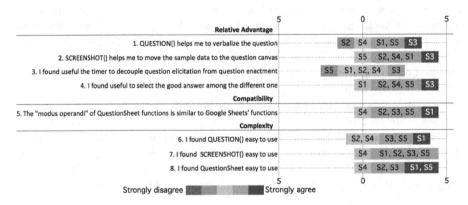

Fig. 7. Diverging stacked bar chart for the satisfaction questionnaire using Likert scales.

---

[5] Studies indicate that how the question is being tagged is the best indicator to predict when an answer is to be expected [7]. Distribution on Fig. 6 was obtained by enacting the query at goo.gl/RgGBE7.

Threats to Validity. A larger number of participants are required to draw any conclusive remarks. As for external validity, *QuestionSheet* could post questions to Q&A platforms other than *StackOverflow* as long as appropriate API-based drivers were provided. On the other hand, our experience can be of interest for frameworks that target home users who can resort to examples to make themselves understood in Q&A forums.

5.3   Lessons Learned

This subsection discusses lessons learned along the model introduced by [5].

Affect. Calefato et al. highlight the importance of a neutral emotional style to promote answers. In a similar vein, we realized subjects were not so aware of the impact of posting questions repeatedly. Specifically, two users posted their questions (or small variations) more than once. *QuestionSheet* might turn both *StackOverflow* too transparent and question posting too easy, with the resulting risk of "question spam". This makes true the famous quote: *"Everything Should Be Made as Simple as Possible, But Not Simpler"*. Users ignored the importance of reputation when interacting with social networks, and the eventuality of being banned, if a proper etiquette is not observed. Hence, *QuestionSheet* needs to be revised as for restricting the number of questions per sheet, or preventing questions which have already been posted at *StackOverflow*.

Presentation Quality. Calefato et al. underscore the importance of attaching code snippets and examples to questions. This remark was even more important in our setting: non-native English speaker. Two subjects noticed that without examples they would have not been able to describe their questions in a narrative way using English. One subject indicated that, though she knew *Stack Overflow*, she was afraid of posing question due to her limited use of English. Here, templates and examples lower the entry bar for home users.

Time Slice. Users were pleasantly surprised for the promptness of the answers. As a subject puts it "it is almost like Google!". This promptness might be well due to the presentation quality of the questions (i.e. neutral language, example-based). But it might also be influenced by the medium complexity of the questions. Being home users, the expectations are for questions to be of medium complexity hence, easily answerable by more proficiency answerers. At this respect, delaying question posting till the evenings or weekends might make sense when there might be a need to attract a large number of answerers. However, medium-complexity questions might well be equally addressed by a smaller population without waiting till the weekend. If this were the case, it would certainly challenge the need of the built-in timer in *QuestionSheet*.

6   Conclusions

This work tackles how to spread the benefits of Q&A platforms to home users. To this end, we look into assisted inline question posting where users post their questions without leaving their working environment. *QuestionSheet* tests this out for *Google Sheets*. Offering Q&A facilities from within *Google Sheets* not only avoids platform switches and facilitates focus, but also accounts for in-place example construction to be used for question clarification, which in turn helps to engage answerers. Informative evaluation is being conducted through five use cases. Though conclusive statements cannot yet be drawn, first insights seem to suggest that in-place scaffolds might help to combat the digital divide between home users and more technical savvy ones as for enjoying the benefits of Q&A platforms. Yet, the benefits of educating users to formulate questions goes beyond the askers themselves. Further investigation should look at how "good questions" enhance effective knowledge-sharing behavior that will eventually lead to the creation of long-lasting value pieces of knowledge in Q&A platforms. Along Israelmore Ayivor quote *"To get answers, ask questions; but to get good answers, ask good questions"*, we could conclude that for Q&A platforms to become a repository of good answers, first you need good questions.

Acknowledgements. This work is co-supported by the Spanish Ministry of Education, and the European Social Fund under contract TIN2014-58131-R. Contell enjoys a doctoral grant from the University of the Basque Country.

References

1. Bacchelli, A., Ponzanelli, L., Lanza, M.: Harnessing Stack Overflow for the IDE. In: Proceedings of the 3rd International Workshop on Recommendation Systems for Software Engineering, pp. 26–30 (2012)
2. Bazelli, B., Hindle, A., Stroulia, E.: On the personality traits of StackOverflow users. In: Proceedings of the 29th IEEE International Conference on Software Maintenance (ICSM), pp. 460–463 (2013)
3. Bosu, A., Corley, C.S., Heaton, D., Chatterji, D., Carver, J.C., Kraft, N.A.: Building reputation in StackOverflow: an empirical investigation. In: Proceedings of the 10th Working Conference on Mining Software Repositories (MSR), pp. 89–92. IEEE Press (2013)
4. Calefato, F., Lanubile, F., Marasciulo, M.C., Novielli, N.: Mining successful answers in Stack Overflow. In: Proceedings of the 12th IEEE/ACM Working Conference on Mining Software Repositories (MSR), pp. 430–433 (2015)
5. Calefato, F., Lanubile, F., Novielli, N.: How to ask for technical help? Evidence-based guidelines for writing questions on stack overflow. Inf. Soft. Technol. **94**(C), 186–207 (2018)
6. Frei, B.: Paid crowdsourcing: current state & progress toward mainstream business use, report (2009). https://www.smartsheet.com
7. Goderie, J., Georgsson, B.M., van Graafeiland, B., Bacchelli, A.: Eta: estimated time of answer predicting response time in Stack Overflow. In: Proceedings of the 12th IEEE/ACM Working Conference on Mining Software Repositories (MSR), pp. 414–417 (2015)
8. Harper, F.M., Raban, D.R., Rafaeli, S., Konstan, J.A.: Predictors of answer quality in online Q&A sites. In: Proceedings of the SIGCHI Conference on Human Factors in Computing Systems, pp. 865–874 (2008)

9. Ko, A.J., Myers, B.A.: Human factors affecting dependability in end-user programming. In: Proceedings of the 1st Workshop on End-User Software Engineering (2005)
10. Korvela, H.: Plz urgent help needed!1!! - aspects of on-line knowledge sharing in end-user development support. In: Proceedings of the 19th Americas Conference on Information Systems (AMCIS) (2013)
11. Korvela, H., Back, B.: The impact of skills and demographics on end-user developers' use of support. In: Proceedings of the 18th Americas Conference on Information Systems (AMCIS) (2012)
12. Korvela, H., Packalén, K.: On-line support - a virtual treasure trove for end-user developers in small organisations? In: Proceedings of the 15th Americas Conference on Information Systems (AMCIS) (2009)
13. Mamykina, L., Manoim, B., Mittal, M., Hripcsak, G., Hartmann, B.: Design lessons from the fastest Q&A site in the west. In: Proceedings of the SIGCHI Conference on Human Factors in Computing Systems, pp. 2857–2866 (2011)
14. Moffitt, K.: Cases on the human side of information technology. In: End-User Computing at BRECI: The Ordeals of a One-Person IS Department, pp. 330–342. IGI Global (2006)
15. Nasehi, S.M., Sillito, J., Maurer, F., Burns, C.: What makes a good code example?: A study of programming Q&A in StackOverflow. In: Proceedings of the 28th IEEE International Conference on Software Maintenance (ICSM), pp. 25–34 (2012)
16. Quinn, A.J., Bederson, B.B.: AskSheet: efficient human computation for decision making with spreadsheets. In: Proceedings of the 17th ACM Conference on Computer Supported Cooperative Work and Social Computing (CSCW), pp. 1456–1466 (2014)
17. Sawadsky, N., Murphy, G.C.: Fishtail: from task context to source code examples. In: Proceedings of the 1st Workshop on Developing Tools and Plug-ins (TOPI), pp. 48–51 (2011)
18. StackOverflow: How do I ask a good question? http://stackoverflow.com/help/how-to-ask. Accessed 23 Mar 2018
19. Tagg, J.: Discovering ideas handbook: use specifics and examples (2003). https://www2.palomar.edu/users/jtagg/handbook/specific.htm. Accessed 23 Mar 2018
20. Thom-Santelli, J., Yuen, S., Matthews, T., Daly, E.M., Millen, D.R.: What are you working on? Status message Q&A in an enterprise SNS. In: Proceedings of the 12th European Conference on Computer Supported Cooperative Work (ECSCW), pp. 313–332 (2011)
21. Tornatzky, L.G., Klein, K.J.: Innovation characteristics and innovation adoption-implementation: a meta-analysis of findings. IEEE Trans. Eng. Manage. EM **29**(1), 28–45 (1982)
22. Treude, C., Barzilay, O., Storey, M.D.: How do programmers ask and answer questions on the web? In: Proceedings of the 33rd International Conference on Software Engineering (ICSE), pp. 804–807 (2011)
23. Wenger, E.: Communities of Practice and Social Learning Systems. The Systems Thinker (1998)
24. Xiao, L., Farooq, U.: Supporting end user development in community computing: requirements, opportunities, and challenges. J. Commun. Inf. **9**, 1 (2013)
25. Yang, J., Hauff, C., Bozzon, A., Houben, G.: Asking the right question in collaborative Q&A systems. In: Proceedings of the 25th ACM Conference on Hypertext and Social Media, pp. 179–189 (2014)

News Recommendation with CF-IDF+

Emma de Koning, Frederik Hogenboom, and Flavius Frasincar[✉]

Erasmus University Rotterdam,
PO Box 1738, 3000 DR Rotterdam, The Netherlands
370761ek@student.eur.nl, {fhogenboom,frasincar}@ese.eur.nl

Abstract. Traditionally, content-based recommendation is performed using term occurrences, which are leveraged in the TF-IDF method. This method is the defacto standard in text mining and information retrieval. Valuable additional information from domain ontologies, however, is not employed by default. The TF-IDF-based CF-IDF method successfully utilizes the semantics of a domain ontology for news recommendation by detecting ontological concepts instead of terms. However, like other semantics-based methods, CF-IDF fails to consider the different concept relationship types. In this paper, we extend CF-IDF to additionally take into account concept relationship types. Evaluation is performed using Ceryx, an extension to the Hermes news personalization framework. Using a custom news data set, our CF-IDF+ news recommender outperforms the CF-IDF and TF-IDF recommenders in terms of F_1 and Kappa.

Keywords: News recommender sytems · CF-IDF · CF-IDF+

1 Introduction

With the continuously growing amount of information on the Internet, distinguishing relevant from irrelevant matters becomes increasingly difficult. What is supposed to be a trivial task for humans, is exceptionally hard for machines because of the difficulties that need to be faced while automatically comprehending the information content. A common type of information that is searched for on the Web is news, i.e., information describing recent events that is or is not yet (partially) known to a user. Articles on news sites are usually already categorized, but this classification is often too course, and does not match the fine-grained user interests. Instead of treating all visitors equally, an automatic recommendation system could aid individual users in selecting interesting information. Determining the user's preferences would be helpful when optimizing browsing experience on news Web sites, for instance by presenting interesting articles on top, or by filtering RSS feeds on interesting contents.

Recommender systems determining the importance of news are not merely useful for enhancing user experiences, but are of utmost importance in decision support scenarios where fast and accurate news provision is needed for reliable

© Springer International Publishing AG, part of Springer Nature 2018
J. Krogstie and H. A. Reijers (Eds.): CAiSE 2018, LNCS 10816, pp. 170–184, 2018.
https://doi.org/10.1007/978-3-319-91563-0_11

decisions, such as in (algorithmic) trading. In addition, such systems are important to news providing companies with revenue models based on advertisements. Their earnings are usually based on page clicks, which can be maximized by presenting the most relevant items to a user. The visitor is more likely to stay on such sites, yielding more clicks and thus more advertisement earnings.

A vast amount of research has been conducted related to algorithms and systems for news recommendation [4–6,10,13,14]. In general, we can distinguish between three basic types of (news) recommendation systems: content-based recommenders that recommend news items to users according to the content of the news items, collaborative filtering recommenders that focus on the articles similar users are interested in, and hybrid recommenders that combine the two previous methods. Therefore, the main difference between content-based recommenders and collaborative filtering recommenders lies in the focus on user-article and user-user similarities. In our efforts, the scope will be limited to content-based recommenders. In these recommenders the emphasis is on measuring the similarity between the user profile, built on the content of previously read news articles, and the content of an unread news article. There are two types of content-based recommenders, i.e., traditional and semantic recommenders. The former type is term-based, whereas the latter one is concept-based. The focus in this research will mainly be on semantic recommenders, as these provide for a better and more intuitive representation of news items.

In previous work, Goossen et al. [10] show that using the semantic recommendation method Concept Frequency - Inverse Document Frequency (CF-IDF) significantly improves the performance over a recommender using the traditional Term Frequency - Inverse Document Frequency (TF-IDF). However, a major shortcoming of CF-IDF is that it does not take into account the various semantic relationships between concepts, like the superclass-of, subclass-of, or domain-specific relationships, which could be useful to provide an accurate representation of news items. The goal of this paper is to extend the CF-IDF weighting technique by employing the semantic relationships of concepts from a domain ontology. The proposed recommendation method, CF-IDF+, is implemented in Ceryx [5], an extension to the Hermes news personalization service for building recommender systems [3]. The performance of the CF-IDF+ will be evaluated against the TF-IDF baseline and various semantic-based counterparts like CF-IDF.

The remainder of this paper is organized as follows. First, we discuss related work in Sect. 2. Next, we introduce our framework and implementation in Sects. 3 and 4, respectively. Section 5 presents the performance evaluation of our algorithm, and we conclude in Sect. 6.

2   Related Work

The current body of literature contains many descriptions of profile-based recommender systems, differing in their approaches for news recommendation. Most importantly, they implement different similarity measures for computing the similarity between a news item and the user profile.

2.1   Term-Based News Recommendation

The most popular traditional content-based recommendation approach is the TF-IDF weighting scheme. A common approach in comparing documents is to use the TF-IDF weighting scheme together with the cosine similarity. It is a statistical method to determine the relative importance for each word within a document in a set of documents.

The TF-IDF weight is defined as a two-step computation, based on the term frequency (TF) and inverse document frequency (IDF). Term frequency indicates the importance of term t_i in document d_j: frequent terms are more likely to be relevant for the topic of the document. The inverse document frequency captures the general importance of a term in a set of documents. The more a term appears in all the documents, the less relevant it is to the topic of a single document, as the term is too generic. The TF-IDF is subsequently computed as the product of TF and IDF.

2.2   Semantics-Based News Recommendation

There are several semantic-based methods available for news recommendation. Some are based on sets of synonyms (synsets), while others are based on onto-logical concepts.

Synsets. The Synset Frequency - Inverse Document Frequency (SF-IDF) [5] weighting method compares the words from unread documents with the words from documents in the user profile. In the comparison, the semantic similarity is obtained by making use of the WordNet online lexical database. This is a database for the English language with over 166,000 senses (i.e., words with their associated parts-of-speech and senses). The advantage of using synsets over words is that the ambiguity of the words can be eliminated by performing word sense disambiguation.

The SF-IDF weighting method outperforms the traditional TF-IDF, but a shortcoming of this semantics-based method is that they do not allow for vari-ous semantic relationships between synsets, thus providing only a limited under-standing of news semantics [13]. The SF-IDF recommender does not take into account inter-synset relationships, like hyponymy, antonymy, troponymy, and synonymy, while this is very important for the interpretability of a document.

Moerland et al. [13] extend SF-IDF to the SF-IDF+ weighting method, by accounting for the semantic relationships between synsets. This method is eval-uated on a set of financial news articles and it outperforms TF-IDF and SF-IDF in terms of F_1-scores.

Ontological Concepts. Concept Frequency - Inverse Document Frequency (CF-IDF) is based on the traditional TF-IDF method, but merely considers key concepts instead of using all the terms within a document [10]. The TF-IDF recommendation method considers all terms in an article to be useful for understanding the content, while this might lead to 'noise' terms in the weighted vector of terms. These are terms that do not give additional information about the content of the article. Instead of a weighted vector of terms, CF-IDF considers the news article as a weighted vector of concepts, where concepts are derived from a domain ontology. Using concepts instead of terms assures that there are no noisy terms which can obfuscate the algorithm's outcomes, making CF-IDF a more intelligent recommender.

3 Framework

The Hermes news personalization framework introduced in [3] is a news personalization service supporting different content-based recommenders. Hermes operates on RSS feeds of user-rated news items. Recommendations are catered using terms, synsets, or concepts from an internal knowledge base for classification.

3.1 Hermes Extensions

The Hermes News Portal has two extensions, i.e., Athena [11] and Ceryx [5]. Athena supports both the classic TF-IDF and the semantics-based recommendation method CF-IDF, assigning words in news articles to different domain concepts, which are stored in an internal knowledge base. A graph-based user interface allows for visualizing inter-concept connections, and can also be used by users to select interesting concepts to manually build a profile. Automatic user profile construction is also supported by analyzing the articles that are read by the user. The Ceryx extension of Athena provides additional support for synset-based recommenders, such as SF-IDF. In our recent efforts, the newly proposed CF-IDF+ is added to the list, which additionally utilizes semantically related concepts and optimizes concept relationship type weights for enhanced performance.

3.2 User Profile Representation

A user profile is represented by a content vector (with terms, concepts, or synsets) of all user-read news articles. The contents are determined differently depending on the recommender. While the TF-IDF recommender merely analyzes every term in a news item after initial stop word removal, the CF-IDF and CF-IDF+ recommenders look at domain concepts found in the news items instead of using all the terms in the text of the news items.

The Hermes framework employs the vector space model, which is interpreted differently by each recommender. Concept-based recommenders use a domain ontology to retrieve the concepts in a news item. This ontology holds a set C of i concepts and their relations, i.e., $C = \{c_1, c_2, c_3, \ldots, c_i\}$. The user profile U consist of j concepts, where $U = \{c_1^u, c_2^u, c_3^u, \ldots, c_j^u\}$ and $c^u \in C$. Concept c^u is linked to k news articles a_i in which it is found, and is denoted as

$c^u = \{a_1^u, a_2^u, a_3^u, \ldots, a_k^u\}$. Hence, for a semantic recommender, an article a is considered to be a set of l elements representing the number of concepts c present, i.e., for CF-IDF:

$$a_{\text{CF-IDF}} = \{c_1^a, c_2^a, c_3^a, \ldots, c_l^a\}, c^a \in C. \tag{1}$$

The traditional TF-IDF recommender has a different interpretation, as it regards article a as a set of m elements representing the number of terms t appearing in article a:

$$a_{\text{TF-IDF}} = \{t_1^a, t_2^a, t_3^a, \ldots, t_m^a\}. \tag{2}$$

Then, we can consider weights w^u and w^a for the user profile and article, respectively, as item scores. Computational details for the weight of a single item (i.e., term or concept) in a specific article for TF-IDF, CF-IDF, and CF-IDF+ can be found in Sect. 3.3. Together, terms and concepts and their respective computed weights comprise the user profile U as a vector V^U, and an unread news item a as vector V^a.

The TF-IDF recommender computes a weight w^u for each term in U and the CF-IDF recommender computes a weight w^u for every concept in U:

$$V_{\text{TF-IDF}}^U = \{\langle\, t_1^u, w_1^u\,\rangle, \ldots, \langle\, t_m^u, w_m^u\,\rangle\,\}, \tag{3}$$

$$V_{\text{CF-IDF}}^U = \{\langle\, c_1^u, w_1^u\,\rangle, \ldots, \langle\, c_j^u, w_j^u\,\rangle\,\}. \tag{4}$$

When computing recommendations, the TF-IDF and CF-IDF recommender convert unread news item a into vector $V_{\text{CF-IDF}}^a$, containing the terms or concepts found in the news items and their corresponding weights, respectively:

$$V_{\text{TF-IDF}}^a = \{\langle\, c_1^a, w_1^a\,\rangle, \ldots, \langle\, c_m^a, w_m^a\,\rangle\,\}, \tag{5}$$

$$V_{\text{CF-IDF}}^a = \{\langle\, c_1^a, w_1^a\,\rangle, \ldots, \langle\, c_l^a, w_l^a\,\rangle\,\}. \tag{6}$$

A similarity measure, such as the cosine similarity, can subsequently be applied to the user profile and news item vectors, resulting in a ranked recommendation based on similarity.

3.3   Weight Computation

In order to ensure a general understanding of the computation of CF-IDF+ weights, we briefly discuss TF-IDF and CF-IDF weight computations. In TF-IDF computations, we are interested in the product of term frequencies tf and inverse document frequencies idf. The term frequency is defined as the number of times a term t_i occurs in document d_j, $n_{i,j}$, divided by the total number of occurrences of all terms in the document, whereas the inverse document frequency indicates the importance of term t in all news items and is computed by dividing the total number of documents $|D|$ by the number of documents in which term t_i can be found:

$$\text{tf}_{i,j} = \frac{n_{i,j}}{\sum_k n_{k,j}}, \tag{7}$$

$$\text{idf}_i = \log \frac{|D|}{|\{d : t_i \in d\}|}, \tag{8}$$

$$\text{tf-idf}_{i,j} = \text{tf}_{i,j} \cdot \text{idf}_i. \tag{9}$$

The similarity between the user profile vector and the vector representation of a news item is subsequently calculated using the cosine similarity measure, i.e.:

$$\mathrm{sim}_{\mathrm{TF\text{-}IDF}} = \frac{P \cdot U}{\|P\|\|U\|}, \tag{10}$$

where the user profile vector is represented by P, and U is the vector representation with the terms in the unread news item. If this similarity score is greater than the cut-off value, the unread news item is recommended to the user. The cut-off value can be any number between 0 and 1. Lower cut-off values will result in more recommended news items (high recall) with a higher likelihood of them being non-interesting (low precision), whereas higher values enforce high similarities and thus yield higher precision and lower recall. Therefore, the cut-off value is an indication of the degree of the user's preference.

The procedures of the Concept Frequency - Inverse Document Frequency recommender [10] are very similar to the TF-IDF recommender, yet operate on concept vectors instead of term vectors. Each news item contains zero or more concepts, which are defined in the domain ontology of Hermes. To obtain these concepts, Hermes employs a Natural Language Processing engine that uses several techniques like tokenization, part-of-speech tagging and word sense disambiguation. Similar to TF-IDF frequencies and inverse document frequencies are computed for concept c_i, i.e., cf and idf$'$, respectively:

$$\mathrm{cf}_{i,j} = \frac{n_{i,j}}{\sum_k n_{k,j}}, \tag{11}$$

$$\mathrm{idf}'_i = \log \frac{|D|}{|\{d : c_i \in d\}|}, \tag{12}$$

$$\mathrm{cf\text{-}idf}_{i,j} = \mathrm{cf}_{i,j} \cdot \mathrm{idf}'_i. \tag{13}$$

The similarity between the user profile vector and the vector representation of the unread news item is computed by the cosine similarity measure introduced before. Again the unread news article U is only recommended if the similarity measure exceeds the cut-off value.

The usage of concepts instead of terms causes the recommender to deal with only the most 'important' terms in an article, making it more effective. Even when considering the additional computation time caused by necessary NLP steps, processing times are much lower for CF-IDF when compared to TF-IDF, because the number of elements that need to be processed is much lower.

Despite its benefits, an earlier mentioned drawback of CF-IDF is that it does not take into account concept relationship types. These types provide valuable information about the content of a news item, that can thus be exploited to retrieve interesting news items that discuss related concepts. The CF-IDF+ recommender also processes the news items into a concept vector representation, but additionally verifies whether the concept is a class or an instance. For a class, its direct super- and subclasses are retrieved from the domain ontology. For an instance, its domain relationships are identified from the domain ontology.

For our subsequent procedures, we extend the original set of concepts C to also include the concepts that are related through various semantical relationships of the included concepts, and define $C(c) = \{c\} \bigcup_{r \in R(c)} r(c)$, where

c denotes a concept in the news item, $r(c)$ denotes a concept that is related to concept c by relationship r and $R(c)$ represents the set of relationships of concept c. Next, we define two sets of extended concepts, i.e., the unread news item U and the user profile R as $U = \{C(u_1), C(u_2), \ldots, C(u_m)\}$, and $R = \{C(r_1), C(r_2), \ldots, C(r_n)\}$, respectively, where $C(u_m)$ denotes the m-th extended concept of U, and $C(r_n)$ represents the n-th extended concept of R. By extended concept we mean all the concepts obtained by following relationships in the domain ontology from the current concept.

The calculation of the CF-IDF+ weight is very similar to the original CF-IDF, but additionally introduces weights per semantic relationship. This has also proven to be very effective in a word sense disambiguation context, by weighting edges (relations) of network representations of semantic lexicons [15]. In CF-IDF+, we focus on domain ontologies, and identify three different weights: one for the superclasses, one for the subclasses and one for the domain relationships. CF-IDF+ is computed as follows:

$$\text{cf-idf}_{i,j,r} = \text{cf}_{i,j} \cdot \text{idf}'_i \cdot w_r. \tag{14}$$

The weight of semantic relationship r needs to be optimized for each cut-off value, for instance through genetic algorithms or brute-force iterative processes (e.g., explicit enumeration). If a concept is found multiple times (direct occurrence in text and indirect occurrence by means of a relationship), we retain only its highest weight value.

4   Implementation

In our endeavours, we have implemented CF-IDF+ in Ceryx, which is an extension to the Java-based Hermes News Portal and its Athena plug-in for recommender systems. The tool makes use of various Semantic Web technologies, operates on user profiles, and processes news items from RSS feeds. The core of the news portal is an OWL domain ontology that is constructed by domain experts, allowing for semantics-based operations on news messages. The ontology contains a set of commonly used, well-known, financial entities such as companies, products, currencies, etc., and these concepts have associated lexical representations. The ontology consists of 65 classes, 18 object properties, 11 data properties, and 1, 167 individuals.

News items are classified using the GATE natural language processing software [8] and the WordNet [9] semantic lexicon. The semantics-based methods additionally make use of the Stanford Log-Linear Part-of-Speech Tagger [16], Lesk Word Sense Disambiguation [12], and the Alias-i's LingPipe 4.1.0 [1] Named Entity Recognizer.

4.1   User Interface

The Ceryx implementation features a tabbed interface in which the user is able to browse through all available news items in the system, which are extracted from RSS feeds. Whenever an item is clicked, the corresponding news message is opened in the Web browser and the user profile is updated. Additionally, the user is able to select a recommendation method. Retrieved recommendations are sorted by relevance to the user's profile. Last, we have implemented a testing environment that compares the various news recommenders implemented in Ceryx. This environment supports XML input files describing user profiles (i.e., lists with the URIs of the read news items and whether the news items are considered as interesting by the user), and displays common evaluation measures, such as precision and recall, based on randomly selected training and test sets. Cut-off testing is also supported, and is preliminary implemented as an iterative process with increments of 0.01 from 0 to 1.

4.2 TF-IDF

For each news item, the TF-IDF news recommender gathers all words (terms), and removes stop words using a predefined list. Subsequently, term counts are computed within each news item and over all news items, which are used for computing term frequencies and inverse document frequencies. Before the news items are recommended to the user, the similarity scores for all the unread news items are MIN-MAX normalized to take values between 0 and 1. If the similarity score exceeds the user-specified cut-off value, the unread news item will be recommended.

4.3 CF-IDF

In order to retrieve concepts from the news items, Ceryx utilizes an advanced Natural Language Processing engine that matches the lexical representations stored in the ontology and subsequently performs word sense disambiguation. When the concepts are obtained from the news items, the CF-IDF recommender operates similarly to the TF-IDF recommender. The CF-IDF recommender counts the number of concept appearances in each news items to obtain the concept frequencies, as well as the number of times a concept is found in all news items to collect the inverse document frequencies. Again, CF-IDF weights are computed and similarities are MIN-MAX normalized. As before, if the similarity score exceeds the user-specified cut-off value, the unread news item will be recommended.

4.4 CF-IDF+

The implementation of CF-IDF+ is an extension to the CF-IDF implementation and additionally determines whether identified concepts are classes or instances, so that related concepts are retrieved. The CF-IDF values of these related concepts are calculated by multiplying the CF-IDF of the original concept with a weight depending on the relationship type between the original concept and the related concept. The CF-IDF+ values are then stored in a vector for the news item or the user profile. When a related concept is already part of the vector, we verify whether the CF-IDF+ value of the related concept is greater than the original CF-IDF+ of the related concept, and choose the highest CF-IDF+ value. Next, the similarity score between the user profile and each news item is computed by using the cosine similarity measure. Again, similarity scores are MIN-MAX normalized between 0 and 1, and if the similarity score exceeds the user-specified cut-off value, the unread news item will be recommended.

5 Evaluation

In order to demonstrate the effectiveness of CF-IDF+, we evaluate the performance of its implementation within the Ceryx plugin of the Hermes news recommendation system against real-world data. The remainder of this section discusses the experimental setup, the optimization of the semantic relationship weights and their properties, and the evaluation of the global performance of CF-IDF+ compared to various other recommenders, respectively.

5.1 Data

In our experiments, we make use of a data set containing 100 real-life news articles, which are collected from a Reuters news feed on news about technology companies. We distinguish between 8 topics of interest and ask 3 domain experts to rate each news item for appropriateness with respect to the selected topics, as described in Table 1, i.e., 'Asia or its countries', 'Financial markets', 'Google or its competitors', 'Internet or Web services', 'Microsoft or its competitors', 'National economies', 'Technology', and 'United States'. We maintain an Inter-Annotator Agreement (IAA) of at least 2 out of 3 reviewers in order for a news item to be added to a user profile for one of the specific subjects. Table 1 described the number of interesting (I+) and non-interesting (I−) items per topic, and additionally shows the fraction of interesting news items indicated as interesting by all domain experts (IAA+), the fraction of non-interesting news items indicated as non-interesting by all experts (IAA−) and the average of these two fractions (IAA).

For evaluation purposes, we split our data randomly into a 30% training, a 30% validation, and a 40% test set. These sets are used for creating user profiles, optimizing relationship type weights, and performance measurement, respectively. Based on true/false positives/negatives, we evaluate performance for various recommendation cut-off values by computing accuracy, recall, precision, and F_1 scores. Weights are optimized using an iterative procedure with a step size of 0.1 between 0 and 1, while aiming to maximize F_1 scores.

5.2 Semantic Relationships Weights

Per cut-off value, ranging between 0 and 1 with an increment of 0.01, we optimize the weights of the superclass, subclass, and domain semantic relationships. The mean and variance of these weights (w_{super}, w_{sub}, and w_{rel}, respectively) are displayed in Table 2. Generally, concepts retrieved through domain

Table 1. Overview of topics and their associated number of (non-)interesting news items, accompanied by their inter-annotator agreements.

Topic	I+	I−	IAA+	IAA−	IAA
Asia or its countries	21	79	100%	98%	99%
Financial markets	24	76	75%	68%	72%
Google and its competitors	26	74	100%	95%	97%
Internet or Web services	26	74	96%	92%	94%
Microsoft or its competitors	29	71	100%	96%	98%
National economies	33	67	94%	85%	90%
Technology	29	71	86%	87%	87%
United States	45	55	87%	84%	85%
Average	29	71	92%	88%	90%

Table 2. Mean and variance of the weights for the semantic relationships.

	w_{super}	w_{sub}	w_{rel}
μ	0.330693	0.161386	0.500000
σ^2	0.126149	0.069594	0.147400

relationships seem to be of more importance than sub- and superclasses, and concepts retrieved through superclasses are of more importance than concepts retrieved through subclasses. Subclasses have a tendency to be too specific, while superclasses give more general information. Domain relationships on the other hand define properties of the original concept, and are hence more likely to be more valuable for extending concepts. For example, the concept 'Windows', an instance in the ontology, has the domain relationships hasUpdate 'Fall Creators Update' and isProducedBy 'MSFT'. These related concept are closely linked to the original concept 'Windows'. Its superclass 'Operating System' is too generic, and its subclass 'Windows Mobile' is too specific.

5.3 Experimental Results

Next, we evaluate the experimental results based on the optimized weights for each cut-off value. The precision, recall, and F_1 scores for the CF-IDF+, CF-IDF, and TF-IDF recommenders are plotted against the cut-off values in Figs. 1a–c. In terms of F_1, overall, the CF-IDF+ outperforms the original CF-IDF method, although for the cut-off values between 0 and 0.3, the difference between CF-IDF+ and CF-IDF is rather small, and at times, both concept-based recommenders are outperformed by TF-IDF.

The performance of the CF-IDF+ recommender is much more stable over all cut-off values, than the other two recommenders. The high F_1 scores are a result of a much higher recall, as CF-IDF(+) precision is lower than TF-IDF

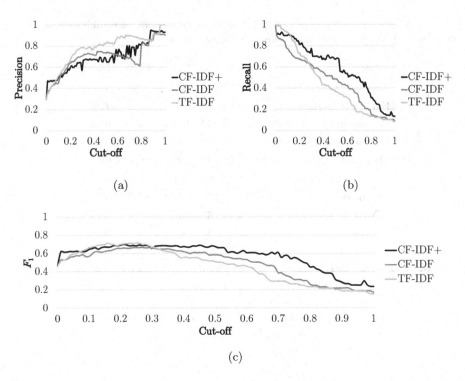

Fig. 1. Global precision, recall, and F_1 scores for the CF-IDF+, CF-IDF, and TF-IDF recommenders for varying cut-off values.

precision. CF-IDF+ clearly outperforms the TF-IDF and CF-IDF recommenders on the recall measure, except for very low cut-off values, when TF-IDF obtains a higher recall. The CF-IDF+ method has a lower precision than the TF-IDF recommender and a relatively similar precision to the CF-IDF recommender. Despite the, at times, disappointing precision scores, the potential benefits of CF-IDF+ remain evident. For low cut-off values, the CF-IDF+ recommender has a higher precision than TF-IDF due to the inherent noise removal by considering concepts rather than terms. The limited ontology quality (i.e., the coverage of concepts and their lexical representations) comes into play for higher cut-off values. Naturally, CF-IDF+'s lower precision could be remedied by improving the employed natural language processing pipeline, more precisely the word sense disambiguation (currently based on the rather basic adapted Lesk algorithm [2]).

Next, we evaluate the Receiver Operating Characteristic (ROC) curves for the CF-IDF+, CF-IDF, and TF-IDF recommenders, which plot recall against the percentage of uninteresting rated news items defined as interesting by the recommender, i.e., the False Positive Rate (FPR) that is computed as $1 -$ specificity (fall-out), while varying the cut-off value. The ROC curves in Fig. 2 show that the CF-IDF+ generally outperforms the CF-IDF recommender, and performs comparably to the TF-IDF recommender. The performances are also underlined

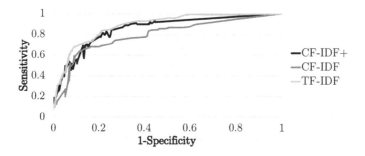

Fig. 2. ROC curves for the CF-IDF+, CF-IDF, and TF-IDF recommenders.

by the Area Under the Curve (AUC). For the ROC curves in Fig. 2, the AUC of the CF-IDF+, CF-IDF, and TF-IDF recommenders are 0.85072, 0.79162, and 0.88274, respectively.

The Kappa statistic [7] measures whether the proposed classifications made by the recommender are better than random guessing. The closer the statistic is to 1, the more classification power a recommender has. Figure 3 shows that the Kappa statistic for CF-IDF+ is significantly higher than for CF-IDF and TF-IDF. For the cut-off values between 0.08 and 0.30, the TF-IDF performs slightly better than the CF-IDF+ and CF-IDF recommenders, but overall, the CF-IDF+ has more classification power than the CF-IDF and TF-IDF recommenders.

The last curve that helps us to assess the performance of our CF-IDF+ recommender is the Precision-Recall (PR) curve. In general, a higher precision is associated with a lower recall, and vice versa. Figure 4 demonstrates that for the 3 recommenders, this also holds. However, even though TF-IDF was outperformed based on F_1, this is clearly not the case judging by the PR curves. The TF-IDF recommender outperforms both the CF-IDF+ and CF-IDF recommenders. The higher F_1 scores can be explained by the fact that most points of CF-IDF+ and CF-IDF are centered around the middle of the graph, resulting in higher F_1 scores due to the characteristics of the F_1 calculation. TF-IDF produces a more

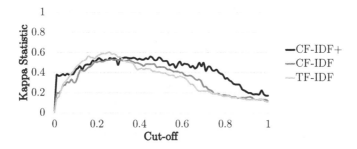

Fig. 3. Kappa statistics for the CF-IDF+, CF-IDF, and TF-IDF recommenders for varying cut-off values.

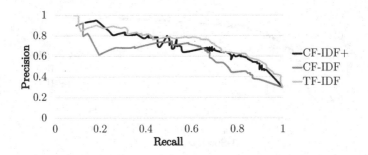

Fig. 4. PR curves for the CF-IDF+, CF-IDF, and TF-IDF recommenders.

evenly distributed set of precision and associated recall values. Overall, CF-IDF+ still performs notably better than the CF-IDF recommender. For example, for a recall of approximately 0.2, the CF-IDF method has a precision of about 0.6, while the CF-IDF+ has a precision higher than 0.9. Therefore, the CF-IDF+ recommender balances precision and recall in a better way than the CF-IDF recommender.

6 Conclusions

In our endeavours, we have aimed to improve the semantics-based CF-IDF (news) recommendation method that operates on ontologies, by additionally taking into account semantic inter-concept relationships, hereby extending the search space. For ontological classes, direct sub- and superclasses are included in the similarity analyses, and for ontological instances, concepts which are related by domain relationships are also taken into consideration. Additionally, we have optimized the semantic relationship weights based on the global F_1-scores.

When comparing our CF-IDF+ recommender to the original CF-IDF and baseline TF-IDF recommenders, we learned from the ROC curve, the Kappa statistic, and from precision, recall, and F_1-scores, that the CF-IDF+ shows a significant improvement over the original CF-IDF recommender. The CF-IDF+ recommender also performs better than the TF-IDF recommender in terms of recall, F_1, and the Kappa statistic. However, the TF-IDF recommender did have a better balance between fall-out and sensitivity, and precision and recall.

We envision various directions for future work. First, a more fine-grained (learning) approach to semantic relationship weight optimization would provide additional insights and could possibly enhance CF-IDF+ performance. Furthermore, we would like to investigate a larger collection of relationships. Now, we have considered the direct super- and subclasses, but hypothetically, non-direct super- and subclasses of concepts could be valuable as well. Last, a more thorough and powerful evaluation based on a larger set of news items would further underline the strong performance of CF-IDF+.

Acknowledgement. The authors would like to thank Tim Vos for his support and the fruitful discussions on this topic.

References

1. Alias-i: LingPipe 4.1.0 (2017). http://alias-i.com/lingpipe
2. Banerjee, S., Pedersen, T.: An adapted lesk algorithm for word sense disambiguation using WordNet. In: Gelbukh, A. (ed.) CICLing 2002. LNCS, vol. 2276, pp. 136–145. Springer, Heidelberg (2002). https://doi.org/10.1007/3-540-45715-1_11
3. Borsje, J., Levering, L., Frasincar, F.: Hermes: a semantic web-based news decision support system. In: 23rd Annual ACM Symposium on Applied Computing (SAC 2008), pp. 2415–2420. ACM (2008)
4. Capelle, M., Hogenboom, F., Hogenboom, A., Frasincar, F.: Semantic news recommendation using WordNet and Bing similarities. In: Shin, S.Y., Maldonado, J.C. (eds.) 28th Symposium on Applied Computing (SAC 2013), The Semantic Web and its Application Track, pp. 296–302. ACM (2013)
5. Capelle, M., Moerland, M., Frasincar, F., Hogenboom, F.: Semantics-based news recommendation. In: Akerkar, R., Bădică, C., Dan Burdescu, D. (eds.) 2nd International Conference on Web Intelligence, Mining and Semantics (WIMS 2012). ACM (2012)
6. Capelle, M., Moerland, M., Hogenboom, F., Frasincar, F., Vandic, D.: Bing-SF-IDF+: a hybrid semantics-driven news recommender. In: Wainwright, R.L., Corchado, J.M., Bechini, A., Hong, J. (eds.) 30th Symposium on Applied Computing (SAC 2015), Web Technologies Track, pp. 732–739. ACM (2015)
7. Cohen, J.: A coefficient of agreement for nominal scales. Educ. Psychol. Meas. **20**(1), 37–46 (1960)
8. Cunningham, H., Maynard, D., Bontcheva, K., Tablan, V.: GATE: a framework and graphical development environment for robust NLP tools and applications. In: 40th Anniversary Meeting of the Association for Computational Linguistics (ACL 2002), pp. 168–175. Association for Computational Linguistics (2002)
9. Fellbaum, C.: WordNet: An Electronic Lexical Database. MIT Press, Cambridge (1998)
10. Goossen, F., IJntema, W., Frasincar, F., Hogenboom, F., Kaymak, U.: News personalization using the CF-IDF semantic recommender. In: Akerkar, R. (ed.) International Conference on Web Intelligence, Mining and Semantics (WIMS 2011). ACM (2011)
11. IJntema, W., Goossen, F., Frasincar, F., Hogenboom, F.: Ontology-based news recommendation. In: Daniel, F., Delcambre, L.M.L., Fotouhi, F., Garrigós, I., Guerrini, G., Mazón, J.N., Mesiti, M., Müller-Feuerstein, S., Trujillo, J., Truta, T.M., Volz, B., Waller, E., Xiong, L., Zimányi, E. (eds.) International Workshop on Business intelligencE and the WEB (BEWEB 2010) at 13th International Conference on Extending Database Technology and 13th International Conference on Database Theory (EDBT/ICDT 2010). ACM (2010)
12. Jensen, A.S., Boss, N.S.: Textual similarity: comparing texts in order to discover how closely they discuss the same topics. Bachelor's thesis, Technical University of Denmark (2008)
13. Moerland, M., Hogenboom, F., Capelle, M., Frasincar, F.: Semantics-based news recommendation with SF-IDF+. In: Camacho, D., Akerkar, R., Rodríguez-Moreno, M.D. (eds.) 3rd International Conference on Web Intelligence, Mining and Semantics (WIMS 2013). ACM (2013)

14. Ostuni, V.C., Noia, T.D., Sciascio, E.D., Mirizzi, R.: Top-N recommendations from implicit feedback leveraging linked open data. In: 7th ACM Conference on Recommender Systems (RecSys 2013), pp. 85–92. ACM (2013)
15. Sussna, M.: Word sense disambiguation for free-text indexing using a massive semantic network. In: Bhargava, B., Finin, T., Yesha, Y. (eds.) 2nd International Conference on Information and Knowledge Management (CIKM 1993), pp. 67–74. ACM (1993)
16. Toutanova, K., Klein, D., Manning, C.D., Singer, Y.: Feature-rich part-of-speech tagging with a cyclic dependency network. In: Human Language Technology Conference of the North American Chapter of the Association for Computational Linguistics (HLTNAACL 2003), pp. 252–259 (2003)

The Cloud and Data Services

Model-Driven Elasticity
for Cloud Resources

Hayet Brabra[1,2]($\boxtimes$), Achraf Mtibaa[3], Walid Gaaloul[1],
and Boualem Benatallah[4]

[1] Telecom SudParis, UMR 5157 Samovar, Universite Paris-Saclay, Paris, France
`hayet.brabra@telecom-sudparis.eu`
[2] FSEG, Miracl Laboratory, Universiy of Sfax, Sfax, Tunisia
[3] ENETCOM, Miracl Laboratory, Universiy of Sfax, Sfax, Tunisia
[4] UNSW, Sydney, Australia

Abstract. Elasticity is a key distinguishing feature of cloud services. It represents the power to dynamically reconfigure resources to adapt to varying resource requirements. However, the implementation of such feature has reached a level of complexity since various and non standard interfaces are provided to deal with cloud resources. To alleviate this, we believe that elasticity features should be provided at resource description level. In this paper, we propose a Cloud Resource Description Model (cRDM) based on State Machine formalism. This novel abstraction allows representing cloud resources while considering their elasticity behavior over the time. Our prototype implementation shows the feasibly and experiments illustrate the productivity and expressiveness of our cRDM model in comparison to traditional solutions.

Keywords: Elasticity · Cloud resources · State machine
Orchestration

1 Introduction

Elasticity is one of the main assets characterizing the cloud services. It is achieved through invocation of reconfiguration actions that run as a result of events, allowing a controller to automatically (re)configure cloud resources. However, exploiting such feature poses a great complexity due to the proliferation of tools that offer heterogeneous resource orchestrations and elasticity services [6]. Existing cloud orchestration and elasticity solutions rely on procedural programing languages (e.g., most of them are based on low-level scripting) to support the elasticity of cloud resources [9,15,17,18]. Prominent examples include: Puppet, Docker, Cloudify, AWS AutoScaling and IBM AutoScale [4,5,18]. This diversity implies that cloud programmers are forced to be aware of different low-level cloud service APIs, command line syntax and procedural programming constructs, to describe and control the elasticity of cloud services. Moreover, this

© Springer International Publishing AG, part of Springer Nature 2018
J. Krogstie and H. A. Reijers (Eds.): CAiSE 2018, LNCS 10816, pp. 187–202, 2018.
https://doi.org/10.1007/978-3-319-91563-0_12

problem worsens as the variety of cloud services and the variations of application resource requirements and constraints increase [15,17]. Not only that, the emergence of federated cloud makes this problem too unreasonably complicated.

A more effective solution should allow users to specify their resources and elasticity features regardless the technical specifications of any cloud provider or orchestration tool. We believe that elasticity features should be provided at resource description level. Instead of relying on low-level scripting mechanisms or provider-specific rule engines, we argue that models and languages for describing cloud resources should be endowed with intuitive constructs that can be used to specify a range of flexible elasticity mechanisms.

Motivated by these considerations, we propose, in this paper, a cloud Resource Description Model (cRDM) to describe cloud resources and their elasticity. Our model is based on a new abstraction that we call Cloud resource requirement State Machine (C-SM), which allows representing cloud resources while considering their elasticity behaviour. We adopt state-machine formalism as it provides refreshing graphical notations in contrast with existing largely text-based solutions. Indeed, state-based model has been commonly used to model the behaviour of systems due to the fact that it is simple and intuitive. In this model, states will be used to characterize application specific resource requirements, when they are needed while transitions between states are triggered when certain conditions are satisfied. Transitions automatically trigger controller actions to perform the desired resource (re-)configurations to satisfy the requirements of target states. Our model has been prototypically implemented and evaluated using experiments with real cloud use cases illustrating its productivity and expressiveness in comparison to traditional solutions.

The paper is structured as follows: Sect. 2 articulates the motivations of our model. In Sects. 3 and 4, we present our model. Section 5 describes our implementation and evaluation. In Sects. 6 and 7, we present related work and conclusion.

2   Limitations in Current Cloud Description and Elasticity Solutions

In this section, we investigate through a motivating example, specific limitations among existing cloud resources description and elasticity solutions.

Motivating Scenario. Consider a cloud user wants to specify resource requirements and constraints for deploying an e-commerce application, which consists of a Mongo database, a NodeJS server and a Java script application (Fig. 1(a)). The user selects Amazon web services (AWS) to deploy this application and specifies that she needs 5 virtual machine (VM) instances to be hosted in for one year from Monday, 1st of January 2018. Each instance has 8 GB RAM and 4 GHz CPU. As QoS Constraint, the user would like that the availability for each instance must be at least 99%. Moreover, when the average CPU usage for 5 min is greater than or equal 80%, she wants 5 more instances to be added from AWS. In contrast, these 5 instances should be removed whenever the average CPU usage is less than 20%. However, during the business spikes every weekend, whenever

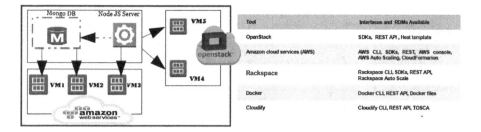

Fig. 1. (a) Cloud Resources for deploying an E-commerce application; (b) List of available cloud description and orchestration Interfaces

the application reaches 10 instances in AWS, she wants to horizontally scale out into another cloud like Openstack by adding 10 instances. Accordingly, the user first needs to describe her application, required resources and dependencies between them. Secondly, she needs to specify elasticity policies that control the application at runtime. Finally, if she wants to use cloud resources from other providers such Rackspace or Openstack, she requires to carry out some adaptations in the application and elasticity policies to make it compatible with the target management interfaces. Additionally, she can use an orchestration tool such Cloudify or Docker to support the multi-providers deployment. However, as shown in Fig. 1(b), AWS, Openstack, Rackspace, Docker and Cloudify provide heterogeneous resource description models (RDMs) and management interfaces and rely on low-level script-based APIs.

Based on these observations, we concluded that existing cloud resources description and elasticity solutions (1) are rarely transparent and adaptive to support the management of resources across various providers; (2) oblige users to acquire new expertise in multiple RDMs and different elasticity mechanisms; and (3) lead to a costly environments and potential vendor lock-in as exploiting resources from a new provider demands extensive programming effort [9].

3 Identifying Basic Cloud Resources Abstractions

We identify the required modeling abstractions that allow users to specify their cloud resources and their elasticity policies. To do so, we first performed a domain analysis on cloud resources from several providers, including AWS, Openstack, OCCI and TOSCA. We selected these services as they represent the range of different types of cloud services available: commercial offer, open source implementation, and open standard. Then, we analyzed the different RDMs used by the cloud orchestration tools, including, Cloudify and Docker and the elasticity solutions used in these tools and cloud solutions cited above. Intentionally, we want a model that is very simple, so that it could help us start from a minimal base and progressively extend it as needed. In this section, we focus on the basic abstractions to be included when using cloud resources from one provider.

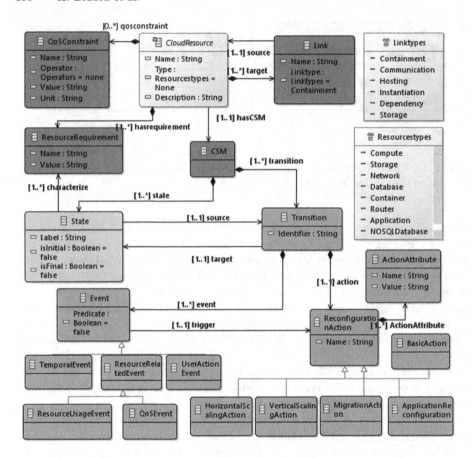

Fig. 2. UML class diagram for the Cloud Resource Description Model

3.1   Abstractions Overview

Figure 2 represents the conceptual UML model illustrating the main modeling abstractions of our cRDM. For illustration purpose, we rely on the motivating example previously described, which consists in deploying an E-commerce application both on AWS and OpenStack. As shown in Fig. 2, CloudResource is the root concept, which is described by a Name that indicates a resource name and Type that indicates its type (e.g. Compute, Database, etc.) and a set of other optional attributes (e.g. Description). Moreover, it is associated by four main concepts. **Resource Requirement** includes the name-value pairs describing the requirement attributes associated with the resource, such as CPU, RAM, etc. **Link** represents relationship between cloud resources. It is defined by a Name (e.g., hosted-in), Linktype (e.g. Containment, Communication, Hosting, etc.) and source and target participants resources. **QoS Constraint** represents a constraint on a particular QoS metric. It is defined through Name that indicates the QoS metric, Operator that indicates the comparison operator, Value that indicates the metric value, and Unit that indicates the measurement unit. *C-SM* is a Cloud Requirement State Machine that aims at capturing the elasticity behavior of a cloud resource over its life cycle.

Example. Figure 3 shows the corresponding cRDM instance of the motivation example. It consists of four cloud resources instantiated from the CloudResource concept: Node-JS, NodeBookshop, Mongo-Database and VM1. For example, VM1 is a virtual machine defined by the name vmcompute and type Compute and represents a host for Node-JS and Mongo-Database resources. This was accomplished by hosted-in links between these resources. Besides, the VM1 is associated with VM1-requirements, VM1-QoS and VM1-CSM. VM1-requirements indicates that VM1 should have 4 GB RAM, 4 GHz CPU, 5 instances and is acquired from AWS provider. VM1-QoS indicates that VM1 availability should be at least 99%. VM1-CSM specifies the elasticity behaviour related to VM1.

3.2   States

A state has a string label and two attributes (isInitial and isFinal) indicating whether the state is a final or initial. It is also associated with the ResourceRequirement concept to characterize the application-specific resource requirements in that state. As illustrated in the Fig. 3, VM1-CSM instance consists of four states S1, S2, S3 and S4 that VM1 may go through during its life cycle where S1 is the initial state and S4 represent the final state. Each state is annotated with resource requirements that should be satisfied under it. For example, the state S1 can only be reached if 5 instances of VM1 from AWS have been started.

3.3   Transitions

We use transitions to express the elasticity behaviour of a cloud resource. Each transition is labeled with a re-configuration policy. A reconfiguration policy aims at automatically triggering a reconfiguration action as result of events to satisfy the requirements of the target state. Conceptually, as shown in Fig. 2, each transition is specified by Identifier that indicates the transition identifier, source that indicates the source state and target that indicates the target state and consists of one or more Event and one Reconfiguration Action. Thus, cloud users can annotate the transitions with those events and reconfiguration actions to model reconfiguration policies of cloud resources without referring to any low-level scripting mechanisms or provider specific policy engines.

3.3.1   Events

We distinguish three event types to trigger a reconfiguration action: Temporal Events, Resource Related Events and User Action Events (see Fig. 2). Each event is specified via a logic predicate expressed using the common grammar used in the traditional state machine [16].

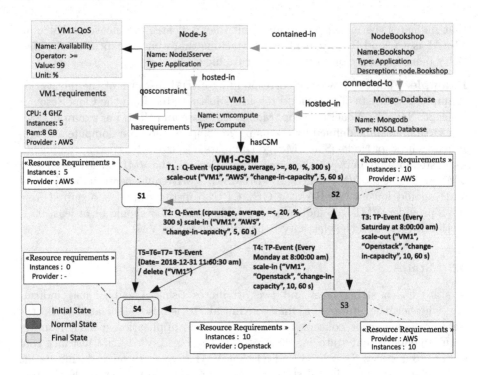

Fig. 3. cRDM instance provided by the cloud user for the use case

Temporal Events. Actions may require that certain temporal events occurred, to be executed. We identify two patterns that these events can take: specified date and periodicity patterns as they are the most used in practice [1,2]. The specified date pattern specifies that an action needs to be executed at a specified date. It defined through a predicate called TS-Event (c) with c defined as **c::= (Date = D)** | **c ∧ c** | **c ∨ c**, where Date is a clock, and D is an absolute date expressed as yy-mm-dd hh-mm-ss am/pm format. As shown in Fig. 3, TS-Event (Date=2018-12-31 11:60:30 am) within the transition T5, T6 and T7 represents a temporal event expressed using this pattern, which will be triggered on 2018-12-31 at 11:60:30 am. Furthermore the periodicity pattern specifies that a certain action needs to be executed following a certain periodicity rule over time, which is defined using TP-Event (p) predicate with p is defined as **p::=D** | **p ∧ p** | **p ∨ p**, where D is a date defined following a periodicity rule and expressed using the following EBNF notations.

```
<D>::= 'Every' <weakday>'-'{<weakday>'-'} 'at' <Time>
|'Everyday' 'at' <Time> 'Except' <weakday>'-'{<weakday>'-'}
<weakday>::= Monday| Tuesday |....| Sunday
<Time>::= <Hour>':'<Minute>':'<Second>'am'|'pm'
...........
```

For example, as shown in Fig. 3, TP-Event (Every Saturday at 8:00:00 am) within T3 represents a temporal event, expressed using the above notations, which shall be triggered every Saturday at 08:00:00 am.

Resource Related Events. Actions need to be executed once certain resource metrics reach a predefined threshold. This type of event includes two sub-types of event that we call Resource Usage Events and QoS Events. Both event sub-types are defined using the predicate Q-event(q, fc, op, tr, u, w), where q is a metric, fc precises the metric evaluation way (average, maximum, etc.), op $\in \{=, \neq, <,$ etc.$\}$, tr is a threshold value, u is the used unit, and finally w defines a time window during which the metric could be evaluated. Resource Usage Events are defined using metrics related to the resource usage such as CPU usage, RAM usage, etc. While QoS events are expressed through QoS metrics such as availability, response time, etc. As shown in Fig. 3, the user defines their resource related events in terms of CPU usage, such as Q-Event (cpuusage, average, $>=$, 80, %, 300 s) within T1, which checks whether the CPU usage average is greater than or equal 80% for 5 min across all VM instances.

User Action Events. Actions are executed at the behest of a cloud user. For instance, a cloud user can demand to set manually the capacity of the VM resource. These events are defined through a predicate called U-Event(c) over a set of messages M, with c defined as **c::=(message=e) | c ∧ c | c ∨ c**, where message is an incoming message from a user and e $\in$ M. For example, U-Event (message=Stop) will be triggered when we receive from user a stop message.

3.3.2   Reconfiguration Actions

They specify how a cloud resource should behave when certain events occurs. To identify them, we examined the reconfiguration and elasticity mechanisms analyzed on research surveys [4,5] and proposed by cloud providers and orchestration tools. We organize them into five categories: horizontal scaling, vertical scaling, migration, application reconfiguration and basic actions. Each reconfiguration action is defined by a name and a set of attributes defining the required inputs to execute this action. In the following, we choose JSON schema to illustrate each action.

Horizontal Scaling (HS): represents the possibility to scale out and in by adding or removing instances (e.g. VM). As shown in Fig. 4(a), HS action can have scale-in or scale-out name and contains 4 attributes. The resource-target represents the resource name that will be adjusted. The adjustment-type specifies the adjustment way which can be change-in-capacity (Add/Remove the given number of resource instances), exact-capacity (Set the current number of resource instances) or percent-change-in-capacity (Add/Remove a given percentage to the instances number). The adjust specifies the adjustment value. Finally, the cooldown indicates the time period during which no other actions on the same resource will be taken.

Vertical Scaling (VS): aims at scaling up and down of resources such as processing, memory. As shown in Fig. 4(b), VS action can have scale-up or

```
(a) { "HorizontalScalingAction" : {"Name :"{"type":        (b) { "VerticalScalingAction" : {"Name": {"type" :
"enum['scale-in', 'scale-out']"},                          "enum ['scale-up', 'scale-down']"},
   "ActionAttributes":{                                       "ActionAttributes":{........,
      "resource-target":{"type":"string"},                      "attribute-target":{"type": "string"},
      "adjustment-type":{"type":"enum['exact-capacity',         ...., ....., ....., ........ } }
      'change-in-capacity', 'percent-change-in-
capacity']"},                                              (d) { "ApplicationReconfigurationAction" : {"Name":
      "adjust":{"type": "string"},                         "update"},
      "cooldown":{"type": "number"} } }                       "ActionAttributes":{
                                                                 "resource-target":{"type": "string"},
(c) { "MigrationAction" : {"Name": "migrate"},                   "attribute-target":{"type": "string"}
   "ActionAttributes":{                                          "attribute-value":{"type": "string"} } }
      "target":{"type": "string"},
      "host":{"type": "string", "default": "None"},       (e) { "BasicAction" : {"Name": {"type" : "enum ['start',
      "type":{"type": "enum ['Cold', 'Hot']"},            'stop', 'delete', 'restart']"},
      "cooldown":{"type": "number"} } }                      "ActionAttributes":{
                                                                 "resource-target": {"type": "string"} } }
```

Fig. 4. A JSON schema describing the main attributes for each reconfiguration action

scale-down name and has same attributes of HS action. Moreover, we add the attribute-target to indicate the attribute name (e.g. CPU, RAM) to be modified.

Migration: includes two migration types: VM Migration and Application migration. In this category, we identify only one action which is a Migration action. As shown in Fig. 4(c), a migrate action has migrate name and contains a set of attributes: The target represents the VM or application component name that will be migrated. The host which can be filled by the host name (e.g. The node name in case of a VM migration or the VM name in case of an application component migration). In some cases, it can be filled by None, so the controller action must choose the appropriate host automatically. The type indicates the migration type: Cold or Hot [4].

Application Reconfiguration (AR): consist of changing specific application aspects such as DB recovery policy. As shown in Fig. 4(d), this action has as name update and contains a set of attributes: the resource-target represents the application component name; the attribute-target indicates the attribute name to be modified and the attribute-value indicates the new value to be assigned.

Basic Actions: regroup the basic actions applied on a cloud resource including: start, stop, restart and delete. As shown in Fig. 4(e), each basic action should contain a Name that should be start, restart, stop, or delete and resource-target as attribute that indicates the resource to be manipulated.

Example. As shown in Fig. 3, the user used both the HS and basic actions: scale-out ("VM1", "AWS", "change-in-capacity", 5, 60 s) within T1 represents an instance from HS action, which allows to add for VM1 resource 5 instances from AWS provider, where 60 s represent a clowdown period that must be respected after triggering this action. delete (VM1) within T5, T6, T7 represents a basic action that aims at removing permanently the resource VM1.

4   Supporting Multi-providers Abstractions

We now present how the above abstractions can be extended in order to support the multi-provider scenario. More specifically, we need to identify required events leading a user to acquire cloud resources from a new provider and correspondingly extend the reconfiguration actions. In fact, once a service has been deployed in a cloud resource from a specific provider, different situation can occur at runtime: (1) Service can be scaled manually (User Action Events) or dynamically (Resource Related or Temporal Events) from a new provider; (2) Service can be migrated to another provider at the behest of its user (User Action Events); (3) Service can be migrated in case that the current provider does not respect the QoS constraints (Resource Related Events); (4) finally, service can be migrated to another provider when this provider offers better utility than the previous one. Consequently, our basic resource model should be extended to support these new considerations. To support (1), (2) and (3) we have to extend the reconfiguration actions (i.e. horizontal and vertical scaling, migration, etc.) by adding the property provider as an attribute to these actions. For instance, the scale-out ("VM1", "Openstack", "change-in-capacity", 10, 60 s) within T3 shows HS action from a new provider which is OpenStack. Furthermore, to support (4), a new type of events should be defined along with the above events, that we call Market related events. As Market related events depend on QoS and resource properties, we define it as one of the Resource Related Event.

Market Related Events. Events can be triggered whenever there is a cloud resource offer providing QoS or any other property (e.g., price) that better satisfies the user needs. We expressed it using the predicate M-Event (q, r, op, tr, u, p), where q is the QoS or resource property, r is the cloud resource, and op, tr and u have the same definition within Q-event. While p indicates the new provider which can be filled by the provider name such Openstack, or by any, so, the controller action must automatically choose the most appropriate provider.

5   Implementation and Evaluation

5.1   Proof of Concept

We built a proof-of-concept (POC) prototype for cRDM, called *cRDM Core*[1], which supports users to describe and configure their cloud resources by exploiting the underlying cloud orchestration and providers solutions. Figure 5 shows an overview of our prototype architecture. cRDM core is the central part in this architecture, consisting in: *cRDM editor, cRDM Validation, cRDM Generation* and *Elasticity controller.* The *cRDM editor* implemented using Sirius technology[2] and Java to provide a drag-and-drop interface enabling user to graphically instantiate from cRDM model the corresponding cRDM instance. This instance is then simply serialized as JSON/XMI file. *cRDM Validation* exploits this file to

---

[1] http://www-inf.it-sudparis.eu/SIMBAD/tools/Cloud-RDM/.
[2] https://www.obeodesigner.com/en/product/sirius.

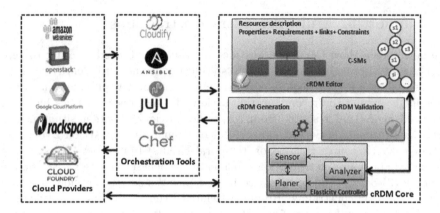

Fig. 5. Architecture overview

check the user cRDM instance consistency by verifying both its syntax and structure. The *cRDM Generation* proceeds to generate all required files for ensuring the deployment task. This generation is based on a model-driven generation technique and a set of connectors that serve to interpret the high-level descriptions related to cloud resources and identify their low-level scripts and commands required to manage and configure them. While the generation phase is important, it is out of the scope of this paper due to the space limitation. Finally, the *Elasticity Controller* is implemented in Java and allows the execution of elasticity policies. From the defined C-SM machines, the elasticity controller detects the needed information to be monitored and identifies the appropriate reconfiguration actions to adapt the cloud resources at runtime.

5.2 Evaluation

We conducted two experiments using the implemented prototype to evaluate the productivity and expressiveness of our cRDM in comparison with traditional solutions. All experiment objects and results have been published online (see footnote 1).

5.2.1 Experimentation 1

Experimental setup. We evaluated the productivity of our cRDM by conducting a user-study with 18 participants from Master students of the university of Paris-Saclay. We test the productivity in terms of the efficiency and the usefulness of cRDM in describing the cloud resources and their elasticity behaviour. The *efficiency* is measured in terms of the time taken to complete the modeling and deployment tasks. The *usefulness* is determined via a questionnaire that asses the participants feedbacks about our cRDM. The questionnaire (see footnote 1) devised into four main parts: Background, Functionality, Usability, Insights/Improvements. The background questions aim at evaluating the participants familiarity with existing cloud resource description and elasticity tools.

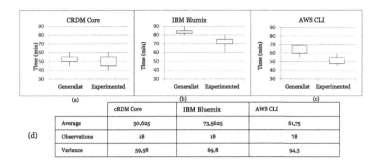

Fig. 6. Time to complete the task; (a), (b) and (c) Time grouped by level of expertise; (d) Average time for all the participants.

The functionality questions verify whether the participants correctly understand the main functionalities of cRDM. The usability questions sought to discover whether the key modeling abstraction offered are easy and intuitive. We asked all participants to model and deploy the application described in our motivation scenario only on AWS provider using our cRDM Core. For quantitative comparison purpose, we asked them to do the same scenario with two provider-specific solutions viz. IBM bluemix platform [2] and AWS CLI [1]. For the sake of analysis, we classified a total of 18 participants into 2 main groups: (1) Generalist: who have average knowledge of cloud tools (12 participants) and (2) Experimented (6 participants): who have a sophisticated understanding of cloud tools.

Evaluation Results. Results of the experiment in Fig. 6(a), (b) and (c) show the time taken using cRDM core, IBM bluemix and AWS CLI for the modeling and deployment tasks. As shown, it was pleasantly surprising that even generalist participants demonstrated a significant reduction in time. More precisely, the time taken to complete these tasks was reduced by 17% in comparison to other solutions. On the whole, as shown in Fig. 6(d), the participants took on average 50 min using our tool, 73 min using IBM bluemix and 61 min using AWS CLI. This demonstrates the efficiency of our cRDM model. In fact, we argue that provider independent resource abstractions and graphical model like state machine to describe the elasticity behaviour significantly improve the time-to-modeling. In contrast, proprietary solutions such AWS CLI and IBM bluemix necessarily demand extensive programming and documentation efforts. More specifically, by using AWS CLI to deploy the requested application, participants are inevitably forced to understand firstly AWS CloudFormation model to create the corresponding template for both describing the required resources and their scaling scripts, which also like AWS CLI provides low-level resource descriptions. Likewise, participants are invited to understand diverse DevOps tools such cloud foundry CLI when using IBM bluemix.

Moreover, to evaluate the cRDM usefulness, we use the Usability section of the questionnaire by asking participants to rate the usability for each abstraction (scale 0–5). We examined the basic cRDM abstractions: Cloud Resource,

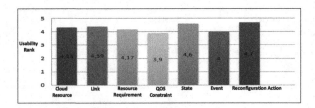

Fig. 7. Usability rate of the main cRDM abstractions

Link, Resource Requirement, QoS Constraint, State, Reconfiguration actions. We observed that the mean score for all abstractions in Fig. 7 is greater than the neutral value of 3 with a noticeable difference. Overall participants reported that our model is a familiar and intuitive, especially C-SM is not far from natural language and allows defining cloud resources elasticity behaviour in a very simple and easy way. Accordingly, giving these observations, we confirm that the key modeling abstractions offered are useful and comprehensible.

5.2.2   Experimentation 2

To evaluate the expressiveness of our cRDM, we used the two well known evaluation techniques in the literature [11]: Comparison of the model with standards or other proposed models; Application of the model to realistic examples or use cases. The former provides a qualitative evaluation, while the latter quantitatively evaluates our model.

Qualitative Evaluation. Herein, we test cRDM expressiveness by evaluating its overall coverage to others RDMs. We choose two RDMs: TOSCA [3] and AWS Cloud Formation [1]. We intentionally choose these solutions as they represent the range of the different types of RDMs available in cloud community: Provider-specific model and open standard. Concretely, we compared the concepts defined in the different models to the ones defined in our cRDM. The results summarized in Table 1(a) show that our cRDM model has a high coverage of TOSCA (80%) concepts related to cloud resources. In contrast, we observe

Table 1. cRDM coverage to (a) RDMs; (b) TOSCA use cases; (c) AWS CF use cases

RDMs	Concepts	Covered	Rate
TOSCA	26	21	80 %
AWS Cloud Formation	9	3	33%
Average			56.5%

(a)

	TOSCA use cases			
	case 1	case 2	case 3	case 4
cRDM instances	25	15	41	125
Generated TOSCA instances	36	18	64	172
TOSCA instances	36	18	64	172
Coverage by the generation	100%	100%	100%	100%

(b)

	AWS CF use cases			
	case 1	case 2	case 3	case 4
cRDM instances	26	36	66	64
Generated AWS CF instances	70	80	102	130
AWS CF instances	70	80	102	130
Coverage by the generation	100%	100%	100%	100%

(c)

a low coverage of AWS Cloud Formation concepts (33%). This is not surprising because we intentionally aimed to hide low-level and technical descriptions related to any provider-dependent solutions. By analyzing AWS Cloud Formation, we reveal that it provides user 9 main concepts to create and configure their cloud resources. However, only three concepts have inevitably to be included in the user template to specify the resources and their properties: Resources, Type and Properties. Therefore, in our cRDM, we support only these three concepts. Others Cloud Formation concepts include Metadata, Packages, Mapping, Transform, are totally omitted in our cRDM with the aim of avoiding any handling with low-level commands and hard coded scripts. Similarly, we only supported TOSCA concepts that are most commonly used in resource descriptions and do not oblige the user to deal with low-level scripting complexity like Interface and Artifact concepts. In the following, we show that the no-support of the low-level scripting at resource description level, does not reduce our cRDM expressiveness.

Quantitative Evaluation. In order to quantitatively evaluate our cRDM expressiveness, we rely on 8 use cases identified in the different chosen models: TOSCA (4 use cases) and AWS Cloud Formation (4 use cases). These use cases are specified as JSON and YAML templates. We use our cRDM core to construct the corresponding cRDM instances for these use cases using our cRDM model. Afterward, we exploit these instances to generate TOSCA and AWS Cloud Formation templates using our generation technique. Finally, we evaluate our cRDM expressiveness in terms of its coverage to use cases, by comparing the concepts instances in generated templates with the ones in use cases templates. Table 1(b) (respectively Table 1(c)) reports, for each TOSCA (respectively AWS CF) use case, the number of obtained instances using cRDM model, the number of obtained instances after the generation, and the number of instances that really exist in TOSCA (respectively AWS CF) use case as well as the coverage percentage of our cRDM after the generation. The obtained results show that we have a complete coverage (100%) of all used use cases, this meaning that all used use cases have been successfully covered when using our cRDM model during the generation. Therefore, we confirm that the no-support of low-level descriptions of cloud resources in our cRDM model does not have any influence on its expressiveness. On the contrary, this leading to a considerable reduction in the modeling complexity. For example, as specified in Table 1(b), to model the first TOSCA use case, we made only 25 instances using our cRDM. While using TOSCA model, this use case is composed of 36 instances since it includes low level and specific descriptions related to each defined cloud resource. To overcome this loss, we exploit the 25 instances created by our cRDM to interpret the required low-level scripts and commands. Therefore, we generate 11 additional instances. This means that the complexity modeling was reduced by 31%

Table 2. MCR rate for TOSCA and AWS CF uses case

	TOSCA use cases				AWS CF use cases			
	Case 1	Case 2	Case 3	Case 4	Case 1	Case 2	Case 3	Case 4
MCR rate	31%	17%	36%	28%	62%	55%	36%	51%

(i.e. $1-(25/36)$). Table 2 reports the modeling complexity reduction rate (MCR rate) for each use case using our cRDM model.

5.3 Threats to Validity

Some potential threats to validity exist in our contribution. First, since we relied on cloud orchestration tools and providers-specific solutions, any change in these solutions oblige us to continually update our connectors to ensure a compliant generation. Second, we validated our model productivity through a cloud use case that supports only the horizontal scaling as it represents the most widespread mechanism in the cloud. Third, our use case has been conducted with 18 students and support only providers-specific solutions such AWS CLI and IBM Bleumix. We believe that a larger number of participants, including professionals and a use case supporting complex elasticity scenarios such as VM/Application migration, need to be considered. Fourth, we validated the productivity and expressiveness of our cRDM model. Our work request a further evaluation, including the correctness of our elasticity controller, which represents a good evidence that the defined state machines behave as expected at runtime. Furthermore, the performance evaluation of the whole system as well as a comparison with orchestration tools like Docker and Cloudify will also be considered.

6 Related Work

Various cloud resource and elasticity description languages and models have been proposed in industry and research. Market-leading providers like AWS CloudFormation [1] and CA AppLogic aim to describe and deploy complete application stacks. They propose provider-specific representations while our approach allows the description of cloud resources and their elasticity policies in a provider-independent way. Modern resource orchestration systems like Puppet, Juju, Ansible and Chef provide scripting-based languages for describing resource configurations over cloud services [18]. However, even sophisticated programmers are regularly forced to understand different low-level cloud APIs to create and maintain complex resource configurations and describe their elasticity policies. In contrast to the above, we contribute in providing high-level modeling abstractions that facilitate the description of cloud resources as well as their elasticity without referring to any low-level languages or providers-specific formats.

In research, to the best of our knowledge, cloud resources and their elasticity description have been studied separately. In [13], the authors introduce a description and deployment model called "Virtual Solution Model" that models a resource as a provider-independent resource configuration. However, this model allows users to describe resource configurations from a single provider for a single deployment. Moreover, various research works have concentrated on using semantic-based languages [6] to describe cloud resources. However, the largest amount of researcher's attention have been largely focused on cloud resources discovery and selection. Although there are a few of research work [8,10] that have recently emerged to deal with the orchestration aspects, none of them consider elasticity policies that are required at control runtime. Other related works have only focused on the definition and implementation of the cloud resource elasticity [7,12,14,19]. They have based on domain specific languages to define the elasticity policies for cloud resources. Our model differs from these works in four main points: (i) We based our model on state machine to describe the elasticity behaviour of a cloud resource, the state machine model considerably simplifies the manner of understanding compared to textual notations provided by the most of the previous work; (ii) We support multiple types of event, in contrast to the above solutions, which support only resource-usage based events; (iii) We support multiple type of elasticity actions, while the previous approaches support only vertical and horizontal scaling; (iv) Finally, our model allows to manage elasticity across multiple clouds, in contrast to these works, which define elasticity rules for scaling resources from only one provider.

7   Conclusion

This paper proposed a novel Cloud Resource Description Model based on a state machine, that provides high-level abstractions to describe cloud resources and their elasticity features. The adoption of state machine to describe the elasticity behaviour hides the actual complex implementation within cloud orchestration and providers tools. The model has been implemented and evaluated using experiments showing significantly its productivity and expressiveness. Future work will focus on evaluating first the correctness of our elasticity controller, and then the performance of the whole orchestration system. Besides, we plan to specify the formal semantics of our cRDM model.

References

1. AWS CLI: AWS Cloud Formation. https://aws.amazon.com/documentation/
2. IBM bluemix platform. https://www.ibm.com/cloud-computing/bluemix/
3. TOSCA. https://www.oasis-open.org/committees/tosca/
4. Al-Dhuraibi, Y., et al.: Elasticity in cloud computing: state of the art and research challenges. IEEE Trans. Serv. Comput. (TSC) (2017)
5. Naskos, A., Gounaris, A., Sioutas, S.: Cloud elasticity: a survey. In: Karydis, I., Sioutas, S., Triantafillou, P., Tsoumakos, D. (eds.) ALGOCLOUD 2015. LNCS, vol. 9511, pp. 151–167. Springer, Cham (2016). https://doi.org/10.1007/978-3-319-29919-8_12
6. Brabra, H., et al.: Semantic web technologies in cloud computing: a systematic literature review. In: IEEE SCC, pp. 744–751 (2016)
7. Copil, G., Moldovan, D., Truong, H.L., Dustdar, S.: SYBL: An extensible language for controlling elasticity in cloud applications. In: International Symposium on Cluster, Cloud, and Grid Computing, pp. 112–119 (2013)
8. Dastjerdi, A.V., et al.: Cloudpick: a framework for QoS-aware and ontology-based service deployment across clouds. Softw. Pract. Exper. 45(2), 197–231 (2015)

9. Weerasiri, D., Barukh, M.C., Benatallah, B., Cao, J.: A model-driven framework for interoperable cloud resources management. In: Sheng, Q.Z., Stroulia, E., Tata, S., Bhiri, S. (eds.) ICSOC 2016. LNCS, vol. 9936, pp. 186–201. Springer, Cham (2016). https://doi.org/10.1007/978-3-319-46295-0_12

10. Dimitris, G.: A semantic framework to support the management of cloud-based service provision within a global public inclusive infrastructure. Int. J. Electron. Commer. 20(1), 142–173 (2015)

11. Horkoff, J., Aydemir, F.B., Li, F.-L., Li, T., Mylopoulos, J.: Evaluating modeling languages: an example from the requirements domain. In: Yu, E., Dobbie, G., Jarke, M., Purao, S. (eds.) ER 2014. LNCS, vol. 8824, pp. 260–274. Springer, Cham (2014). https://doi.org/10.1007/978-3-319-12206-9_21

12. Jrad, A.B., Bhiri, S., Tata, S.: Description and evaluation of elasticity strategies for business processes in the cloud. In: SCC, pp. 203–210 (2016)

13. Konstantinou, A.V., et al.: An architecture for virtual solution composition and deployment in infrastructure clouds. In: International Workshop on Virtualization Technologies in Distributed Computing, pp. 9–18 (2009)

14. Kritikos, K., et al.: SRL: A scalability rule language for multi-cloud environments. In: International Conference on Cloud Computing Technology and Science (2014)

15. Liu, C., Loo, B.T., Mao, Y.: Declarative automated cloud resource orchestration. In: Proceedings of the 2nd ACM Symposium on Cloud Computing (SOCC) (2011)

16. Ponge, J., et al.: Analysis and applications of timed service protocols. ACM Trans. Softw. Eng. Methodol. 19(4), 1–38 (2010)

17. Ranjan, R., Benatallah, B.: Programming cloud resource orchestration framework: operations and research challenges. CoRR (2012)

18. Thomas, D., Wouter, J., Bart, V.: A survey of system configuration tools. In: International Conference on Large Installation System Administration, LISA (2010)

19. Zabolotnyi, R., et al.: SPEEDL-a declarative event-based language to define the scaling behavior of cloud applications. In: IEEE World Congress on Services (2015)

Fog Computing and Data as a Service: A Goal-Based Modeling Approach to Enable Effective Data Movements

Pierluigi Plebani, Mattia Salnitri$^{(\boxtimes)}$, and Monica Vitali

Dipartimento di Elettronica Informazione e Bioingegneria, Politecnico di Milano,
Piazza Leonardo da Vinci, 32, 20133 Milano, Italy
{pierluigi.plebani,mattia.salnitri,monica.vitali}@polimi.it

Abstract. Data as a Service (DaaS) organizes the data management life-cycle around the Service Oriented Computing principles. Data providers are supposed to take care not only of performing the life-cycle phases, but also of the data movements from where data are generated, to where they are stored, and, finally, consumed. Data movements become more frequent especially in Fog environments, i.e., where data are generated by devices at the edge of the network (e.g., sensors), processed on the cloud, and consumed at the customer premises.

This paper proposes a goal-based modeling approach for enabling effective data movements in Fog environments. The model considers the requirements of several customers to move data at the right time and in the right place, taking into account the heterogeneity of the resources involved in the data management.

Keywords: Data movement · Fog Computing · Decision system
Goal-based model

1 Introduction

The adoption of Service Oriented Architectures [18] has changed the way in which capabilities of an information system are offered and consumed. Although a gap exists between the initial expectations of this research domain and the actual adoption [6] (e.g., about automatic service composition), the Cloud Computing paradigm – where everything is offered as a service – has demonstrated that service orientation has a significant value for both consumers and providers. Nevertheless, limited attention has been paid, so far, to the link between the service oriented paradigm and data management. There are some approaches, under the umbrella of the so-called Data Base as a Service (DBaaS, a.k.a., Cloud Databases), concerning how to provide DBMS functionalities according to the Cloud Computing paradigm [1][1]. Actually, data management's scope is wider

---

[1] Available commercial solutions are Microsoft Azure SQL Database, Amazon RDS and Oracle Cloud, to name the few.

© Springer International Publishing AG, part of Springer Nature 2018
J. Krogstie and H. A. Reijers (Eds.): CAiSE 2018, LNCS 10816, pp. 203–219, 2018.
https://doi.org/10.1007/978-3-319-91563-0_13

and Data as a Service (DaaS) aims to take care of all the activities needed to collect, process, store, and publish data, which must be accessible on-demand and regardless of the location where they are stored or from where they are requested. Although DaaS providers are mainly focused on taking care of the activities composing the life-cycle, data movement management is also crucial. For instance, in IoT scenarios, data are mainly generated at the edge of the network (e.g., by sensors), but they are usually moved to the cloud, where a theoretically unlimited amount of resources is available to efficiently store and process the data and to make them available to the customers. Indeed, cloud resources ensure high reliability and scalability, but the network capacity might negatively influence the latency when data movements among resources on the cloud and the edge occur. Thus, the advantage of the fast-processing at the cloud might be wasted resulting in lower quality of service.

The goal of this paper is to support the data management offered through a DaaS paradigm, by enabling effective data movements able to deliver data at the right time, the right place, and with the right quality and format, to satisfy the customer requirements, as conjectured in a preliminary work [8]. To achieve this goal, the proposed solution is based on two main pillars. Firstly, DaaS provisioning adopts the Fog Computing paradigm, which creates a continuum between the resources living on the edge and on the cloud [19] to exploit the advantages of both: data on the edge are closer to where they are generated or consumed (thus, latency can be reduced), while data on the cloud can have more capacity (thus, processing can be more efficient). Secondly, a goal model is used to design a decision system that includes the customers' requirements, the data movement actions that the environment is able to execute, and the effects of the enactment of a data movement on the satisfaction of the requirements. According to these two pillars, the main contributions of this paper are:

- formalization of data movement actions enriched with data transformations (e.g., aggregation, pseudonymization, encryption) for DaaS provisioning in Fog Computing, considering heterogeneous resources both in the edge and in the cloud belonging to different stakeholders (the data provider and its customers);
- context-based selection of valid movement actions and transformation for each cloud provider based on the definition of the storage resources provided by both the provider and the customers;
- extension of a goal-based modeling language for the definition of customers' non-functional requirements with new concepts such as data movement, data transformations, and a representation of the effect of data movements on goals satisfaction;
- application of a goal-based model at run-time for dynamically selecting a proper movement action to fix the violation of goals.

The rest of the paper is organized as follow. Section 2 motivates the proposed approach by introducing a running example. Section 3 analyzes data movement strategies in Fog environments, while Sect. 4 discusses how to use data movement with the goal-based model. Section 5 demonstrates the scalability of the proposed approach. Section 6 discusses how the proposed approach is related with the current state of the art, and, finally, Sect. 7 summarizes the proposed approach and identifies future work directions.

2   Motivating Example

Figure 1 shows an example of DaaS providing a data source about traffic information. The data provider manages some sensors and cameras placed on the highways of both Europe and United States to provide real-time information about the traffic. In particular, traffic data from the West Coast go to a *local data storage* where they are temporarily stored. The same occurs to data coming from the highways in the East Coast, as well as from the highways in Europe. Due to the high number of customers, the data provider relies on two cloud sites, i.e., *US site* and *EU site*, where two *Cloud Data Storage*, that must be maintained consistent, are fed by all the local storages, as customers must have visibility of traffic worldwide.

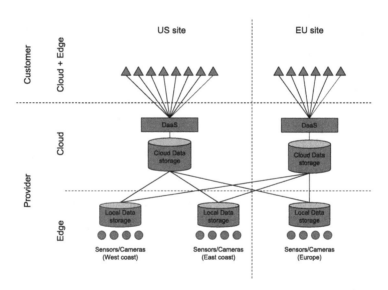

Fig. 1. Traffic information DaaS

Even in this very simple scenario, it is clear how the data continuously move from where data are generated (i.e., cameras and sensors), to the place where data are analyzed, to the data storages where they are saved, to all the final customers' storage devices. To deal with this situation, data providers usually implement solutions aiming to ensure the consistency and the timeliness of the two cloud data storages. In this way, all users see the same data set and the data are provided as soon as possible. Moreover, as the capacity of the local data storages are limited, and in order to ensure a proper timeliness of data offered to the customers, data are periodically moved from the edge to the cloud. This type of solution is actually reasonable only if all users behave in the same way. Actually, we can assume that the traffic information about EU are mainly used by the European citizens, while the US citizens are more interested on the US traffic data. Furthermore, customers can express different requirements in terms of quality of service (QoS), including latency, timeliness, availability, and so on. As a consequence, we want to support the data providers with a solution that enables effective data movements among the data storages driven by the objective of delivering the right data, at the right time, in the right format.

The proposed solution is based on two pillars: Fog Computing-based infrastructure and a goal-based decision model. The Fog Computing paradigm aims to consider the edge and the cloud resources involved in the service provisioning as a seamless environment. As a consequence, cloud resources do not include only the data provider storages and the scalable applications used to make those data available, but also, if any, the cloud resources used at the client side. Similarly, the edge includes not only the resources that are close to where the data are generated, but also the resources directly managed by the client to store and process the data (e.g., mobile devices).

Referring to our example, we can assume that the following resources are considered:

- data provider edge resources: local storages collecting data about traffic from sensors, positioned in EU and US sites (both in west and east coast);
- data provider cloud resources: data storage located in the cloud containing an aggregation of the information coming from the several edge resources, positioned in EU and US sites;
- customer cloud resources: for customers in EU and US, these are storage resources that belong to the clients but can be used by the data provider to store traffic data useful for the customer application;
- customer edge resources: edge resources with limited storage capabilities that can be used as in the previous case for storing useful information (moving or duplicating data in these additional resources).

We can reasonably assume that storages on the cloud and the edge can have different capabilities: they can store either a complete or a portion of the data set relevant for the specific phase of the data life cycle. Referring to our example, storages at the edge of the provider contain the most recent traffic information about a specific set of highways, while the cloud data storages, due to their capabilities, contain the complete set.

Moving to the second pillar, a goal-based modeling language is used to express the non-functional requirements negotiated between a data customer and a data provider for DaaS provisioning. We chose a goal-based modeling language as this is an intuitive approach for the specification of requirements. The adopted language is the Business Intelligent Model (BIM) [13], used to model trees where each level represents a set of subgoals required to satisfy specific properties of the data provisioning, and each goal is associated with one or more metrics used to assess the goal satisfaction.

For each customer, a tree is generated to express the QoS agreed with the DaaS provider. An example of such tree is shown in Fig. 2 where there is one top goal, High quality of Service, that represents the main objective to be achieved. This goal is AND-decomposed into three sub-goals, meaning that all sub-goals must be achieved in order to achieve the top-goal. Each sub-goal is OR-decomposed into two sub-goals. The OR-decomposition specifies that at least one of the sub-goals must be achieved, in order to consider the top-goal achieved. For example, Reliable Service is OR-decomposed as Service available and Service scalable: in order to offer a reliable service, the data provider must offer a service with a defined level of availability or a defined level of scalability. A data provider can enrich the model with as many goals as needed to describe the capability of the DaaS service and refine the goals with the needed AND and OR decompositions. The achievement of goals can be defined using metrics that specify properties to be monitored and conditions that determine when goals are satisfied. For example, in Fig. 2, the Fast data process goal is evaluated with the metric Response time and it is considered achieved when its value is lower than 5 s. The overall figure specifies that the data provider and customer agreed in the provisioning of a service that must be reliable, be fast and will maintain data consistency (first AND-decomposition of the top-goal).

The model is the starting point to monitor the customers satisfaction and to detect possible violations of the agreed QoS. It is worth noticing that the proposed approach can be easily extended to include other requirements (e.g., security, privacy, data quality) by adding an additional top goal connected with AND-decompositions to all these non-functional requirements and their sub-trees. Data movement enactment can be used to avoid violations of the tree, as will be described in the next section.

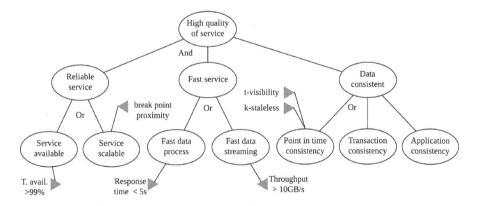

Fig. 2. Example of a goal-based diagram for the specification of QoS requirements

3   Data Movement in Fog Computing

Moving data implies moving portions of the offered data set from a data storage to another in a different location either in the edge or in the cloud. As there might be differences in the way in which the data can be stored, data movement could also require some data transformation. In this section, we classify data movement actions (Sect. 3.1) and discuss their instantiation in a specific context (Sect. 3.2). It is worth noticing that the proposed method is not limited to a specific storage model, even if different models would affect the implementation of the movement actions. The DITAS H2020 project[2] which funded this research is dealing with this issue by providing DaaS independence from the storage model and technology.

3.1   Movement Actions and Transformations

Although the generic term is data movement, the actions to be considered are: the actual movement (M), which consists of deleting the data from the original data storage and move them to a different one, and the duplication (D), where data are copied from a data storage to another while keeping them in the original one. These two classes of actions can be specified at a finer level of detail by considering the location where data are moved. In a heterogeneous environment, where data storages can be placed both in the edge and in the cloud, we have the following scenarios:

- Move/duplicate from cloud to edge (M_{CE}, D_{CE}): data contained in a cloud storage are moved or duplicated in a storage placed in the edge.
- Move/duplicate from edge to cloud (M_{EC}, D_{EC}): data contained in an edge storage are moved or duplicated in a storage placed in cloud.
- Move/duplicate from cloud to cloud (M_{CC}, D_{CC}): data contained in a cloud storage are moved or duplicated in another cloud storage.
- Move/duplicate from edge to edge (M_{EE}, D_{EE}): data contained in an edge storage are moved or duplicated in another edge storage.

Additionally, all the possible classes of movement actions can be subject to an additional data transformations T when data is moved or duplicated from a storage to another. Transformations consist in the manipulation of the content of a data storage and they are requested when the format required by the source and destination data storages are different or when they must be altered for security/privacy reasons. Examples of transformations include:

- aggregation: the content of a data storage is reduced using aggregation operations (e.g., average, maximum, minimum) summarizing several tuples;
- pseudonymization: data are manipulated to substitute identifying fields within a data record with artificial identifiers;
- encryption: the data contained in a data storage are manipulated using encryption algorithms to make them unreadable to unauthorized users.

---

[2] https://www.ditas-project.eu.

For a given movement action, we can have different sets of transformations, which can be either optional or mandatory. Optional transformations can be executed according to the user requirements, whereas mandatory transformations need to be executed every time data are moved from the data storage. Both movement actions and transformations are associated with metadata defining cost and execution time. These metadata are required to select which action to apply given a specific strategy (cost minimization or time minimization).

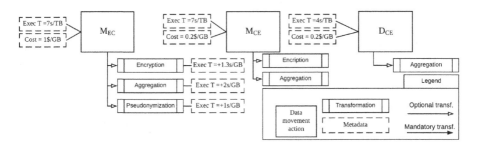

Fig. 3. Data movement actions and transformations example in a fog environment

The BIM modeling language has been extended with the elements described in this section, which create an additional layer. An example is shown in Fig. 3. The extended modeling language allows to specify both data movement actions and transformations. Movement actions are represented as rectangular boxes containing the specification of the movement action (e.g., M_{EC} specifies a movement class from an edge to a cloud storage). Transformations are boxes connected to the movement action to which they are associated. The association can be optional (white arrow) or mandatory (black arrow). Both actions and transformations can be annotated with information about their cost and execution time, represented as dashed rectangular boxes associated to the action or the transformation. For example, in Fig. 3 the data movement action M_{EC} has a cost of 7 s for each Tera Byte moved and of 1\$ for each Giga Byte transferred, while its Encryption data transformation add 1.3 s for each Giga Byte transformed.

While the execution time for a transformation can be obtained by testing the algorithms used for this purpose, the execution time of a movement or duplication is affected by the network latency. For this reason, we assume to have some information about the network capacity from which we can derive the needed information. Since we are dealing here with movement classes, without specifying the storage resources and locations involved, the value that we include in the model represents the average behavior of that class of actions. Distinguishing between classes of resources (edge vs cloud) enables us to better predict the metadata associated to a class since, due to the heterogeneity of the resources, edge and cloud storages will behave differently in terms of execution time and cost. This distinction will become implicit when moving from movement classes to movement instances as discussed in Sect. 3.2.

3.2   From Movement Classes to Movement Instances

The data movement classes described in Sect. 3.1 represent all the possible movement actions applicable in a generic context. When instantiating the model in a specific scenario, instances of these classes have to be defined according to the storage resources available. As stated before, storage resources are made available for different customers by the DaaS. Additional data storages can be made available by the customer, near to where the data will be analyzed.

Referring to the running example, at the edge we have three edge data storages: i.e., US West coast S_{wc}, US East coast S_{ec}, and Europe S_{eu}. Two geographical distributed cloud storages are also available in the US and Europe, S_{US} and S_{EU}.

Finally, each customer can provide an edge storage resource S_{cust}^E in which the data provider can store a subset of the information contained in an edge or cloud storage.

Distribution of data sets affects data management and in the specific case data movement capabilities of a DaaS provider. Indeed, movement or duplication is possible between two data sets only if their schemas are compatible. A movement action instance should be created for each possible combination of data storages. However, a limitation on movement might derive from security and privacy constraints or from policies defined by the provider. As an example, the provider can decide which classes of movement are allowed and which transformation are mandatory for a specific class. As an example, constraints can be expressed on data localization (e.g., it is not possible to move data from a cloud location in Europe and another in US), or security and privacy constraints (e.g., to be moved from the location in which they are produced, data have to be pseudonymized). The model captures these constraints by removing unauthorized movement instances and properly setting the transformations.

Knowing the available data sets, their location, their relations, and the constraints defined by the provider, it is possible to define which are the data movement actions that can be applied between two data sets. The steps for instantiating the data movement actions from the movement class M_{xy}, with x and y indicating the type of resource (i.e., edge or cloud) are the following:

1. generate a movement instance for each possible combination of resources of type x to resources of type y;
2. remove data movement instances based on the constraints defined by the provider;
3. apply constraints on the transformation and change the required transformation from optional to mandatory (leaving optional the other transformations associated to the action);
4. if additional information is given on resources, mutual location and capabilities, recompute metadata; else inherit metadata from the movement class.

An example on how to get movement instances from the M_{EC} movement class (representing movement from edge to cloud) for our running example is shown in Fig. 4 assuming the provider had specified a localization constraint (e.g., data cannot be moved between Europe and US resources), and a pseudonymization

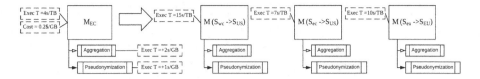

Fig. 4. Data movement actions instances

transformation constraint. In the example all possible combinations of movement instances from edge to cloud are generated, discarding movement actions forbidden due to the localization constraint, and setting mandatory transformations expressed by the provider. As shown, three movement instances are generated from the movement class: from the edge in west and east coast to cloud in the US ($D(S_{wc} \rightarrow S_{US})$ and $D(S_{ec} \rightarrow S_{US})$), and from edge in Europe to cloud in Europe ($D(S_{eu} \rightarrow S_{EU})$). All transformations remain optional with the exception of pseudonymization which changes to mandatory. For the sake of simplicity, inherited metadata are not represented in the figure. Similarly, instances will be generated for movement class D_{EC} and for other allowed movement actions. As can be seen, classes metadata are useful for providing time and cost prediction in unknown contexts. Observations at execution time are used for refining the metadata of both the instance (collecting data about time and cost of the instance execution) and the class (computing the average behavior of all the instances of that class).

4   A Goal-Based Approach for Data Movement Management

The goal model expresses data customer requirements that the provider has to keep satisfied. When a requirement is violated, the model supports the selection of the best data movement action in order to restore goal achievement. To enable this, we need to enrich the goal model, expressing the agreement on QoS between the provider and the customer, taking into account the role of data movements among the data sources available.

4.1   A Goal-Based Modeling Language for Data Movement Management

As the initial goal model only specifies what has been agreed between the provider and the customer in terms of QoS, we propose to extend this model also taking into account the effect of actions, i.e., data movement and transformations over goals. To model the relations between data movement actions or transformations and goals, we use *contribution links*. A *contribution link* specifies that the execution of the action (and transformation) has an impact on the achievement of the goal. Contribution links can have a positive effect (the execution of an action or transformation helps the achievement of a goal) or a negative effect (the execution of an action or transformation hurts the achievement of the linked goal). Contribution links can be defined by the data provider according to its specific platform and data resources (i.e., the data provider knows that duplicating data between two data sets has a negative effect on the consistency of the data).

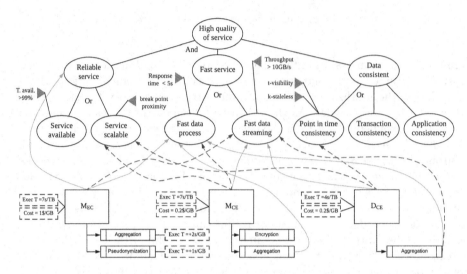

Fig. 5. Example of a diagram for the selection of data movement actions (Color figure online)

Figure 5 shows an example of the goal tree and the data movement actions connected with contribution links, which constitute the proposed extension of the BIM. For example, M_{EC} action is connected with a positive contribution link to Fast data process, since its adoption will improve the metric Response time and, therefore, it will help the achievement of the goal. Similarly, the same action impacts negatively Fast data streaming, since the movement of a data set in the cloud, in this example, will move the data set farther from the sensors that are creating data. For the sake of simplicity, only movement classes are represented in the figure. Movement instances will inherit contribution links from their classes, since the kind of effect (positive or negative) is the same for all the instances. In this work, we represent positive (green) or negative (red) effect with these links. In future work we are planning to assign a quantitative value to these links. In this case, each instance will have a different contribution value. To model this, the data provider and the data customer may customize the contribution links according to expected behavior. Other automatic approaches can be used for setting or refining the contribution links. As an example, in [22], the authors have proposed a reinforcement learning approach to update the knowledge of the effect of a set of actions over a set of goals using a Multi-Armed Bandit inspired algorithm, thus refining the confidence of such link every time the action is enacted.

4.2   Using the Goal-Based Model

Given the requirements of a customer, our extension of BIM is used at *design time* by the data provider, to produce a customized goal-based model, containing the constraints on the requirements relevant for the data customer and the movement actions filtered according to the available resources.

After that, at *runtime*, i.e. when the data customer uses the DaaS, the data provider monitors the goals satisfaction through the associated metrics. When a metric is out of the defined thresholds, an automated *controller*, in charge of managing the DaaS resources, selects to execute the movement action that might improve the current situation. To this aim, the controller analyzes the goal model negotiated with the data customer and selects a set of data movement actions that affect the violated goal.

The goal-based model supports the detection of goal violations. When a metric goes beyond the thresholds defined in a goal model, the linked goal is considered unsatisfied, and the model is analyzed to check, using backward analysis [7,11,21], whether the top goals are satisfied. For example, in Fig. 5, imagine that the goals Service available, Fast data streaming and Transaction consistency are achieved; in this context the top-goal is achieved too. After a while, the Throughput metric goes below 10 GB/s and the goal Fast data streaming becomes unachieved. In this setting, the top goal is not achieved anymore and a data movement action has to be enacted. For further details about goal analyses in BIM please refer to [12].

The action selection is led by the knowledge of the contribution links of both the actions and (if needed) the transformations with the goals. Before this analysis is executed, contribution links on non-leaf nodes are moved to leaf nodes. For example, in Fig. 5, the positive impact propagation of the action Move from edge to Cloud to Reliable service, is propagated to all leaf nodes Service available and Service scalable.

In order to select relevant movement actions, the controller considers all actions and transformations that have a positive impact on the unachieved goal. If a data movement action and one or more of its data transformations have conflicting contribution links on the same goal, we assume the positive contribution link from the transformation is always stronger (it overcomes) than the negative contribution from the data movement action. In the complementary case, the negative transformation link from the transformation nullifies the positive contribution link from the action. For instance, in Fig. 5 the data movement action Move from cloud to edge has a negative impact on Fast data process while its transformation Aggregation has a positive impact, therefore, the combination of the data movement action and transformation have a positive contribution to the goal. The rationale behind this decision lays on the idea that transformations are used to fix the weaknesses of the data movement actions, therefore, even if transformations have negative effects, such effects should never overcome the positive effects of the whole data movement action.

Three possible outcomes are expected: (i) no movement action is selected, meaning that the situation is so critical that none of the possible actions, that can be executed by the data provider, can solve the violation; (ii) one movement action is selected; (iii) multiple movement actions are selected. We do not investigate further the first option since other research work [4, 10] already faced similar problems and can be adopted as solutions for this case. For the third option, the controller considers the metadata associated to movement actions and transformations which express costs and time for enacting the action. Indeed, when several alternative actions are available for fixing a violation, the action selection might be led by the movement strategy selected by the customer. Two main strategies can be expressed: (i) **cost minimization strategy**: the controller selects the action that maximizes the goals satisfaction while minimizing the cost of enactment; (ii) **time minimization strategy**: the controller selects the action that maximizes the goals satisfaction while minimizing the time of enactment. The application of such strategies creates a ranked list of data movement actions.

The decision on which data movement action to apply cannot been taken for a single customer without considering other customers who concurrently access the same data sources. Indeed, the applications sharing the same data sources interfere with each other and a movement action might improve the QoS of one of them while negatively affecting another one. As an example, using the traffic information DaaS, let's consider the situation in which to bring data about traffic in the EU zone nearer to a customer, a movement action moves a subset of them from the cloud storage to a customer's edge storage. This action will improve the Fast service goal of the customer without violating the Data consistent goal. However, another concurrent application using the same data will be affected and its Reliable service goal will be violated.

To avoid interferences, after the selection of a set of candidate actions, the controller might check their effect on the goal trees of other customers. Each action selected is analyzed against all goal trees related to the data source that is being moved. If in at least one goal tree, the action negatively impacts a goal that has no positive contribution links from other movement actions, and therefore no action can be later adopted to restore the goal satisfaction, then the action is moved down in the ranking.

According to where the decision for each customer tree is performed, it is possible to implement the framework in a centralized fashion (global decision and global monitoring) or in a decentralized fashion (distributed monitoring and distributed decision). The first solution is easier in terms of management but the controller is a bottleneck since it has to manage all the customers. The second solution is more scalable but introduce a higher complexity in the coordination of the movement actions.

When the best movement action is selected, the framework will ignore violations that will be signaled in the period immediately after the enactment, in order to avoid oscillations of data sources between two or more locations. More complex mechanisms can be adopted to avoid subtler oscillations of data sources, however this is out of the scope of the paper and will be considered for a future work.

5 Scalability Evaluation

The efficiency of our solution mainly relies on the BIM engine. Thus, an efficient decision making depends on the ability of the BIM engine to produce a result in an acceptable amount of time. Being the complexity of the algorithms for the forward and backward propagation of goal satisfaction implemented in the BIM engine depends on the number of goals, we evaluated the response time of the BIM engine considering a variation of goals from 1 goal to 31 goals[3]. The tests have been executed on a virtual machine with 4 GB of RAM and 2 dedicated 3,3 GHz cores with Linux Ubuntu 16.04 installed. Future work will concentrate on evaluating the effectiveness of the proposed decision making system, as the infrastructure able to move data among cloud and edge storages is under development.

Figure 6 shows results for the backward analysis: on the x-axis there is the number of goals while the y-axis shows the execution time in milliseconds. The dotted line represents the linear regression, which indicates that the execution time increases linearly with the number of nodes. The results, especially in the right side of the chart, may appear distant from the linear regression, however, considering the scale of the chart, the distance can be considered minimal. The maximum execution time with 31 goals is 10 ms, which indicates the software returns the results almost immediately.

The scalability tests for the forward analysis return similar results in terms of execution time. In particular, the linear regression indicates that the execution time remains almost constant. The results are not included in the paper because of lack of space.

Both tests indicate that the backward and forward algorithms are executed in few milliseconds and, therefore, they can be integrated in the software for the decision of the best data movement action at runtime. In order to select the best action, multiple goal trees are evaluated, however, each tree is considered separately, therefore, the analysis can be executed in parallel for each goal tree. Other operations will be executed for the selection of a data movement action, such as the creation of the ranking using the data movement strategies, or the update of the ranking based on the impact of the actions in other goal trees. However, such operations are very fast and they do not impact on the performance of the overall approach.

Since the measured values are very low, they may be influenced by external factors, such as other CPU consuming operations. We solved this possible threat by executing the test 10 times and by excluding the minimum and maximum values.

The scalability test measures the execution time of the forward and backward reasoning software engine, however, other factors should be considered, such as the expressiveness of the modeling language and its usability. Although BIM has been used and validated with many case studies [9,12] we, nevertheless, will perform empirical experiments to evaluate our extension of BIM and the overall framework.

---

[3] The maximum value corresponds to a binary goal tree with depth equals to 4, a size that, from our experience, we believe is much bigger the goal model that will be used for the purposes of this paper.

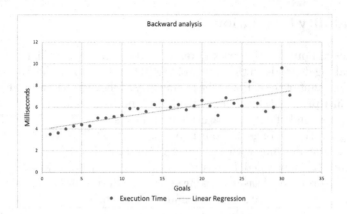

Fig. 6. Scalability tests results

6 Related Work

Initially introduced in the telecommunication domain by Cisco [5], Fog Computing has recently emerged as a hot topic also in the software domain, and especially for data-intensive applications, with the goal of creating a continuum between the resources living on the cloud and the ones living on the edge [19]. The adoption of Fog computing enables an effective data provisioning [20] as data can be moved among different environments in a seamless way. Data movement has been widely studied from different perspectives in the literature to try to reduce the problems arising from the management and use of large quantities of data from different sources and represented in different formats [2,17]. With respect to these approaches, this paper proposes a method for selecting which is the best one to be enacted.

Data movement is also the focus of Content Delivery Networks (CDN) with the aim of geographically distributing a service to ensure high availability and performance. CDNs have been evolving since their first implementation [15] and new solutions also considers deployment on edge facilities[4]. The main limitation of CDNs is that resources used for caching data are predefined and owned and managed by the provider. Moreover, the caching algorithm is only addressing performance and availability optimization of all the users. In the proposed approach, resources are dynamic and can be also controlled by the customers. Also, the data movement policies are driven by the requirements of each specific user and not by a general purpose.

Goal models are used in requirement engineering to specify the objectives of users and applications to be designed. In this paper, we have decided to use BIM [13] as a reference model. However, other approaches are available. In particular, the Goal-oriented Requirement Language (GRL) [3] is a rich modeling language that covers most of the concepts of BIM. However, GRL is a very rich language and may

---

[4] https://aws.amazon.com/cloudfront/.

prevent a correct usage of the method since many concepts of GRL are not used by our method and may confuse users. We, therefore, decided to extend BIM since it contains the minimal set of concepts needed. Yet, GRL may be considered for a future work.

The tree-like structures of goal models can be used to take decisions on which subset of goals to achieve. A great variety of analyses techniques have been proposed for analyzing goal models for this purpose [14, 16]. The satisfaction analyses propagate the satisfaction or denial of goals forward and backward in the goal tree structure. The forward propagation [16] can be used to check alternatives while the backward propagation [7, 11, 21] can be used to understand what are the consequences of a satisfied or denied goal. Such approaches however, were defined for other domains and, therefore, they do not include concepts needed for this paper.

7   Concluding Remarks

This paper proposes a solution to support data provisioning based on a DaaS paradigm in a Fog Computing environment by enabling an effective data movement among the data storages belonging not only to data providers but also to data consumers. Data movement is driven by a goal-based model capturing the agreement between a provider and its customers and can be used to figure out the most suitable data movement strategy. The model is enriched with data movement actions, defined and classified in the paper. We mapped the effect of actions over goals using contribution links that enable the method to be used at runtime for selecting the best action given a goal violation. At this stage, the validation of the approach is limited to the analysis of scalability, which demonstrated a linear increase of the response time with respect to the increase of the number of goals. Additional experimentations are planned in the near future to also demonstrate that the enactment of the data movement is able to improve the satisfaction of the customer requirements. In future work, we are going to refine the existing model by exploring the outcome of dealing with partially satisfied goals, instead of boolean conditions, also enabling the customer to set weights indicating the most relevant requirements. We are also refining the contribution links associating to them a quantitative value expressing the expected impact of the movement on the indicators associated to the goal, similarly to [22]. We are also going to investigate the implementation of controllers to manage multiple goal-based models for supporting multi-client requirements satisfaction.

Acknowledgments. DITAS project is funded by the European Union's Horizon 2020 research and innovation programme under grant agreement RIA 731945.

References

1. Agrawal, D., Abbadi, A.E., Emekci, F., Metwally, A.: Database management as a service: challenges and opportunities. In: Proceedings of IEEE International Conference on Data Engineering, pp. 1709–1716 (2009)
2. Amarasinghe, S.P., Lam, M.S.: Communication optimization and code generation for distributed memory machines. SIGPLAN Not. **28**(6), 126–138 (1993)
3. Amyot, D., Mussbacher, G.: User requirements notation: the first ten years, the next ten years. JSW **6**(5), 747–768 (2011)
4. Aydemir, F.B., Giorgini, P., Mylopoulos, J.: Multi-objective risk analysis with goal models. In: Proceedings of the Research Challenges in Information Science, pp. 1–10. IEEE (2016)
5. Bonomi, F., Milito, R., Zhu, J., Addepalli, S.: Fog computing and its role in the Internet of Things. In: Proceedings of the MCC Workshop on Mobile Cloud Computing, pp. 13–16 (2012)
6. Bouguettaya, A., et al.: A service computing manifesto: the next 10 years. Commun. ACM **60**(4), 64–72 (2017). http://doi.acm.org/10.1145/2983528
7. Chung, L., Nixon, B.A., Yu, E., Mylopoulos, J.: Non-functional Requirements in Software Engineering. International Series in Software Engineering, vol. 5. Springer, New York (2012). https://doi.org/10.1007/978-1-4615-5269-7
8. D'Andria, F., Field, D., Kopaneli, A., Kousiouris, G., Garcia-Perez, D., Pernici, B., Plebani, P.: Data movement in the Internet of Things domain. In: Dustdar, S., Leymann, F., Villari, M. (eds.) ESOCC 2015. LNCS, vol. 9306, pp. 243–252. Springer, Cham (2015). https://doi.org/10.1007/978-3-319-24072-5_17
9. Francesconi, F., Dalpiaz, F., Mylopoulos, J.: Models for strategic planning: applying TBIM to the Montreux Jazz Festival case study. In: 2015 IEEE 9th International Conference on Research Challenges in Information Science (RCIS), pp. 229–238. IEEE (2015)
10. Gembicki, F., Haimes, Y.: Approach to performance and sensitivity multiobjective optimization: the goal attainment method. IEEE Trans. Autom. control **20**, 769–771 (1975)
11. Giorgini, P., Mylopoulos, J., Nicchiarelli, E., Sebastiani, R.: Formal reasoning techniques for goal models. J. Data Seman. **1**(1), 1–20 (2003)
12. Horkoff, J., Barone, D., Jiang, L., Yu, E., Amyot, D., Borgida, A., Mylopoulos, J.: Strategic business modeling: representation and reasoning. Softw. Syst. Model. **13**(3), 1015–1041 (2014)
13. Horkoff, J., Borgida, A., Mylopoulos, J., Barone, D., Jiang, L., Yu, E., Amyot, D.: Making data meaningful: the business intelligence model and its formal semantics in description logics. In: Meersman, R., et al. (eds.) OTM 2012. LNCS, vol. 7566, pp. 700–717. Springer, Heidelberg (2012). https://doi.org/10.1007/978-3-642-33615-7_17
14. Horkoff, J., Yu, E.: Interactive goal model analysis for early requirements engineering. Requir. Eng. **21**(1), 29–61 (2016)
15. Leighton, F.T., Lewin, D.M.: Content delivery network using edge-of-network servers for providing content delivery to a set of participating content providers, 22 April 2003
16. Letier, E., Van Lamsweerde, A.: Reasoning about partial goal satisfaction for requirements and design engineering. ACM SIGSOFT Soft. Eng. Notes. **29**, 53–62 (2004)
17. Lu, P., Zhang, L., Liu, X., Yao, J., Zhu, Z.: Highly efficient data migration and backup for big data applications in elastic optical inter-data-center networks. IEEE Netw. **29**(5), 36–42 (2015)

18. MacKenzie, C.M., Laskey, K., McCabe, F., Brown, P.F., Metz, R.: Reference model for service oriented architecture 1.0. Technical report, OASIS (2006)
19. OpenFog Consortium Architecture Working Group: OpenFog Architecture Overview, February 2016. http://www.openfogconsortium.org/ra
20. Plebani, P., Garcia-Perez, D., Anderson, M., Bermbach, D., Cappiello, C., Kat, R.I., Pallas, F., Pernici, B., Tai, S., Vitali, M.: Information logistics and Fog computing: the DITAS approach. In: Proceedings of the Forum and Doctoral Consortium at CAISE 2017, pp. 129–136 (2017)
21. Sebastiani, R., Giorgini, P., Mylopoulos, J.: Simple and minimum-cost satisfiability for goal models. In: Persson, A., Stirna, J. (eds.) CAiSE 2004. LNCS, vol. 3084, pp. 20–35. Springer, Heidelberg (2004). https://doi.org/10.1007/978-3-540-25975-6_4
22. Vitali, M., Pernici, B., O'Reilly, U.M.: Learning a goal-oriented model for energy efficient adaptive applications in data centers. Inf. Sci. **319**, 152–170 (2015)

An Ontology-Based Framework
for Describing Discoverable Data Services

Xavier Oriol[(✉)] and Ernest Teniente

Universitat Politècnica de Catalunya, Barcelona, Spain
{xoriol,teniente}@essi.upc.edu

Abstract. Data-services are applications in charge of retrieving certain data when they are called. They are found in different communities such as the Internet Of Things, Cloud Computing, Big Data, etc. So, there is a real need to discover how can an application that requires some data automatically find a data-service which is providing it. To our knowledge, the problem of automatically discovering these data-services is still open. To make a step forward in this direction, we propose an ontology-based framework to address this problem. In our framework, input and output values of the request are mapped into concepts of the domain ontology. Then, data-services specify how to obtain the output from the input by stating the relationship between the mapped concepts of the ontology.

Keywords: Data-service · Data-service discovery · Internet of Things

1 Introduction

Data-services are applications whose main concern is to provide data to their client applications. Data-services play a key role in areas like the Internet of Things (IoT), where smart objects obtain data from sensors (or other devices) and then make it available to others by means of data-services. Additionally, smart objects can also be interested in consuming data coming from other external data-services, so that they can use it to take their smart decisions. This idea is being currently exploited in several european projects like, for instance, the BIG IoT European project [1].

In IoT, smart objects should be as autonomous as possible due to the huge amount of devices and data that are constantly added. Hence, they should be able to offer/discover data-services on the fly, without human intervention. E.g., an autonomous car looking for parking in its current street should be able to automatically discover data-services retrieving available parking places of a street.

To make data-services discoverable, the usual strategy is to register data-services in some kind of service-broker, i.e., a marketplace where data-services are publicly offered [2]. Then, smart objects query the service-broker, and the service-broker is responsible to match the request with its data-services. How to perform this matching automatically is still an open problem in IoT [3].

© Springer International Publishing AG, part of Springer Nature 2018
J. Krogstie and H. A. Reijers (Eds.): CAiSE 2018, LNCS 10816, pp. 220–235, 2018.
https://doi.org/10.1007/978-3-319-91563-0_14

In this sense, we propose a framework for specifying data-services so that they can be automatically discovered. To achieve it, we provide unambiguous descriptions of the data-services and the request, together with a mechanism capable of interpreting these descriptions and check whether they match. Our solution is grounded on ontology-based data integration and can be applied in the IoT context, although it can also be used in any other domain involving the discovery of applications retrieving data.

An ontology is a common set of terms (a vocabulary) with semantic relationships among them (e.g. subclass/superclass relationships, etc.) that describes the real world. Figure 1 shows a UML ontology regarding parking concepts.

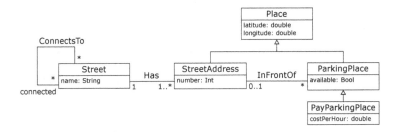

Fig. 1. Ontology for the parking domain in UML

Assume now a data-service receiving as input a *string* and returning a list of *paired numbers*. Clearly, with no description, a service-broker cannot determine what does the data-service compute since it cannot know what does such string stands for, what are the numbers describing, and which is the relationship between the string and the numbers. Hence, automatic discovery is not possible.

However, using the previous ontology we can state that the input are *Streets* (described by *street names* strings) and the output are *Parking places* (described by *latitude/longitude geolocalizations* paired numbers). These descriptions are somehow related to Semantic Web mark-ups provided by, for instance, schema.org, but these mark-ups alone are not enough since we need to know which is the relationship between the input *Street* and the output *Parking places* to know what the service is actually doing.

In our framework, we describe this relationship between the input and the output by defining a new association in the ontology linking their corresponding classes. The contents of this association is defined by means of a constraint (aka ontology axiom), which states what objects from the output are obtained from those in the input using the properties of the ontology. In this manner, we can specify that, for instance, given a street name, a data-service provides exactly the available parking places that are located in front of the street addresses that the street has. All these descriptions can be specified with ontological meta-concepts (concepts/properties/axioms), thus, no need of special new meta-concepts is required (such as operation, or query).

In this way, our framework reduces the capability of storing data-service descriptions to that of storing ontologies, so, any existing tool for managing ontologies can be immediately used by the service-broker. Moreover, we also have that the problem of data-service/request matching is reduced to that of ontology reasoning. More precisely, to relation subsumption which is a very well-known and studied problem in the field of automated ontology reasoning [4,5].

We first define our framework in an abstract way, independently of the particular language used for specifying the ontology. Secondly, as a proof of concept that this framework can already be used in practice, we further develop it in the UML/OCL language with UML/OCL reasoning tools. We argue that using UML/OCL as our ontology languages are a natural choice since (1) they are widely used in software engineering, (2) they are expressive enough to specify all descriptions of our framework, and (3) there exist several reasoning tools that can be used for the service description/request matching. We finally comment on some experiments which show that these current reasoning tools have good response-times when applied in these settings.

In summary, the main contributions of the paper are: (1) *An ontology-based framework to describe discoverable data-services.* We define a set of descriptions to unambiguously specify data-services and, thus, enabling its automatical discovery. These descriptions only involve meta-concepts already existing in modeling languages (e.g., relationships, axioms); (2) *UML/OCL suitability for the framework.* We show how to use UML/OCL for writing these descriptions, and show that UML/OCL already has tool support for solving the discoverability matching problem; and (3) *Efficiency evaluation.* We evaluate the efficiency for matching a service description/request using a UML/OCL reasoning tool.

It is worth to mention that this framework is an adoption of Semantic Web contributions (starting from local-as-view data integration [6]) to the context of Data-Services in IoT using software engineering languages (UML/OCL).

2   Basic Concepts

Data-Service. We refer as data-service to any application that receives some data as input and returns some data as output. During the paper, a data-service might be thought as a smart object providing information on demand.

Data-Request. We refer as data-request to the description of some data-service required by some application. During the paper, a data-request might be thought as the request from some smart object to find another smart object providing its interested information.

Data-Service/Request Matching. We refer as data-service/request matching to the problem of identifying whether some data-request coincides with the description of some existing data-service. For the sake of simplicity, we tackle the problem only in the semantic level. That is, we are not intended to match technological aspects of the service (e.g., SOAP/REST calls, XML/JSON formatting answers, etc.). Thus, we assume that the community using our proposed framework has an initial agreement about the communication technologies involved.

Service-Broker. We refer as service-broker to an application responsible of storing the data-service descriptions and resolving the data-request invoked by means of the data-service/request matching.

Ontology. An ontology is a set of concepts and properties describing some real world domain. Ontology axioms (aka constraints) state conditions over these concepts/properties that hold in the real world. A typical axiom is the *isA* hierarchy between two concepts which states that instances of the first concept are also instances of the second. For our purposes, we require the ontology to include *primary key constraints*, i.e., to state which properties take unique values.

3 Framework Essentials

In the following, we summarize how does the framework specify data-services in an abstract way (i.e., without bounding to any particular language), and show how to discover the service (i.e. how to match a data-service with a data-request).

3.1 Describing Data-Services

The framework assumes the use of one or several ontologies to describe data-services and application data-requests. These ontologies should be stored and maintained by the service-broker in which the data-services are registered.

The basic idea is that, if a service receives as input *Streets* and returns as output its available *Parkings places*, this service can be seen as a new relationship in the ontology between *Streets* and *Parking places*, whose contents is computed by the service. In general, a service that receives concepts $I_1, ..., I_M$ and returns concepts $O_1, ..., O_N$ can be described as a new M+N-ary relationship in the ontology.

The difficulty to describe data-service as a new ontology-relationship is that it does not receive/return objects, but values describing the real-world objects. Indeed, a data-service never returns a *Parking*, but some values describing it (e.g. a pair of integers representing its latitude/longitude). Note, in addition, that different data-services might describe the same object through different properties (e.g. a strings encoding the street address number of the parking). Thus, to specify the input and output of a data-service in terms of a relationship, we first need to map the input and output parameters to the ontology concepts.

Describing Input/Output Concepts. To identify the concepts handled by a data-service, we need to group its parameters and specify which concept of the ontology are they describing. Formally, given the set of parameters *PARAMS*, its powerset $\mathcal{P}(PARAMS)$, and the set of ontology concepts *CONCEPT*, this description corresponds to the input/output mappings:

$$Input_C : \mathcal{P}(PARAMS) \to CONCEPT \qquad Output_C : \mathcal{P}(PARAMS) \to CONCEPT$$

For instance, if some output parameters *lat, long* describe an object *ParkingPlace*, we map {*lat, long*} to *ParkingPlace*.

Moreover, since each parameter represents a specific property of the described concept, we map each parameter into the particular property it represents. Formally, given the parameters *PARAMS*, and the ontology properties *PROPERTIES*, this description corresponds to the maps:

$$Input_P : PARAMS \rightarrow PROPERTIES \qquad Output_P : PARAMS \rightarrow PROPERTIES$$

For instance, the output parameters *lat* and *long* would be mapped to the properties *latitude* and *longitude* from *ParkingPlace*.

The important thing here is that the set of parameters describing an input object should univocally determine the real-world object. In essence, this means that the description should include, at least, one primary key of its corresponding concept. For instance, if some data-service returns properties for describing *ParkingPlaces*, these properties should include, at least, a latitude/longitude pair or a street number address. Note that, without the primary key, we cannot identify which object from the real world do the properties belong to and, therefore, the data-service would not be able to identify the input objects for which computing the data required by the output.

The service-broker should be responsible of checking that the data-service is unambiguously describing their objects. It is worth saying that this condition is not necessarily required by the output objects since we may be interested in anonymizing some kind of sensible data which should not appear in the output.

Describing the Input/Output Relationship. Once we know the input and output concepts to which the parameters refer to, we can define the input/output relationship computed by the service, i.e. the logic of the service. That is, the computation that the data-service performs in terms of a navigation through the ontology input concept to the ontology output concept.

This relationship is specified only in terms of the known basic concepts/properties of the ontology by means of an axiom (also called constraint). Specifically, this axiom defines the contents of the relationship in terms of contents of the input/output concepts and its related properties. Equivalently, we can see this relationship as *derived* from the rest of the ontology terms.

Formally, assuming that our data-service describes n input objects and m output objects, we have to define an axiom with the form:

$$R(I_1, ...I_n, O_1, ..., O_m) \leftrightarrow \phi(I_1, ...I_n, O_1, ..., O_m)$$

where R is the relation computed by the data-service, and ϕ is a statement (usually a first-order formula) that defines the contents of R in terms of the rest of ontology properties.

As an example, assuming a different predicate for each concept and relationship in our ontology, the input/output relationship of a service computing the

ParkingPlaces that are *available* and located *infrontof* some *streetAddress* that has the input *Street* would be specified in logic by means of the following rule:

$$R(s,p) \leftrightarrow Street(s) \wedge HasStreetAddress(s,st) \wedge InFrontOf(st,p)$$
$$\wedge ParkingPlace(p) \wedge IsAvailable(p,a) \wedge a = \mathbf{true}$$

A service-broker might check whether an input/output relationship R description is correct by means of checking whether R is lively. We say that R is lively if there exists a state of the real-world domain I (that is, a valid instantiation of the whole ontology) in which R has some contents. Clearly, if no such real-world domain I exists, this means that this data-service is not out-coming any output from any input, which would mean that such data-service is useless. Checking relationship liveliness is a well-known ontology reasoning task [4,5].

Summary of a Data-Service Description. The description of a data-service is a tuple of the form: $< Input_P, Input_C, Output_P, Output_C, R(x) \leftrightarrow \phi(x)) >$

It is worth noting that our framework is only meant for data-services that retrieve data. Thus, it is not suited for services modifying this data. Moreover, our current proposal does not consider preconditions either, although they could be easily added at the ontological level as new ontological axioms. It is also important to remark that a data service should adapt to the terms in the ontology of the service-broker used (instead of being able to use its own ontology to define them). This limitation is caused because we need data-services and data-service requests to use the same ontological terms to enable its matching.

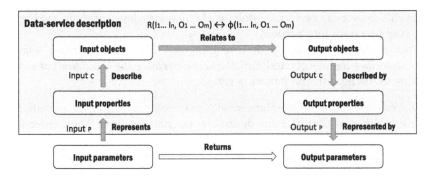

Fig. 2. Data-service description

Finally, we would like to stress that the descriptions proposed in our frame-work are unambiguous and capable of describing services considering several input/output parameters (even describing several input/output objects). The descriptions are unambiguous because, fixed a real-world state and some input values, there is only one valid collection of output values that satisfies them. Intuitively, this is so because all the semantic descriptions stated in Fig. 2 can be composed into a unique function that converts input parameter values into input objects, input objects into output objects, and output objects as output values again. The description can handle services with an arbitrary number of input/output parameters as long as the language we use to state them allows n-ary associations. This is also a motivation for using UML as specification language (instead of OWL/RDF) since it directly supports n-ary associations.

3.2 Discovering Data-Services

To discover a data-service, a data consumer application should make a request to the service-broker so that it can check which data-services match the request.

The essential idea is that the request should be described in the same way as a data-service. I.e., with the input/output mappings that state how parameters describe concepts, and with a new ontology relationship and an axiom defining which input/output object relation is the requester looking for. Intuitively, a data request is described exactly as the hypothetical data-service it requires.

Then, the service-broker can match the request with its data-services by checking which relationships describing the services are subsumed by the relationship in the request. By subsumption, we mean that the contents of the relationship computed by the service is necessarily included in the contents of the one desired by the request. If the service relationship is subsumed by the request, this means that all the objects retrieved by the service are data that the requester wants to find. Thus, our proposal is semantic-based (i.e., independant from the particular syntactic way of writing the request/service in OCL).

Note that we only need to check if a data-service relationship is subsumed by a request relation if they agree on the related input/output concepts. Moreover, such discovery is performed only at the ontology level, without taking the underlying raw data into account.

Given a data-service and a data-request, the distinction between *how the service describes the objects* and *how the service relates the input/output objects* allows us to distinguish the following cases:

1. *Different relationships*: the data-service may compute an input/output relationship different than the one desired by the data-request. Thus, there is no match between the two. E.g.: a data service relates Streets with Parkings, but our data-request looks for a relationship between Streets and Streets.
2. *Same relationship, different descriptions*: it might happen that, at the ontology level, the data-service computes the same input/output relationship as the one requested, but the parameters given to describe these input/output are not the expected by the data-request. E.g.: a data service relates Streets with Parkings, and the data-request looks exactly for this relationship but, however, the service describes parkings by means of latitudes/longitudes and the requester expects street address numbers as output.
3. *Same relationship and same descriptions*: the data-service computes the input/output relation desired by the data-request and the parameters used are those it expects. E.g.: a data service relates Streets with Parkings, the

data-request looks exactly for this relationship, and they both agree on the properties used to describe both concepts.

Obviously, when looking for some data-service matching a data-request, the best case is the third one. However, the second case should not be neglected. In fact, the second case should further dinstinguish whether the descriptions misagreement is because the service provides too much information (which might not be a problem at all), or because it does not provide all the desired information (which might be a problem, or not).

4 Using UML/OCL as Ontology Language

We show now how to use UML/OCL as the ontology language for the framework. UML/OCL is a good candidate language because it is well-known in the software engineering community, it has all necessary elements for encoding the framework descriptions, and there are already some reasoning tools that can be used to solve the matching problem.

In the following, we first discuss the usage of UML/OCL for describing the common ontology to be used by the service-broker. Then, we show how a data-service provider can use UML/OCL to describe a new service. Finally, we show how to discover a data-service by means of a UML/OCL reasoning tool.

4.1 Using UML/OCL to Define the Service-Broker Ontology

The service-broker should provide, at least, one common ontology so that all their published data-services are described over it. That is, all the data-services are described using the terms of the ontology provided by the broker, and assuming all the axioms defined in this ontology.

In Fig. 1 we have shown a UML class diagram describing *Parking* concepts. In this ontology, all parkings are places with a latitude and a longitude, they can be available or not, and some of them have an associated cost per hour. These parkings may be in front of street addresses, which are also places with a latitude and a longitude. These street addresses belong to a street, and the streets can be connected to other streets.

To complete the ontology, the UML class diagram must be complemented with a set of OCL constraints stating conditions that the real-world satisfies. In the following, we show some OCL constraints stating that there are no two places with the same latitude/longitude, nor two streets with the same name, and that fixed one street, there are no two addresses with the same number:

context Street **inv** StreetPrimaryKey: Street.allInstances()->isUnique(name)
context StreetAddress **inv** StreetAddressPrimaryKey: StreetAddress.allInstances()
   ->forAll(a, b | a = b or a.number <> b.number or a.street <> b.street)
context Place **inv** PlacePrimaryKey: Place.allInstances()
   ->forAll(a, b | a = b or a.latitude <> b.latitude or a.longitude <> b.longitude)

4.2   Using UML/OCL to Describe the Service

Once we have the common ontology specified, we can use it to define a service along the framework defined in Sect. 3. That is, by means of describing how the parameters describe input/output concepts, and how the input and output concepts are related at the ontology level. These ideas are similar to [7], where they model queries as UML classes and inputs/outputs as associations to these classes.

Describing Input/Output Concepts Through UML/OCL. To describe the mappings from the parameters to the concepts (i.e., the $Input_C/Output_C$ mappings), we define a new UML class and association for each described concept.

That is, if our parking service has an input parameter *name* that is mapped to the concept *Street*, we add a new UML class in the class diagram with an attribute name and associate this class to *Street*. Figure 3 illustrates this example in the UML class diagram.

In general, we may need to create several of these classes in case our input/output parameters describe several objects of different classes. To keep track of the newly added classes because of describing these mappings, we propose to use two new UML stereotypes: *ServiceInput*, and *ServiceOutput*.

Then, for describing the mappings to the properties (i.e., the $Input_P/Output_P$ mappings), we rely on the usage of OCL constraints. In particular, we define a new OCL invariant for each parameter that, in essence, equates its value to some attribute from the common ontology. Figure 3 also exemplifies how to do so.

To correctly and unambiguously describe the objects through the properties, two instances of parameters with exactly the same values should describe exactly the same object (e.g., two *InputStreetPar* with the name *Broadway* should describe exactly the same real-world *Street*: *Broadway*). Ensuring that two equal instances of parameters cannot describe (i.e., be associated to) the same object is a reasoning task that can be solved by many current UML/OCL reasoning tools [5, 8–10].

Describing the Input/Output Relation Through UML/OCL. Next, we add a new association in the UML class diagram to describe the input/output relationship computed by the service. This new association relates the described input and output concepts. Similarly as before, we propose distinguishing this new association using a UML stereotype: *Service*.

In addition, a new OCL invariant *InputOutputRelationship* establishing the contents of this association has to be specified, as shown in Fig. 3. In our example, the invariant states that all the retrieved output parking places are available parking places from some street address of the given input street.

Intuitively, this invariant specifies the logics of the service in terms of the common ontology. Note that several equivalent ways of specifying the same logics may exist (i.e., by using different OCL operators, or using different OCL

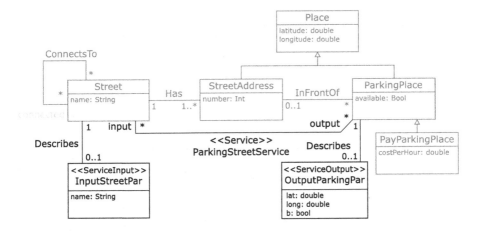

context Input **inv** DescribesStreet:
self.street.name = self.name
context Output **inv** DescribesParkingPlace:
self.parkingplace.latitude = self.lat and self.parkingplace.longitude=self.long and
self.parkingplace.available = self.b
context Street **inv** InputOutputRelationship:
self.output = self.streetAddress.parkingPlace->select(p|p.available)

Fig. 3. Description of a data-service in the UML/OCL ontology

navigations, etc.) and the service designer is free to choose the one he/she likes
the most. The important thing is that all the equivalent ways of defining the same
invariant will match with the same equivalent requests during service discovery.

Finally, it is worth noting that this way of specifying services scales for any
number of input/output described objects (and not only one object for the input,
and one object class for the output). Indeed, if the service relates N objects from
the input (e.g. Street and Place) to M output objects (e.g. StreetAddress and
ParkingPlace), we only need to define an $(N + M)$-ary relation.

There is a special case when $N = 0$ and $M = 1$ because UML has no unary
associations. However, this case can be handled by using a subclass. Indeed,
consider that our parking service returns all parking places that are available
(thus, no input is required, and the output is composed of only 1 object). In this
case, our service is not relating any input to any output, but is considering a
new class of parkings (i.e., the class of those parkings that it is retrieving -e.g.,
the available ones-). So, the service is specified by means of a subclass rather
than an association. Similarly as before, this new subclass should be paired with
the OCL constraint specifying its contents in terms of the common ontology.

4.3 Discovering the Services

Now, the idea is that, when a service-broker receives a data-request, it must
look for those data-services that match the request. More specifically, taking in

account that services/requests logics are, at the end, defined through an association in the UML class diagram, it should look for those Services associations that are subsumed by the Request association.

There is a match between the data-service and the data-request when the association encoding the data-service R_S is subsumed by the association encoding the request R_R. Recall that, an association R_S is subsumed by R_R if the contents of R_S are, for any valid instance of the ontology, a subset of the contents of R_R.

Checking if R_S is subsumed by R_R can be done by analyzing whether the UML schema admits a finite valid instantiation I such that: (1) it satisfies all the integrity constraints of the UML class diagram, and (2) it contains some instance i of R_S not appearing in R_R. If such instantiation does not exist, this means that all instances of R_S are contained for sure in R_R and, thus, R_S is subsumed by R_R. On the contrary, if such instantiation I actually exists, I represents a state of the real world in which the data-service would provide some output i not desired according to the request.

As an example, consider two data-requests described with an association from *Streets* to *ParkingPlace* by means of the following invariants:

context Street **inv** Request1: self.output = self.streetAddress.parkingPlace->select(a|
   not a.oclIsTypeOf(PayParkingPlace))
context Street **inv** Request2: self.output = self.streetAddress.parkingPlace
   ->union(self.connected.streetAddress.parkingPlace)

Intuitively, the first request is asking for parking places that are *free* (in the sense that you do not have to pay for their use), while the second one is asking for parking places in the given street and its connected streets. The first request does not subsume our data-service running example, but the second does.

The first data-request does not subsume our data-service because there might be a real-world state of the domain I in which the data-service computes some output undesired by the data-request. E.g., consider a real-world state I with a Street *Abbey Road* which contains a parking place in the geocoordinates *51.532, 0.177* such that is of kind *PayParkingPlace*. In this state of the domain I, our data-service would retrieve the previous parking, when the data-request is asking to avoid *PayParkingPlaces*. Thus, there is no match between the two.

On the contrary, the second data-request subsumes our service. Indeed, in any real-world state of the domain I, all the contents retrieved by the service are contents desired by the request, although our data-service is limited to only retrieve a subset of it (those that are available and in the given street, thus, ignoring the non-available parkings and the parkings from the connected streets).

In case we were only interested in data-services retrieving exactly all our requested contents, we could additionally check if the data-service also subsumes the data-request. Indeed, if they both mutually subsume each other, then, they are exactly requesting/computing the same relationship.

Reasoning association subsumption is a known problem in the conceptual schema reasoning field and it can be performed by several of their UML/OCL reasoners. E.g., we could use Alloy [8], USE [9], UMLtoCSP [10], or AuRUS [5].

Moreover, these tools are capable of computing the counterexample I that proves when a data-service would compute an undesired output for the data-request.

It is important to note that we are only checking whether the service request and the data-service offered coincide in terms of their intended logics (that is, their semantics). This checking does not take into account the different descriptions/representations that this services might have for their concepts. That is, it only checks whether the service/request matches the intended *Parking*, but it does not check if the way parkings are described also matches (which might be checked with a simple comparison between the actual/expected descriptions). This behavior is the one we already discussed while presenting the framework essentials in Sect. 3.2.

5 Experiments

To show the feasibility of our framework, together with the suitability of current UML/OCL reasoners to solve the matching problem, we have conducted some experiments with real data-services developed in the BIG IoT Project [1], where we participate and that has motivated the ideas proposed in this paper.

In particular, we have used our framework to describe 3 different traffic data-services from BIG IoT: a data-service for finding parking places of a given current location, a data-service for retrieving the current traffic status of a given street, and a third data-service for obtaining the average car speed of some predefined streets. These data-services cover several relevant aspects of our framework: the first one receives and retrieves different concepts (i.e, it receives *Street* and returns *ParkingPlaces*), the second receives and retrieves the same concept (i.e., *Streets*), and the third has no input and only retrieves one concept (i.e., *Streets*).

To define the common ontology for these services we have departed from its current semantic annotation present in the BIG IoT project. In essence, BIG IoT tags the parameters of the services with the ontology property they represent (i.e., it defines the $Input_P/Output_P$ maps) using an extension of the schema.org ontology. Thus, we have built in UML/OCL the fragment of schema.org involved, and added the necessary extensions to define those concepts and properties which are not present in schema.org. Then, we have described all mentioned data-services in terms of it[1], using the constructions specified in Sect. 4.

Then, we have used the AuRUS [5] UML/OCL reasoner for evaluating the execution times for: (1) checking the correctness of the descriptions, (2) matching data-services with data-requests. We have run all experiments over a Windows 8 in an Intel i7-4710HQ up to 3.5 GHz machine with 8 GB of RAM.

Table 1 shows the results when checking the correctness of a data-service description. The first two checks are aimed at determining whether the input and output parameters, respectively, univocally identify one ontology object. Moreover, we have checked the *liveliness* of the data-service relationship that allows computing the output from the input. That is, we have analyzed whether there

---

[1] Descriptions available at http://www.essi.upc.edu/~xoriol/obf-files/.

Table 1. Checking data-service correct descriptions execution times

	Checking input	Checking output	Checking service liveliness
ParkingService	531 ms	485 ms	546 ms
TrafficStatusService	515 ms	505 ms	531 ms
CarAverageSpeedService	-	516 ms	531 ms

is at least one possible real world domain instantiation I where the described data-service would return, at least, one instance.

Table 2 shows the results for matching the services with some requests. In particular, we have evaluated the matching execution time against three kinds of data-requests: one that was equal to the intended service, another that was subsumed by the service, and a third one that was subsuming the request. It is important to note that for the *CarAverageSpeedService* we could not define a request for exactly matching the data-service. This is because this service only returns data for those streets for which it has a sensor (which is not a selection criteria that can be described through the ontology).

Table 2. Matching data-services with data-requests

	Same request	Subsumed request	Subsumed service
ParkingService	203 ms	203 ms	156 ms
TrafficStatusService	125 ms	203 ms	125 ms
CarAverageSpeedService	-	218 ms	125 ms

As it can be seen, all the performed tests behaved satisfactorily and were executed in ms. It is also worth to remark that those tests aimed at performing data-service matching (i.e. the ones in Table 2) took always below 0.3 s. For these reasons, we believe that our proposed framework works properly for the automated discovery of data-services and it is also feasible in practice since it offers an efficient way to implement the data-service matching problem.

6 Related Work

Our work is based on the *local-as-view* approach to data integration as defined by [6]. In this approach, data-services and requests are defined as queries over a common ontology which describes the real-world. In this manner, the problem of data-service/request matching reduces to database query containment. However, and differently from [6], we distinguish among the concepts that the data-service deals with, from the properties used to describe these concepts. Thus, a data-request or data-service computing exactly the same conceptual relation will match in our framework, whereas they will not match in [6] if they use different properties to refer to the same concepts.

This distinction between concepts and their descriptions also appears in OWL-S [11]. OWL-S is, in essence, an ontology for describing data-services. The main idea is that any data-service can be described through instantiating the OWL-S concepts and properties. Other approaches working similarly are SAWSDL [12] and WSMO/WSML [13] according to a recent survey [14]. However, in the context of IoT, the difficulty to learn these languages has already been claimed to be a barrier to provide semantic descriptions for sensors [15].

We argue that these frameworks are difficult to use because they require instantiating very particular concepts/properties defined over them. In contrast, our framework is purely based on the observation that services can be seen as relationships. Thus, we can model services as any other relationship (without learning new vocabularies). In MOF terms, whereas approaches such as OWL-S or WSMO/WSML look at particular services as instances of the M0 level, M1 as the schema used to describe the service, and M2 as the language that enables so; in our framework, the data retrieved by the service belongs to M0, M1 corresponds to the particular service that relates the data, and M2 is the usual modelling language to specify the relationship (in our case, UML).

It can be argued that we could have chosen OWL instead of UML as our modelling language for defining the services. However, regarding the syntax, UML brings as natural support for specifying n-ary associations (which are mandatory to define services associating more than 2 objects through its input/output), while OWL does not. Under the point of view of semantics, UML/OCL is interpreted under the close-world assumption, thus considering only finite real-world states. In contrast, OWL is interpreted under the open-world assumption, thus considering infinite real-world states. As a consequence, it might happen that two services that compute exactly the same input/output relation are determined to be equal under a UML specification, but to be different in OWL. However, we think it is not realistic to consider infinite real-world states (states with an infinite number of *Parkings*, *Streets*, etc.). It is not clear how approaches based on pure OWL (such as SSWAP [16]) handle these cases.

Regarding the context of IoT, the current proposals that we know for semantically describing IoT sensors to enable their discovery are not fully automatic [1,17,18]. The approach presented in [17] is meant for helping human users to find their desired semantically-described sensors through a GUI and, thus, it implements manual discovery. In the case of [1], and despite pursuing automatic discovery, for the moment its unique way to describe the input/output relation is by means of a single tag annotation, so, its discovery process is reduced to checking the coincidence of such tag, which is a syntactic check rather than a semantic one. On the other side, the proposal in [18], although it can be fully automated, is in essence based on selecting sensors according to non-functional criteria (e.g. time-response, availability, cost) rather than semantics.

7   Conclusions and Future Work

We have presented a framework to specify IoT data-services so that they can be automatically discovered. Our framework specifies a data-service in terms of how it describes its input and output objects (that is, which properties of the objects does it speak about), and how it relates the input to the output objects.

We have seen that these descriptions can be written in UML/OCL. That is, both input and output can be modelled as UML classes related to the concepts they refer to, and the data-service logics can be seen as a new UML association between such classes whose contents are specified by means of a constraint. In this way, the service matching problem is reduced to relation subsumption.

In the experiments, we have demonstrated that our framework could be used to describe and discover three real IoT data-services that are currently being implemented in the BIG IoT European Project. Moreover, we have also shown that current UML/OCL reasoners can be used to solve the matching problem, and obtaining execution times below 0.3 s. As future work, we plan to study how to deal with different data formats and how to chain/orchestrate several data-services to accomplish a data-request.

Acknowledgements. Work partially supported by the BIG IoT project (European Commission's Horizon 2020 research and innovation programme) and by the Spanish Ministerio de Economía, Industría y Competitividad, under project TIN2017-87610-R.

References

1. Bröring, A., Schmid, S., Schindhelm, C., Khelil, A., Käbisch, S., Kramer, D., Phuoc, D.L., Mitic, J., Anicic, D., Teniente, E.: Enabling IoT ecosystems through platform interoperability. IEEE Softw. **34**(1), 54–61 (2017)
2. Papazoglou, M.P., van den Heuvel, W.: Service oriented architectures: approaches, technologies and research issues. VLDB J. **16**(3), 389–415 (2007)
3. Perera, C., Zaslavsky, A.B., Christen, P., Georgakopoulos, D.: Context aware computing for the internet of things: a survey. IEEE Commun. Surv. Tutorials **16**(1), 414–454 (2014)
4. Berardi, D., Calvanese, D., De Giacomo, G.: Reasoning on UML class diagrams. Artif. Intell. **168**(1–2), 70–118 (2005)
5. Rull, G., Farré, C., Queralt, A., Teniente, E., Urpí, T.: AuRUS: explaining the validation of UML/OCL conceptual schemas. Softw. Syst. Model. **14**(2), 953–980 (2015)
6. Levy, A.Y., Rajaraman, A., Ordille, J.J.: Querying heterogeneous information sources using source descriptions. In: VLDB 1996, Proceedings of 22th International Conference on Very Large Data Bases, pp. 251–262 (1996)
7. Olivé, A., Raventós, R.: Modeling events as entities in object-oriented conceptual modeling languages. Data Knowl. Eng. **58**(3), 243–262 (2006)
8. Cunha, A., Garis, A.G., Riesco, D.: Translating between alloy specifications and UML class diagrams annotated with OCL. Softw. Syst. Model. **14**(1), 5–25 (2015)
9. Kuhlmann, M., Hamann, L., Gogolla, M.: Extensive validation of OCL models by integrating SAT solving into USE. In: Bishop, J., Vallecillo, A. (eds.) TOOLS 2011. LNCS, vol. 6705, pp. 290–306. Springer, Heidelberg (2011). https://doi.org/10.1007/978-3-642-21952-8_21

10. Cabot, J., Clarisó, R., Riera, D.: UMLtoCSP: a tool for the formal verification of UML/OCL models using constraint programming. In: 22nd IEEE/ACM International Conference on Automated Software Engineering (ASE 2007), pp. 547–548 (2007)

11. Martin, D.L., Burstein, M.H., McDermott, D.V., McIlraith, S.A., Paolucci, M., Sycara, K.P., McGuinness, D.L., Sirin, E., Srinivasan, N.: Bringing semantics to web services with OWL-S. World Wide Web **3**, 243–277 (2007)

12. Kopecký, J., Vitvar, T., Bournez, C., Farrell, J.: SAWSDL: semantic annotations for WSDL and XML schema. IEEE Internet Comput. **11**(6), 60–67 (2007)

13. de Bruijn, J., Lausen, H., Polleres, A., Fensel, D.: The web service modeling language WSML: an overview. In: Sure, Y., Domingue, J. (eds.) ESWC 2006. LNCS, vol. 4011, pp. 590–604. Springer, Heidelberg (2006). https://doi.org/10.1007/11762256_43

14. Klusch, M., Kapahnke, P., Schulte, S., Lécué, F., Bernstein, A.: Semantic web service search: a brief survey. KI **30**(2), 139–147 (2016)

15. Song, Z., Cárdenas, A.A., Masuoka, R.: Semantic middleware for the internet of things. In: Internet of Things (IOT), IoT for a green Planet, Proceedings (2010)

16. Gessler, D.D., Bulka, B., Sirin, E., Vasquez-Gross, H., Yu, J., Wegrzyn, J.: iPlant SSWAP (simple semantic web architecture and protocol) enables semantic pipelines for biodiversity. In: Semantics for Biodiversity (S4BioDiv), p. 101 (2013)

17. Perera, C., Vasilakos, A.V.: A knowledge-based resource discovery for internet of things. Knowl.-Based Syst. **109**, 122–136 (2016)

18. Perera, C., Zaslavsky, A.B., Christen, P., Compton, M., Georgakopoulos, D.: Context-aware sensor search, selection and ranking model for internet of things middleware. In: IEEE 14th International Conference on Mobile Data Management, vol. 1, pp. 314–322 (2013)

Process Discovery

How Much Event Data Is Enough?
A Statistical Framework for Process Discovery

Martin Bauer[1], Arik Senderovich[2], Avigdor Gal[2], Lars Grunske[1],
and Matthias Weidlich[1(✉)]

[1] Humboldt-Universität zu Berlin, Berlin, Germany
{bauermax,grunske,weidlima}@informatik.hu-berlin.de
[2] Technion – Israel Institute of Technology, Haifa, Israel
sariks@technion.ac.il, avigal@ie.technion.ac.il

Abstract. With the increasing availability of business process related event logs, the scalability of techniques that discover a process model from such logs becomes a performance bottleneck. In particular, exploratory analysis that investigates manifold parameter settings of discovery algorithms, potentially using a software-as-a-service tool, relies on fast response times. However, common approaches for process model discovery always parse and analyse all available event data, whereas a small fraction of a log could have already led to a high-quality model. In this paper, we therefore present a framework for process discovery that relies on statistical pre-processing of an event log and significantly reduce its size by means of sampling. It thereby reduces the runtime and memory footprint of process discovery algorithms, while providing guarantees on the introduced sampling error. Experiments with two public real-world event logs reveal that our approach speeds up state-of-the-art discovery algorithms by a factor of up to 20 .

Keywords: Process discovery · Log pre-processing · Log sampling

1 Introduction

Process mining emerged as a discipline that targets analysis of business processes based on event logs [1]. Such logs are recorded by information systems during the conduct of a process, such that each event denotes the timestamped execution of a business activity for a particular instance of the process. One of the essential tasks in process mining is process discovery—the construction of a model of the process from an event log.

In recent years, a plethora of algorithms have been proposed for process discovery [2]. These algorithms differ along various dimensions, such as the imposed assumptions on the notion of an event log (e.g., atomic events vs. interval events [3,4]), the applied representational bias (e.g., constructing transition

© Springer International Publishing AG, part of Springer Nature 2018
J. Krogstie and H. A. Reijers (Eds.): CAiSE 2018, LNCS 10816, pp. 239–256, 2018.
https://doi.org/10.1007/978-3-319-91563-0_15

systems [5], Petri-nets [6], or BPMN models [7]), the handling of log incompleteness and noise (e.g., heuristics to balance over-fitting and under-fitting [8] or filter noise [9]), and the formal guarantees given for the resulting model (e.g., ensuring deadlock-freedom [10]).

Regardless of the specific algorithm chosen, efficiency becomes an increasingly important requirement for process discovery, due to several reasons. First, the increasing pervasiveness of data sensing and logging mechanisms led to a broad availability of large event logs in business contexts, reaching up to billions of events [11]. Hence, the manifesto of the IEEE Task Force on Process Mining concludes that *'additional efforts are needed to improve performance and scalability'* [12]. Second, discovery algorithms typically have multiple parameters that need to be configured. Since these parameters have a large impact on the resulting model [13], discovery becomes an exploratory analysis rather than a one-off procedure. Third, companies started to offer software-as-a-service solutions for process mining, see Signavio[1] or Lana-Labs[2], which are trouble-some to use if analysis needs to be preceded by an upload of very large logs.

Acknowledging the need for more efficient process discovery, divide-and-conquer strategies enable the decomposition of a discovery problem into smaller ones [14,15]. Also, models for distributed computation may be exploited to increase efficiency [16,17]. All these techniques fundamentally consider *all* available data, conducting a *complete* scan of the event log. Yet, in practice, we observe that a small fraction of a log may already enable discovery of a high-quality model. This holds in particular as logs are not trustworthy and discovery algorithms apply heuristics to cope with incomplete information and noise in the data. Hence, small deviations between models discovered from a complete log and a partial log, respectively, may be considered a result of the inherent uncertainty of the process discovery setting. Following this line, the challenge then becomes to answer the following question: *How to systematically determine how much of data of an event log shall be considered when discovering a process model?*

In this paper, we set out to answer the above question with a framework for statistical pre-processing of an event log. More specifically, this paper contributions are as follows:

(1) A statistical framework for process discovery that is based on an incremental pre-processing of an event log. For each trace, we assess whether it yields new information for the process at hand. By framing this procedure as a series of binomial experiments, we establish well-defined bounds on the error introduced by considering only a sample of the log.

(2) We instantiate this framework for control-flow discovery. That is, we show how discovery algorithms that exploit the directly-follows graph of a log such as the Inductive Miner [9,10] can benefit from this framework.

(3) We handle the enrichment of discovered process models with information on execution times, e.g., to estimate the cycle time of a process. That is, we elaborate on different variants to incorporate the respective information in our framework.

---

[1] https://www.signavio.com/.

[2] https://lana-labs.com/.

In the remainder, we first introduce preliminary notions and notations (Sect. 2), before presenting our statistical framework for process discovery and its instantiation (Sect. 3). An evaluation of our techniques with two real-world datasets reveals that they indeed reduce the runtime and memory footprint of process discovery significantly (Sect. 4). We close with a discussion of related work (Sect. 5) and conclusions (Sect. 6).

2 Preliminaries

This section defines event logs, as well as directly-follows graphs and process trees as two process modelling formalisms, and elaborates on their discovery from event logs.

Event Logs. Following [4,18], we adopt an interval-based notion of an event log. Let $\mathcal{A}$ be a set of activity identifiers (activities, for short) and $\mathcal{T}$ be a totally ordered time domain, e.g., given by the positive real numbers. Then, an event e recorded by an information system is characterised by at least the following information: an activity $e.a \in \mathcal{A}$; a start time $e.s \in \mathcal{T}$; and a completion time $e.c \in \mathcal{T}$ with $e.s \leq e.c$. We say that an event e is instantaneous, if $e.s = e.c$. Furthermore, let $\mathcal{E}$ denote the universe of all possible events. A single exe-

Trace	Activity	Start	Complete
1	R: Receive Claim	9:05	9:05
1	F: Fetch Previous Claim	9:05	9:10
1	P: Plausibility Check	9:08	9:20
1	U: Update Claim Status	9:20	9:22
1	U: Update Claim Status	9:46	9:50
2	R: Receive Claim	9:51	9:51
3	R: Receive Claim	9:55	9:55
2	P: Plausibility Check	9:57	10:03
2	F: Fetch Previous Claim	10:01	10:06
3	F: Fetch Previous Claim	10:12	10:17
3	P: Plausibility Check	10:13	10:22
3	U: Update Claim Status	10:22	10:25

Fig. 1. Example event log.

cution of a process, called a trace, is modelled as a set of events, $\xi \subseteq \mathcal{E}$. However, no event can occur in more than one trace.

An event log is modelled as a set of traces, $L \subseteq 2^{\mathcal{E}}$. To illustrate this notion, Fig. 1 depicts the log of an example claim handling process. It contains three traces, each comprising events that denote the execution of particular activities. Each event has a start and completion time. As a short-hand, we define $A_L = \{a' \in \mathcal{A} \mid \exists\, \xi \in L, e \in \xi : e.a = a'\}$ as the set of activities referenced by events in the log. Using the activity identifiers in Fig. 1, we have $A_L = \{R, F, P, U\}$ for the example.

Process Models. A rather simple formalism to capture the behaviour of a process, which is widely used in process mining tools in practice, is a directly-follows graph (DFG). A DFG is a directed graph $G = (V, E)$. Vertices V denote the activities of a process. Directed edges $E \subseteq V \times V$ model that the target activity can be executed immediately after the source activity in a process instance. A plain DFG may be extended by marking some vertices $V_<, V_> \subseteq V$ as start and completion ver-

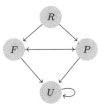

Fig. 2. Example DFG.

tices. Then, a path in the DFG from a start to a completion vertex represents a possible execution sequence of the process. An example DFG is given in Fig. 2. With $V_< = \{R\}$ and $V_> = \{U, F\}$, it defines that $\langle R, F, U \rangle$ would such an execution sequence.

Based on DFGs, richer types of models that feature explicit concepts, e.g., for repetitive behaviour and concurrent execution can be constructed. In this work, we consider the formalism of a process tree [10, 18]. In a process tree, leaf nodes denote activities or a specific silent activity τ. Non-leaf nodes are control-flow operators, such as sequence ($\rightarrow$), exclusive choice ($\times$), concurrency ($\wedge$), interleaving ($\|$) and structured loops ($\circlearrowleft$). Given a process tree, a set of execution sequences of activities is constructed recursively. For a leaf node, this set contains a single execution sequence, consisting of the respective activity (the sequence is empty for the silent activity). For non-leaf nodes, semantics is induced by a function that joins the execution sequences of the subtrees of the node. The process tree given in Fig. 3, e.g., defines that $\langle R, P, F, U, U \rangle$ would be an execution sequence of the process.

Fig. 3. Process tree.

Models such as process trees may also be annotated with additional information, such as times or costs of activity execution. Adopting the model from [4], for instance, each activity is assigned a duration in terms of a cumulative distribution function (CDF). Semantics of such a timed process tree are no longer given in terms of execution sequences of activities, but in terms of events, which, as defined above, associate activity executions with start and completion times. To this end, an execution sequence of the untimed tree is enriched by constructing the start time for each activity execution based on the completion time of the previous activity, while the completion time is then determined by drawing a duration from the respective CDF.

Process Discovery. Process discovery constructs a process model from an event log. With $\mathcal{L}$ and $\mathcal{M}$ as the universe of event logs and process models, respectively, a discovery algorithm can thus be seen as a function $\rho : \mathcal{L} \rightarrow \mathcal{M}$.

As mentioned earlier, a large number of specific algorithms have been proposed in recent years [1, 2]. Referring to the above formalisms for process models, a DFG may trivially be constructed from a log L, once a total order $\succ_\xi$ has been defined for the events of each trace $\xi \in L$. For an interval-based log, this order may be derived from either the start times or the completion times (breaking ties, if needed). Then, a DFG is discovered as $(A_L, \succ)$ with $\succ = \bigcup_{\xi \in L} \succ_\xi$. This simple construction of a DFG may be adapted through frequency-based techniques for noise filtering [5]. The DFG in Fig. 2 is obtained from the log in Fig. 1, when ordering events by their completion time.

Process trees may be discovered by various variants of the Inductive Miner. In its basic version, this algorithm relies on the DFG and iteratively identifies cuts, specific sets of vertices, in the graph to build a process tree [10]. The tree in Fig. 3 would be discovered for the log in Fig. 1, using the DFG in Fig. 2. Variants of the miner target robustness against noise [9] or exploit start and completion times to distinguish interleaved from concurrent execution [18] and detect delays between activities [4].

3 A Statistical Framework for Process Discovery

This section introduces our approach to statistical pre-processing of an event log. By reducing a log's volume through sampling of traces until no new information is encountered, the efficiency of process discovery algorithms is improved. Below, we first give an overview of the general framework (Sect. 3.1), before we instantiate it for control-flow discovery (Sect. 3.2) and extraction of performance information (Sect. 3.3).

3.1 Statistical Pre-processing of Event Logs

The general idea of our approach is to avoid a full scan of an event log during process discovery. This is to cope with the phenomena of traces that contain highly redundant information. Specifically, even though each event in a trace constitutes original information, process discovery typically works only on abstractions of events and traces. As a consequence, many events and traces are considered to be equivalent or highly similar by a discovery algorithm. Hence, it is reasonable to assume that process discovery based on a representative subset of traces (and thus events) of a log can be expected to yield a highly similar result compared to the model obtained by processing a complete log. Based on this assumption, we propose the following statistical approach to reason on the 'discovery sufficiency' of a given subset of traces.

A Statistical View on Log Sampling. When parsing a log trace-by-trace, some traces may turn out to be similar to those already encountered. As motivated above, this similarity is assessed in terms of the information used by a discovery algorithm. On a conceptual level, this is captured by a trace abstraction function $\psi : 2^{\mathcal{E}} \to \mathcal{X}$ that extracts from a trace, information of some domain $\mathcal{X}$ that is relevant for discovery. This information may, for example, relate to the occurrence of activities, their ordering dependencies, or quantitative information, and needs to be instantiated for a specific discovery algorithm.

Based on the trace abstraction function, one can define a random Boolean predicate $\gamma(L', \xi)$ that captures whether a trace $\xi \in 2^{\mathcal{E}}$ provides new information with respect to an event log $L' \subseteq 2^{\mathcal{E}}$. Here, in a strict sense, trace ξ provides novel information solely if its abstraction is not part of those jointly derived for all traces in L':

$$\gamma(L', \xi) \Leftrightarrow \psi(\xi) \notin \bigcup_{\xi' \in L'} \psi(\xi'). \tag{1}$$

In some cases, however, the abstraction function of a trace yields information of rather fine granularity, e.g., numerical values that represent the time or cost of activity execution. If so, it is reasonable to also consider a relaxed notion of new information. Assuming that the distance between the abstract information provided by traces can be quantified by a function $d : \mathcal{X} \times \mathcal{X} \to \mathbb{R}_0^+$ and given a relaxation parameter $\epsilon \in \mathbb{R}_0^+$, we therefore also consider the following predicate:

$$\gamma^{\epsilon}(L', \xi) \Leftrightarrow d\left(\psi(\xi), \bigcup_{\xi' \in L'} \psi(\xi')\right) > \epsilon. \tag{2}$$

In other words, if the observed abstraction $\psi(\xi')$ is ϵ far from L', then $\psi(\xi')$ adds new information beyond L'.

Next, we define the *discovery sufficiency* property for an event log $L' \subseteq L$, with respect to an abstraction ψ, a relaxation parameter ϵ, and some probability measure δ.

Definition 1 (Discovery Sufficiency). *An event log $L' \subseteq L$ is called (δ, ϵ, ψ)-discovery sufficient, if for a newly sampled trace $\xi : \xi \in (L \setminus L')$, it holds that:*

$$p_{\gamma}(L', \xi) = P(\gamma(L', \xi) = 1) < \delta, \tag{3}$$

with P being a standard probability measure.

Essentially, discovery sufficiency of L' requires that the chance of a newly sampled trace to add new information beyond L' is bounded by δ. Our goal is to come up with a (δ, ϵ, ψ)-discovery sufficient log L' based on the original log L. When it is clear from the context, we shall use discovery sufficiency without its parameters.

We now formulate a statistical hypothesis testing procedure to assess when *enough* information has been extracted from the log, thereby giving a criterion to terminate the construction of L'. Then, we shall present an algorithm that is based on the procedure.

Given $L' \subseteq L$, we would like to test whether it is discovery sufficient. We assume under the null hypothesis that L' is sufficient; or not, under the alternative hypothesis. Next, we consider an independent and identically distributed (i.id.) sample of N traces $\xi_1, \ldots, \xi_N$ from L, such that $\xi_i \neq L'$. We denote $p = p_i = p_{\gamma}(L', \xi_i)$ the probability that ξ_i will yield new information, i.e., the probability that $\gamma(L', \xi_i) = 1$. By the i.id. assumption, the probability p_i to have new information in ξ_i is the same for all ξ_i. Hence, under the null hypothesis, it holds $p = 0$; under the alternative, we hypothesise $p > 0$. Further, let k be the number of traces of the N samples that bring new information. Under our assumptions, k is binomially distributed, with parameters N and $1 - p$.

To assure that L' is discovery sufficient, we wish to bound the probability $p_{\gamma}(L', \xi)$ by δ for new ξ. In statistical terms, we wish to bound the probability that the null hypothesis is not rejected, given that the alternative is true for L'. Furthermore, we want to provide a statistically significant answer, with significance level α. To ensure α, and bound $p_{\gamma}(L', \xi)$ with δ, we must select an appropriate sample size N_{min}, which is referred to as *minimum sample size* [19]. Using the normal approximation to the binomial distribution, the minimal sample size is given by,

$$N_{min} \geq \frac{1}{2\delta}\left(-2\delta^2 + z^2 + \sqrt{z}\right), \tag{4}$$

with z corresponding to the realisation of a standardised normal random variable for $1 - \alpha$ (one-sided hypothesis test). For $\alpha = 0.01$ and $\delta = 0.05$, for example, we obtain $N_{min} \geq 128$. After seeing 128 traces without new information, sampling can be stopped knowing with 0.99 confidence that the probability of finding new information in the remaining log is less than 0.05. We apply this idea in what follows.

Algorithm 1: Statistical Framework for Process Discovery

---

input : L, an event log,
$\qquad\quad$ α, a significance value,
$\qquad\quad$ δ, a similarity value,
$\qquad\quad$ γ, a predicate that holds true, if a trace provides new information,
$\qquad\quad$ ρ, a discovery algorithm.
output : $\hat{m}$, a process model.

1 $\;L', \hat{L} \leftarrow \emptyset$; $\qquad$ /* The sampled logs, overall and for current experiment */
2 $\;i \leftarrow 0$; $\qquad\qquad$ /* The number of current iterations without new information */
$\;\;$ /* Determine the number of failed trails to be observed, based on Eq. 4 */
3 $\;N \leftarrow 1/(2\delta)\left(-2\delta^2 + z^2 + \sqrt{z}\right)$ with z of $(1 - \alpha)$;

4 **repeat** $\quad$ /* Repeat until N traces without new information have been seen */
5 $\quad\;\xi \leftarrow select(L \setminus \hat{L})$; $\qquad\qquad\qquad$ /* Sample a single trace */
6 $\quad\;$**if** $\gamma(\xi, L')$ **then** $\qquad\qquad$ /* If ξ provides new information */
7 $\quad\quad\;\;i \leftarrow 0$; $\qquad\qquad\qquad\qquad\qquad$ /* Reset the counter */
8 $\quad\quad\;\;\hat{L} \leftarrow \emptyset$; $\qquad\qquad$ /* Reset sampled log for current experiment */
9 $\quad\;$**else**
10 $\quad\quad\;\;i \leftarrow i + 1$; $\qquad\qquad\qquad\qquad$ /* Increment the counter */
11 $\quad\quad\;\;\hat{L} \leftarrow \hat{L} \cup \{\xi\}$; $\quad$ /* Add trace to sampled log for current experiment */
12 $\quad\;$**end**
13 $\quad\;L' \leftarrow L' \cup \{\xi\}$; $\qquad\qquad$ /* Add trace to overall sampled log */
14 **until** $i \geq N \vee \hat{L} = L$;
15 **return** $\hat{m} = \rho(L')$; $\qquad$ /* Return model discovered from overall sampled log */

Statistical Framework for Process Discovery. The above observations are exploited as formalised in Algorithm 1. This statistical framework for process discovery takes as input an event log, a significance level α, a sufficiency bound δ, a predicate to determine whether a trace provides new information, and a discovery algorithm. After initialisation, in line 3, the algorithm computes the number of trails that need to fail, i.e., the number of consecutive traces that do not provide new information, according to Eq. 4. Then, it samples traces from the log (line 4–line 14). For each trace, it determines whether new information has been obtained (line 6). If so, an iteration counter is reset. Once the counter indicates that N traces without new information have been sampled, the procedure terminates by applying the mining algorithm to the sampled log.

From a technical point of view, the framework in Algorithm 1 is independent of the chosen discovery algorithm, which is applied only once a sampled log has been obtained. However, it relies on a predicate to decide whether a trace provides new information over a set of traces. This predicate, in turn, shall be devised with respect to the properties of a trace that are exploited by the discovery algorithm. Consequently, the effectiveness of the approach is not only affected by the configuration of Algorithm 1 in terms of statistical significance and sufficiency bound, but depends also on the definition of the respective predicate. How often it holds true is influenced by the relation of the richness of the employed trace abstraction and the variability of the event data to the overall size of the original log. With richer abstractions and higher variability, more traces will provide new information, so that the size of the sampled log approaches the size of the original log.

3.2   Instantiation for Control-Flow Discovery

Next, we turn to the instantiation of the above framework for control-flow discovery, which yields a process model comprising a set of activities along with their execution dependencies. According to the model introduced in Sect. 3.1, this requires us to define a suitable trace abstraction function. Based thereon, using predicate $\gamma(L', \xi)$ of Eq. 1, it is determined, whether a trace ξ provides new information over a log L'.

As a particular example, we consider the Inductive Miner [10] for the definition of this predicate. As detailed in Sect. 2, this miner constructs a DFG and then, deterministically, builds a process tree from this graph. Hence, a trace provides new information, if the sets of vertices or edges of the DFG change. Considering also explicit start and completion vertices (not exploited by the Inductive Miner, but useful when employing the DFG directly), we define a trace abstraction function as follows. Given a total order of events in a trace (see Sect. 2), the function yields a tuple of the contained activities, their order, and the minimal and maximal elements induced by that order:

$$\psi_{IM}(\xi) \mapsto \left(A_{\{\xi\}}, \succ_\xi, \min_{\succ_\xi}(\xi), \max_{\succ_\xi}(\xi) \right) \tag{5}$$

Lifting union of sets and containment of elements in sets to tuples of sets, the predicate $\gamma(L', \xi)$ based on the trace abstraction function ψ_{IM} holds true, if, intuitively, trace ξ shows a new activity, a new order dependency, or new information about activities starting or completing the process.

As an example, consider the information extracted by the above function from the first two traces of the log in Fig. 1. Here, $\psi_{IM}(\xi_1) \cup \psi_{IM}(\xi_2)$ yields the following information in terms of known activities, their order dependencies, and the minimal and maximal elements: $(\{R, F, P, U\}, \{(R, F), (R, P), (F, P), (P, F), (P, U), (U, U)\}, \{R\}, \{F, U\})$. We note that this information is sufficient to construct the DFG in Fig. 2. Hence, adding the third trace of the example log does not provide any new information, as $\psi_{IM}(\xi_3) = (\{R, F, P, U\}, \{(R, F), (F, P), (P, U)\}, \{R\}, \{U\})$.

When aiming at process discovery with other variants of the Inductive Miner, the above trace abstraction may also be applied, even though the respective discovery algorithms exploit further information. For instance, the Inductive Miner Infrequent [9] incorporates relative frequencies of edges in the DFG to cope with noise in event logs. On the one hand, these frequencies may be neglected during sampling from the event log (applying the above abstraction) and only be derived from the selected sample of traces. In that case, the guarantee in terms of δ-similarity of the sampled log would not cover these frequencies, though. A different approach, thus, is to consider the relative edge frequencies in the abstraction, so that they influence the predicate to assess whether a trace contains new information. As these relative frequencies can be expected to change slightly with each sampled trace, the relaxed predicate, see Eq. 2, needs to be applied. Then, the relative edge frequencies stabilise during sampling and traces that incur only minor changes in these frequencies are not considered to provide new information.

3.3   Instantiation for Performance Discovery

We complement the above discussion with an instantiation of the framework that incorporates additional aspects of a trace, beyond execution dependencies. That is, we consider the extraction of performance details in terms of the cycle time, i.e., the time from start to completion of the process. However, further aspects of a trace, such as costs induced by a trace or the involved roles, may be integrated in a similar manner.

The cycle time of a process is commonly captured as a numerical value at a rather fine granularity. Consequently, observations of the cycle time of particular traces will show some variability—if at all, only very few traces will have the same cycle time. According to the model presented in Sect. 3.1, this suggests to rely on the relaxed predicate $\gamma^\epsilon(L', \xi)$, see Eq. 2. It allows for a certain deviation, bound by ϵ, of the information provided by ξ from the one known already from L', while still considering ξ as providing no new information. Without this relaxation, all traces with different cycle times would add new information, even if the absolute difference of these times is negligible. To incorporate information on cycle time and instantiate predicate $\gamma^\epsilon(L', \xi)$, we need to define a respective trace abstraction function and a distance function, as detailed below.

Model-Based and Activity-Based Cycle Time Approximation. The estimation of the average cycle time of a process based on a log may be approached in different ways. One solution is to compute the average over the cycle times observed for all traces, considering each trace as an independent observation. Another solution is to compute the cycle time analytically based on the average durations of activities. The latter approach builds a performance annotated process model such as the timed process tree discussed in Sect. 2, fitting a distribution for the duration of an activity based on all observations.

Either approach is realised in our framework by specific trace abstraction functions. If the cycle time is considered per trace, referred to as *model-based approximation*, the abstraction captures the time between the start of the first event and the completion time of the last event of a trace (again, $\succ_\xi$ is the total order of events in trace ξ):

$$\psi_{tCT}(\xi) \mapsto \{e_2.c - e_1.s\} \quad \text{with} \quad e_1 = \min_{\succ_\xi}(\xi) \quad \text{and} \quad e_2 = \max_{\succ_\xi}(\xi). \quad (6)$$

If the cycle time is captured on the level of activities (*activity-based approximation*), the trace abstraction function yields the observed durations per activity:

$$\psi_{aCT}(\xi) \mapsto \bigcup_{a' \in A_{\{\xi\}}} \left\{ \left(a, \operatorname*{avg}_{e \in \xi, e.a = a'} (e.c - e.s) \right) \right\} \tag{7}$$

For the first trace in the log in Fig. 1, these abstractions yield $\psi_{tCT}(\xi_1) = 45$ (minutes) for the model-based approximation, and $\psi_{aCT}(\xi_1) = \{(R, 0), (F, 5), (P, 12), (U, 3)\}$ for activity-based approximation, where $(U, 3)$ stems from two events related to U in ξ_1.

Difference of Cycle Time Approximations. Using one of the above trace abstractions, instantiation of the predicate $\gamma^\epsilon(L', \xi)$ further requires the definition of a distance function, to assess whether the information provided by ξ deviates with less than ϵ from the information provided by L'. To this end, we compare the cycle times computed based on the measurements from L' and computed based on L' and ξ. Specifically, while adding traces to L', the cycle time computation is expected to converge, so that the difference will continuously fall below ϵ at some point.

We illustrate the definition of this difference for the trace abstraction function ψ_{tCT}. Then, given a trace ξ, the singleton set with its cycle time $\psi_{tCT}(\xi) = CT_\xi$, and the cycle times of all traces in the sampled log $\bigcup_{\xi \in L'} \psi_{tCT}(\xi) = CT_{L'}$, this difference is measured as the change in the average of cycle times, if trace ξ is incorporated:

$$d(CT_\xi, CT_{L'}) \mapsto \left| \operatorname*{avg}_{ct \in CT_{L'}} (ct) - \operatorname*{avg}_{ct \in CT_{L'} \cup CT_\xi} (ct) \right|. \tag{8}$$

If this difference is smaller than ϵ, the cycle time measurement provided by trace ξ is negligible and, thus, the trace is considered to not provide any new information. It is worth to mention that sequence of average cycle time measurements obtained when adding traces to L' corresponds to a Cauchy-sequence, as the sequence converges and the space of possible cycle times is a complete metric space. Hence, it can be concluded, for any chosen ϵ, that after a certain number of traces have been incorporated, adding additional traces will not increase the difference above ϵ again.

Referring to our running example, the log in Fig. 1, consider the case of having sampled the first and second trace already. Then, with $L' = \{\xi_1, \xi_2\}$, we obtain $CT_{L'} = \{45, 15\}$. Trace ξ_3 with $CT_{\xi_3} = \{30\}$ does not add new information, as the average of values in $CT_{L'}$ and the average of values in $CT_{L'} \cup CT_{\xi_3}$ are equivalent.

The definition of the difference function for the case of ψ_{aCT} uses the same principle, applied to individual activities. For each activity duration, we assess, whether the average of values observed in L' is changed by more than ϵ, if the values in ξ are incorporated.

Inspecting the third trace ξ_3 of the log in Fig. 1, again, it would not provide new information: Activity R is instantaneous and F has the same duration

(5 min) in all traces. For activities P and U, the durations in ξ_3 correspond to the averages of durations observed in traces ξ_1 and ξ_2.

Using the above functions means that the effectiveness of sampling depends on the selection of parameter ϵ and the variability of cycle times (or activity durations) in the log. A high value for ϵ means that the information provided by more traces will be considered to be negligible, leading to a smaller sampled log as the basis for discovery. We later explore this aspect in our experimental evaluation.

4   Experimental Evaluation

This section reports on an experimental evaluation of our statistical framework for process discovery. Using two real-world datasets, we instantiated the framework with the Inductive Miner Infrequent, a state-of-the-art algorithm, for the construction of timed process trees. Our results indicate that the runtime of the algorithm can be reduced by a factor of up to 20. Below, we elaborate on datasets and the experimental setup (Sect. 4.1), before turning to the actual results (Sect. 4.2).

4.1   Datasets and Experimental Setup

Datasets. We relied on two event logs that have been published as part of the Business Process Intelligence (BPI) Challenge and are, therefore, publicly available.

BPI-2012 [20] is a log of a process for loan or overdraft applications at a Dutch financial institute. It contains 262,200 events of 13,087 traces.

BPI-2014 [21] is the log of an incident management process as run by Rabobank Group ICT. For the experiments, we employed the event log of incidence activities, which comprises 343,121 events of 46,616 traces.

Discovery Algorithms. For our experiments, we adopted the Inductive Miner Infrequent (IMI) [9] as a discovery algorithm. This choice is motivated as follows: IMI discovers a process tree and, thus, supports common control-flow structures, such as exclusive choices and concurrency. Second, IMI features a frequency-based handling of noise in event logs, making it suitable for the application to real-world event logs.

In the experiments, the IMI is applied to construct a process tree using a 20% noise filtering threshold. In addition, we collect the performance details needed to quantify the cycle time of the process from the log. Hence, our framework is instantiated with a combination of two trace abstraction functions, one focusing on control-flow information (Eq. 5) and one extracting performance details. Regarding the latter, we employ the model-based cycle time approximation (Eq. 6) as a default strategy. In one experiment, however, we also compare this strategy with the activity-based approximation.

When sampling the event log, we set $\alpha = 0.01$ and $\delta = 0.99$, so that the information loss is expected to be small. Due to the consideration of performance details, we use the relaxed version of the predicate to indicate new information (Eq. 2). Hence, the relaxation parameter ϵ of this predicate becomes a controlled variable in our experiments.

Measures. We measure the effectiveness of our statistical pre-processing by the size of the sampled log, i.e., the *number of traces*, and relate it to the size of the original log.

As the effectiveness of sampling is an analytical, indirect measure for the efficiency of our approach, we also quantify its actual efficiency as implemented in ProM [22]. To this end, we measure the total *runtime* (in ms) and total *memory footprint* of the complete discovery procedure, contrasting both measures with the plain IMI implementation.

To explore a potential loss of information incurred by our approach, we also compare the quality of the models obtained with IMI and our approach. Here, the control-flow dimensions is considered by measuring the *fitness* [23] between the model and the event log, while we consider performance information in terms of the *approximated cycle time*.

For all measures, we report the mean average of 100 experimental runs, along with the 10th and 90th percentiles.

Experimental Setup. We implemented our approach based on the IMI implementation within ProM [22]. Specifically, we developed two ProM plugins, one targeting the direct application of our approach by users and one conducting the experiments reported in the remainder of this section. Both plugins are publicly available at github.[3]

Data on the efficiency of our statistical framework for process discovery has been obtained on a PC (Dual-Core, 2.5 GHz, 8 GB RAM) running Oracle Java 1.8.

4.2 Experimental Results

The first experiment concerned the effectiveness of the statistical pre-processing of the event log. For both datasets, Fig. 4 illustrates the number of traces that are included in the sampled log by the statistical version of IMI (denoted by sIMI), when varying the relaxation parameter ϵ. As a reference point, the flat blue line denotes the size of the complete event log as used by the plain version of the IMI. Choosing the smallest value for the relaxation parameter means that virtually every trace provides new information, so that pre-processing has no effect. For any slightly higher value, however, the sampled log is significantly smaller than the original one. Already for ϵ values of around 20 (corresponding to deviations in performance details of 20 min for a process that runs for 9 days on average), the sampled log contains only a small fraction of all traces. This provides us

---

[3] https://github.com/Martin-Bauer/StatisticalInductiveMinerInfrequent.

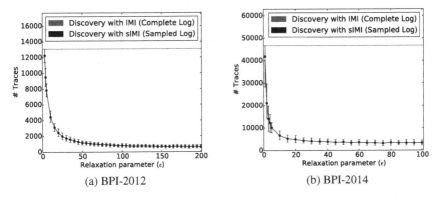

Fig. 4. Effectiveness of process discovery: # Traces vs. relaxation parameter. (Color figure online)

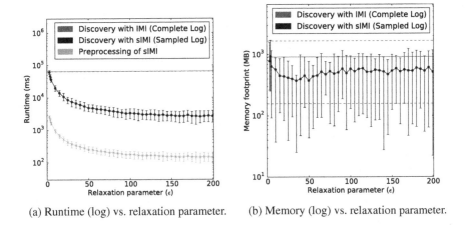

(a) Runtime (log) vs. relaxation parameter.   (b) Memory (log) vs. relaxation parameter.

Fig. 5. Efficiency of process discovery for the BPI-2012 event log. (Color figure online)

with evidence that our approach can indeed reduce the volume of event data in process discovery significantly.

The drastic reduction of the size of the event log immediately improves the actual efficiency of process discovery. Figure 5 shows the runtime and memory footprint measured when running the statistical and the plain version of the IMI in ProM for the BPI-2012 dataset (BPI-2014 yields the very same trends). Runtime drops drastically when increasing the relaxation parameter (note the logarithmic scale). Compared to the runtime of the plain algorithm, we observe speed-ups by a factor of 20. Figure 5a further illustrates that the time spent in the pre-processing of the event log is relatively small, compared to the overall runtime of the discovery procedure.

In terms of memory consumption, the statistical variant of the IMI is more efficient than the plain one, see Fig. 5b. While there is a large variability in the

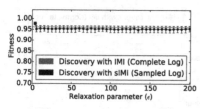
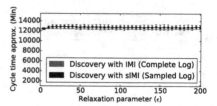

(a) Fitness vs. relaxation parameter.   (b) Cycle time approx. vs. relaxation parameter.

Fig. 6. Quality of the resulting models in terms of fitness and cycle time approximation for the BPI-2012 event log.

measurements, we see that this variability is also observed for the implementation of the plain IMI in ProM (blue dashed lines denote the 10th and 90th percentiles). However, the statistical variant of the IMI has a consistently smaller memory footprint.

Next, Fig. 6 illustrates results on the quality of the obtained models. For the BPI-2012 dataset, the fitness measured between the discovered model and the event log is shown in Fig. 6a, also in relation to the fitness of the model mined by the IMI for the complete log. The results demonstrate that, while the exact same fitness value is not guaranteed to be obtained due to noise filtering by the IMI, the observed deviations are arguably negligible. For the BPI-2014 dataset (not visualised), the plain IMI and the statistical variant achieve a fitness value of 1.0 in all configurations and experiment runs.

Figure 6b further demonstrates, for the BPI-2012 dataset, that also the cycle times approximations obtained with the models discovered from the sampled log are relatively close to the one derived from the complete log, even for very large relaxation parameters. The same trend also materialised for the BPI-2014 dataset.

Finally, we turn to the two strategies for cycle time estimation, see Sect. 3.3. For the BPI-2012 dataset, Fig. 7 compares the model-based strategy with the activity-based one in terms of the sampling effectiveness. Both variants show the same trend. It turns out, though, that the activity-based strategy yields a smaller log for the same ϵ value. This is explained by the fact that this strategy considers performance details at a more fine-granular level. That is, deviations of size ϵ need to be observed on average for all activities, instead of the complete trace, in order to mark a trace as containing new information.

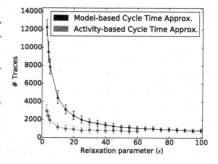

Fig. 7. Cycle time approx. strategies.

5   Related Work

Having discussed some traditional approaches to process discovery in Sect. 1, below, we review literature on incremental and online process discovery, and sampling methods.

Incremental and Online Discovery. In the Big Data era, it is essential to consider process discovery from streams of data. Mining procedural and declarative process models from event streams was first proposed in [24], and [25], respectively. These works propose novel discovery algorithms for settings where continuously arriving data cannot be fully stored in memory. Subsequently, a more generic event processing architecture was proposed by [26], thus allowing to re-use existing discovery algorithms such as the α-miner [27], and the inductive miner [9,10], on streams of events. In this paper, we apply our statistical sampling method to finite historical data, which is large enough to incur high performance costs, yet exists as a batch.

While most online methods for process discovery rely on incremental processing, the latter is relevant also in the context of event logs. For instance, it has been argued that process discovery based on region theory and Petri net synthesis shall be conducted incrementally, to avoid full re-computation of the model when a new log becomes available [28]. Incremental techniques are also motivated by performance considerations. As mentioned earlier, divide-and-conquer strategies to decompose discovery problems have been explored for mining based on directly follows graphs [14,15] as well as for region-based approaches [29]. Unlike our approach, however, these methods always consider the complete event log.

Sampling Sequences. The statistical part of our approach is based on sampling from sequence databases, i.e., datasets that contain traces, similarly to process logs. In process discovery, it has been suggested to sample Parikh vector of traces from a log to increase processing efficiency [30]. Yet, this approach does not give sample size guarantees, as detailed in our work. In the related field of specification mining, the notion of k-confidence has been introduced to capture the probability that a log is complete [33], yet without providing statistical guarantees. The latter was addressed by Busany and Maoz [31], who developed theoretical lower bounds on the number of samples required to perform inference. The approach was instantiated for the k-tails algorithm [32] and mining of Markov Chains. Our work adopts this general idea, yet focuses on trace abstractions used by common process discovery algorithms.

6   Conclusion

In this paper, we developed an approach to increase the efficiency of common process discovery algorithms. We argued that, instead of parsing and analysing all available event data, a small fraction of a log can be sufficient to discover a process model of high quality. Following this line, we presented a statistical framework for process discovery. In a pre-processing step, a sampled log is obtained, while still providing guarantees on the introduced error in terms of the probability of missing information. We instantiated this framework for a state-of-the-art algorithm for control-flow discovery, namely the Inductive Miner Infrequent, and also showed how the framework is used when discovering performance details about a process. Our experiments with two publicly available real-world event logs revealed that the pre-processing indeed yields a drastic decrease in runtime and reduces the memory footprint of a ProM-based implementation of the respective mining algorithm. At the same time, the resulting models show only minor deviations compared to the model mined from the complete log.

Note that the presented instantiation of our framework, in terms of trace abstraction functions, is applicable for a broad range of mining algorithms. Our abstraction for control-flow information relies on directly-follows graphs (DFG), a model employed, e.g., by the family of α-algorithms [34] or mining based on activity dependencies [35]. Similarly, the abstraction for performance information captures what is incorporated by common performance-aware variants of heuristic discovery, e.g., as implemented in Disco.[4] For algorithms that extract further information from traces, the respective abstractions need to be adapted. For instance, for delay-aware process discovery as introduced in [4], the abstraction would include interval relations between activities.

References

1. van der Aalst, W.M.P.: Process Mining - Data Science in Action. Springer, Heidelberg (2016)
2. Augusto, A., Conforti, R., Dumas, M., Rosa, M.L., Maggi, F.M., Marrella, A., Mecella, M., Soo, A.: Automated discovery of process models from event logs: review and benchmark. CoRR abs/1705.02288 (2017)
3. Wen, L., Wang, J., van der Aalst, W.M.P., Huang, B., Sun, J.: A novel approach for process mining based on event types. J. Intell. Inf. Syst. 32(2), 163–190 (2009)
4. Senderovich, A., Weidlich, M., Gal, A.: Temporal network representation of event logs for improved performance modelling in business processes. In: Carmona, J., Engels, G., Kumar, A. (eds.) BPM 2017. LNCS, vol. 10445, pp. 3–21. Springer, Cham (2017). https://doi.org/10.1007/978-3-319-65000-5_1
5. Weijters, A.J.M.M., van der Aalst, W.M.P.: Rediscovering workflow models from event-based data using little thumb. Integr. Comput. Aided Eng. 10(2), 151–162 (2003)
6. Solé, M., Carmona, J.: Process mining from a basis of state regions. In: Lilius, J., Penczek, W. (eds.) PETRI NETS 2010. LNCS, vol. 6128, pp. 226–245. Springer, Heidelberg (2010). https://doi.org/10.1007/978-3-642-13675-7_14
7. Conforti, R., Dumas, M., García-Bañuelos, L., Rosa, M.L.: BPMN miner: automated discovery of BPMN process models with hierarchical structure. Inf. Syst. 56, 284–303 (2016)
8. van Zelst, S.J., van Dongen, B.F., van der Aalst, W.M.P.: Avoiding over-fitting in ILP-based process discovery. In: Motahari-Nezhad, H.R., Recker, J., Weidlich, M. (eds.) BPM 2015. LNCS, vol. 9253, pp. 163–171. Springer, Cham (2015). https://doi.org/10.1007/978-3-319-23063-4_10

---

[4] https://fluxicon.com/disco/.

9. Leemans, S.J.J., Fahland, D., van der Aalst, W.M.P.: Discovering block-structured process models from event logs containing infrequent behaviour. In: Lohmann, N., Song, M., Wohed, P. (eds.) BPM 2013. LNBIP, vol. 171, pp. 66–78. Springer, Cham (2014). https://doi.org/10.1007/978-3-319-06257-0_6

10. Leemans, S.J.J., Fahland, D., van der Aalst, W.M.P.: Discovering block-structured process models from event logs - a constructive approach. In: Colom, J.-M., Desel, J. (eds.) PETRI NETS 2013. LNCS, vol. 7927, pp. 311–329. Springer, Heidelberg (2013). https://doi.org/10.1007/978-3-642-38697-8_17

11. van der Aalst, W.M.P.: Data scientist: the engineer of the future. In: Mertins, K., Bénaben, F., Poler, R., Bourrières, J.-P. (eds.) Enterprise Interoperability VI. PIC, vol. 7, pp. 13–26. Springer, Cham (2014). https://doi.org/10.1007/978-3-319-04948-9_2

12. van der Aalst, W.M.P.: Process mining manifesto. In: Daniel, F., Barkaoui, K., Dustdar, S. (eds.) BPM 2011, Part I. LNBIP, vol. 99, pp. 169–194. Springer, Heidelberg (2012). https://doi.org/10.1007/978-3-642-28108-2_19

13. Solé, M., Carmona, J.: Region-based foldings in process discovery. IEEE Trans. Knowl. Data Eng. **25**(1), 192–205 (2013)

14. van der Aalst, W.M.P., Verbeek, H.M.W.: Process discovery and conformance checking using passages. Fundam. Inform. **131**(1), 103–138 (2014)

15. Leemans, S.J.J., Fahland, D., van der Aalst, W.M.P.: Scalable process discovery with guarantees. In: Gaaloul, K., Schmidt, R., Nurcan, S., Guerreiro, S., Ma, Q. (eds.) CAiSE 2015. LNBIP, vol. 214, pp. 85–101. Springer, Cham (2015). https://doi.org/10.1007/978-3-319-19237-6_6

16. Wang, S., Lo, D., Jiang, L., Maoz, S., Budi, A.: Scalable Parallelization of Specification Mining Using Distributed Computing, pp. 623–648. Morgan Kaufmann, Boston (2015)

17. Evermann, J.: Scalable process discovery using map-reduce. IEEE TSC **9**(3), 469–481 (2016)

18. Leemans, S.J.J., Fahland, D., van der Aalst, W.M.P.: Using life cycle information in process discovery. In: Reichert, M., Reijers, H.A. (eds.) BPM 2015. LNBIP, vol. 256, pp. 204–217. Springer, Cham (2016). https://doi.org/10.1007/978-3-319-42887-1_17

19. Fleiss, J.L., Levin, B., Paik, M.C.: Statistical Methods for Rates and Proportions. Wiley, New York (2013)

20. Van Dongen, B.: BPI Challenge 2012 (2012). https://doi.org/10.4121/uuid:3926db30-f712-4394-aebc-75976070e91f

21. Van Dongen, B.: BPI Challenge 2014 (2014). https://doi.org/10.4121/uuid:c3e5d162-0cfd-4bb0-bd82-af5268819c35

22. Verbeek, E., Buijs, J.C.A.M., van Dongen, B.F., van der Aalst, W.M.P.: Prom 6: the process mining toolkit. In: BPM Demos. CEUR, vol. 615. CEUR-WS.org (2010)

23. van der Aalst, W.M.P., Adriansyah, A., van Dongen, B.F.: Replaying history on process models for conformance checking and performance analysis. Wiley Interdisc. Rew. Data Mining Knowl. Discov. **2**(2), 182–192 (2012)

24. Burattin, A., Sperduti, A., van der Aalst, W.M.P.: Control-flow discovery from event streams. In: IEEE CEC, pp. 2420–2427. IEEE (2014)

25. Burattin, A., Cimitile, M., Maggi, F.M., Sperduti, A.: Online discovery of declarative process models from event streams. IEEE TSC **8**(6), 833–846 (2015)

26. van Zelst, S.J., van Dongen, B.F., van der Aalst, W.M.P.: Event stream-based process discovery using abstract representations. CoRR abs/1704.08101 (2017)

27. Van der Aalst, W., Weijters, T., Maruster, L.: Workflow mining: discovering process models from event logs. IEEE TKDE **16**(9), 1128–1142 (2004)
28. Badouel, E., Schlachter, U.: Incremental process discovery using petri net synthesis. Fundam. Inform. **154**(1–4), 1–13 (2017)
29. Solé, M., Carmona, J.: Incremental process discovery. TOPNOC **5**, 221–242 (2012)
30. Carmona, J., Cortadella, J.: Process mining meets abstract interpretation. In: Balcázar, J.L., Bonchi, F., Gionis, A., Sebag, M. (eds.) ECML PKDD 2010. LNCS (LNAI), vol. 6321, pp. 184–199. Springer, Heidelberg (2010). https://doi.org/10.1007/978-3-642-15880-3_18
31. Busany, N., Maoz, S.: Behavioral log analysis with statistical guarantees. In: ICSE, pp. 877–887. ACM (2016)
32. Biermann, A.W., Feldman, J.A.: On the synthesis of finite-state machines from samples of their behavior. IEEE Trans. Comput. **21**(6), 592–597 (1972)
33. Cohen, H., Maoz, S.: Have we seen enough traces? In: ASE. IEEE CS, pp. 93–103 (2015)
34. Wen, L., van der Aalst, W.M.P., Wang, J., Sun, J.: Mining process models with non-free-choice constructs. Data Min. Knowl. Discov. **15**(2), 145–180 (2007)
35. Song, W., Jacobsen, H., Ye, C., Ma, X.: Process discovery from dependence-complete event logs. IEEE TSC **9**(5), 714–727 (2016)

Process Discovery from Low-Level Event Logs

Bettina Fazzinga[2], Sergio Flesca[1], Filippo Furfaro[1], and Luigi Pontieri[2(✉)]

[1] DIMES, University of Calabria, Rende, Italy
{flesca,furfaro}@dimes.unical.it
[2] ICAR-CNR, Rende, Italy
{fazzinga,pontieri}@icar.cnr.it

Abstract. The discovery of a control-flow model for a process is here faced in a challenging scenario where each trace in the given log L_E encodes a sequence of low-level events without referring to the process' activities. To this end, we define a framework for inducing a process model that describes the process' behavior in terms of both activities and events, in order to effectively support the analysts (who typically would find more convenient to reason at the abstraction level of the activities than at that of low-level events). The proposed framework is based on modeling the generation of L_E with a suitable Hidden Markov Model (HMM), from which statistics on precedence relationships between the hidden activities that triggered the events reported in L_E are retrieved. These statistics are passed to the well-known *Heuristics Miner* algorithm, in order to produce a model of the process at the abstraction level of activities. The process model is eventually augmented with probabilistic information on the mapping between activities and events, encoded in the discovered HMM. The framework is formalized and experimentally validated in the case that activities are "atomic" (i.e., an activity instance triggers a unique event), and several variants and extensions (including the case of "composite" activities) are discussed.

Keywords: Process discovery · Log abstraction · Bayesian reasoning

1 Introduction

Thanks to the diffusion of automated process management and tracing platforms, many process logs (i.e., collections of execution traces) are now available. Log data can be used to analyze and improve a process, by possibly using process mining techniques [2], and in particular *process discovery* (PD) techniques. PD aims at inducing a model that compactly describes the behavior of the process in terms of ad-hoc formalisms (such as *Workflow Nets*, *Heuristics Nets*, etc.), where temporal dependencies between the actions performed during the process enactments are represented. In order to make the model usable, the activities described in it should correspond to some high-level view of the steps of the

© Springer International Publishing AG, part of Springer Nature 2018
J. Krogstie and H. A. Reijers (Eds.): CAiSE 2018, LNCS 10816, pp. 257–273, 2018.
https://doi.org/10.1007/978-3-319-91563-0_16

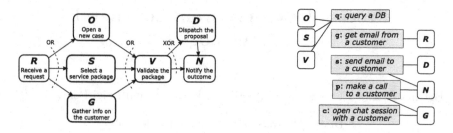

Fig. 1. (*a*) The activity flow of the service-activation process; (*b*) The mapping between high-level activities (upper-case symbols) and low-level events (lower-case symbols).

process, with which analysts are familiar. However, all PD techniques require each event reported in the log to coincide with (or be univocally mapped to) these high-level activities. Unfortunately, often this assumption does not hold in practice. As a matter of fact, in the logs of many real systems, the recorded events just represent low-level operations, with no clear reference to the business activities that were carried out through these operations, as shown in the following example.

Example 1. Consider the case of a phone company, where a *service activation* process is carried out. The process performs the flow of activities depicted in Fig. 1(*a*), and each activity requires the execution of a low-level operation. The mapping between activities and operations is *many to many*, as depicted in Fig. 1(*b*): different executions of the same activity can result in different low-level operations, and, vice versa, the same operation can be the result of performing different activities. For instance, activity G can be accomplished by performing either c or p. Analogously, the event of performing p can be generated by the execution of either activity G or N.

Each enactment of the process is monitored by a tracing system, which will store a log trace consisting of a sequence of low-level events, capturing each the execution of a low-level operation. Thus, a process instance consisting in the execution of sequence $R\,S\,G\,V\,N$ may be recorded in the log as the traces g q p q s q, or g q c q p q, depending which low-level event is triggered by each activity.   □

It is worth noticing that a situation like the one sketched in the example above is not rare when analyzing the logs of knowledge-intensive processes and of legacy applications based on messaging and/or document management systems (e.g., SCM/PDM systems), as well as the events triggered by human activity detection systems (as in the Lifelogging analysis scenario of [22]). More generally, the relevance of a scenario where the tracing system provides low-level event logs rather than high-level activity logs is witnessed by several recent research works aiming at defining process mining tools for these kinds of logs. In particular, some "event abstraction" techniques [5,6,12,13,17,22] have been proposed for translating low-level traces into sequences of high-level activities, so that "traditional" process mining tools [2] working with activity logs can be eventually exploited on the translated logs. Unfortunately, most of these techniques require some knowledge of several aspects of the process behavior (e.g., procedural/declarative process models [5,12,17], event-oriented disambiguation rules [6], activity-annotated example traces [22]), that is not available in typical process discovery settings. Moreover, as these techniques often cannot automatically dissolve every possible ambiguity in the event-activity mapping, an expert's intervention is required to eventually bring the original log at the abstraction level of the activities.

We address the PD problem in this complex setting, and introduce a framework that, starting from a low-level event log L_E, produces a *two-level process model* W^2, where the process behavior is described at both the abstraction levels of activities and events. The proposed framework can work in the presence of a limited domain knowledge, where only the alphabets of the types of activities and events are required to be specified, along with the indication of a candidate mapping between activity types and event types (some preliminary knowledge on activity dependencies can be also exploited by the framework, even if it is not mandatory). In particular, our approach is structured according to the following steps (depicted in detail in Fig. 3):

1: learn a stochastic model capable of reproducing the generation of the input log L_E: this yields a *Hidden Markov Model* (HMM), where the observations encode the low-level events, and the underlying hidden states the activities. The HMM's structure is built on the basis of what is known about the activity/event mapping and the activity dependencies, while its parameters are learned using L_E's traces as training sequences;

2: extract statistics from the stochastic model: statistics on precedence relationships between the hidden activities that triggered the events reported in L_E are obtained by looking into the HMM resulting from step 1;

3: produce a process model in terms of activities: the statistics retrieved at step 2 are given as input to a minor modification of the well-known *Heuristics Miner* algorithm (whose standard implementation would extract these statistics from an activity-aware log), that produces a model of the process at the abstraction level of activities;

4: augment the process model with the events: the control-flow model produced at step 3 is enriched by associating each activity with a probability distribution (pdf) over the low-level events that the activity execution can trigger (see Fig. 2). This pdf is built using the emission probabilities of the learned HMM.

The framework can straightforwardly embed other PD techniques, in place of *Heuristics Miner*, and is both formalized and tested in the case that the activity/event mapping is many-to-many, but the activities are "simple" (i.e. the execution of an activity triggers exactly one event)—as also assumed in [7,21]. In a

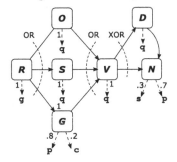

Fig. 2. A 2-level model

setting (like that of log abstraction works [5,6,12,17]) featuring "composite" activities, one might preprocess the log with event clustering methods, as suggested in [7]. Anyway, for the sake of generality, Sect. 6 discusses how our framework can be extended to handle directly logs of processes with composite activities.

Plan of the Paper. The next section introduces preliminary notions and notations, and recalls the fundamentals of Hidden Markov Models and *Heuristics Miner*. Section 3 introduces our framework (Sects. 3.1–3.4 regard steps 1–4 above). Section 4 reports the experimental evaluation we performed, Sect. 5 discusses some related work and Sect. 6 discusses some limitations and possible extensions of our framework.

2   Preliminaries

Logs, Traces, Processes, Activities and Events. A log is a set of *traces*. Each trace Φ describes a process instance at the abstraction level of basic *low-level events*, each generated by the execution of a *high-level activity*. That is, a process instance w is the execution of a sequence $A_1, \ldots, A_m$ of activities; in turn, the execution of each activity A_i generates an event E_j; hence, the trace Φ_w describing w is a sequence of events $E_1, \ldots, E_m$. Activities will be denoted with upper-case letters, possibly adorned with subscripts, and events specifically with letter E, possibly adorned with subscripts.

We assume that we are given a low-level event log L_E, whose traces can have different lengths (we denote as T the maximum length of a trace in L_E). We also assume that the alphabets $\mathcal{A}$ of activities and $\mathcal{E}$ of events are given, along with a *candidate mapping* $\mu : \mathcal{A} \to 2^{\mathcal{E}}$ associating each activity A with a superset $\mathcal{E}'$ of the set of events that can be the result of an execution of A. As it will be clear later, μ is used as a starting point and will be refined using a learning procedure exploiting L_E, thus resulting in a probabilistic mapping between activities and events (see Sect. 6 - point e), where possible ways to preliminarily obtain a mapping μ are discussed.

Observe that we allow different activities to result in the same event, and different events to be the result of the execution of the same activity (this corresponds to a rather general scenarios, where shared functionalities are allowed).

Hidden Markov Models (HMMs) and the Baum-Welch (BW) Algorithm. A *Hidden Markov Model* (HMM) is a statistical Markov model in which the system being modeled is a Markov process with unobserved (i.e. hidden) states, meaning that only the observations are visible. Specifically, an HMM is a tuple $\langle Q, \pi, O, \alpha, \beta \rangle$, where: $Q = q_1, \ldots, q_N$ is a set of states, $\pi = \pi_1, \ldots, \pi_N$ is an initial probability distribution over the states (where each π_i represents the probability that q_i is the initial state of the Markov chain), α is an $N \times N$ transition matrix, where each α_{ij} represents the probability of moving from state q_i to state q_j, $O = o_1, \ldots, o_K$ is a set of possible observations, and β is a $K \times N$ matrix, where each β_{ij} (also called *emission probability*) expresses the probability that the state q_j generates the observation o_i.

A classical problem concerning HMMs is their training, that is the learning of the parameters of an HMM given a sequence of observations $\boldsymbol{\omega}$. The standard algorithm for HMM training is the *Baum-Welch* algorithm (BW), a special case of the *Expectation-Maximization* (EM) algorithm. BW iteratively refines the transition and emission probabilities of the HMM, with the aim of maximizing the likelihood that the sequence of observations $\boldsymbol{\omega}$ is generated. Specifically, it starts with initializing the α and β matrices, and then, at each iteration, it performs the expectation and the maximization step. The expectation step uses α and β to compute, for each time point, (i) an $N \times N$ matrix ξ_t, s.t. each cell $\xi_t(i, j)$ contains the probability of being in state q_i at time t and in q_j at $t + 1$, and (ii) a vector γ_t of size N, s.t. each $\gamma_t(i)$ is the probability of being in q_i at t. Then, the maximization step starts from ξ_t and γ_t and recomputes α, β, and π, whose new values are given as input of the next iteration. At the end of the iterations (whose number is an input parameter), the algorithm outputs α and β.

The main reason for adopting BW is that it is the *de facto* standard learning algorithm for HMMs, at least when the HMM is not "very large" (BW's complexity is $O(N \cdot |\omega| \cdot Q^{\max})$, where $Q^{\max}$ is the maximum state outdegree). In our scenario, the number of HMM states quadratically depends on the number of activities, that is typically not large, so that the feasibility is not jeopardized. Anyway, BW is easy to parallelize, and efficient sampling-based variants exist that can deal with settings yielding a very large number of states (as it may happen with the extension to composite activities – see Sect. 6). More details on BW and its variants can be found in [19].

Process Discovery Algorithms: Heuristics Miner. Among the many techniques for inducing a control-flow model from an activity log L (see, e.g., [1, 2, 4, 23]), we here consider the simple and popular *Heuristics Miner* [24] algorithm, which has been widely used in real-life process mining projects, owing to its fastness and robustness to noise. The relevance of *Heuristics Miner* is also witnessed by recent efforts to extend it to deal with large logs [10, 11] and online discovery settings [8]. However, as discussed in Sect. 6, the framework is orthogonal to the PD technique invoked at one of its steps. In particular, the possibility of using techniques guaranteeing the correctness of the returned model, that is not guaranteed by *Heuristics Miner*, is worth mentioning.

Heuristics Miner relies on the measures $|A > B|$ and $|A \gg B|$, indicating how many times the sub-sequences AB and ABA occur in the log, respectively. Furthermore, it considers also how many times each activity A appears at the beginning/end of a trace, mainly to decide if A is a starting/final activity. On the basis of these statistics, *Heuristics Miner* builds a process representation in the form of a "Heuristics Net".

In Sect. 6, it is discussed how further statistics that can be optionally used by *Heuristics Miner* can be inferred in our framework.

3   A Framework for Process Discovery

In brief, the problem to be addressed is that of inferring a model of the process behavior from a log L_E describing, at the abstraction level of low-level events, the actions performed during the process enactments. As an output, we intend to provide a model W^2 describing the control-flow at both the abstraction levels of events and activities, so that typical analysts (who are generally more familiar with high-level activities, rather than with low-level events) are allowed to reason through W^2 on the behavioral aspects of the process from different standpoints.

The starting point is the well-established algorithm *Heuristics Miner*. As recalled in Sect. 2, *Heuristics Miner* is a popular technique for inferring a process model from a log, in the case that the log and the desired model are at the same level of abstraction (i.e., both refer to high-level process activities). Our proposal is a framework that uses the algorithmic core of *Heuristics Miner* as an intermediate step, as shown in Fig. 3. Basically, the process for inferring a model works as follows.

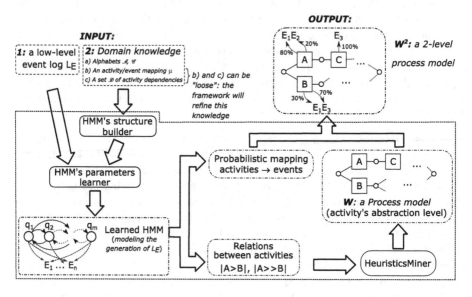

Fig. 3. The phases of our approach for inferring a model from a low-level event log

The input consists of a low-level event log L_E along with some domain knowledge, covering: (*a*) the alphabets $\mathcal{A}$, $\mathcal{E}$, and (*b*) the candidate activity/event mapping μ, and (*c*) a set $\mathcal{D}$ of dependencies between activities, such as "*A always/never precedes/is preceded by B*". Points (*b*) and (*c*) can be specified even "loosely": μ is not required to be tight (as discussed in Sect. 2), and $\mathcal{D}$ can be incomplete (in fact, in our experiments, $\mathcal{D} = \emptyset$). This aspect will be further addressed in Sect. 6 (point *a*), after having made clearer the details of the framework. The knowledge encoded by (*a*), (*b*), (*c*) is used to define the structure of an HMM, that is in turn given as input, along with the log L_E, to the BW algorithm. This way, an HMM reproducing the generation of the sequences of events in L_E is obtained. The structure of the HMM given to BW is specifically designed so that it can encode different forms of dependencies between activities: this enables the result of BW to be used to extract the statistics on the precedence relations between the activities that are needed by *Heuristics Miner* to infer a model.

Then, *Heuristics Miner* is run on these statistics, and a high-level model W of the process is obtained. In other words, we circumvent the unavailability of an activity-aware log, from which the "standard" *Heuristics Miner* would extract statistics on the mutual dependencies exhibited by the activities, with two steps: (1) we learn an HMM modeling the sequences of events recorded in L_E as the *observed* counterparts of the sequences of *hidden* activities executed during the process enactments; (2) we extract the needed activity-level statistics from this HMM, and use them to feed *Heuristics Miner*.

Now, the model W returned by *Heuristics Miner* describes the control-flow at the abstraction level of the activities. Thus, as a final step, we augment W with a description of the behavior exhibited at the abstraction level of events. In particular, the augmented model W^2 returned by the framework is obtained by suitably embedding into W the probabilistic mapping between activities and events encoded in the learned HMM.

Let us now discuss some major aspects of our process discovery approach in detail.

3.1 Structure of the HMM

The semantics of HMMs suggests a natural way of modeling the generation of a low-level event log, as the result of executing sequences of high-level activities. In fact, the scenario is the following: (1) what is observed (i.e., the log) is a set of traces, each consisting of a sequence of events; (2) what generate the events are the activities, that provide a high-level description of each step of the process, but are not explicitly represented in the log. These arguments back the use of an HMM's structure where the observations are the events in $\mathcal{E}$, and the hidden states are the activities in $\mathcal{A}$ (see Fig. 4(a)).

However, a major requirement of the HMM in our context is that it must allow the information needed by *Heuristics Miner* to be easily extracted, once the parameters of the same HMM have been learned. Since *Heuristics Miner* needs to know $|A \gg B|$ for each pair of activities A, B, the above-discussed naïve model is unsuitable, as it does not represent sequences of three activities (in fact, states represent single activities and arcs pairs of consecutive activities). Hence, a more suitable model is that of associating each hidden state q with a pair of activities $A^{pre} A^{cur}$ (referred to as $q.A^{pre}$ and $q.A^{cur}$), where A^{cur} represents the current activity, and A^{pre} the activity performed immediately before A^{cur}. For representing the starting of a process (corresponding to an activity execution preceded by no other activity), we use states q where $q.A^{pre}$ is the "null" activity

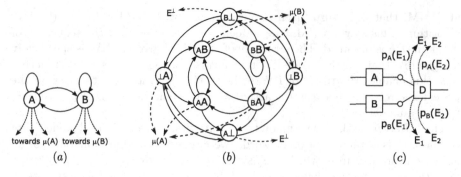

Fig. 4. A naïve HMM (*a*) and an HMM with memory of the previous state (*b*). Associating an activity with a pdf over the resulting events in W^2 (*c*)

(denoted as "$\perp$"). Analogously, a state q where $q.A^{cur} = $ "$\perp$" indicates the end of a process instance. In particular, emissions and transitions are set as follows:

i. every state q has one emission for each $E \in \mu(a.A^{cur})$. As a special case, we assume that the "fake" activity $\perp$ has the unique possible outcome $\mu(\perp) = \{E^\perp\}$, where $E^\perp$ is an event registered at the end of each trace to indicate the termination of the corresponding process instance;

ii. transitions are put from a state q_1 to a state q_2 if and only if $q_1.A^{cur} = q_2.A^{pre}$ and if the possibility that $q_1.A^{cur}$ precedes $q_2.A^{cur}$ in a process execution is not forbidden by the dependencies in $\mathcal{D}$. Herein, the transition from a state q with $q.A^{cur} = $ "$\perp$" to a q' with $q'.A^{pre} = $ "$\perp$" corresponds to moving from the end of a process instance to the start of another process instance.

Before explaining how to train the HMM (thus, how to set the transitions' and emissions' probabilities), and how to derive $|A > B|$ and $|A \gg B|$ from it, let us show an example, in Fig. 4(*b*), with two activities and three events, assuming $\mathcal{D} = \emptyset$ (as done in our tests). In this figure and in what follows, a state q with $q.A^{pre} = A$ and $q.A^{cur} = B$ is denoted as $q = {}_AB$, where the smaller font refers to the previous activity.

3.2   Training the HMM

The HMM's parameters are learned using L_E as set of training sequences. In particular, we use the BW algorithm, that is invoked on a starting configuration of the HMM where:

(1) for each state, the outgoing transitions are equiprobable, as well as the emissions;

(2) the pdf π indicating the initial states is uniform over the states q with $q.A^{pre} = \perp$, and assigns 0 to all the other states.

Once the parameters of the HMM have been learned, the HMM will have a subset of the states, of the transitions and of the emissions of the starting configuration. For instance, if the BW algorithm has learned that an execution of A, when preceded by B, is never followed by C, then the transition from the state BA to AC is deleted. Analogously, if BW has learned that A is never followed by B, then the state AB and all of its ingoing and outgoing transitions are deleted, along with its emissions. Thus, the learning step refines the domain knowledge used to define the HMM's structure, as it can learn dependencies not initially specified in $\mathcal{D}$ and make the initial candidate mapping μ probabilistic. In fact, for each activity A, every $E \in \mu(A)$ will be associated with an emission probability representing the likelihood that E is the outcome of A (a 0 emission probability means removing E from the set of candidate events).

3.3 Extracting the Statistics Needed by *Heuristics Miner* from the HMM

From the learned HMM, the statistics to be given as input to *Heuristics Miner* are evaluated as follows. First of all, we assume that the BW algorithm returns, along with the probabilities of the transitions and the emissions, also the vector $\gamma = \gamma_1, \ldots, \gamma_T$, where each γ_i is a pdf assigning each state q the probability of being the actual state of the process at time i, considering all the process enactments encoded in L_E.[1] Then, for each pair of activities A, B, the number of times that an execution of A is followed by an execution of B is estimated as the sum of the probabilities that at the different steps the process is in the state AB of the HMM, that is: $|A > B| = \Sigma_{i \in [1..T]} \gamma_i(AB)$.

Analogously, the number of occurrences of ABA is estimated by summing the probabilities that, at each step, the process has performed activity B after having performed activity A, and then will perform A again. This means taking probabilities at the various steps that the process is in state AB and then moves to BA, that is: $|A \gg B| = \Sigma_{i \in [1..T]} \gamma_i(AB) \cdot a_{(AB)(BA)}$ where $a_{(AB)(BA)}$ is the entry of the transition matrix associated with the transition between the states AB and BA.

3.4 Invoking *Heuristics Miner* and Building a Two-Level Process Model W^2

The values of $|A > B|$ and $|A \gg B|$, computed for each activity pair, are passed as argument to *Heuristics Miner*, that uses them to build a process model.

The output of *Heuristics Miner* is a model W of the process at the abstraction level of activities (in particular, in our experiments we leveraged an implementation of the algorithm available in ProM 6.7 that returns a *Heuristics Net*).

Starting from W, a two level process model W^2 is composed, that augments W by associating each activity A occurring in W with a pdf over the events into which the execution of A can result (see Fig. 3 for an example of shapes of W

---

[1] As recalled in Sect. 2, γ is computed by the standard BW algorithm to perform the maximization step. In our implementation, γ is not disposed when the algorithm ends, as it is useful as a statistics on the states of the process.

and W^2). This pdf is suitably extracted from the emission probabilities in the learned HMM. In particular, for each activity A in W, the emission probabilities that are associated with A in W^2 are those of the states q in the HMM with $q.A^{cur} = A$ and such that $q.A^{pre}$ is an activity that precedes A in W. For instance, consider the case that $\mathcal{A} = \{A, B, C, D\}$ and that, according to W, activity D can be preceded by A or B, but not by C, and it cannot be the initial activity. Then, the emission pdf associated with D will incorporate the emission probabilities of states AD and BD, but not that of CD. This is depicted in Fig. 4(c), where the pdf of D is "split" into $p_A(E)$ and $p_B(E)$, denoting the probabilities that D triggers the event E when the execution of D is preceded by A and B, respectively. The pdfs of A and B are not distinguished based on the previous activity, meaning that the event produced by these activities does not depend on the previous activity.

Observe that it can happen in practice that a state q is present in the HMM, but the models W and W^2 do not allow the possibility that $q.A^{pre}$ precedes $q.A^{cur}$. In fact, *Heuristics Miner* may infer that $q.A^{pre} \Rightarrow q.A^{cur}$ does not hold since the cardinality of $|A > B|$, although greater than 0, is "small". In this case, *Heuristics Miner* concludes that the occurrence of the subsequence AB in some process enactment is symptomatic of noise and thus it must not be encoded in the resulting process model.

4   Experiments

The Datasets. The empirical validation was conducted over noisy datasets generated from the process models $W_1 = $ "*parallel5*", $W_2 = $ "*a12*", $W_3 = $ "*herbstFig3p4*", $W_4 = $ "*herbstFig6p18*", $W_5 = $ "*herbstFig6p41*", selected from a popular set of benchmark logs used in many previous works (e.g., [18,23]), and featuring different control-flow constructs (sequences, choices, parallelism, short loops, and invisible tasks).

For each W_i, we generated 20 low-level event logs, each denoted as L_{ij}^s, with $s \in \{1.2, 1.4, 1.6, 1.8, 2.0\}$ and $j = [1..4]$, where s is a parameter denoting the average number of activities onto which an event can be mapped (i.e., the average number of activities that possibly result in the same event). Thus, s measures the amount of shared functionalities and, intuitively enough, gives a measure of the uncertainty to be dealt with (in particular, $s = 1$ means that each event corresponds to one activity exactly).

Specifically, each L_{ij}^s was generated as follows. First, we used the model W_i (that contains references to high-level activities only) to generate a noise-free activity log NFL_i^A conforming to W_i. Then, a noisy activity log L_i^A was obtained by perturbing each trace in NFL_i^A by randomly switching pairs of consecutive activities, or replacing an activity occurrence with another activity, or deleting an activity occurrence. Perturbations were applied with probability 2% for each step in each trace. After obtaining the (noisy) activity log L_i^A, we generated 20 mappings, denoted as μ_{ij}^s (with $s \in \{1.2, 1.4, 1.6, 1.8, 2.0\}$ and $j = [1..4]$), between the activities in W_i and a set of events with the same cardinality as $|\mathcal{A}|$.

Specifically, for each value of s, we generated 4 different mappings $\mu_{i1}^s, \ldots, \mu_{i4}^s$, all ensuring that, on the average, each event is mapped to s activities. Observe that, since $|\mathcal{E}| = |\mathcal{A}|$, s measures also the average number of events that are the possible outcomes of the same activity. Starting from μ_{ij}^s, we obtained a probabilistic mapping $\tilde{\mu}_{ij}^s$, by interpreting the events in μ_{ij}^s as the possible outcomes of a random variable following a Zeta distribution (in particular, the events in $\tilde{\mu}_{ij}^s(A)$ were associated with a progressive rank k, and their probabilities were set proportional to $1/k$). Finally, from the activity log L_i^A, for each $s \in \{1.2, 1.4, 1.6, 1.8, 2.0\}$ and $j = [1..4]$, we obtained an event log L_{ij}^s as follows: for each occurrence of an activity A in L_i^A, an event E was sampled from $\tilde{\mu}_{ij}^s(A)$, and the occurrence of A was replaced with E. This way, we obtained 20 event logs for each model W_i.

Measuring the Effectiveness. To validate our framework, we performed the following tests, for each log L_{ij}^s. First, we ran our prototype on L_{ij}^s and obtained a two-level model $W^2(L_{ij}^s)$. Then, the effectiveness was evaluated by measuring how far $W^2(L_{ij}^s)$ is from the actual model originating L_{ij}^s, that is the combination of the activity model W_i and the mapping $\tilde{\mu}_{ij}^s$. This was accomplished using three metrics: $\mathcal{F}$, $\mathcal{P}$, and Err.

$\mathcal{F}$ is a *fitness* metric returning a sort of "recall" score quantifying the capability of W^2 to capture the behavior of the ground-truth model W_i manifested in the (noise-free) activity log NFL_i^A. The metric is evaluated by using the fitness calculation method in [20][2], to compare the log NFL_i^A and a Petri-net representation of the activity level of $W^2(L_{ij}^s)$, i.e., the model obtained from $W^2(L_{ij}^s)$ by disregarding events' emissions.

$\mathcal{P}$ is instead and empirical metric of *precision*, computed in a symmetric way by first generating an activity log L' (of 5000 traces) from a Petri-net representation of the activity level of $W^2(L_{ij}^s)$, and then computing a fitness score (still using the method of [20]) for the ground-truth model W_i (represented as a Petri net) and L'.

Finally, Err provides an *error* measurement for the emission probabilities represented in $W^2(L_{ij}^s)$, computed as the maximum difference between the probability associated with an emission in $W^2(L_{ij}^s)$ and that in $\tilde{\mu}_{ij}^s$.

The Setting. The HMM to be learned was constructed according to what said in Sect. 3.1, assuming $\mathcal{D} = \emptyset$ (thus, benefiting from no knowledge of the process behavior in terms of activity dependencies) and assuming the presence of a domain expert providing as candidate mapping the actual sets of possible outcomes of every activity (however, no information on the probability associated with each event was assumed, thus the emission probabilities were set uniformly

---

[2] This method mainly relies on a non-blocking replay (where tokens can be put artificially into any place of the net when the latter cannot reproduce a trace step), and penalizes both unexpected events and improper completion (based on how many tokens were created artificially and left unconsumed, respectively). It can be hence applied to a not sound process model (as were those returned by *Heuristics Miner* on some of the low-level logs produced with $s > 1.6$).

and had to be refined by BW). The benefits that could arise from using a non-empty $\mathcal{D}$ are discussed in Sect. 6.

Our results are easily reproducible, as our prototype uses a standard implementation of the BW algorithm, and the implementations of algorithm Heuristics Miner and of the fitness algorithm [20] available in ProM 6.7 and in ProM 5.2, respectively—precisely, when computing each fitness score we set a maximum search depth of 1 over the invisible transitions (if any) of the net. Moreover, we used the default parameter setting for *Heuristics Miner*, except for the parameters *DependencyDivisor* and *relativeToBest* that were set equal to 20 (corresponding to about 0.4% the number of analyzed log traces) and 10%, instead of 1 and 5%, respectively.

The results were obtained using only 2 iterations of the BW algorithm, as we noticed that in most cases further iterations did not improve the quality of the discovered process models substantially—in fact, allowing BW to perform more steps can help improve the learned HMM, and refine the dependency statistics for *Heuristics Miner*, but such refinements may be unnecessary to reckon the structure of the process well enough.

Discussion of the Results. The results in Table 1 show that our framework is rather effective in finding a model that is representative of the behavior of the processes analyzed. In particular, the closeness of the two-level model to the actual processes' behavior is witnessed by the high values of $\mathcal{F}$, $\mathcal{P}$, and $1 - Err$. As expected, these measures are affected by s: the higher s (i.e., the average number of activities that could generate the same event), the lower $\mathcal{F}$, $\mathcal{P}$, and $1 - Err$. In fact, intuitively enough, larger values of s correspond to dealing with a higher level of uncertainty when mapping each log event to the actual activity generating it, as the average number of candidate activities is s. Interestingly, the lower efficacy of the framework at the highest value of s is more evident for the models W_3 and W_4, that exhibit a more complex structure than W_1, W_2, and W_5 (containing choices and parallelism, but not mixed with loops). All the results in Table 1 must be read considering that invoking *Heuristics Miner* on the activity logs $L_1^A, \ldots, L_5^A$ (from which all the low-level event logs were derived) returned the correct models $W_1, \ldots, W_5$. This means that the values of $\mathcal{F}$ less than 1 are due to the uncertainty introduced by the many-to-many mapping between activities and events.

Figure 5 shows the sensitivity of the run times vs. various parameters. For every run, the time spent by *Heuristics Miner* was negligible compared with that taken by the BW algorithm; thus we do not distinguish between the time shares of these two components.

The diagram in Fig. 5(a) reports, for each $X \in \{1000, 2000, 3000, 4000, 5000\}$, the mean values of the run times over all the event logs $L_{i1}^s(X)$ (for all $s \in \{1.2, 1.4, 1.6, 1.8, 2.0\}$ and $i \in [1..5]$), where $L_{i1}^s(X)$ consists of the first X traces in L_{i1}^s. This diagram shows that the run time is linear with the number of traces, which is in line with the fact that BW is linear in the number of observations. The diagram in Fig. 5(b) reports the average run times over all the L_{i1}^s logs grouped by s, and shows that the computation time does not depend on s.

Table 1. Effectiveness of the framework (using 2 invocations of BW algorithm). For each model W_i and each value of s, the averages of the results obtained over the logs $L^s_{i1}, \dots, L^s_{i4}$ are shown.

	$W_1(parallel5)$			$W_2(a12)$			$W_3(herbstFig3p4)$			$W_4(herbstFig6p18)$			$W_5(herbstFig6p41)$		
	$\mathcal{F}$	$\mathcal{P}$	Err	$\mathcal{F}$	$\mathcal{P}$	Err	$\mathcal{F}$	$\mathcal{P}$	Err	$\mathcal{F}$	$\mathcal{P}$	Err	$\mathcal{F}$	$\mathcal{P}$	Err
$s=1.2$	0.97	0.97	2%	0.90	0.92	2%	0.93	0.89	1%	0.96	0.97	2%	0.92	0.94	3%
$s=1.4$	0.98	0.96	5%	0.88	0.81	6%	0.94	0.92	4%	0.95	0.96	2%	0.96	0.81	5%
$s=1.6$	0.95	0.97	6%	0.89	0.77	8%	0.88	0.84	5%	0.90	0.83	3%	0.94	0.88	4%
$s=1.8$	0.92	0.88	8%	0.88	0.72	9%	0.84	0.77	8%	0.90	0.86	4%	0.92	0.86	6%
$s=2.0$	0.89	0.92	9%	0.88	0.75	9%	0.78	0.80	10%	0.83	0.72	8%	0.88	0.79	8%

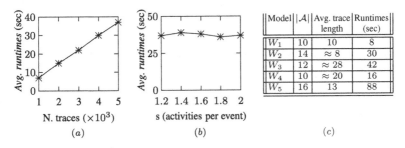

		Model	$|\mathcal{A}|$	Avg. trace length	Runtimes (sec)
		W_1	10	10	8
		W_2	14	≈ 8	30
		W_3	12	≈ 28	42
		W_4	10	≈ 20	16
		W_5	16	13	88

(a)   (b)   (c)

Fig. 5. Run times vs. number of traces (a), activities per events (b) and models (c)

This result is obvious for *HM* (that does not deal with the events). For *BW*, it follows from the fact that its computation complexity mainly depends on the number of states and observation types, and not on their mutual correlations. Finally, the table in Fig. 5(c) reports the average run times vs. the models used to produce the logs, and characteristics of both the models and logs.

5 Related Work

The problem of discovering a control-flow model has been deeply addressed in the last years, and a wide variety of solutions have been proposed (see [2,4,23] for recent surveys). A method for inducing a dependency graph was first presented in [3]. Similar graph-like models also underly subsequent methods [18,24], which also introduced mechanisms for describing the semantics of join/split nodes. In most of these methods, including algorithm *Heuristics Miner* [24], activity dependencies are derived from pairwise ordering statistics over the activities, extracted from the log, collectively referred to as "log abstractions".

By contrast, in other works the discovery problem was conceptually stated as a search over more expressive classes of process models, such as Petri nets (as done in [25]), or block-structured workflow models (as done in [15]). This search has been faced using various strategies, ranging from ad-hoc algorithms, to approximated optimization methods. An approach that was shown quite effective (cf. [4]) is the *Inductive Miner* algorithm [15], which founds on iteratively refining a grammar-like representation of the process, while recursively splitting the events in the log.

However, as empirically observed in [23], *Heuristics Miner* often manages to produce effective and readable process models, compared to many later approaches, and to reach a good trade-off between effectiveness (owing to its capability to deal with both complex constructs and noise) and fastness/scalability. In particular, the usage of compact log abstractions (quadratic in the number of activities, and independent of the log size), that can be computed in linear time (possibly in a parallelized way), makes it a natural candidate to implement scalable process discovery approaches [10], and even to deal with streaming logs [8]—by contrast, most current process discovery methods need to keep the log in main memory [16]. To the best of our knowledge, this feature is shared with only one advanced process discovery algorithm [16], where the search procedure of Inductive Miner [15] is not performed by recursing on the log, but rather on a matrix storing directly-follow measurements (the same as *Heuristics Miner*).

All the methods above cannot be applied to our problem setting, as they assume that each log event refers (or can be deterministically mapped) to a process activity.

Other related approaches are those addressed at interpreting low-level event logs [5,6,12,13,17,22], and eventually translating event traces into sequences of high-level activities. In principle, such techniques could be used as a pre-processing tool, to enable the application of activity-oriented process discovery algorithms to the abstracted version of a low-level log. However, most of these techniques need to be provided with explicit and rather deep knowledge on the process' behavior, such as: procedural/declarative process models (as in [5,12,17]), event-oriented disambiguation rules (like the attribute and event context conditions in [6]), documents containing activity descriptions [5,6], activity-annotated traces for training a sequence-labelling model (as in [22]). Moreover, since this does not usually suffice to resolve all ambiguities, the intervention of an expert [5,6] is needed to distill the "right" translation.

6   Conclusions: Discussion of Limitations and Extensions

Our framework for performing the process discovery task starting from a low-level event log has been proved effective in returning accurate two-level process models when tested over logs adhering to different real-life models taken from repositories popular in the process mining community. In what follows, we discuss some critical points, limitations and possible extensions of our approach.

(a) Domain knowledge: how much? Our experimental results suggest that the framework can be effective even in the presence of a rather limited amount of domain knowledge: models of good quality were obtained despite $\mathcal{D}$ was assumed empty. Obviously, there are limits to what can be learned in the absence of knowledge of the process behavior. For instance, activities O, S, V in Fig. 1 are indistinguishable from their low-level counterparts (i.e., the event q), thus

no mechanism is likely to effectively solve the ambiguity when interpreting an occurrence of q in the log. However, in this example, the knowledge of the dependency "V *always precedes* D *or* N" (that can be straightforwardly encoded in the HMM's structure) can dramatically reduce the uncertainty, so that an accurate model becomes obtainable (in fact, taking into account this dependency, only the last occurrence of q in a trace could be interpreted as the outcome of V). The possibility of exploiting knowledge from a domain expert (in the spirit of [14]) also places the proposed framework as the core of an iterative scheme, where, after a model is produced, the expert is called for validating it and possibly enriching $\mathcal{D}$ with dependencies that s/he believes can help solve ambiguities. This progressive refinement would also address possible limits of the high-level PD algorithm embedded in the framework, that might yield models that are unsound or with deadlocks, and thus need some revising (as may happen with *Heuristics Miner*, that does not guarantee the correctness of the produced model).

(b) Dealing with composite activities. The extension to the case that activities are composite is straightforward if no interleaving is allowed between the sequences of events of two activities (but interleaving sequences of activities are still allowed). In this case, each state AB can be represented by three states $start_{AB}$, $intermediate_{AB}$, and $final_{AB}$ with the events in $\mu(B)$ as possible emissions. The general case where the sequences of events of two activities A and B can be interleaved can still be modelled, by putting into each state a bitstring encoding the set of activities that are still "active". The point is that this can dramatically increase the size of the HMM, thus more scalable variants of the BW algorithm (for instance, sampling-based ones) must be tried.

(c) Dealing with non-free choices. In order to allow *Heuristics Miner* to discover non-free choices in the process behavior, we need to extract statistics of the form $|A \ggg B|$, representing how many times activity B followed activity A in the log, at any distance. These statistics cannot be obtained directly from the HMM, but can be inferred by (i) running the Viterbi algorithm over the trained HMM and every trace of the log, in order to obtain the most probable sequence of activities for each trace, and (ii) computing $|A \ggg B|$ by analyzing each of the so obtained sequences of activities.

(d) Exploiting PD algorithms other than Heuristics Miner. Techniques such as α [1] and $\alpha+$ [9] can be straightforwardly used in place of *Heuristics Miner*, as the information they need as input is obtainable from our trained HMM and the values of γ. The same holds for the "single-pass" scalable version of *Inductive Miner* [16]. As a matter of fact, we have conducted some preliminary experiments using *Inductive Miner* over the data sets generated from the process models W_1 and W_2. Interestingly, we obtained models with higher fitness but lower precision (compared to those obtained with *Heuristics Miner*). However, this result must be read along with the fact that the models produced by *Inductive Miner* were all sound (while some of those returned by *Heuristics Miner* were not). Further experiments are needed to investigate the sensitivity of this behavior to the parameters of *Inductive Miner* and to the process features.

PD techniques taking as input information different from the statistics taken by *Heuristics Miner* can be also embedded in the framework: after learning the HMM, the event traces can be converted into activity traces by means of the Viterbi algorithm, and then the required information can be extracted from the converted log.

(e) Obtaining the candidate mapping via semi-automatic approaches. The initial candidate mapping μ, that is refined by the learning phase into a probabilistic event/activity mapping, can be provided by a domain expert. Otherwise, existing event-to-activity matching techniques, like those in [5,6], can help distillate a small subset of all possible mappings between activities and events (at the level of types), by leveraging background knowledge on the process behavior and/or textual descriptions available for both the activities and the events. By only considering the ⟨activity, event⟩ pairs returned by these techniques as admissible mappings, it is possible to initialize more precisely the emission probabilities of the HMM model, as emissions supported by no evidence (but otherwise considered as candidate mappings) can be preliminarily discarded.

References

1. van der Aalst, W., Weijters, T., Maruster, L.: Workflow mining: discovering process models from event logs. IEEE TKDE **16**(9), 1128–1142 (2004)
2. van der Aalst, W.M.P.: Process Mining: Discovery, Conformance and Enhancement of Business Processes, 1st edn. Springer, Heidelberg (2011). https://doi.org/10.1007/978-3-642-19345-3
3. Agrawal, R., Gunopulos, D., Leymann, F.: Mining process models from workflow logs. In: Schek, H.-J., Alonso, G., Saltor, F., Ramos, I. (eds.) EDBT 1998. LNCS, vol. 1377, pp. 467–483. Springer, Heidelberg (1998). https://doi.org/10.1007/BFb0101003
4. Augusto, A., Conforti, R., Dumas, M., La Rosa, M., Maggi, F.M., Marrella, A., Mecella, M., Soo, A.: Automated discovery of process models from event logs: review and benchmark. arXiv preprint arXiv:1705.02288 (2017)
5. Baier, T., Di Ciccio, C., Mendling, J., Weske, M.: Matching events and activities by integrating behavioral aspects and label analysis. Softw. Syst. Model., 1–26 (2017)
6. Baier, T., Mendling, J., Weske, M.: Bridging abstraction layers in process mining. Inf. Syst. **46**, 123–139 (2014)
7. Baier, T., Rogge-Solti, A., Weske, M., Mendling, J.: Matching of events and activities - an approach based on constraint satisfaction. In: Frank, U., Loucopoulos, P., Pastor, Ó., Petrounias, I. (eds.) PoEM 2014. LNBIP, vol. 197, pp. 58–72. Springer, Heidelberg (2014). https://doi.org/10.1007/978-3-662-45501-2_5
8. Burattin, A., Sperduti, A., van der Aalst, W.M.: Control-flow discovery from event streams. In: 2014 IEEE Congress on Evolutionary Computation (CEC), pp. 2420–2427. IEEE (2014)
9. de Medeiros, A.K.A., van Dongen, B.F., van der Aalst, W.M.P., Weijters, A.J.M.M.: Process mining: extending the α-algorithm to mine short loops. Technical report, University of Technology, Eindhoven (2004). bETA Working Paper Series, WP 113

10. Evermann, J.: Scalable process discovery using map-reduce. IEEE Trans. Serv. Comput. **9**(3), 469–481 (2016)
11. Fazzinga, B., Flesca, S., Furfaro, F., Masciari, E., Pontieri, L.: A compression-based framework for the efficient analysis of business process logs. In: Proceedings of 27th International Conference on Scientific and Statistical Database Management, SSDBM 2015, pp. 6:1–6:12 (2015)
12. Fazzinga, B., Flesca, S., Furfaro, F., Masciari, E., Pontieri, L.: Efficiently interpreting traces of low level events in business process logs. Inf. Syst. **73**, 1–24 (2018)
13. Fazzinga, B., Flesca, S., Furfaro, F., Pontieri, L.: Online and offline classification of traces of event logs on the basis of security risks. J. Intell. Inf. Syst. **50**(1), 195–230 (2018)
14. Greco, G., Guzzo, A., Lupia, F., Pontieri, L.: Process discovery under precedence constraints. ACM Trans. Knowl. Discov. Data **9**(4), 32:1–32:39 (2015)
15. Leemans, S.J.J., Fahland, D., van der Aalst, W.M.P.: Discovering block-structured process models from event logs - a constructive approach. In: Colom, J.-M., Desel, J. (eds.) PETRI NETS 2013. LNCS, vol. 7927, pp. 311–329. Springer, Heidelberg (2013). https://doi.org/10.1007/978-3-642-38697-8_17
16. Leemans, S.J.J., Fahland, D., van der Aalst, W.M.P.: Scalable process discovery with guarantees. In: Gaaloul, K., Schmidt, R., Nurcan, S., Guerreiro, S., Ma, Q. (eds.) CAISE 2015. LNBIP, vol. 214, pp. 85–101. Springer, Cham (2015). https://doi.org/10.1007/978-3-319-19237-6_6
17. Mannhardt, F., de Leoni, M., Reijers, H.A., van der Aalst, W.M.P., Toussaint, P.J.: From low-level events to activities - a pattern-based approach. In: La Rosa, M., Loos, P., Pastor, O. (eds.) BPM 2016. LNCS, vol. 9850, pp. 125–141. Springer, Cham (2016). https://doi.org/10.1007/978-3-319-45348-4_8
18. Medeiros, A.K., Weijters, A.J., Aalst, W.M.: Genetic process mining: an experimental evaluation. Data Min. Knowl. Disc. **14**, 245–304 (2007)
19. Rabiner, L.R.: A tutorial on hidden Markov models and selected applications in speech recognition. In: Waibel, A., Lee, K.F. (eds.) Readings in Speech Recognition, pp. 267–296. Morgan Kaufmann (1990)
20. Rozinat, A., van der Aalst, W.M.: Conformance checking of processes based on monitoring real behavior. Inf. Syst. **33**(1), 64–95 (2008)
21. Senderovich, A., Rogge-Solti, A., Gal, A., Mendling, J., Mandelbaum, A.: The ROAD from sensor data to process instances via interaction mining. In: Nurcan, S., Soffer, P., Bajec, M., Eder, J. (eds.) CAiSE 2016. LNCS, vol. 9694, pp. 257–273. Springer, Cham (2016). https://doi.org/10.1007/978-3-319-39696-5_16
22. Tax, N., Sidorova, N., Haakma, R., van der Aalst, W.M.P.: Event abstraction for process mining using supervised learning techniques. In: Bi, Y., Kapoor, S., Bhatia, R. (eds.) IntelliSys 2016. LNNS, vol. 15, pp. 251–269. Springer, Cham (2018). https://doi.org/10.1007/978-3-319-56994-9_18
23. Weerdt, J.D., Backer, M.D., Vanthienen, J., Baesens, B.: A multi-dimensional quality assessment of state-of-the-art process discovery algorithms using real-life event logs. Inf. Syst. **37**(7), 654–676 (2012)
24. Weijters, A., van Der Aalst, W.M., De Medeiros, A.A.: Process mining with the heuristics miner-algorithm. Technische Universiteit Eindhoven, Technical report WP 166, pp. 1–34 (2006)
25. van der Werf, J.M.E.M., van Dongen, B.F., Hurkens, C.A.J., Serebrenik, A.: Process discovery using integer linear programming. Fundamenta Informaticae **94**, 387–412 (2009)

Detection and Interactive Repair of Event Ordering Imperfection in Process Logs

Prabhakar M. Dixit[1,2]([✉]), Suriadi Suriadi[2], Robert Andrews[2], Moe T. Wynn[2], Arthur H. M. ter Hofstede[2], Joos C. A. M. Buijs[1], and Wil M. P. van der Aalst[1,2]

[1] Eindhoven University of Technology, Eindhoven, The Netherlands
{p.m.dixit,j.c.a.m.buijs,w.m.p.v.d.aalst}@tue.nl
[2] Queensland University of Technology, Brisbane, Australia
{s.suriadi,r.andrews,m.wynn,a.terhofstede}@qut.edu.au

Abstract. Many forms of data analysis require timestamp information to order the occurrences of events. The *process mining* discipline uses historical records of process executions, called event logs, to derive insights into business process behaviours and performance. Events in event logs must be ordered, typically achieved using timestamps. The importance of timestamp information means that it needs to be of high quality. To the best of our knowledge, no (semi-)automated support exists for detecting and repairing ordering-related imperfection issues in event logs. We describe a set of timestamp-based indicators for detecting event ordering imperfection issues in a log and our approach to repairing identified issues using domain knowledge. Lastly, we evaluate our approach implemented in the open-source process mining framework, ProM, using two publicly available logs.

Keywords: Event log imperfection · Event ordering · Data quality

1 Introduction

Process mining [1], a form of data mining, utilises historical records of process executions to derive insights into business process performance and behaviour. Central to all process mining techniques is the "event log". An event log is a set of records of processing steps (events) carried out during multiple process instances (cases). Each event is minimally characterised by attributes identifying the process instance to which it belongs *(case id)*, the process step that was carried out *(activity)* and an attribute, typically a *timestamp*, that allows the events in the case to be ordered. Event records may optionally contain other attributes such as the person that carried out the process step.

The saying *garbage in - garbage out* alludes to the fact that, for all forms of data-to-information transformations, process mining included, poor quality input data

© Springer International Publishing AG, part of Springer Nature 2018
J. Krogstie and H. A. Reijers (Eds.): CAiSE 2018, LNCS 10816, pp. 274–290, 2018.
https://doi.org/10.1007/978-3-319-91563-0_17

leads to poor quality information. Many forms of data analyses, especially time-series analyses, e.g., stock market prediction and weather forecasting, need quality timestamp information. Likewise, within the domain of process mining, timestamps are the principal means for ordering activities within cases. Correctly ordering activities within a case is critical to deriving accurate process models (discovery), to determining whether activities were carried out in accordance with organisational expectations (conformance) and to determine whether activities and cases are executed efficiently (performance). Thus, it is clear that the quality of process mining results is contingent on the quality of timestamp data. For instance, consider a situation, where activities in an event log are drawn from two different information systems, one of which records timestamps at only day level granularity, while the other records timestamps to the milli-second. The effects on process mining analysis (see Fig. 1 for an example), include discovered process models which do not reflect the actual ordering of events.

The importance of timestamps in process mining analysis is well recognised [2]. A variety of authors have identified timestamp-related data quality issues and their impact on process mining [1,3–6]. As incorrect ordering of events can have

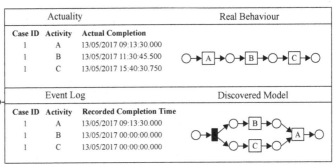

Fig. 1. Effect of inaccurate timestamps on event ordering.

adverse effects on the outcomes of process mining analysis, we provide a set of relevant indicators to detect ordering related problems in an event log to pinpoint those activities that *might* be incorrectly ordered.[1] Next, we let the user repair the event log by interactively injecting domain knowledge directly into the event log, with the help of process fragments, as well as analyze the impact of the repaired log. While tools for automated and user-driven detection and repair of time-oriented data exist, e.g. TimeCleanser [7], to our knowledge, this is the first technique that provides consolidated detection and repair mechanisms to deal with data quality-related ordering issues in event logs.

In this paper, Sect. 2 describes the *indicators* of event ordering imperfection data quality issues, Sect. 3 outlines our approach to detect, repair, and analyze the impact of repair of the event log, and Sect. 4 describes our implementation. Section 5 evaluates the tool against two publicly-accessible data sets. Related work is provided in Sect. 6, followed by conclusion in Sect. 7.

---

[1] These indicators may be specific to event logs; however, some of the indicators that we propose are generic enough to be applicable to other time-ordered data.

2   Indicators of Event Ordering Imperfection

The approach proposed in this paper aims to enable domain experts to detect and repair event-ordering imperfection that arises due to timestamp-related issues. As a first step, we need to be able to *locate exactly where timestamp issues may exist in a log* by recognising characteristics commonly associated with event ordering imperfection. From existing literature that describe data quality issues in event logs [1,2,4,5,8] in particular, and time-oriented data in general [3], we abstract three classes of indicators that may be used to locate event ordering issues. It is not the intention of this paper to provide a comprehensive list of indicators of event ordering imperfection; rather, our aim here is to highlight the importance of recognising the various indicators of event ordering imperfection from an event log as a starting point for our log repair approach. Note that the existence of these indicators does not automatically mean that the log has event ordering issues. However, the ability to highlight those activities will assist domain experts in selecting the best repair actions, if necessary.

Granularity. One of the indicators of event ordering imperfection is the existence of either coarse timestamp granularity (e.g. *imprecise timestamp* [1,8]) or mixed timestamp value granularity (e.g. event log includes events from multiple systems where each system records timestamps differently [2,3]). The mixture of varying timestamp granularity may result in events being ordered incorrectly. For example, an event *'Seen by Doctor'* may be recorded at day-level granularity, e.g. '05 Dec 2017 00:00:00'. Within the same case, another event *'Register Patient'* may have second-level granularity, e.g. '05 Dec 2017 19:45:23'. The ordering of these two events will be *'Seen by Doctor'* followed by *'Register Patient'*, which is incorrect as it should have been the other way around.

Order Anomaly. Locating events affected by ordering imperfection can also be performed by identifying events exhibiting unusual temporal ordering, e.g. duplicate entry of exactly the same event [3]. Given an activity a_1 that was recorded twice (due to a user mistakenly double-clicking a button on his/her screen), we may observe an unusual (infrequent) directly-followed temporal order $a_1 \rightarrow a_1$, highlighting a potential event ordering imperfection. Issues related to missing events [1,4,8] and incorrect timestamps due to events being recorded post-mortem [1] or due to manual entry [8] can also be detected by learning if there exist other forms of unusual temporal orderings between activities.

Statistical Anomaly. Extracting more generic statistical anomalies, such as learning the temporal position of a particular activity in the context of other activities, or the distribution of timestamp values of all events in a log, may indicate the existence of timestamp-related problems. For example, when a log is comprised of events from multiple systems, there may be more than one way in which timestamps are formatted, which may lead to the 'misfielded' or 'unanchored' timestamp problem [3,5] whereby timestamp values are interpreted incorrectly. A common situation is when timestamp values formatted as DD/MM/YYYY are interpreted as MM/DD/YYYY. In this situation, one may see imbalance distribution of the 'date' values of all events in the log whereby all events will have the date value between 1–12 only, and none for 13–31. There are other indicators of event ordering problems that can be detected through statistical anomaly, such as the use of batch processing [3,5] and multiple timezone problem [3].

3   Approach

We start with an event log and adopt an iterative approach (see Fig. 2) to addressing event order imperfection, cycling through (i) automated, indicator-based issue detection, (ii) user-driven repair, (iii) impact analysis, and (iv) log update. In this section we show, through examples, how our indicators are used to detect potential event-order issues, how identified issues are repaired using domain knowledge infused process-fragments and an alignment strategy. Lastly we describe a set of metrics for assessing the impact of the repair actions on the log which can guide the user in accepting or rejecting the repairs.

3.1   Detection

In this section, we discuss the three detection strategies employed in our tool to detect the three indicators presented earlier in Sect. 2. The output of detection is a consolidated list of possible time-oriented data quality issues, containing the actual problem and description, the affected activity(ies) and the number of instances affected. Our technique doesn't require any input, other than the event log, for detection of the issues. It should be noted that the aim of our detection technique is to provide hints on the probable ordering related issues in the event log, and

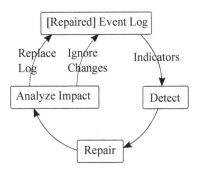

Fig. 2. Detect/repair approach

the user may choose to ignore, even *whitelist*, some of the detected items from the list. These whitelisted items will not be shown in the subsequent detection cycles.

Table 1. Sample granularity table populated by scanning an event log.

Activity	Year	Month	Day	Hour	Minute	Second
a	0	0	0	1	2	500
b	0	0	0	450	20	0
c	0	0	0	0	5	350
d	0	0	0	2	7	448
⋮	⋮	⋮	⋮	⋮	⋮	⋮

Granularity based detection makes use of a table where rows correspond to activities and columns correspond to the granularity of timestamps. Each cell represents the frequency (by finest recording granularity) of a particular activity in the event log. Table 1 gives an example of such a table, which clearly indicates that activity b is usually recorded at a coarser granularity (Hour) compared to the other activities. Every such b is added to the list of detected issues.

Ordering based detection is conceptually similar to the challenge faced by the process discovery techniques, which try to deduce the correct order of activities using an event log. Here we use pairwise causal relations between activities to *guess* the correct order of activities based on frequency thresholds. In order to achieve this, we populate two tables: one containing the directly *follows* relations and the other containing the directly *precedes* relations between the activities in the event log.

Table 2. An example *follows relations* table snapshot (left) and *precedes relations* table snapshot (right).

→	a	b	c	d	e	f
⋮	⋮	⋮	⋮	⋮	⋮	⋮
c	3	0	0	150	200	25
⋮	⋮	⋮	⋮	⋮	⋮	⋮

←	a	b	c	d	e	f
a	0	3	3	450	0	0
⋮	⋮	⋮	⋮	⋮	⋮	⋮
f	0	0	25	0	0	0

Each cell contains the number of times an activity from a corresponding row, was directly followed/preceded by an activity from the corresponding column in the event log. For example, in the *follows* relations of Table 2, activity c is followed by activities a, d, e and f: 3, 150, 200 and 25 times respectively. The discrepancies in the ordering w.r.t. each activity are analyzed. For example, consider the activity c in Table 2 and a (user defined) threshold value of 0.8. The first step is to filter out the smallest set of activities which are directly followed by c resulting in at least 80% of the total occurrences of c.

In the *follows* table (Table 2), these are activities d and e. Next, for every remaining activity with non-zero frequency (a and f) not within the threshold, we check the corresponding *precedes* relations table with the threshold values for infrequent non zero activity relations. In the case of a, these would be activities b and c. Since both the directly follows and precedes relations ($c \rightarrow a$ and $a \leftarrow c$) are infrequent, we conclude that there is an issue in ordering between activities a and c, and thus these activities are added to the detected issues list. However, for activity f, activity c is within the threshold, i.e., all occurrences of f are preceded by c. Since activity f is highly infrequent (compared to c), the ordering between c and f is assumed to be correct. Hence by considering both the directly *follows* and *precedes* relations, we *detect* only those infrequencies that are ordering related. A similar approach is followed for eventually *follows* and *precedes* relations to explore long term infrequent relations.

Statistical anomaly can be used to detect event ordering imperfection. In this paper, we investigated one type of statistical anomaly useful in detecting event ordering issues due to *mis-fielding* (see Sect. 2). For example, data extracted as MM/DD/YY from the system might be incorrectly processed as DD/MM/YY format in the event log. In this scenario, one would only witness days in the range 1–12 and not in the range 13–31. In order to evaluate such issues, we employ a directional statistical technique - Kuiper's test. Kuiper's test is especially useful in the scenarios when the data is circular. That is, while analyzing the hourly distribution of activity distribution, timestamp values 23:59 h, and 00:01 would be considered close. For more details about the Kuiper's test we refer the reader to [9] (p. 99). We employ Kuiper's test to compare the distribution of each activity with all the other activities in the event log, at 6 levels: hour of the day, minute of the hour, second of the minute, day of the week, day of the month and month of the year. The activity whose distribution is statistically significantly different ($p = 0.05$) compared to all the other activities, would then show up in the detection list.

3.2   Repair

Insights from the detection phase are coupled with domain knowledge to repair the event log. A single activity is repaired at a time. The repair workflow consists of four steps as shown in Fig. 3 and explained in the subsections that follow.

Fig. 3. The four steps of the process repair workflow

Modeling process fragments lets the user specify domain knowledge as a free choice Petri net fragment via an interactive Petri net editor. Minimally, the Petri net contains only the activity which should be repaired. The user can additionally include "other activities" in the Petri net with respect to which the activity should be repaired. For example, consider that the original expected model of Fig. 1 is unavailable. However, the user is aware that activity C should happen only after activity A. From the detection phase, the user would know that there is a problem with the granularity of activity C. Hence the user can decide to repair the activity C in relation to activity A, by modeling a simple sequential Petri net. It should be noted that in many real life scenarios, the event log contains multiple activities and the user may not be aware of relationship between *all* the activities. Hence we would like to support the user in modeling partial fragments of Petri nets, explicitly specifying the relations between activities that the user is aware of. Modeling via Petri nets allows the possibility to include complex graphical fragments such as concurrency, choices, loops, duplications and introducing *silent* activities. Duplicate activities in the Petri net are uniquely numbered. After modeling the Petri net fragment, the next step is configuration of the repair options.

Table 3. The repair activity configuration options.

Activity	Description
Activity to repair	Specify the activity to be repaired from the Petri net fragment. For duplicate occurrences in the Petri net, select the (uniquely numbered) activity instance
Add events?	Specify if new activities should be artificially inserted in the event log. If set to true, the technique may insert artificial events (corresponding to the activity to be repaired) in the event log
Remove events?	Specify if activities should be removed from the event log. If set to true, the technique may remove existing events (corresponding to the activity to be repaired) from the event log

Repair Configuration. The repair configuration allows the user to specify the settings for correcting the ordering of a particular activity, and consists of: (i) the repair activity configuration, and (ii) the repair context configuration. The repair activity configuration deals with the activity to be repaired and is shown in Table 3. The repair context configuration provides the contextual information of how to correct an activity w.r.t. another activity, using the so-called repair actions. Each *repair action* consists of:

(i) **Anchor**: The timestamp of the activity to be repaired would be corrected based on a so-called anchor activity. Anchor activity is an activity from the Petri net which cannot be the same as the activity to be repaired. Alternatively, the anchor could be the start of a case (first activity) or the end of a case (last activity).

(ii) **Position**: Select whether the activity to be repaired should occur *before* the anchor or *after* the anchor activity.

(iii) **Value**: Specify the value of the new timestamp of the activity to be repaired in relation to the anchor. This could be an absolute value, e.g. 4 h, or a mean/median value of all the conforming relations from the event log between the anchor activity and the activity to be repaired per the modeled Petri net.

Multiple repair actions can be specified for repairing an activity. The order of repair actions determines the sequence in which repairs are performed in the event log, such that if the first action fails, the second one is applied and so on.

Perform Repairs. The actual repair is performed using the outcome from the alignments based conformance technique [10]. We demonstrate the use of alignments strategy to perform the actual repair of the event log with an example. Consider the ordering of an activity b is incorrect in an event log and needs to be fixed. Lets assume that the correct ordering of activity b in relation to two activities a and c is a sequential relation: $a \rightarrow b \rightarrow c$, i.e. activity a should be followed by b, which should be followed by c. This relation can easily be modeled using a Petri net. As a first step, the original event log is duplicated. Let's call this duplicated event log L_D. This duplicated event log and the Petri net fragment are used as input for the alignments. The alignments based conformance technique finds an optimal navigation path in a Petri net model for a particular sequence of activities. It does so by assigning a *cost* to each activity, which is then used for determining the correct ordering with an objective of minimizing the overall cost. All the activities which are not present in the Petri net fragment are assigned a cost of zero. In our example, as b is the activity to be repaired, it is assumed that the positioning of activity b is inaccurate compared to a and c. Hence a and c are assigned higher costs compared to b, i.e. it is preferred to align a or c at the expense of b. Now consider there is a case in the event log L_D, such that all the three activities (a, b and c) happen once and on the same day, however b happens at 08:00, a happens at 09:00 and c happens at 10:00. It is easy to see that this case does not fit the *expected* sequential behavior ($a \rightarrow b \rightarrow c$). The outcome of alignments for such a case would thus be:

log sequence	b	a	$>>$	c
model sequence	$>>$	a	b	c

In the alignments example above, activities a and c exhibit *synchronous* behavior per the Petri net fragment and the event log. The first occurrence of b (red) is a *move on log*, indicating a behaviour in the event log that cannot be replayed in the net. The second b (gray) is a *move on model*, indicating an expected occurrence of b as per the Petri net fragment which did not occur in the event log. Now, the ordering of b is corrected using each repair action from the context configuration box as follows:

1. For every *move on model* of b in the alignment of a particular case, change the timestamp of an event corresponding to a *move on log* of b for that case. The new timestamp is set based on the current repair action. A timestamp of an event can be updated only once for a particular repair action. It should be noted that, in case of Petri net patterns such as *loops*, the nearest *synchronous* anchor activity is chosen (in case the anchor is not case start or end). In our example above, the timestamp of the event corresponding to the first b is changed based on the configured repair action.

2. If *add events* is set to true: Let n be the difference between the number of *move on model* of b and the number of *move on log* of b for a particular case. If n is non-zero positive number, then add n number of artificial events of activity b, based on the chosen repair action.

3. If *remove events* is set to true: Let n be the difference between the number of *move on log* of b and the number of *move on model* of b for a case. If n is

non-zero positive number, then remove n number of events of activity b whose timestamp values were not changed (in step (a)), and whose alignment step corresponds to *move on log*.

4. Perform alignments on the repaired event log, and retrace and replicate all those cases with correct ordering of activity b in the event log L_D. Create a new event log using L_D containing only those cases which have incorrect ordering of activity b. Perform alignments on this newly created event log, and repeat the steps (a)–(d) until all the repair actions are performed. Alternatively, stop if the positioning of activity b for all the cases is correct.

3.3 Impact Analysis

Having repaired the event log, the user is presented with information about the impact of the repair on the event log. The user can choose to replace the current working copy of the event log with the repaired log (L_D), or choose to ignore the repaired event log. The metrics presented to the user to analyze the impact are: (i) edit distance: Levenshtein distance between the repaired event log and the original event log, (ii) fitness [10] of the original event log and the repaired event log for the activities from the Petri net fragment, (iii) the total number of cases impacted, (iv) the total number of events added, (v) the total number of events removed, and (vi) the total number of events with a changed timestamp value.

4 Implementation

The proposed techniques have been implemented in ProM[2], and can be used from the "Interactive Process Mining" package in the nightly build version of ProM. Figure 4 shows the detection screen of our tool. The user is presented with all the detected quality issues (along with their description, frequencies etc.) in a tabular format (panel A). On selecting an issue from the detection table, the user is provided with the filtered event log (panel B) which contains only the events relevant for the detected issue. Furthermore, the user is also provided with an aggregated histogram (panel C) showing the distribution of all the affected events based on the type of the detected issue.

Figure 5 shows the repair screen of our tool. The user models a Petri net interactively by adding one activity at a time (panel A). This Petri net specifies the relation between the activity to be repaired with other activities. The user specifies the repair strategy in repair configuration view (panel B). Upon performing the repair the impact is presented to the user (panel C) which can be used for the decision making of either keeping or discarding the changes.

---

[2] http://www.processmining.org/prom/start.

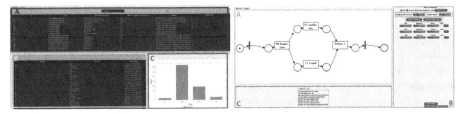

Fig. 4. The detection tab showing the three detection panels: (A) the list of detected issues, (B) the event log view, and (C) the graph view showing the distributions for the selected issue.

Fig. 5. The repair tab showing the three repair panels: (A) the Petri net modeled interactively by the user, (B) the repair configuration view, and (C) the impact of a (sequence of) repair(s).

5   Evaluation

Detecting and repairing event ordering anomalies involve re-structuring the event log to correct the control flow aspect. Hence for each event log, we demonstrate the detectability of ordering related issues in the event log, followed by repair of the event logs using the insights from detection along with domain knowledge. The final outcome is evaluated based on the process models discovered using the repaired event log. The event logs used are: (i) Sepsis event log - wherein the process model discovered from a repaired event log is compared with the ground truth process model, (ii) BPIC 2015 event log - wherein the outcomes of process discovery techniques are compared before and after the event log repair. From [11] and the reports submitted to BPIC 2015, we know that both the event logs have quality issues pertaining to event ordering. However, the exact problems and repairs were unknown and/or unavailable.

As detailed in Sect. 3, access to *domain knowledge* is essential in our approach. For the evaluation of our approach using the Sepsis log, domain knowledge was acquired through a process analyst who had extensive consultation with relevant domain experts [11]. For the evaluation using the BPIC 2015 log, domain knowledge was acquired by consulting the reports submitted to BPIC 2015. We acknowledge that our evaluation may be limited by the absence of *domain experts* who would be able to confirm the validity of the repaired log and to shed light on any peculiar, but still valid, phenomenon within the process being analysed that may be lost due to the repair action(s) applied. Nevertheless, even without access to domain experts, our approach is capable of improving the quality of the log as evidenced by the ability to generate a process model that is closer to the known ground-truth process model (for the Sepsis log) and of better process model quality (for the BPIC 2015 challenge log). The details of the evaluation of our approach are provided below.

5.1   Sepsis

The Sepsis log[3] [11] (1,050 patients/cases and 16 activities) contains the Sepsis treatment process of patients in a Dutch hospital. The "ground truth" process model (see Fig. 6) was hand-drawn in consultation with domain experts (see [11] for details).

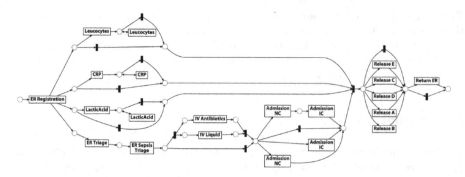

Fig. 6. Original Sepsis model for the event log as modeled in [11].

The detection outcome of our tool indicated that 96% of the issues were granularity and ordering related and the remaining 4% were issues arising due to statisical anomalies. Among the ordering related issues, almost 50% (14 out of 28) of the *direct* ordering problems had at least one of the three activities: *IV Antibiotics, IV Liquid* and/or *Admissions NC*, which hinted at a possible ordering problem with regards to these activities.

Here we give an example of the steps followed to repair one of the activities (*Admissions NC*). The Petri net fragment shown in Fig. 7 was modeled based on the domain knowledge about Sepsis protocols from the literature. In order to repair the activity *Admissions NC*, the position of *Admissions NC* was configured to be after (the mean duration of all the fitting cases) *IV Antibiotics* or *IV Liquids* for the mis-aligned cases. In total, the timestamp ordering of six activities was corrected (see Table 4). After repairing the time values (and removing duplicates) of the 6 activities, Inductive miner-incomplete [12] was used to automatically discover a process model from the final repaired event log (Fig. 9).

It should be noted that the process model (see Fig. 8) discovered from the original event log using inductive miner-incomplete includes much parallelism and allows self-loops for almost all the activities. Upon comparing the process models in Figs. 6 and 9, it is quite evident that by using our tool to repair the event log in just 6 steps, we were able to come very close to the process model discovered manually by the authors of [11]. We were able to achieve this despite the assumption of *absence* of a complete normative model to begin with. Instead, we looked at the *problem* activities from the detection phase, and used

---

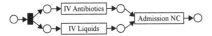

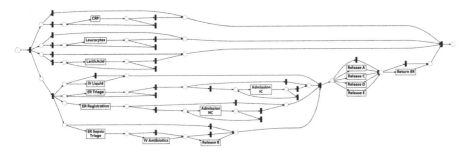

Fig. 7. *Admission NC* should be optional (see Fig. 6), but according to the available domain knowledge it is only known that if *Admission NC* occurs, it should always occur (only once) after *IV Antibiotics* and *IV Liquid.*

Table 4. Activities repaired on Sepsis log.

Activity repaired	Edit distance	% traces impacted
ER Triage	15	0.9
ER Sepsis Triage	46	2.3
IV Liquid	124	6.2
IV Antibiotics	22	1.1
Admission NC	407	32.3
Admission IC	17	1.2

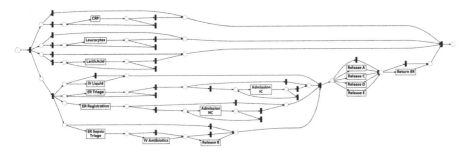

Fig. 8. Sepsis model discovered from the original event log using the Inductive miner-incomplete discovery algorithm.

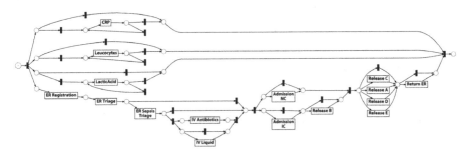

Fig. 9. Sepsis model discovered from the repaired event log using the Inductive miner-incomplete discovery algorithm.

the clinical protocols pertaining to such activities as-is, specified by Petri net fragments, in order to repair the event log.

5.2   BPIC 2015

We also evaluated our approach against the BPI Challenge 2015[4] event logs which deal with building permit applications of five municipalities in the Netherlands. We use the event log from one of the municipalities (municipality 1) which

---

[4] http://www.win.tue.nl/bpi/2015/challenge.

contains almost 400 activities. To make the results comprehensible, we filter the log to select the top-25 frequent activities belonging to the main application process and focus on almost 700 *closed* cases.

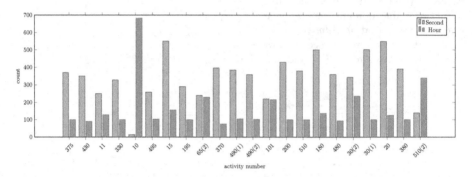

Fig. 10. Number of events per activity with hourly vs. second-granularity.

Our tool's detection technique indicated a likely granularity problem (50% of all the issues) It is clear that granularity problems can influence ordering, and hence there were a high number of ordering related issues too (47%), compared to 3% of statistical anomaly issues.

Among granularity issues, more than 10 activities had multi-granular times-tamp values (see Fig. 10) and many activities had high number of events registered at *hourly* granularity resulting in the loss of ordering information in the event log. Therefore, we repair the event log in order to manually correct the ordering of top 10 activities with high number of hourly granularity.

The activities in the filtered event log consists of the following naming structure: 01_HOOFD_### (a few activities have an additional _## at the end). From the available domain knowledge, it is known that typically the last three digits from 01_HOOFD_### denote the ordering sequence of activities (however this may not always be the case). We use this information to *repair* the positioning of activities which have a high number of events with hourly granularity, in relation to the activities which have high number of events with seconds granularity. Upon doing this repeatedly, we increased the granularity of almost 45% of the events which initially had hourly

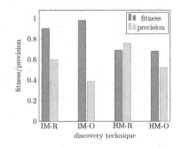

Fig. 11. Fitness and precision (original log). IM/HM: Inductive miner infrequent/Heuristic miner algorithms. Leading -R/-O denote repaired and original logs used for discovery.

granularity. Because we do not have *ground truth* for the BPIC 2015 log, we use the process mining quality dimensions of fitness (i.e., the faithfulness of the model to the log) and precision (i.e., the extent to which behaviours not seen in the log are allowed by the model) to evaluate the results. Generally, a process model with higher fitness and precision is desirable [1]. We discover process models using two state of the art process discovery algorithms (Heuristic miner [13] and Inductive miner [12]) using the original (non-repaired) event log at default settings. Next, we use the same algorithms to discover process models using the repaired event log. All the four discovered process models are evaluated against the original (non-repaired) event log to assess using the fitness [10] and precision [14] dimensions (see Fig. 11). From the detection phase it was clear that there were granularity issues (and thus event ordering issues) in the event log, by repairing the event log using the domain knowledge available, we significantly improved the precision of the models thereby restricting the model from allowing too much unobserved behavior, with little or no impact on the fitness of the model with the event log.

6 Related Work

In this section we consider notions of data quality in general, data quality for event logs (timestamps in particular) and discuss work related to detection and repair of event order-related data quality issues. Data quality has a long history of active research (see [15–17]). Data quality is generally described as being a multi-dimensional concept with dimensions such as accuracy/correctness, completeness, understandability, currency and precision [18,19] being frequently mentioned. Data quality for event logs, its impact on process mining and the particular importance of timestamps for event data was first considered in [2]. In line with general data quality discussion, consideration of event log quality also uses multi-dimensional frameworks and adopts similar dimensions. For instance, Mans et al. [6] describe event log quality as a two-dimensional spectrum with the first dimension concerned with the level of abstraction of the events and the second concerned with the accuracy of the timestamp (in terms of its (i) *granularity*, (ii) *directness of registration* (i.e., the currency of the timestamp recording) and (iii) *correctness*) thus making *explicit* the importance of these quality dimensions for proper temporal ordering of events.

We note that many process mining techniques *implicitly* detect quality issues in the event log. For example, conformance checking technique by [10] matches expectation (modeled by a Petri net) with reality (as per event logs) to detect conforming and deviating behavior. Thus the order related data quality issues would surface as deviating behavior. However such techniques *require* a process model to do the analysis. Our approach does not require ground truth during the detection phase. Many process discovery algorithms (for e.g. [12,13,20]) implicitly try to detect noise in the event log, which can sometimes be attributed to event ordering imperfections. However, the decisions and detection's made during the discovery phase are implicitly incorporated in the discovered process model, but not explicitly presented to the user. Our approach explicitly presents the user with the individual anomalies related to event ordering imperfections.

Some techniques from the literature automatically quantify the quality of an event log. [21] is a package in R which provides aggregated information about the structuredness and behavioralness of an event log. Similarly [22] discusses various metrics to quantify the overall quality of the event log. However, these techniques typically provide a global measure of data quality and do not pin point the exact list of issues within the event log. Our approach focuses on providing the user with a list of time related quality issues in the event log.

Unlike process discovery and performance analysis, repair of data quality issues in event logs is almost unexplored territory in the field of process mining. The authors of [23,24] describe techniques to correct the positioning of events based on (timed) Petri nets and probabilities derived from alignments. In [25] the authors describe a Petri net decomposition based heuristic repair strategy for efficient event log repair. In reality however, a de-facto standard process model is seldom available. Compared to these approaches which assume presence of a process model, our approach does not require an end-to-end process model for event log repair. Our approach only requires sufficient domain knowledge to recognise the sources of timestamps problems from the detected timestamp quality indicators and to then construct appropriate process fragments.

In [26,27], the authors describe a denial constraints based approach and a temporal constraints based approach resp. to automatically repair the timestamps of events in the event log. Again, these approaches require the users to pre-specify the constraints, whereas in our case the user can first analyze the detected issues, and then choose to act upon them. In [7] a visual guided approach for repairing time related issues in event log is discussed. Contrary to both constraints based approaches and visualizations driven approaches to event log repair, our approach allows the user to flexibly specify the domain knowledge using graphical process fragments, which typically are more intuitive and allow specification of complex process behavior such as concurrency, loops, choices, duplication of activities etc. In [28] a technique is proposed to automatically remove the "noisy" behavior from the event log without any user involvement. However, in our case we present such probable issues to the user, instead of automatically removing them from the event log. The user can temporarily repair the issues, analyze the impact and optionally make the changes permanent.

7   Conclusion

We have presented an integrated technique for detecting and repairing event ordering imperfections in an event log. The indicators of event ordering imperfections were discussed followed by a comprehensive strategy to detect such indicators. The proposed repair approach allows the user to develop a global truth (designed as a process fragment), which can be enforced locally on each case in the event log. This makes our repair approach far easier compared to a case by case repair strategy. Furthermore, intuitive graphical process fragments can be interactively designed by users for easy incorporation of domain knowledge. We applied our detection and repair techniques on two real life event logs.

We were able to show that we can detect anomalies and repaired them, leading to better process models being discovered. This work mostly focused on the data quality issues from a control flow perspective of process mining. In the future, we would like to address the performance and compliance perspective, along with exploring other data quality issues in process mining.

Acknowledgment. The contributions to this paper of R. Andrews, M.T. Wynn and A.H.M. ter Hofstede were supported through ARC Discovery Grant DP150103356.

References

1. van der Aalst, W.M.P.: Process Mining - Data Science in Action, 2nd edn. Springer, Heidelberg (2016). https://doi.org/10.1007/978-3-662-49851-4
2. van der Aalst, W., Adriansyah, A., de Medeiros, A.K.A., Arcieri, F., Baier, T., Blickle, T., Bose, J.C., van den Brand, P., et al.: Process mining manifesto. In: Daniel, F., Barkaoui, K., Dustdar, S. (eds.) BPM 2011. LNBIP, vol. 99, pp. 169–194. Springer, Heidelberg (2012). https://doi.org/10.1007/978-3-642-28108-2_19
3. Gschwandtner, T., Gärtner, J., Aigner, W., Miksch, S.: A taxonomy of dirty time-oriented data. In: Quirchmayr, G., Basl, J., You, I., Xu, L., Weippl, E. (eds.) CD-ARES 2012. LNCS, vol. 7465, pp. 58–72. Springer, Heidelberg (2012). https://doi.org/10.1007/978-3-642-32498-7_5
4. Bose, J.C., Mans, R.S., van der Aalst, W.M.P.: Wanna improve process mining results? IEEE CIDM **2013**, 127–134 (2013)
5. Suriadi, S., Andrews, R., ter Hofstede, A.H.M., Wynn, M.T.: Event log imperfection patterns for process mining: towards a systematic approach to cleaning event logs. Inf. Syst. **64**, 132–150 (2017)
6. Mans, R.S., van der Aalst, W.M.P., Vanwersch, R.J.B., Moleman, A.J.: Process mining in healthcare: data challenges when answering frequently posed questions. In: Lenz, R., Miksch, S., Peleg, M., Reichert, M., Riaño, D., ten Teije, A. (eds.) KR4HC/ProHealth -2012. LNCS (LNAI), vol. 7738, pp. 140–153. Springer, Heidelberg (2013). https://doi.org/10.1007/978-3-642-36438-9_10
7. Gschwandtner, T., Aigner, W., Miksch, S., Gärtner, J., Kriglstein, S., Pohl, M., Suchy, N.: TimeCleanser: a visual analytics approach for data cleansing of time-oriented data. In: i-KNOW, p. 18. ACM (2014)
8. Mans, R.S., van der Aalst, W.M.P., Vanwersch, R.J.B.: Data quality issues. Process Mining in Healthcare. SBPM, pp. 79–88. Springer, Cham (2015). https://doi.org/10.1007/978-3-319-16071-9_6
9. Mardia, K.V., Jupp, P.E.: Directional Statistics. Wiley, Chichester (1999)
10. Adriansyah, A., van Dongen, B.F., van der Aalst, W.M.P.: Towards robust conformance checking. In: zur Muehlen, M., Su, J. (eds.) BPM 2010. LNBIP, vol. 66, pp. 122–133. Springer, Heidelberg (2011). https://doi.org/10.1007/978-3-642-20511-8_11
11. Mannhardt, F., Blinde, D.: Analyzing the trajectories of patients with sepsis using process mining. RADAR+EMISA **1859**, 72–80 (2017)
12. Leemans, S.J.J., Fahland, D., van der Aalst, W.M.P.: Discovering block-structured process models from event logs containing infrequent behaviour. In: Lohmann, N., Song, M., Wohed, P. (eds.) BPM 2013. LNBIP, vol. 171, pp. 66–78. Springer, Cham (2014). https://doi.org/10.1007/978-3-319-06257-0_6

13. van der Aalst, W.M.P., Weijters, A.J.M.M., Maruster, L.: Workflow mining: discovering process models from event logs. IEEE Trans. Knowl. Data Eng. **16**(9), 1128–1142 (2004)
14. Muñoz-Gama, J., Carmona, J.: A fresh look at precision in process conformance. In: Hull, R., Mendling, J., Tai, S. (eds.) BPM 2010. LNCS, vol. 6336, pp. 211–226. Springer, Heidelberg (2010). https://doi.org/10.1007/978-3-642-15618-2_16
15. Wang, R.Y., Storey, V., Firth, C.: A framework for analysis of data quality research. IEEE Trans. Knowl. and Data Eng. **7**(4), 623–640 (1995)
16. Batini, C., Palmonari, M., Viscusi, G.: Opening the closed world: a survey of information quality research in the wild. In: Floridi, L., Illari, P.(eds.) The Philosophy of Information Quality. SL, vol. 358, pp. 43–73. Springer, Cham (2014). https://doi.org/10.1007/978-3-319-07121-3_4
17. Laranjeiro, N., Soydemir, S.N., Bernardino, J.: A survey on data quality: classifying poor data. In: PRDC, pp. 179–188. IEEE (2015)
18. Wand, Y., Wang, R.Y.: Anchoring data quality dimensions in ontological foundations. Commun. ACM **39**(11), 86–95 (1996)
19. ISO/IEC FDIS 25012: Software engineering - software product quality requirements and evaluation - data quality model (2008)
20. Günther, C.W., van der Aalst, W.M.P.: Fuzzy mining – adaptive process simplification based on multi-perspective metrics. In: Alonso, G., Dadam, P., Rosemann, M. (eds.) BPM 2007. LNCS, vol. 4714, pp. 328–343. Springer, Heidelberg (2007). https://doi.org/10.1007/978-3-540-75183-0_24
21. Swennen, M., Janssenswillen, G., Jans, M., Depaire, B., Vanhoof, K.: Capturing process behavior with log-based process metrics. In: SIMPDA (2015)
22. Kherbouche, M.O., Laga, N., Masse, P.A.: Towards a better assessment of event logs quality. In: 2016 IEEE SSCI, pp. 1–8, December 2016
23. Rogge-Solti, A., Mans, R.S., van der Aalst, W.M.P., Weske, M.: Improving documentation by repairing event logs. In: Grabis, J., Kirikova, M., Zdravkovic, J., Stirna, J. (eds.) PoEM 2013. LNBIP, vol. 165, pp. 129–144. Springer, Heidelberg (2013). https://doi.org/10.1007/978-3-642-41641-5_10
24. Rogge-Solti, A., Mans, R.S., van der Aalst, W.M.P., Weske, M.: Repairing event logs using timed process models. In: Demey, Y.T., Panetto, H. (eds.) OTM 2013. LNCS, vol. 8186, pp. 705–708. Springer, Heidelberg (2013). https://doi.org/10.1007/978-3-642-41033-8_89
25. Song, W., Xia, X., Jacobsen, H.A., Zhang, P., Hu, H.: Heuristic recovery of missing events in process logs. In: ICWS, pp. 105–112, June 2015
26. Chu, X., Ilyas, I.F., Papotti, P.: Holistic data cleaning: putting violations into context. In: IEEE ICDE, pp. 458–469, April 2013
27. Song, S., Cao, Y., Wang, J.: Cleaning timestamps with temporal constraints. Proc. VLDB Endow. **9**(10), 708–719 (2016)
28. Conforti, R., Rosa, M.L., ter Hofstede, A.H.M.: Filtering out infrequent behavior from business process event logs. IEEE Trans. Knowl. Data Eng. **29**(2), 300–314 (2017)

Fusion-Based Process Discovery

Yossi Dahari[1], Avigdor Gal[1], Arik Senderovich[1(✉)], and Matthias Weidlich[2]

[1] Technion – Israel Institute of Technology, Haifa, Israel
{dahari,sariks}@technion.ac.il, avigal@ie.technion.ac.il
[2] Humboldt-Universität zu Berlin, Berlin, Germany
matthias.weidlich@hu-berlin.de

Abstract. Information systems record the execution of transactions as part of business processes in event logs. Process mining analyses such event logs, e.g., by discovering process models. Recently, various discovery algorithms have been proposed, each with specific advantages and limitations. In this work, we argue that, instead of relying on a single algorithm, the outcomes of different algorithms shall be fused to combine the strengths of individual approaches. We propose a general framework for such fusion and instantiate it with two new discovery algorithms: The Exhaustive Noise-aware Inductive Miner (exNoise), which, exhaustively searches for model improvements; and the Adaptive Noise-aware Inductive Miner (adaNoise), a computationally tractable version of exNoise. For both algorithms, we formally show that they outperform each of the individual mining algorithms used by them. Our empirical evaluation further illustrates that fusion-based discovery yields models of better quality than state-of-the-art approaches.

1 Introduction

Information systems monitor and support business processes in real-time, while recording process transactions into event logs [1]. Process mining strives for analysis of business processes based on such event logs [2]. Specifically, various algorithms for the discovery of process models from event logs have been developed in recent years. Each of these algorithms strikes different trade-offs in process discovery, e.g., related to the scalability of the approach or the accuracy and complexity of the resulting models [3,4]. Specifically, due to differences in how noise and incompleteness in the event log is handled and which representational bias [5] is adopted, algorithms differ in terms of the behavioural structures that they recognise in an event log and, thus, represent in a process model.

In this work, we argue that the results of process discovery can be improved by combining several of the existing algorithms and suggest to adopt the idea of ensemble methods from domains such as a statistics and machine learning. That is, by combining the strengths of different discovery algorithms, the resulting model shall have higher quality than a model that would be discovered by any of the base algorithms in isolation. Such an effect would stem from one algorithm outperforming other algorithms on *parts* of the event log, whereas for other parts a different algorithm turns out to be superior.

© Springer International Publishing AG, part of Springer Nature 2018
J. Krogstie and H. A. Reijers (Eds.): CAiSE 2018, LNCS 10816, pp. 291–307, 2018.
https://doi.org/10.1007/978-3-319-91563-0_18

To realise this idea, we present an approach for process discovery, referred to as FuseDisc, that is based on the notion of model fusion. The FuseDisc framework takes as input an event log and a set of discovery algorithms, and produces a process model by fusing results from several discovery algorithms. The fused model should then yield an improvement over the models that would have been discovered by the algorithms in isolation. This improvement is assessed using common quality measures for process discovery, which evaluate different aspects of the resulting models [6], such as its ability to replay the event log or the extent of generalisation beyond the behaviour recorded in the log. We capture instantiations of the FuseDisc framework that come with guarantees on the improvement with respect to a quality measure using the notion of a *proper fusion*.

Following the above idea, we develop two novel algorithms to discover process trees using the FuseDisc framework: First, the Exhaustive Noise-aware Inductive Miner, exNoise for short, employs a set of discovery algorithms, which are given as different variants of the Inductive Miner [7,8]. These variants in differ in their handling of noise in the event log: By applying different thresholds in noise filtering they are more or less aggressive in discarding infrequent behaviour. Second, given the high run-time complexity of exNoise, we also introduce the Adaptive Noise-aware Inductive Miner, adaNoise, which is a computationally tractable and heuristics-driven version of exNoise. That is, adaNoise uses a pre-defined discovery quality measure as a black-box heuristic.

Both algorithms are shown to generalise the Inductive Miner Infrequent [8], a variant of the Inductive Miner that is robust against noise. It relies on a single, global threshold for noise filtering that has to be configured upfront. Our algorithms, in turn, select the appropriate intensity of noise handling locally, i.e., per part of the event log.

Below, we summarise our contributions, while providing the structure of this paper after the definition of preliminaries in the next section:

- We propose FuseDisc, a framework for process discovery based on model fusion and define the notion of a proper fusion (Sect. 3).
- Using this framework, we propose two discovery algorithms, exNoise and adaNoise (Sect. 4). We prove both algorithms to be proper fusions and elaborate on their computational complexity.

Section 5 evaluates our algorithms empirically, using both synthetic logs and a real-world healthcare log. Our results indicate that the proposed algorithms improve over each of the base discovery methods, in terms of combined quality measures. Finally, Sects. 6 and 7 review related work and present concluding remarks, respectively.

2   Preliminaries

This section introduces preliminaries for process discovery. We formalise the notion of an event log (Sect. 2.1), before turning to process trees, their automated discovery from event logs, and common quality measures for process discovery (Sect. 2.2).

2.1   Event Logs

We adopt a common model of event logs that is grounded in sequences of activity labels that denote the executions of activities as part of a single process instance, also known as case. Let $\mathcal{A}$ be a universe of *activity labels* (*activities* for short). A *trace* $\sigma = \langle a_1, \ldots, a_n \rangle \in \mathcal{A}^*$ is a finite sequence of activities. The universe of traces is denoted by $\mathcal{T}$. An *event log* $L : \mathcal{T} \to \mathbb{N}$ is a multi-set of traces. We write $\mathcal{L}$ for the universe of event logs. Furthermore, let $|\sigma|$ denote the length of a trace, and $|L|$ be the number of unique traces in the event log. The set representation $\bar{L} \in \mathcal{L}$ of a log L is the set of traces that occur at least once in L, which is defined as $\bar{L} = \{\sigma \in \mathcal{T} \mid L(\sigma) > 0\}$.

For example, $L = \{\langle a, b, c \rangle^3, \langle b, c \rangle^1\}$ is an event log that comprises three traces $\langle a, b, c \rangle$ of length three and one trace $\langle b, c \rangle$ of length two. Its set representation contains two traces and is given as $\bar{L} = \{\langle a, b, c \rangle, \langle b, c \rangle\}$.

Furthermore, we define a trace operator set $\bigoplus$ with $\oplus \in \bigoplus$ being a function that maps n traces into a new trace, i.e., $\oplus : \mathcal{T}^n \to \mathcal{T}$. For example, the concatenation operator $\oplus_\to \in \bigoplus$ concatenates two given traces. Let $\sigma_1 = \langle a, b, c \rangle$ and $\sigma_2 = \langle d, e \rangle$ be two traces. Then, it holds that $\sigma_1 \oplus_\to \sigma_2 = \langle a, b, c, d, e \rangle$.

A partition of an event log L into sublogs with respect to a set of operators $\bigoplus$ is denoted by $\pi_\bigoplus(L)$. Formally it can be written as follows:

$$\pi_\bigoplus(L) = \left\{ L_1, \ldots, L_M \in \mathcal{L} \mid \bar{L} = \bigoplus_{i=1}^M \bar{L}_i \right\} \subseteq 2^\mathcal{L}.$$

That is, we may reconstruct the set representation $\bar{L}$ of the log by applying a sequence of (possibly different) operators from $\bigoplus$ on the set representations of $L_1, \ldots, L_M$.

2.2   Process Trees and Their Discovery

In this work, we focus on the combination of discovery algorithms that adopt the same representational bias, but differ in their handling of noise in event logs. Specifically, we consider algorithms that discover process trees, such as the Inductive Miner [7] and the Evolutionary Tree Miner [9].

Process Trees. A process tree represents a process as a rooted tree, in which the leaf nodes are activities and all non-leaf nodes are control-flow operators, see [7]. Common control-flow operators include sequences of activities ($\to$), exclusives choice ($\times$), concurrency ($\wedge$), and structured loops ($\circlearrowright$). Process trees are defined recursively, as follows. Let $\Phi = \{\to, \times, \wedge, \circlearrowright\}$ be a set of *operators* and $\epsilon \notin \mathcal{A}$ be the *silent activity*. Then, $a \in \mathcal{A} \cup \{\epsilon\}$ is a process tree; and $\phi(T_1, \ldots, T_n)$, $n > 0$, with $T_1, \ldots, T_n$ being process trees and $\phi \in \Psi$ being an operator, is a *process tree* ($n > 1$ if $\phi = \circlearrowright$). The universe of process trees is denoted by $\mathcal{M}_T$. The semantics of a process tree T is defined by a set of traces, which is also constructed recursively: A function $\nu : \mathcal{M}_T \to 2^\mathcal{T}$ assigns a set of traces to a process tree. Trivially, $\nu(a) = \{\langle a \rangle\}$ for $a \in \mathcal{A}$ and $\nu(\epsilon) = \{\langle\rangle\}$. The interpretation of an operator $\phi \in \Phi$ is grounded in a language join function $\phi_l : 2^T \times \ldots \times 2^T \to 2^T$. Then, the semantics of a process tree $\phi(T_1, \ldots, T_n)$ is defined as $\nu(\phi(T_1, \ldots, T_n)) = \phi_l(\nu(T_1), \ldots, \nu(T_n))$. For instance, the traces induced by the exclusive choice operator $\times_l(L_1, \ldots, L_n)$ are given by the union of the traces of its children $\bigcup_{1 \leq i \leq n} L_i$. See [7] for formalisations of all operators in Φ.

A process tree for the aforementioned example log L could be one that describes a sequence of a choice between a and a silent activity ϵ, followed by activity b, and then c:

Process Tree Discovery. With $\mathcal{L}$ as the universe of event logs and $\mathcal{M}$ as the universe of process models, e.g., $\mathcal{M}_T$ for all process trees, we capture the essence of process discovery as follows:

Definition 1 (Discovery Algorithm). *A discovery algorithm is a function* $\gamma : \mathcal{L} \to \mathcal{M}$*, i.e.,* γ *produces a process model from an event log.*

A prominent approach for discovering process trees from event logs is the *Inductive Miner* (IM) [7,8]. To balance between over- and under-fitting, it is *parametrised* by a noise filtering technique that uses a predefined threshold $\tau \in [0,1]$ [8]. The workings of the Inductive Miner are summarised as follows. The algorithm recursively applies a *select* function $\eta : \mathcal{L} \times [0,1] \to \Phi \times 2^{\mathcal{L}}$, which, given an event log and a noise threshold, selects a trace operator and partitions the log based on the selected operator.

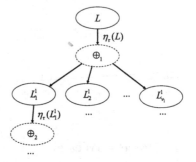

Fig. 1. Schematic view of the IM.

An overview of the algorithmic steps is given in the bipartite tree of Fig. 1. The algorithm starts by operating the select function η on the given event log, L, using a fixed noise threshold. A log partitioning operator $\oplus$ is returned (dashed circle), along with its corresponding logs, $L_1^1, \ldots, L_{v_1}^1$ (solid circles). The select function is then applied to every resulting event log L_i^1 recursively, until a base case of a log containing only single-activity traces is reached. The algorithm is guaranteed to terminate [7].

We observe that for the Inductive Miner, trace operators $\bigoplus$ are strongly coupled with process tree operators Φ. In fact, partitioning based on a log operator $\oplus$ yields a corresponding tree operator, denoted $\phi_\oplus \in \Phi$. Hence, the construction of a process tree from the result of the recursive application of η is uniquely defined: The respective operators are appended according to the bipartite tree shown in Fig. 1.

An alternative approach for mining process trees is rooted in the concept of searching for an improved model, with respect to some quality criteria. A well-known representative of such an approach is the *evolutionary tree miner* [9], which applies genetic mining to elicit the best model. However, this approach has a major advantage: Even though any returned solution is guaranteed to be of high quality, there is no guarantee that such a solution will be found in a finite period of time—a common drawback of genetic algorithms. As shown in experiments in [8], this is a limitation of practical relevance. The miner turned out to not discover any process tree within reasonable amounts of time for several real-world event logs.

Quality Measures for Process Discovery. Once a process model has been discovered from an event log, its quality shall be assessed. To this end, it has been argued that there is a common set of evaluation dimensions that shall be considered with according measures [6] With $\mathcal{L}$ and $\mathcal{M}$ as the universe of event logs and process models, respectively we capture these quality measures as follows:

Definition 2 (Discovery Quality Measure). *A discovery quality measure is a function $\psi : \mathcal{L} \times \mathcal{M} \to \mathbb{R}^{0+}$.*

Applied to a log L and a model T (a process tree in our case), a measure ψ potentially quantifies several dimensions of the relation between the log and the model. Also, the definition of the measure does not necessarily mean that the process tree T has been discovered from L. Instead, it could be a normative model that represents the process.

In the remainder, we focus on the following three dimensions: Fitness (*can the model represent all of the behaviour that is observed in the log?*), precision (*does a model allow only for the behaviour observed in the log?*), and generalisation (*does a model generalise beyond the behaviour that was observed in the log?*) [6]. Technically, we consider the three measures jointly, based on a weighted scoring function:

$$\psi_{Score}(L,T) = \omega_{Fit}\psi_{Fit}(L,T) + \omega_{Prec}\psi_{Pre}(L,T) + \omega_{Gen}\psi_{Gen}(L,T),$$

with $\sum_i \omega_i = 1, \omega_i \in [0,1]$. In our experiments, we shall test various values of ω_i. Note that we omitted simplicity due to the selected representational bias. That is, the algorithms used in the remainder construct process trees with uniquely labelled leaf nodes, and hence we assume that they achieve maximal simplicity.

3   Process Discovery Based on Model Fusion

Different discovery algorithms, understood also in the sense of varying configurations of a single algorithm, have particular strengths and limitations. Hence, discovery shall not be restricted to the *selection* of the most suitable algorithm for an event log, but rely on a *combination* of various algorithms, selecting a suitable one for each specific part of a log. For obvious reasons, such a combination should not provide worse results compared to applying any of the base algorithms to the complete log. In such a case, the overhead of considering multiple algorithms in the first place would not be justified.

To realize the above idea, we define FuseDisc, a framework for process discovery that is based on model fusion. An instance of this framework is given by a set of process discovery algorithms and a quality measure. The latter is used to determine the suitability of the algorithms for particular parts of the log, thereby guiding how the algorithms are combined. An instance of the framework is *proper*, if indeed the combined application of several algorithms leads to results that are at least as good as those obtained with any individual algorithm, in terms of the given quality measure.

Definition 3 (Fusion-based Process Discovery (FuseDisc); Proper Fusion). *Let* $\Gamma = \{\gamma_1, \ldots, \gamma_n\}$ *be a finite set of process discovery algorithms. Given an event log* L, *and a discovery quality measure* ψ, *a fusion-based discovery (FuseDisc) algorithm* $\gamma^{\langle \Gamma, \psi \rangle}$ *produces a process model using the discovery algorithms in* Γ, *potentially guided by* ψ. *Such an algorithm is called proper, if and only if it holds that*

$$\psi\left(L, \gamma^{\langle \Gamma, \psi \rangle}(L)\right) \geq \psi\left(L, \gamma(L)\right), \ \forall \ \gamma \in \Gamma. \tag{1}$$

Clearly, a trivial solution to achieve properness of a FuseDisc algorithm would be to define it as $\gamma^{\langle \Gamma, \psi \rangle} = \mathrm{argmax}_{\gamma \in \Gamma} \psi(L, \gamma(L))$. However, this trivial algorithm is an uninteresting one, since it does not stand a chance to satisfy Eq. 1 in a strict manner. In practice, the expectation is that a FuseDisc algorithm would yield a strictly better model. Yet, for the sake of flexibility, we do not enforce this requirement in Definition 3. We later show how empirical arguments demonstrate the actual benefit of a FuseDisc algorithm.

4   Inductive Mining with Adaptive Noise Filtering

This section introduces two discovery algorithms based on model fusion. To this end, Sect. 4.1 outlines how to instantiate the FuseDisc framework using a divide-and-conquer scheme. Following this idea, Sect. 4.2 presents exNoise, a specific discovery algorithm that relies on the Inductive Miner and is proven to yield a proper fusion. Due to its exponential runtime complexity, Sect. 4.3 then introduces adaNoise, a greedy and thus more efficient version of exNoise.

4.1   Fusion-Based Discovery Based on a Divide-and-Conquer Scheme

The idea of the FuseDisc framework as defined in Sect. 3 is to combine a set of discovery algorithms Γ in the construction of a process model for a given event log. While this combination shall be guided by some discovery quality measure ψ, the framework does not enforce any assumptions on *how* to combine the algorithms from Γ.

In this section, we argue that one way of organising the combination of different discovery algorithms is by means of a divide-and-conquer scheme. That is, the given event log is decomposed into sub-logs, and the algorithms from Γ are applied to each of them. This results in a set of sub-models per sub-log. Guided by the measure ψ, such sub-models are composed again to obtain a single model, which represents the result of fusion-based discovery.

Specifically, a log L is split into a set of sub-logs $L_1, \ldots, L_m$. For each sub-log L_i, $1 \leq i \leq m$, either the algorithms from $\Gamma = \{\gamma_1, \ldots, \gamma_n\}$ are applied to create sub-models $M_i^1, \ldots, M_i^n$, or the split is applied again, splitting L_i into $L_{i,1}, \ldots, L_{i,m'}$. In the composition step, in turn, the model for the sub-logs $L_1, \ldots, L_m$ is obtained by selecting one sub-model M_i^j, derived by applying discovery algorithm γ_j to sub-log L_i, for each of the sub-logs L_i and composing them into a single model.

The above idea requires Γ to contain discovery algorithms, so that the resulting models can be composed correctly in a hierarchical manner. In the remainder, we populate Γ with configurations of the Inductive Miner, which is motivated as follows:

- The Inductive Miner internally splits the event log hierarchically into sub-logs and associates a tree operator to each split. This tree operator provides an immediate means to compose the models obtained for the respective sub-logs.
- The Inductive Miner further guarantees the absence of behavioural anomalies, such as deadlocks. Consequently, behavioural anomalies cannot be introduced as part of the composition of models obtained for sub-logs.

The above points illustrate that the definition of fusion-based discovery by means of a divide-and-conquer scheme is closely related to notions of model compositionality: The representational bias adopted by the discovery algorithms in Γ must enable correct composition of a model from sub-models. As an example, one may also consider populating Γ with different discovery algorithms that construct a Petri-net [10]. Then, composition may be approached based on existing notions of Petri-net composition and refinement [11]. However, most existing discovery algorithms constructing Petri-nets, e.g., the α-algorithms [12] or the ILP-miner [13] lack guarantees on both the structure of the resulting net and the absence of behavioural anomalies. This makes model composition challenging in the general case.

4.2   The exNoise Algorithm

The *Exhaustive Noise-aware Inductive Miner*, exNoise, is a FuseDisc algorithm that is parametrised by a process quality measure ψ and by defining $\Gamma = \{\gamma_1, \ldots, \gamma_n\}$ as a set of n variants of the Inductive Miner Infrequent, see Sect. 2.2. However, each algorithm γ_i applies a different threshold τ_i for noise filtering.

The idea of the algorithm is depicted in Fig. 2a. Adopting the above divide-and-conquer scheme, the log is split into sub-logs hierarchically. For each sub-log (solid circle), exNoise considers the n trace operators that originate from the application of the n algorithms in Γ to the sub-log. For example, starting at L, the exNoise runs all *select* functions η_i of the respective algorithms γ_i on the event log. Each of these functions will determine a particular trace operator $\oplus_i$. Next, the algorithm considers all log partitions $\pi_{\oplus_1}, \dots, \pi_{\oplus_n}$ that stem from these trace operators in a recursive manner, until a base case is reached. A path of operators from the root of the graph to its leafs corresponds to a process tree, as every trace operator $\oplus$ is associated with a tree operator $\phi_\oplus$.

By means of the above procedure, exNoise constructs a search space of process tress. From this set, denoted by $\mathcal{T}_{ex}$, exNoise then selects a tree T_{ex}^* as follows:

$$T_{ex}^* = \underset{T \in \mathcal{T}_{ex}}{\mathrm{argmax}} \ \psi(L, T). \tag{2}$$

Note that the Inductive Miner Infrequent, as discussed in Sect. 2, is a special case of exNoise. It corresponds to an instantiation of exNoise with Γ containing solely a single variant of the discovery algorithm with a single, pre-defined noise filtering threshold.

To show properness of exNoise, we note that the result of applying any of the considered algorithms in isolation, is contained in the search space of exNoise.

Proposition 1. *Let $\mathcal{T}_{ex}$ be the set of process trees explored by* exNoise. *Then, it holds that $\gamma(L) \in \mathcal{T}_{ex}, \forall \ \gamma \in \Gamma$.*

Proof. We prove the proposition by construction. Let $\gamma \in \Gamma$ be the Inductive Miner Infrequent with noise threshold τ. Every selection of a trace operator in exNoise includes the option of applying η. Choosing this option for every level of the recursion is a feasible solution to exNoise. Hence, it holds that $\gamma(L) \in \mathcal{T}_{ex}$. □

Corollary 1. *The* exNoise *algorithm is proper.*

Proof. We assume by negation that there exists an algorithm $\gamma \in \Gamma$, such that $\psi(L, T_{ex}^*) < \psi(L, \gamma(L))$, which contradicts the properness condition. By combining Proposition 1, $\gamma(L) \in \mathcal{T}_{ex}$ and Eq. 2, $T_{ex}^* = \mathrm{argmax}_{T \in \mathcal{T}_{ex}} \psi(L, T)$, however, we get that $\psi(L, T_{ex}^*) \geq \psi(\gamma(L))$. This contradicts with the above negative assumption. □

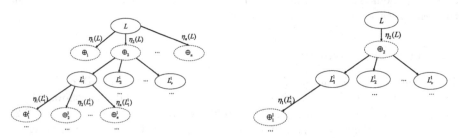

(a) Exhaustive Noise-Aware IM Algorithm.   (b) Adaptive Noise-Aware IM Algorithm.

Fig. 2. Two fusion-based discovery algorithms.

We now turn the focus on the computational complexity of the exNoise. Denote by $v = \max(|L|, |\sigma_{max}|)$, with $\sigma_{max} = \text{argmax}_{\sigma \in \bar{L}}|\sigma|$, the maximum between event log size, i.e., the number of traces, and the length of L's longest trace.

For every event log, exNoise considers $n = |\Gamma|$ select functions (dashed nodes of the bipartite tree in Fig. 2a), and creates a log partitioning for each of them. In the worst-case, this partitioning step needs v calculations: The worst-case may result either from a horizontal partition of the event log into $|L|$ separate event logs (every event log is a single trace of the originating log), or from a vertical partitioning of σ_{max} into traces with single activity (e.g., due to the sequence operator). Therefore, for every level in the search space (dashed circles in Fig. 2a), exNoise performs at most $v \times n$ steps.

Since we recursively select *every* trace operator, the time complexity of exNoise is $\mathcal{O}((v \times n)^k)$, with k being the depth of the bipartite tree. In the worst case, the latter is $k = v$, since the maximum number of recursive splits, again, depends on the length of the longest trace in the originating log and the size of the original event log [7]. Hence, the complexity of exNoise is $\mathcal{O}((v \times n)^v)$.

In practice, event logs may become very large, see [14], so that one cannot guarantee to have a bound on v. Hence, we consider the time complexity of exNoise to be exponential in the size of the log, so that exNoise quickly becomes intractable in practice.

4.3   The adaNoise Algorithm

Given the above results, the *Adaptive Noise-aware Inductive Miner*, adaNoise, is a quality-aware and greedy version of the exNoise. Specifically, it attempts to overcome the computational complexity of exNoise by moving from an exhaustive to a heuristic search. To this end, it incorporates the given quality measure ψ not only to select the best combination of discovery algorithms for a log partitioning. Rather, the measure directly guides the exploration of combinations of discovery algorithms. Similarly to exNoise, the adaNoise algorithm is parametrised by a quality measure ψ and a set Γ of n algorithms, given as variants of the Inductive Miner Infrequent.

We explain the intuition of adaNoise by means of Fig. 2b. Compared to the approach taken by the Inductive Miner (Fig. 1), adaNoise applies selection based on a (locally) optimal discovery algorithm γ^*, where optimality is measured using ψ. Since we consider a set of inductive mining algorithms, this corresponds to the (local) optimisation of the noise filtering threshold. In each step of deriving the next log partitioning, adaNoise chooses the select function η of the algorithm γ^*, such that

$$\gamma^* = \underset{\gamma \in \Gamma}{\text{argmax}}\ \psi(L, \gamma(L)), \tag{3}$$

with L being the log on which η is applied in the respective step.

While proceeding according to this heuristic search, adaNoise further constructs the process models for each intermediate step. To do so, it considers the operators that were selected prior to the current log L (i.e., the path from the root of the bipartite tree to L), concatenated with the result of applying each of

the possible discovery algorithms to the current log L. The resulting n intermediate models are added to a solution set $\mathcal{T}_{ada}$, and their ψ values are maintained
for future use.

The trees in Fig. 2 illustrate the difference between exNoise and adaNoise. For
every log, adaNoise proceeds with one trace operator and recursively splits the
log with a single select function. In contrast, exNoise considers all trace operators
for all logs.

If adaNoise reaches leaf nodes, the recursion is stopped, the corresponding
process tree is computed, and added to the solution set $\mathcal{T}_{ada}$. Once leaf nodes
have been reached for all logs, adaNoise returns T^*_{ada}, as follows:

$$T^*_{ada} = \underset{T \in \mathcal{T}_{ada}}{\mathrm{argmax}} \ \psi(L, T). \tag{4}$$

Similar to exNoise, the Inductive Miner Infrequent is a special case of adaNoise,
when Γ is a singleton set containing only one configuration of the algorithm,
with a single noise filtering threshold. Next, we provide theoretical guarantees
for adaNoise.

Proposition 2. *Let $T_\Gamma = \{\gamma(L) | \gamma \in \Gamma\}$ be the set of process trees discovered
by the algorithms in Γ to an event log L. Then, it holds that $T_\Gamma = \{\gamma(L) | \gamma \in \Gamma\} \subseteq \mathcal{T}_{ada}$.*

Proof. The result is due to the computation of intermediate models for every
event log that is considered for partitioning. Given a log L, adaNoise always
obtains $\gamma(L)$ for all $\gamma \in \Gamma$ and these intermediate models are added to $\mathcal{T}_{ada}$. □

Corollary 2. *The adaNoise algorithm is proper.*

Proof. By Proposition 2, the solution space $\mathcal{T}_{ada}$ contains T_Γ. Then, due to Eq. 4,
we derive that the following holds true: $\psi(L, T^*_{ada}) \geq \psi(L, \gamma(L)), \forall \ \gamma \in \Gamma$. □

Note that $\psi(L, T^*_{ada})$ is not guaranteed to attain $\psi(L, T^*_{ex})$, as the greedy search
in adaNoise may miss out on solutions that are found by exNoise.

The adaNoise algorithm attempts to provide a computationally feasible alternative to the exhaustive search of exNoise. Yet, its computational complexity
depends on the performance of ψ. Specifically, the algorithm computes the
heuristic n times at every level, which results in $n \times v$ operations, since, in
the worst case, we consider v event logs per level. Furthermore, the maximal
depth of the bipartite tree is v. This yields a time complexity of $\mathcal{O}(n \times v^2)$. The
computation of ψ at each level factorizes this expression. Hence, if ψ is exponential in v (e.g., computing ψ based on alignments [15]), adaNoise has exponential
runtime. In our experiments, however, we show that, in practice, adaNoise runs
efficiently, even if ψ has exponential time complexity.

5   Evaluation

We evaluate the FuseDisc framework by comparing adaNoise against the plain
Inductive Miner Infrequent (IMi). The latter is indeed dominated by adaNoise,

with the difference being large, when focusing on precision as a quality dimension. This is explained by the fact that, unlike IMi, our adaNoise algorithm is able to improve precision, even when weighted against fitness and generalization.

5.1 Benchmarks and Experimental Setup

We run experiments on three benchmarks. First, we tested the algorithms with synthetic event logs from two BPM (2016–2017) discovery contest benchmarks.[1] The two benchmarks comprise 10 event logs from 10 corresponding process models. These models include complex control-flow constructs, and log phenomena such as recurrent activities, loops, inclusive choices, etc. Each log contains 1000 traces, which we split into training-validation-test sets, as described below. The third benchmark is a real-world hospital log, which comprises one month of event data from an outpatient cancer clinic in the US. The dataset comprises about 25000 treatment paths ($\sim$1000 patients per day) that consist of a total 68800 events. Log behaviour includes parallelism, loops, and exclusive choices.

For the fusion, we considered the Inductive Miner Infrequent (IMi) [8] with noise filtering thresholds of increasing size, with step size being 0.05. In other words, we set $\Gamma = \{\gamma_1, \ldots, \gamma_n\}$ with the threshold of γ_i being set to $\tau_i = 0.05 \times (i - 1)$. We always compared the results obtained with adaNoise against the best base miner, i.e., we compare against $\gamma^* = \mathrm{argmax}_{\gamma \in \Gamma} \psi(L, \gamma(L))$. As such, we consider the most challenging baseline in our evaluation, even though, in practice, there would be no means to know a-priori which of the baseline algorithms performs best.

In our experiments, we measured three quality metrics, namely fitness, precision and generalization, along with the total score. For the evaluation, the adaNoise algorithm was implemented as a ProM plugin, using the *EDU-ProM* [16] variant of ProM.

To measure fitness we employ measures that are grounded in the notion of an *alignment* between model and log [15]. Specifically, the alignment is defined based on steps $(x, y) \in \mathcal{A}^{\perp} \times \mathcal{A}^{\perp}$, where $\mathcal{A}^{\perp} = \mathcal{A} \cup \{\perp\}$ is constructed from the universe of activities and a symbol $\perp \notin \mathcal{A}$. A step (x, y) is legal if $x \in \mathcal{A}$ or $y \in \mathcal{A}$ and is interpreted such that an alignment is said to 'move in both' traces $((x, y) \in \mathcal{A} \times \mathcal{A})$, 'move in first' $(y = \perp)$, or 'move in second' $(x = \perp)$. Given two traces σ, σ', an alignment is a sequence of legal steps. Each step is assigned a cost and a common cost model assigns unit cost if either $x = \perp$ or $y = \perp$; zero cost if $x = y$; and infinite cost if $x \neq y$. An alignment is a sequence of steps and the alignment cost is the sum of the costs of its steps. To obtain the fitness score, costs per log trace are aggregated, weighted by the trace frequency in the log, and normalized by the maximal possible cost.

To measure precision, we rely on a log traversal technique proposed in [17]: Every trace in the log is mapped to a state in the model, and *escaping edges* (traces and activities allowed by the model, while missing in the log) are computed. The precision score is defined as a function of the ratio of escaping edges over all allowed activities.

---

[1] See http://bit.ly/2BgjmIl and http://bit.ly/2a8z2T3.

Generalization was measured similarly to the method that was used in past discovery contests, assessing the model's capability to allow for legal behaviour that was not observed in the event log. Formally, let σ_0 be a legal trace which was never observed in L. Denote by R a trace replay function that given a trace σ and a process tree T, returns 1, if σ can be replayed on T, and 0 otherwise. Since σ_0 is random, generalization is written in expectation as $\mathbb{E}R(\sigma_0, T)$. This corresponds to the probability that T is able to parse a new legal trace σ_0. A generalization measure ψ_{Gen} should estimate $\mathbb{E}R(\sigma_0, T)$ explicitly from data. To this end, we use the *validation set* approach, which is a standard method to quantify the generalization error in the Machine Learning literature [18].

The event log L is partitioned into a *training set* L_T and a *validation set* L_V. We align every σ in L_V on T using R using a standard tree and log alignment procedure [6]. The generalization is given by the following unbiased estimator for $\mathbb{E}R(\sigma_0, T)$:

$$\psi_{Gen}(L, T) = \widehat{\mathbb{E}R(\sigma_0, T)} = \frac{1}{|L_V|} \sum_{i=1}^{|L_V|} R(\sigma_i, T).$$

Clearly, higher values of $\psi_{Gen}(L, T)$ imply greater generalization.

Our experiments were divided into two parts. First, we considered the synthetic datasets, setting ω_i of ψ_{Score}, see Sect. 2, to be uniform. Then, we altered the weight of ω_{Prec} to demonstrate the major advantage of fusion-based discovery. Second, we validated our approach on a real-world hospital log, testing in particular the sensitivity of adaNoise to the weights in the definition of ψ_{Score}.

5.2   Results

Before assessing the qualitative results, we note that even when considering an alignment-based calculation of ψ for fitness and precision, adaNoise had reasonable response times. Discovery took 80 s on average (stdev 60 s) for the synthetic data, and 27 s (stdev 7.8 s) for the real-life log. While this is less efficient than the plain IMi that runs in less than one second on average, we argue that the dominance of adaNoise in terms of discovery quality justifies this difference in run-time.

Part I: Synthetic Logs. Table 1 shows the results for the two BPM contest benchmarks. Specifically, we compare the discovery quality of the best algorithm in Γ (denoted IMi), to the quality of models created by the adaNoise algorithm for the BPM 2016 and BPM 2017 benchmarks. Furthermore, we present the results for all three quality measures, along with the total score measure obtained with a uniform weighting of the three measures. For every miner, IMi and adaNoise, we present the results for the each of the 10 individual logs per benchmark.

First, the theoretical guarantee formalised as Corollary 2 is visible the results. The total quality score obtained with adaNoise, is always higher (or equal) to the one observed for IMi. Furthermore, adaNoise better balances between precision and the other two measures. It is a known property of the IMi to sacrifice precision by adding generality, in the form of 'flower' constructs, while preserving

Table 1. Results obtained for the BPM 2016/2017 contest event logs.

Benchmark	Measure	Miner	L_1	L_2	L_3	L_4	L_5	L_6	L_7	L_8	L_9	L_{10}
BPM 2016 Contest	Fitness	IMi	0.97	1.0	0.99	0.99	0.98	1.0	1.0	1.0	0.95	0.86
		adaNoise	0.97	0.99	0.99	0.99	0.99	0.99	1.0	0.82	0.95	0.77
	Precision	IMi	0.84	0.53	0.46	0.96	0.22	0.37	0.22	0.46	0.90	0.61
		adaNoise	0.84	0.56	0.53	0.96	0.21	0.46	0.22	0.92	0.90	0.96
	Generalization	IMi	0.96	1.0	0.99	0.99	0.98	1.0	1.0	1.0	0.96	0.83
		adaNoise	0.96	0.99	0.99	0.99	0.99	1.0	1.0	0.83	0.96	0.78
	Total Score	IMi	0.92	0.84	0.82	0.98	0.73	0.79	0.74	0.82	0.94	0.77
		adaNoise	0.92	0.85	0.84	0.98	0.73	0.82	0.74	0.86	0.94	0.84
BPM 2017 Contest	Fitness	IMi	1.0	0.99	0.98	0.88	0.98	1.0	0.97	0.90	0.99	0.82
		adaNoise	0.95	0.99	0.97	0.86	0.98	0.95	0.99	0.90	0.99	1.0
	Precision	IMi	0.34	0.87	0.57	0.66	0.62	0.77	0.57	0.96	0.35	0.79
		adaNoise	0.50	0.87	0.66	0.88	0.62	0.86	0.56	0.96	0.35	0.55
	Generalization	IMi	1.0	0.98	0.98	0.87	0.99	1.0	0.98	0.88	0.99	0.82
		adaNoise	0.94	0.98	0.96	0.85	0.99	0.94	0.99	0.88	0.99	1.0
	Total Score	IMi	0.78	0.95	0.84	0.81	0.86	0.92	0.84	0.918	0.78	0.81
		adaNoise	0.80	0.95	0.86	0.86	0.86	0.92	0.85	0.918	0.78	0.85

fitness [8]. These results show that the adaNoise algorithm indeed follows the FuseDisc paradigm by compensating for the weaknesses of one algorithm by using other algorithms from Γ. Specifically, adaNoise flexibly tunes the noise thresholds that shall be applied to different parts of the event log. This yields improved results, without sacrificing precision.

We further vary the weight of precision for two selected event logs, namely L_8 of either benchmark. Figure 3 illustrates obtained results with both algorithms. Here, the vertical axis represents ψ_{Score}, the total quality score, while the horizontal axis corresponds to the weight assigned to precision, with the remaining weights being set uniformly to the complement of the precision weight. We observe that as the importance of precision grows, adaNoise's dominance increases, as the IMi is insensitive to ω_{Prec}.

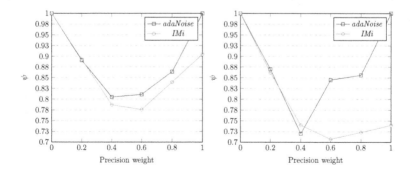

Fig. 3. Details results for the 8th event logs of the BPM 2016/2017 benchmarks.

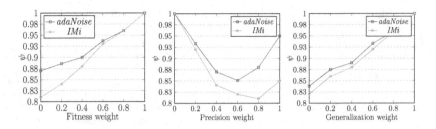

Fig. 4. Exploring the model quality with varying weights.

Part II: Hospital Log with Varying Weights. Next, we explored the sensitivity to the weights in the definition of model quality with the real-life hospital log. Figure 4 shows the overall score ψ_{Score} as function of changes in the respective weights. For example, in Fig. 4a, when the fitness weight is set to 0, the other two weights are set uniformly to 1/2. We observe throughout the results that adaNoise dominates the IMi, as formalised as a guarantee in Corollary 2. Furthermore, the largest improvement is observed when the emphasis is on precision. This is expected, since, as mentioned above, IMi tends to sacrifice precision, but is not guided by a comprehensive quality measure. A consequence of this behaviour is that, when the weight of precision is 0, the two algorithms yield the same value for ψ_{Score}. For virtually all other configurations, however, adaNoise provides a considerable improvement over the result of IMi in terms of model quality.

6 Related Work

Our approach falls in the area of process discovery, see a recent review by Augusto et al. [4]. We instantiated our framework with the Inductive Miner, introduced by Leemans et al. in [7] and later extended with a noise filtering mechanism [8]. The latter filters an event log representation (e.g., a directly follows graph) using a *single* noise threshold. Similar noise filtering strategies have been incorporated in many other discovery algorithms, e.g., those extracting simple heuristic nets, see [19,20]. Another example is the Split-Miner [21] to generate BPMN models, which reduces noise by several threshold-based pruning techniques, while ensuring appropriate levels of fitness. We argue that our fusion-based algorithms are able to overcome the limitations of such individual configurations and tune the noise filtering strategy for each of the event log.

We instantiated the FuseDisc framework with a divide-and-conquer scheme that partitions an event log. This is similar to splitting a log into passages for process discovery [22]. Yet, there is a fundamental difference: Our methods split the log to optimise accuracy in terms of fitness, precision, and generalization, whereas passage-based discovery is purely driven by performance considerations. In fact, it requires the selection of a particular discovery algorithm. Once this choice is taken, the model discovered from the complete log or obtained by combining the results discovered for the sub-logs are expected to be equivalent. While one may try to optimise accuracy, in the spirit of exNoise and adaNoise, for each sub-log obtained from minimal passages, this would require a careful selection of considered discovery algorithms to achieve a correct composition. Without that, applying [22] in the scheme defined in Sect. 4.1, would not even guarantee connectedness of the resulting model.

Discovery based on the theory of regions, see [24], may also give hints on how to split a log for fusion-based discovery. However, the regions in a transition system representing a log may be overlapping, so that it is unclear how the composition of models obtained for the respective sub-logs would have to be done.

Divide-and-conquer schemes may speed-up the computation of a alignments [23]. However, unlike our methods, this decomposes a model into fragments, not an event log.

Our approach is also related to discovery based on trace clustering. Respective approaches typically proceed in two steps: (1) traces of an event log are clustered based on a pre-defined similarity measure, and (2) a model is derived for each of the clustered sub-logs. A large number of such trace clustering methods has been proposed in the literature, see [25, 26], and references within. Recently, trace clustering has also been conducted such that the impact of clustering on the discovery result is taken into account to improve the discovered models [27]. Model improvement by trace clustering can further be achieved by slicing of discovered models [28], i.e., models are split based on identified trace clusters. The main difference of these works on trace clustering and our approach is that trace clustering partitions a log *horizontally*, i.e., per trace. Our FuseDisc framework, however, targets *vertical* and *horizontal* partitioning of traces, selecting a suitable algorithm for the log partitioning obtained with particular trace operators.

Finally, search-based process discovery methods, e.g., the Genetic Miner [29] and the Evolutionary Tree Miner [30] are related, as they explore a space of models when attempting to find an optimal one, given a set of quality criteria. Yet, these algorithms are not guaranteed to terminate in finite time. Recently, it was also suggested to consider the question of conformance between a log and a model as a search problem [31]. In our work, we avoid a direct search in the space of process models, yet adopt search ideas when partitioning the event log and combining different discovery algorithms.

7   Conclusion

To improve the quality of discovered process models, we introduced FuseDisc, a framework that combines the results of multiple discovery algorithms by fusing process models obtained for parts of an event log. Within this framework, we defined the notion of properness, which guarantees that the fused result will be at least as good as the results of all base algorithms. We then argued that fusion-based discovery may exploit divide-and-conquer schemes and presented two specific algorithms, namely exNoise and adaNoise. We showed that they are proper, and discussed their computational complexity. To evaluate our approach, we ran experiments using synthetic and real-world event logs. The results illustrate that the presented techniques improve over a state-of-the-art discovery algorithm, the Inductive Miner Infrequent, in terms of combined quality measures that include fitness, precision, and generalization.

References

1. Dumas, M., Van der Aalst, W.M., Ter Hofstede, A.H.: Process-Aware Information Systems. Wiley, New York (2005)
2. van der Aalst, W.M.: Process Mining: Data Science in Action. Springer, Heidelberg (2016). https://doi.org/10.1007/978-3-662-49851-4
3. Dumas, M., García-Bañuelos, L.: Process mining reloaded: event structures as a unified representation of process models and event logs. In: Devillers, R., Valmari, A. (eds.) PETRI NETS 2015. LNCS, vol. 9115, pp. 33–48. Springer, Cham (2015). https://doi.org/10.1007/978-3-319-19488-2_2
4. Augusto, A., Conforti, R., Dumas, M., Rosa, M.L., Maggi, F.M., Marrella, A., Mecella, M., Soo, A.: Automated discovery of process models from event logs: review and benchmark. CoRR abs/1705.02288 (2017)
5. van der Aalst, W.M.P.: On the representational bias in process mining. In: WET-ICE, pp. 2–7. IEEE CS (2011)
6. Buijs, J.C., van Dongen, B.F., van der Aalst, W.M.: Quality dimensions in process discovery: the importance of fitness, precision, generalization and simplicity. IJCIS **23**(01), 1440001–1440040 (2014)
7. Leemans, S.J.J., Fahland, D., van der Aalst, W.M.P.: Discovering block-structured process models from event logs - a constructive approach. In: Colom, J.-M., Desel, J. (eds.) PETRI NETS 2013. LNCS, vol. 7927, pp. 311–329. Springer, Heidelberg (2013). https://doi.org/10.1007/978-3-642-38697-8_17
8. Leemans, S.J.J., Fahland, D., van der Aalst, W.M.P.: Discovering block-structured process models from event logs containing infrequent behaviour. In: Lohmann, N., Song, M., Wohed, P. (eds.) BPM 2013. LNBIP, vol. 171, pp. 66–78. Springer, Cham (2014). https://doi.org/10.1007/978-3-319-06257-0_6
9. Buijs, J.C.A.M., van Dongen, B.F., van der Aalst, W.M.P.: Mining configurable process models from collections of event logs. In: Daniel, F., Wang, J., Weber, B. (eds.) BPM 2013. LNCS, vol. 8094, pp. 33–48. Springer, Heidelberg (2013). https://doi.org/10.1007/978-3-642-40176-3_5
10. Reisig, W.: Understanding Petri Nets - Modeling Techniques, Analysis Methods, Case Studies. Springer, Heidelberg (2013). https://doi.org/10.1007/978-3-642-33278-4
11. Suzuki, I., Murata, T.: A method for stepwise refinement and abstraction of petri nets. J. Comput. Syst. Sci. **27**(1), 51–76 (1983)
12. Wen, L., van der Aalst, W.M.P., Wang, J., Sun, J.: Mining process models with non-free-choice constructs. Data Min. Knowl. Discov. **15**(2), 145–180 (2007)
13. van der Werf, J.M.E.M., van Dongen, B.F., Hurkens, C.A.J., Serebrenik, A.: Process discovery using integer linear programming. Fundam. Inf. **94**(3–4), 387–412 (2009)
14. Leemans, S.J.J., Fahland, D., van der Aalst, W.M.P.: Scalable process discovery with guarantees. In: Gaaloul, K., Schmidt, R., Nurcan, S., Guerreiro, S., Ma, Q. (eds.) CAISE 2015. LNBIP, vol. 214, pp. 85–101. Springer, Cham (2015). https://doi.org/10.1007/978-3-319-19237-6_6

15. Van der Aalst, W., Adriansyah, A., van Dongen, B.: Replaying history on process models for conformance checking and performance analysis. Wiley Interdiscip. Rev. Data Min. Knowl. Disc. **2**(2), 182–192 (2012)
16. Dahari, Y., Gal, A., Senderovich, A.: EDU-ProM: ProM for the classroom. In: BPM Demo. CEUR, vol. 1920. CEUR-WS.org (2017)
17. Muñoz-Gama, J., Carmona, J.: A fresh look at precision in process conformance. In: Hull, R., Mendling, J., Tai, S. (eds.) BPM 2010. LNCS, vol. 6336, pp. 211–226. Springer, Heidelberg (2010). https://doi.org/10.1007/978-3-642-15618-2_16
18. Friedman, J., Hastie, T., Tibshirani, R.: The Elements of Statistical Learning. Springer, New York (2001). https://doi.org/10.1007/978-0-387-21606-5
19. Weijters, A.J.M.M., Ribeiro, J.T.S.: Flexible heuristics miner (FHM). In: CIDM, pp. 310–317. IEEE (2011)
20. Cnudde, S.D., Claes, J., Poels, G.: Improving the quality of the heuristics miner in ProM 6.2. Expert Syst. Appl. **41**(17), 7678–7690 (2014)
21. Augusto, A., Conforti, R., Dumas, M., Rosa, M.L.: Split miner: discovering accurate and simple business process models from event logs. In: ICDM (2017)
22. van der Aalst, W.M.P., Verbeek, H.M.W.: Process discovery and conformance checking using passages. Fundam. Inform. **131**(1), 103–138 (2014)
23. Munoz-Gama, J., Carmona, J., van der Aalst, W.M.P.: Single-entry single-exit decomposed conformance checking. Inf. Syst. **46**, 102–122 (2014)
24. Carmona, J., Cortadella, J., Kishinevsky, M.: New region-based algorithms for deriving bounded petri nets. IEEE Trans. Comput. **59**(3), 371–384 (2010)
25. Weerdt, J.D., vanden Broucke, S.K.L.M., Vanthienen, J., Baesens, B.: Active trace clustering for improved process discovery. IEEE TKDE **25**(12), 2708–2720 (2013)
26. Chatain, T., Carmona, J., van Dongen, B.: Alignment-based trace clustering. In: Mayr, H.C., Guizzardi, G., Ma, H., Pastor, O. (eds.) ER 2017. LNCS, vol. 10650, pp. 295–308. Springer, Cham (2017). https://doi.org/10.1007/978-3-319-69904-2_24
27. Sun, Y., Bauer, B., Weidlich, M.: Compound trace clustering to generate accurate and simple sub-process models. In: Maximilien, M., Vallecillo, A., Wang, J., Oriol, M. (eds.) ICSOC 2017. LNCS, vol. 10601, pp. 175–190. Springer, Cham (2017). https://doi.org/10.1007/978-3-319-69035-3_12
28. Ekanayake, C.C., Dumas, M., García-Bañuelos, L., La Rosa, M.: Slice, mine and dice: complexity-aware automated discovery of business process models. In: Daniel, F., Wang, J., Weber, B. (eds.) BPM 2013. LNCS, vol. 8094, pp. 49–64. Springer, Heidelberg (2013). https://doi.org/10.1007/978-3-642-40176-3_6
29. van der Aalst, W.M.P., de Medeiros, A.K.A., Weijters, A.J.M.M.: Genetic process mining. In: Ciardo, G., Darondeau, P. (eds.) ICATPN 2005. LNCS, vol. 3536, pp. 48–69. Springer, Heidelberg (2005). https://doi.org/10.1007/11494744_5
30. Buijs, J.C.A.M., van Dongen, B.F., van der Aalst, W.M.P.: A genetic algorithm for discovering process trees. In: CEC, pp. 1–8. IEEE (2012)
31. Rogge-Solti, A., Senderovich, A., Weidlich, M., Mendling, J., Gal, A.: In log and model we trust? a generalized conformance checking framework. In: La Rosa, M., Loos, P., Pastor, O. (eds.) BPM 2016. LNCS, vol. 9850, pp. 179–196. Springer, Cham (2016). https://doi.org/10.1007/978-3-319-45348-4_11

Decisions and the Blockchain

On the Relationships Between Decision Management and Performance Measurement

Bedilia Estrada-Torres[✉], Adela del-Río-Ortega, Manuel Resinas,
and Antonio Ruiz-Cortés

Depto. de Lenguajes y Sistemas Informáticos, Universidad de Sevilla, Sevilla, Spain
{iestrada,adeladelrio,resinas,aruiz}@us.es

Abstract. Decision management is of utmost importance for the achievement of strategic and operational goals in any organisational context. Therefore, decisions should be considered as first-class citizens that need to be modelled, analysed, monitored to track their performance, and redesigned if necessary. Up to now, existing literature that studies decisions in the context of business processes has focused on the analysis of the definition of decisions themselves, in terms of accuracy, certainty, consistency, covering and correctness. However, to the best of our knowledge, no prior work exists that analyses the relationship between decisions and performance measurement. This paper identifies and analyses this relationship from three different perspectives, namely: the impact of decisions on process performance, the performance measurement of decisions, and the use of performance indicators in the definition of decisions. Furthermore, we also introduce solutions for the representation of these relationships based, amongst others, on the DMN standard.

Keywords: Performance measurement · Performance indicators
Decision management · DMN · PPINOT

1 Introduction

Decisions are a key aspect of every business and its processes. Traditionally, decisions have been modelled either inside business process models or through decision logic using business rules or decision tables, amongst others. Recently, the Decision Model and Notation (DMN) standard [16] has been released with the aim of providing constructs to model decisions and decoupling decisions from process models. DMN can be used to model human decision-making, to identify requirements for automated decision-making and to implement those decisions.

Optimal decision making, and decision management as a more general concept, is of utmost importance for the achievement of strategic and operational

---

This work has been partially supported by the European Commission (FEDER), the Spanish and the Andalusian R&D&I programmes (grants TIN2015-70560-R (BELI) and P12-TIC-1867 (COPAS)).

© Springer International Publishing AG, part of Springer Nature 2018
J. Krogstie and H. A. Reijers (Eds.): CAiSE 2018, LNCS 10816, pp. 311–326, 2018.
https://doi.org/10.1007/978-3-319-91563-0_19

goals in any organisational context. Therefore, decisions should be considered as first-class citizens that need to be modelled, measured, analysed, monitored to track their performance, and redesigned if necessary [8]. Similarly, Nura et al. [15] argue that currently, decisions are based on quantitative and qualitative proofs that can be measured by means of statistical methods for the former or using techniques like benchmarking or balance scorecard for the latter. In addition, they claim that by means of decision measurement organisations can set targets and get feedback on the progress made towards their objectives.

Regarding the analysis of a decision, several approaches agree on the importance of differentiating the *quality of the decision* that is judged by the process followed to reach the decision; and the *quality of its outcome* and the associated consequences [7,10,13]. According to those authors, a good decision does not guarantee a good outcome because of uncertainty presented in the decision process; and just looking at the decision outcome does not provide information about the quality of the decision. Most scenarios found in the literature evaluate decisions on the basis of the knowledge and preferences of the decision makers, such as in [2,7]; and few information is taken from evidences in an objective manner, or is related to the process in which the decision takes place. Decisions are also studied in the context of business processes. However, authors have focused on the modelling of decisions and the analysis of the definition of decisions themselves, in terms of accuracy, certainty, consistency, covering and correctness [6,11,20]; using performance values to define decision rules [5] or providing languages for the definition of those decisions [17]; but to the best of our knowledge, no prior integrated work exists that analyses the relationship between decisions modelled in DMN and process performance and that evaluates decision performance itself based on data from event logs.

In this paper, we identify and analyse this relationship from three different perspectives. First, we analyse the impact of decisions on process performance and how this information can help the decision-making process. Second, we focus on the performance measurement of decisions themselves based on evidences gathered from the process execution. Finally, we analyse how process performance measures and indicators can be used for the definition of decisions. Furthermore, we also introduce solutions for the representation of these relationships: the concept of *decision performance indicator* (DPI) is proposed; PPINOT Metamodel [22] and the DMN standard are extended and integrated to propose a formal alternative for the measurement of decision performance; and the inclusion of performance information in decisions is improved providing more expressiveness by using PPINOT. Our proposal is applied to decisions made within the context of business processes and, ideally, modelled using DMN or a similar notation. Therefore, these decisions are usually made several times because they are repeated in each process instance.

The rest of the paper is organised as follows. Section 2 introduces the DMN standard in a real scenario. Section 3 analyses the relationship between decisions and performance measurement, whose representation using DMN and PPINOT is shown in Sect. 4. Section 5 presents the related work. Section 6 contains a brief discussion of the proposal. Finally, Sect. 7 concludes the paper.

2 Decision Model and Notation in a Running Example

DMN [16] is a standard for describing and modelling repeatable decisions within organisations, which provides a readily understandable notation by business and IT users and ensures that decision models can be automated and interchangeable. Two levels are used in DMN to model and describe decisions. For the first one, the requirement level, decision requirement diagrams (DRD) are used. They comprise a set of elements, their connection rules and a corresponding notation. For the second one, the decision logic level, the Friendly Enough Expression Language (FEEL) is provided for defining and assembling decision tables, calculations, if/then/else logic, etc. In addition, a notation for decision logic (boxed expressions) is provided to graphically represent those expressions and to show their relationship with elements of a DRD.

Figure 1 shows a DMN model based on real decisions made as part of the IT incident management process of a public organisation whose identity or characteristics cannot be revealed for privacy reasons. The IT Department receives and records incidents in one of its information systems. Incidents are resolved by agents external to the organisation, so before resolving them it is necessary to determine their priority level and the resource responsible for resolving it. By way of example, we focus on decisions related to the priority setting of IT incidents. The model was built with information provided by the organisation and the related data presented along the paper were taken from event logs of the aforementioned information system. An incident can be classified with the highest (P1), the intermediate (P2) or the lowest priority level (P3). This level determines the resource allocation and the total time allowed for the resolution of each incident. The priority is assigned considering two values: First, the level of *impact* (major, high, medium or low), which is a measure of the criticality of an incident often equal to the extent to which an incident leads to degradation of agreed service levels, and usually considers the number of people or systems affected. Second, the *urgency* (high, medium or low), which refers to the necessary speed for resolving an incident of a certain impact.

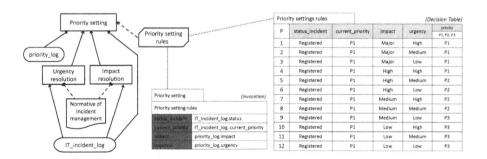

Fig. 1. Decisions in an IT Incident management process modelled using DMN.

A *decision* denotes the act of determining an output from a number of inputs, using decision logics. In our example, the main decision "Priority setting" requires information from two decisions ("Urgency resolution" and "Impact resolution") and from *Input data* ("priority_log" and "IT_incident_log"). Solid arrows represent *Information Requirements* of decisions. "Priority setting rules" represents a so-called *Business Knowledge*, which encapsulates business know-how in the form of business rules or other formalisms. The invocation of a *Business Knowledge* by a decision logic of a decision is made by means of *Knowledge Requirements* depicted as a dashed arrow. Finally, "Normative of incident management" is a knowledge source that denotes an authority. In our case, the internal documentation establish relationship between systems and users, which must be taken into account to determine the level of impact and urgency.

Two tables, called *Boxed Expressions*, conform the notation provided by DMN to represent logic decisions. *Invocation* is a container for the parameter bindings that provides the context for the evaluation. *Decision tables* are a tabular representation of a set of related input and output expressions, organised into rules indicating which output applies to a specific set of input entries. In Fig. 1, both tables are included. The decision table shows an excerpt of all decision rules that can be defined to evaluate the priority of an incident. The decision table only shows rules for those incidents whose current priority is P1. Four input values are required to make a decision. The *status_incident* describes the current state of the incident selected: "Registered" refers to unresolved recorded incidents. The *current_priority* is P1, but in a complete table the value could be *null* (to do the first assignation), P1, P2 or P3; this is required because an incident priority may be evaluated and changed more than once. The *impact* and *urgency* are values stemming from previous decisions.

3 Analysis of the Relationship Between Decisions and Performance Measurement

There is a well-known relationship between decisions and business processes [1,3,4,12], which has been analysed from different angles. However, although process performance is a relevant part of business process management, its relationship with decisions and decision models has not been analysed in depth. In this section, we identify and analyse this relationship from three perspectives, namely: the impact of decisions on process performance, the performance measurement of decisions themselves based on evidences gathered from the process execution, and the use of process performance indicators on the definition of decisions.

3.1 Impact of Decisions on Process Performance

Decisions are a key part of business processes and, as such, they can have an impact on their performance [14] that can be observed through their impact on their process performance indicators (PPIs). PPIs are quantifiable metrics that allow an evaluation of the efficiency and effectiveness of business processes [21]. Specifically, we can say that *a decision has an impact on a PPI* if the PPI value changes depending on the output selected in the decision. This fact is already acknowledged by DMN, which explicitly allows modelling the impact of decisions in performance indicators [16] by means of the relation `impactedPerformanceIndicator`. Thus, this relation enables the definition of a set of decisions and the performance indicators impacted by them. Figure 2 depicts an example of this relationship, in which PPI-1 and PPI-2 are identified as being impacted by *Decision-1* and *Decision-2*, respectively. This example also allows one to identify indirect impacts between decisions and PPIs: *Decision-1* has an indirect impact on PPI-2 because *Decision-2* has an impact on PPI-2 and *Decision-2* depends on *Decision-1*.

Based on relationships like those shown in Fig. 2, and on data about the execution of processes that allows the computation of PPIs, it is possible to obtain insights about the impact of decisions on process performance. Next, we detail two ways in which this information can be exploited.

Fig. 2. Example of relationships between decisions and PPIs

Warn About Potential Performance Issues. PPIs include a target value that allows one to determine whether they are fulfilled or not. Those PPIs that are not being fulfilled enable the identification of process improvement areas. Therefore, a description of relationships, as shown in Fig. 2, can be used to identify all those decisions that have a direct or indirect impact on PPIs that are not being fulfilled. The result is a set of decisions that might be causing a negative impact on the performance of the process. These decisions should then be analysed by domain experts to determine whether they are actually having a negative impact on the performance or not.

Insights About How a Decision Impacts a PPI. The relationship between decisions and PPIs can also be useful to identify how a specific decision value (outcome) may impact a PPI. This can be done by computing the PPI for each outcome of the decision and comparing the values obtained for them. Table 1 gathers data related to PPI-1, which is defined as the total time spent by an IT Department to solve an IT incident and whose target value is *less than 30 h*. This table shows how the priority decision impacts on PPI-1. The depicted data correspond to 476 PPI-1 values calculated monthly during November 2017, using PPINOT tool suite. Although the priority value of an incident may change several times, for our example, only the last priority value registered was considered. Specifically, only those cases where priority value after the last evaluation

Table 1. Influence of output decision in process performance - Part I

PPI: PPI-1	Target value:	< 30 h (monthly)	Total instances: 476
From	To - Decision *output*	Value - time	*Instances*
P1	P1	21,7 h	308
P1	P2	48,6 h	111
P1	P3	7,9 h	57

remained P1 or changed from P1 to P2 or to P3 were taken into account. *Value* column represents the PPI value for those cases in which the output of the decision is the one specified in the first column of the table (h stands for hours). *Instance* column represents the number of instances whose current decision is P1 and changes to another one.

Several interesting conclusions can be derived from this table. First, it is clear that the value of PPI-1 depends on the decision analysed because it changes significantly depending on the decision made, especially if the decision is to change from P1 to P2. Second, contrary to expectations, changing from P1 to P3 does not significantly increase the resolution time required, but the opposite. Possibly, this is because this decision is made when the incident has been thoroughly analysed, the causes have been identified and its priority has not been identified as critical. Third, it provides insights about actions that can be taken to improve the performance of PPI-1. Specifically, we can try to reduce the time it takes when the decision is P2.

3.2 Performance Measurement of Decisions

Decision measurement has been recognised as an important aspect within the organisation, because it helps to identify the progress made towards their objectives [15] and because the quality and speed of decisions may influence the success within an organisation [14]. Statistical methods or benchmarking and balance scorecard techniques are used to measure decisions [15]. Certain approaches, such as [2, 7, 14], suggest measures to assess decisions, but most of them are based on information provided directly by participants using surveys and interviews about users opinions or preferences, and not on objective data.

PPIs quantify the performance of a process in an objective manner, calculating their values with data generated within the process. Decisions have been decoupled from business processes to provide flexibility in the process management. With the aim of quantifying decision performance, in this paper we propose the concept of *decision performance indicator* (DPI) as analogous indicators that can also be computed from data gathered from the process execution. A DPI can relate to both the activities of the process where the decision is made and the decision elements (inputs, outcomes or other decisions, for example).

Considering the four performance dimensions defined in the Devil's quadrangle: time, cost, quality and flexibility, we discuss how the first three dimensions can be addressed using DPIs. Concerning *time* and *cost*, they can be applied in the same way as in the activities of a business process since, in this context, a decision is usually considered an activity. Therefore, one can measure for instance the elapsed time to make a decision or the cost of making the decision by considering the resources involved and the time spent in the decision. In our IT incident management scenario, a DPI that measures time could be "the total time spent to make a priority assignment decision should be less than 10 min."

Theories on the *quality* of decisions emphasize the differentiation between the *quality of the decision* as a process and the *quality of the output* [10] because the output may depend on external factors that are not under the control of the decision-maker. Factors that may influence the quality of decisions are [10]: *Decision makers*, who have their own knowledge and experiences; *the frame* as the possible ways of seeing the decision or its resolution contexts; the set of viable *alternatives* created for each frame; the decision maker's *preferences*; the *information* that we know; and the *logic* by which the decision is made. As it usually happens with quality, it cannot be measured directly, but it is possible to find proxies for it. In this way, in our IT incident management scenario we could define DPIs to evaluate decision makers such as "at least 80% of the people that make the priority escalation decision must be senior developers," DPIs to evaluate information, like "at least 95% of incidents must have the root cause field filled," or DPIs to evaluate the quality of the output, like "at least the 85% of incidents must not have changes in their priority".

3.3   Use of Performance Indicators on the Definition of Decisions

A decision involves a set of inputs and rules that are evaluated to determine an output. Inputs usually represent literal values or information provided by input data [16], as shown in Fig. 1. However, in some cases, decisions could require information provided by measures or indicators related to the performance of the process [17]. For instance, in the context of our example, let us suppose that it is necessary to inform the management staff when the time an IT incident is in Priority 1, or the number of evaluation iterations for the IT incident exceed some predefined thresholds. Under these conditions a decision table can be constructed, but in this case its inputs are a set of *performance measures* that need to be evaluated to determine the need to send an alert or not and the type of it. Therefore, a mechanism to be able to include process performance information into decisions is necessary to support these use cases. Furthermore, it would also be possible to consider not only the value of a performance measure, but its prediction [5]. For instance, one could also consider the predicted total resolution time as an input for the priority decision.

4   The Decision Performance in PPINOT

In this section we propose an integrated approach for representing the relationships described above that extends the DMN standard and PPINOT, an existing solution for PPI management.

4.1   PPINOT

PPINOT [22] is a metamodel for the definition and modelling of PPIs, which
has high expressiveness, allows the definition of PPIs in an unambiguous and
complete way, facilitates traceability between business process elements and PPIs
and promote the fulfillment of SMART criteria (specific, measurable, achievable,
relevant and time bounded). In PPINOT a PPI is defined by means of a set of
attributes: *goals*, indicating its relevance; a *target* value to be reached; a *scope*,
which defines the subset of instances to be considered during its calculation; a
set of *human resources* involved and a *measure definition* that specifies how the
PPI is computed.

Measure definition is a complex attribute that can be one of three types: *base
measures* that measure time, count, state conditions or data over a single process
instance; *aggregated measures* to aggregate one base measure over several process instances; and *derived measures* used for the calculation of a mathematical
function over other measures (aggregated or not). Measures are connected with
processes by *Conditions* that indicate how and when to take values from the
process, and *DataContentSelections* to obtain an attribute of a data object.

Furthermore, PPINOT also comes with a template-based notation that pro-
vides a set of linguistic patterns to allow user-friendly definitions of PPIs [23].
Table 2 shows linguistic patterns for the definition of measures based on the
PPINOT Metamodel (see Fig. 3). It has been extended to allow definitions
of PPIs considering decisions concepts. Gray boxes in the figure comprise the
PPINOT extension that will be explained in following subsections.

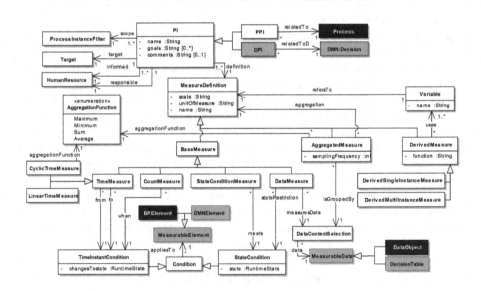

Fig. 3. Excerpt of the PPINOT metamodel

Table 2. L-patterns for measure definitions in PPINOT

Measure	Linguistic patterns
TimeMeasure ::=	*LinearTimeMesure* | *CycleTimeMeasure*
LinearTimeMesure ::=	*the duration between the* [first] *time instant*[s] *when* *<Event₁> and* [*the last time instant*] *when* *<Event₂>*
CycleTimeMeasure ::=	*the* {*total* | *maximum* | *minimum* | *average* | *...*} *duration between the pairs of time instants when* *<Event₁> and <Event₂>*
Event ::=	*<BP element type> <BP element name> becomes <BP element state>* | *<BP event name> is triggered*
CountMeasure ::=	The number of times *<Event>*
ConditionMeasure ::=	*<StateCondM>* | *<DataPropertyCondM>*
StateCondM ::=	*<BP element type> <BP element name>* [that] {[*is*] [*not*] *currently* | *has* [*not*] *finished*} [in state] *<BP element state>*
DataPropertyCondM ::=	[*<dataobjectstate>*] *<data object name> that satisfies: <condition on data object properties>*
DataMeasure ::=	*the value of* [*property*] *<data object property name> of* [*dataobject*] *<data object name>*
DeriverdMeasure ::=	*the function <expression over $x_1...x_n$>, where* {*<xᵢ> is <MeasureForDerᵢ>*}$_{i=1..n}$
MeasureForDer ::=	*TimeMeasure* | *CountMeasure* | *ConditionMeasure* | *DataMeasure* | *AggregatedMeasure*
AggregatedMeasure ::=	*the* {*sum* | *maximum* | *minimum* | *average* | *...* } *of <MeasureForAgg>* [*groupedby*[*property*] *<data object property name>of* [*dataobject*] *<data object name>*]
MeasureForAgg ::=	*TimeMeasure* | *CountMeasure* | *ConditionMeasure* | *DataMeasure* | *DerivedMeasure*

*BP - Business process

4.2   PPINOT on the Impact of Decisions on Process Performance

The DMN metamodel considers a relationship between classes *Decision* and *PerformanceIndicator* to specify the list of indicators impacted by the decision. However, as PPIs are not DMN's objective, *PerformanceIndicator* class does not provide specialised attributes to clearly and unambiguously define PPIs; which could provide errors and inconsistencies in calculating and analysing information. In this sense, we propose to integrate DMN with PPINOT to benefit from the aforementioned characteristics for the definition of PPIs. This is carried out by refining the class *DMN:PerformanceIndicators* in the DMN metamodel with the class *PPINOT:PPI*. This means, for instance, that these PPIs modelled

Table 3. Influence of output decision in process performance - Priority times

From	To - Decision *output*	Value - time	*Instances*	Total
P2	P1	51.6 h	156	
P2	P2	23.5 h	848	1242
P2	P3	29.3 h	238	
P3	P1	94,3 h	201	
P3	P2	131,5 h	440	98695
P3	P3	195.6 h	98054	

with PPINOT can be automatically computed and analysed by means of tools developed on the basis of the PPINOT metamodel [22]. This automation could facilitate the calculation of, for example, the values presented in Table 1.

Similar to Table 1, Table 3 gathers PPI values of the average time spent in solving an incident that changed priority from P2 or P3 to another priority value. In this case, the results show data computed using PPINOT tool suite from a total of 99937 instances of our scenario event log. It can be seen that the change from priority P2 to P1 did not reduce the resolution time, probably because resources allocated to solve incidents with P1 are not enough or because the analysis before solving P1 incidents is more exhaustive and requires more time. For incidents with P3, the behaviour is the expected: if priority changes from P3 to P1 the resolution time is less than if it changes from P3 to P2.

4.3   Definition of Decision Performance Indicators in PPINOT

Based on the PPINOT metamodel, we argue that the performance of decisions can be measured in three dimensions. First, we may be interested in measuring the output of a decision in terms of how many times each output occurs. Second, we may be interested in measuring the time a decision (or a part of a decision) takes. Obviously these two dimensions could be combined (e.g., the average time a decision takes grouped by its output). And third, we may want to know the quality of the information involved in the decision, because in some cases, a decision is made without all the information originally required, due to the high cost for obtaining that information (in terms of time or resources assigned). For example, if *urgency* value is not known, a policy could be defined for the *Priority setting* decision in which if the *current_priority* is null or P3 and *impact* is *major*, the incident priority is automatically changed to P1.

We propose an extension of the PPINOT metamodel that allows the modelling and definition of DPIs over DMN elements, taking into account the three dimensions identified. Just like a PPI definition and based on it, we define a DPI as *a quantifiable metric that allows the evaluation of the efficiency and effectiveness of decisions*, and it shares its attributes with a PPI: target, scope, etc. Figure 3 shows the PPINOT metamodel including new elements to support the measurement of decisions, depicted in light grey.

In the metamodel, a performance indicator (*PI*) is related to all indicator attributes. From PI, *PPI* and *DPI* are derived. *PPI* is defined over business processes and *DPI* over DMN decisions. The measure definition attribute can be instantiated as one of a set of measures. In the context of decisions, several of those measures can be reused. First, both *TimeMeasure* and *CountMeasure* are related to a *Condition*, specifically a *TimeInstantCondition*. For a *TimeMeasure* two time instants are required: the starting (*from*) and ending (*to*) points when values will be taken. The *CountMeasure* requires a time instant to indicate the moment *when* the value is taken. The *TimeInstantCondition* can be applied to a *MeasurableElement* that can be instantiated as a *BPElement* for a PPI definition, or as a *DMNElement* for a *DPI* definition. *DMNElements* can be: decision, business knowledge, input data or decision tables. *DataMeasure* measures the value of an attribute of a *DataObject* or a *DMNElement* in this case, being possible to specify a condition to obtain this information. Both *AggregatedMeasure* to consider several instances and *DerivedMeasure* to define formulas, can also be reused for decisions. By way of example, a DPI could be defined as *the average time spent to assign an IT incident priority*, whose scope is comprised by all instances and its target is set to less than 4 hours. On the basis of PPINOT templates [23] and linguistic patterns (see Table 2), the PPI is defined in Table 4. The *LinearTimeMeasure* pattern specifies its measure definition.

Several patterns can be used to define other types of DPIs. For example, a *CountMeasure* and its linguistic pattern can be used to define a DPI to count the number of decisions made or to know the number of times that a specific outcome was selected in a decision. By combining a *DataMeasure* and a *DerivedMeasure* it is possible to identify decisions with particular values such as "urgency" *null* and "current_prioriy" *P3*. Thus, the quality of information involved in a decision can be evaluated, because it is possible to identify decisions whose output has been selected with a lack of information. We have not found evidence about the possibility of using a *StateCondition* in the context of the performance measurement of decisions, reason why it was not included in the PPINOT metamodel extension.

Table 4. Example of DPI definition using template and linguistic patterns

DPI-1	Average time spent to assign an IT incident priority
Process	IT incident management
Goals	BG01 - Reduce IT incident resolution time
Measure definition	The DPI value is calculated as *the duration between the time instant when Activity 'IT incident registration' becomes active and Decision 'Priority Setting' output becomes assigned*
Target	Less than 4 h
Scope	All instances
Responsible	Head of IT Department
Informed	IT Manager

4.4   Using PPINOT PPIs in the Definition of Decisions

In order to provide greater flexibility and expressiveness on the definition of decisions, we propose the use of PPIs in the definition of decisions by means of the extension of DMN boxed expressions with linguistic patterns of PPINOT.

Boxed invocations use business parameters to indicate the business knowledge model to be used, and each parameter is accompanied by a binding expression that indicates the value assigned to the parameter for the purpose of evaluating a business knowledge model invoked. In our proposal, a *boxed invocation* is extended to include *measure definitions* as parameters, expressing them through PPINOT linguistic patterns. The value of this measure will be evaluated in a decision table, where according to the value obtained for each measure, a particular output can be selected. In the same way, a *boxed invocation* can also be extended to include PPIs as parameters. In this case, the extension incorporates: a scope, to indicate instances involved in the calculation; the target, to specify the expected PPI value; and a measure definition that indicates how to calculate the PPI. The evaluation in a decision table consists of identifying if the PPI target value is achieved or not. Decision tables are not extended because they support literals, values and expressions that can also be used to evaluate new parameters. Unlike the previous subsection, here we focus on PPI definitions and the complete set of *MeasureDefinitions*, because we want to measure process performance and use this information as an input to make decisions.

Let's suppose a scenario in which two types of alerts, *time_alert* and *workload_alert*, can be sent depending on specific values obtained from the process (see Fig. 4.a). *Boxed expressions* receive PPINOT measure definitions as inputs. The *Invocation* (on the left) lists measures involved in the decision "Decide alert": *execution_time* and *number_it_incidents_received*. The column "type" indicates whether the measure is applied over a single instance or over a set of them using an aggregation operation. Traditional attributes of *Boxed Invocations* can also be used, such as *priority_type* to indicate the incident current priority. The *decision table* (on the right) contains the same measures as the invocation. Rules are defined to select a type of alert, according to the knowledge of experts and depending on the value established for each one. This *decision table* is implemented using traditional FEEL expressions. The second example shown in Fig. 4.b depicts a decision that involves two PPIs. Here, the *invocation table* is extended to include the scope, the target value and its measure definition. The *decision table* maintains its structure but in this case, the rules to evaluate PPI values are based on true-false values to indicate whether the PPI target value has been reached. In invocation, as many measures or PPIs as needed can be included as inputs.

5   Related Work

Relations between business processes and decisions, and specifically decisions modelled in DMN, have been addressed in different approaches. For example, [1] proposes the integration of processes and decision modelling using BPMN

a) Measures

Decide alert					Decide alert rules					*(Decision table)*
Decide alert rules										**ALERT**
measure_name	**type**	**measure_definition**			**P**	**priority_type**	**execution_ time**	**number_it_ incidents_received**		**time_alert, workload_alert**
execution_time	Single instance	the duration between the time instant when *Activity IT incident registration* becomes *active* and when *Activity IT incident resolution* becomes *completed*			1	P1	>= 20 hours	-		time_alert
					2	P1	>= 35 hours	2000		workload_alert
number_it_ incidents_ received	Aggregated (all instanes)	The number of times *DataObject IT_incident* becomes *registered*.			3	P2	>= 45 hours	-		time_alert
					4	P2	>= 70 hours	3000		workload_alert
priority_type	Priority_log.priority_type				5	P3	>=120 hours	-		time_alert
					6	P3	-	2500		workload_alert

b) PPIs

Decide alert						Decide alert rules			*(Decision table)*
Decide alert rules									**ALERT**
PPI_name	**type**	**scope**	**Target**	**measure_definition**		**P**	**execution_ time**	**number_it_ incidents_ received**	**time_alert, workload_alert**
execution_time	Single instance	all	< 20 hours	the duration between the time instant when *Activity IT incident registration* becomes *active* and when *Activity IT incident resolution* becomes *completed*		1	True		time_alert
						2	True	True	workload_alert
number_it_incidents_ received	Aggregated (all instanes)	all	2000	The number of times *DataObject IT_Incident* becomes *registered*.		3	False	-	-

Fig. 4. Example of PPINOT measures (a) and PPIs (b) DMN box expressions.

and DMN, [4] derives decision models from business processes, [12] relates processes and decisions by means of a set of integration scenarios, or [3] presents frameworks for adjusting decision models dynamically according to the business process environment and ensure SLA compliance, to name a few. However, process performance is not considered in the context of these relationships.

Other proposals focus on the quality of the logic expressed in decision tables. To this end, measures such as certainty, consistency and covering are proposed to evaluate a set of decision-rules extracted from a complete [18], incomplete [19] or an ordered decision table [20]. In the same vein, [11] proposes an algorithm for measuring rule set consistency evaluating similarity between different rule sets; and [6] proposes algorithms for correctness checking tasks over DMN tables. However, they do not evaluate each decision instance, but the decision model expressed as decision tables.

More related to our proposal are [5,9], which are related to the impact of decisions in process performance. Specifically, [9] derives decision criteria formulated as decision rules based on experience gained through past process executions, although they do not consider the specifics of the DMN standard. Concerning [5], the authors propose a formal framework to derive decision models from event logs using DMN and BPMN and taking into account predictions of PPIs. However, they are concerned with obtaining decision models instead of helping to understand the consequences of each decision. Regarding the performance measurement of decisions, [2] addresses the quality of customer decisions using measures mostly based on preferences of decision makers and not on objective data taken from the process. Finally, the use of performance indicators in the definition of decisions is dealt with in [17], which proposes a query language to extract information from process or task instances that allows definitions of measures in boxed invocation.

6   Discussion

The analysis of the performance-decision relationships and the application of the integrated approach introduced for representing them using DMN and PPINOT in our scenario have shown improvements on the current state-of-the-art in three main directions. First, the analysis of the impact of decisions in performance complements the work of [5,9] by focusing on helping to understand the impact of decisions explicitly modelled using notations such as DMN and has proven to be very useful in our scenario. It also helps to identify the more problematic decisions to be analysed (i.e. those related to the PPIs with undesired values), preventing an important waste of time and effort that would be put on analysing hundreds of decisions otherwise. Finally, the integration of PPINOT in DMN to represent this relationship makes it easier the automated computation of PPIs. Second, with the novel concept of DPI proposed and the specific mechanism with templates and patterns provided, the performance of decisions can be defined and measured based on the data generated along the execution of the process. This opens a new way of measuring the performance of decisions that can be used to complement or replace more traditional approaches [2,7] that measure the performance of decisions using more subjective data such as surveys and do not take process information into account. And third, the use of the complete toolbox provided by PPINOT for the management of PPIs in DMN boxed expressions means an enrichment of the type of information that can be currently added as decision criteria in proposals such [17], whose definition language is less expressive than PPINOT; or [5] that does not focus on the definition of metrics.

Along the application of our approach to the evaluation scenario, some limitations were nevertheless identified. Concerning the impact of decisions in performance, although some tasks like the computation of Tables 1 and 3 can be automated, it remains, however, a domain-expert task to propose how to improve a decision in the light of the gathered insights. To do so, it would be interesting to explore the use of machine learning techniques to support this task. Furthermore, the computation of Tables 1 and 3 could also be extended so that the "priority" is not considered a single value, but the set of priorities the incident was assigned during its life. Finally, regarding DPIs, most of the limitations stem from deficient information sources. In order to define and compute measures, PPIs and DPIs, certain information must be available in the corresponding logs. However, this is not always the case. For instance, the information about the instant when a decision starts to be considered or the experience of a decision-maker are often missing. Therefore, if we need to measure the duration of a decision or the influence a user's experience has on a decision output, it will be necessary to cross-check information from different sources, in case this information is actually recorded somewhere. Nevertheless, the quick pace at which BPMSs are improving their support to decision management makes us think that more and better information concerning decisions will be available soon, mitigating this issue.

7 Conclusions and Future Work

This paper seeks to improve the understanding of the relationship between decision management and performance measurement. The main conclusion of our analysis is that decision management can be significantly enriched by considering performance indicators and their relationship with decisions in three different perspectives: the impact of decisions on process performance, the measurement of the performance of decisions and the use of PPIs and performance measures as inputs for decisions. Some advantages of explicitly defining these relationships have been encountered, such as the provision of important insights regarding possible dysfunctional decisions from a performance point of view or the identification of possible actions to be taken to improve the performance. Furthermore, in this paper we also outline how these relationships can be modelled and supported by extending PPINOT. However, performance relationships could be represented using other performance notations or techniques as it has been partially shown in papers such as [17]. The future work is aimed at developing an exhaustive evaluation of each of the relationships identified between decisions and performance measurement and extending PPINOT tools to facilitate the modelling and management of DPIs.

References

1. van der Aa, H., Leopold, H., Batoulis, K., Weske, M., Reijers, H.A.: Integrated process and decision modeling for data-driven processes. In: Reichert, M., Reijers, H.A. (eds.) BPM 2015. LNBIP, vol. 256, pp. 405–417. Springer, Cham (2016). https://doi.org/10.1007/978-3-319-42887-1_33
2. Aksoy, L., Cooil, B., Lurie, N.H.: Decision quality measures in recommendation agents research. J. Interact. Mark. **25**(2), 110–122 (2011)
3. Batoulis, K., Baumgraß, A., Herzberg, N., Weske, M.: Enabling dynamic decision making in business processes with DMN. In: Reichert, M., Reijers, H.A. (eds.) BPM 2015. LNBIP, vol. 256, pp. 418–431. Springer, Cham (2016). https://doi.org/10.1007/978-3-319-42887-1_34
4. Batoulis, K., Meyer, A., Bazhenova, E., Decker, G., Weske, M.: Extracting decision logic from process models. In: Zdravkovic, J., Kirikova, M., Johannesson, P. (eds.) CAiSE 2015. LNCS, vol. 9097, pp. 349–366. Springer, Cham (2015). https://doi.org/10.1007/978-3-319-19069-3_22
5. Bazhenova, E., Weske, M.: Deriving decision models from process models by enhanced decision mining. In: Reichert, M., Reijers, H.A. (eds.) BPM 2015. LNBIP, vol. 256, pp. 444–457. Springer, Cham (2016). https://doi.org/10.1007/978-3-319-42887-1_36
6. Calvanese, D., Dumas, M., Laurson, Ü., Maggi, F.M., Montali, M., Teinemaa, I.: Semantics and analysis of DMN decision tables. In: La Rosa, M., Loos, P., Pastor, O. (eds.) BPM 2016. LNCS, vol. 9850, pp. 217–233. Springer, Cham (2016). https://doi.org/10.1007/978-3-319-45348-4_13
7. Dowding, D., Thompson, C.: Measuring the quality of judgement and decision-making in nursing. J. Adv. Nurs. **44**(1), 49–57 (2003)
8. Dumas, M.: Decision Management with DMN in Practice (2016). http://bpmtips.com/decision-management-with-dmn-in-practice/#Dumas

9. Ghattas, J., Soffer, P., Peleg, M.: Improving business process decision making based on past experience. Decis. Support Syst. **59**, 93–107 (2014)
10. Howard, R.A., Abbas, A.E.: Foundations of Decision Analysis. Prentice-Hall, Englewood Cliffs (2015)
11. Huysmans, J., Baesens, B., Vanthienen, J.: A new approach for measuring rule set consistency. Data Knowl. Eng. **63**(1), 167–182 (2007)
12. Janssens, L., et al.: Consistent integration of decision (DMN) and process (BPMN) models. In: CAiSE Forum, pp. 121–128 (2016)
13. Keren, G., De Bruin, W.B.: On the assessment of decision quality: considerations regarding utility, conflict and accountability. In: Thinking: Psychological Perspectives on Reasoning, Judgment and Decision Making, pp. 347–363 (2003)
14. Negulescu, O., Doval, E.: The quality of decision making process related to organizations effectiveness. Procedia Econ. Finance **15**, 858–863 (2014)
15. Nura, A.A., Osman, N.H.: A toolkit on effective decision making measurement in organizations. Int. J. Humanit. Soc. Sci. **2**(4), 296–303 (2012)
16. Object Management Group: Decision Model and Notation (DMN) V1.1 (2016)
17. Perez-Alvarez, J.M., et al.: Process instance query language to include process performance indicators in DMN. In: Enterprise Distributed Object Computing Workshop (EDOCW), pp. 1–8 (2016)
18. Qian, Y., et al.: Measures for evaluating the decision performance of a decision table in rough set theory. Inf. Sci. **178**(1), 181–202 (2008)
19. Qian, Y., et al.: On the evaluation of the decision performance of an incomplete decision table. Data Knowl. Eng. **65**(3), 373–400 (2008)
20. Qian, Y., et al.: Evaluation of the decision performance of the decision rule set from an ordered decision table. Knowl. Based Syst. **36**, 39–50 (2012)
21. del Río-Ortega, A., et al.: Visual PPINOT: a graphical notation for process performance indicators. Bus. Inf. Syst. Eng. tbd(tbd) (2017)
22. del Río-Ortega, A., et al.: On the definition and design-time analysis of process performance indicators. Inf. Syst. **38**(4), 470–490 (2013)
23. del Río-Ortega, A., et al.: Using templates and linguistic patterns to define process performance indicators. Enterp. Inf. Syst. **10**(2), 159–192 (2016)

DMN Decision Execution
on the Ethereum Blockchain

Stephan Haarmann$^{(\boxtimes)}$, Kimon Batoulis, Adriatik Nikaj, and Mathias Weske

Hasso Plattner Institute, University of Potsdam, Potsdam, Germany
stephan.haarmann@student.hpi.de,
{kimon.batoulis,adriatik.nikaj,mathias.weske}@hpi.de

Abstract. Recently blockchain technology has been introduced to execute interacting business processes in a secure and transparent way. While the foundations for process enactment on blockchain have been researched, the execution of decisions on blockchain has not been addressed yet. In this paper we argue that decisions are an essential aspect of interacting business processes, and, therefore, also need to be executed on blockchain. The immutable representation of decision logic can be used by the interacting processes, so that decision taking will be more secure, more transparent, and better auditable. The approach is based on a mapping of the DMN language S-FEEL to Solidity code to be run on the Ethereum blockchain. The work is evaluated by a proof-of-concept prototype and an empirical cost evaluation.

Keywords: Blockchain · Interacting processes · DMN

1 Introduction

Business processes are an acknowledged means to describe working procedures in organizations [1]. Activities, their ordering, the data they work on, and organizational responsibilities can be represented in business processes, for instance using the Business Process Model and Notation (BPMN) [2]. Recently, BPMN has been accompanied by the Decision Model and Notation (DMN) [3] standard to capture decision structure and decision logic, thereby achieving a separation of concerns of process and decision logic [4]. More and more companies use both standards to describe business processes and decisions taken by them. Process enactment architectures are enhanced by decision engines that are capable of executing decision logic during process executions.

While we currently witness a strong uptake of decision management in industry, decision models are almost exclusively used to represent internal decisions of a company, such as whether a specific credit can be granted or not. However, many decisions are linked to more than one process and are part of interacting business processes. Such processes have been subject to behavioral analysis [1,5].

To address the security and transparency needs of interacting business processes, recently blockchain technology was proposed as an enactment platform

© Springer International Publishing AG, part of Springer Nature 2018
J. Krogstie and H. A. Reijers (Eds.): CAiSE 2018, LNCS 10816, pp. 327–341, 2018.
https://doi.org/10.1007/978-3-319-91563-0_20

for interacting business processes [6]. Blockchains enable the execution of interacting processes through software code to be run on a blockchain – so-called smart contracts – even if the participants do not trust each other.

While this work introduces blockchain technology to execute interacting processes, it does not address decisions that are taken by them. To close this gap, in this paper we propose an approach to execute DMN decision models on an Ethereum blockchain, thereby improving security, transparency, and auditing of decisions taken by multiple interacting business processes. The approach is based on a mapping of DMN decision models expressed in the language S-FEEL to Solidity, the programming language of the Ethereum blockchain [7].

The remainder of this paper is structured as follows. In Sect. 2 we provide an overview of DMN and of relevant aspects of blockchain technology. Section 3 contains the mapping of DMN decision models to Ethereum smart contracts. The approach is evaluated with respect to the blockchain specific costs in Sect. 4. An overview of related work is provided in Sect. 5. We discuss our research in Sect. 6.

2   Background

DMN decision models express requirements and logic of decisions. While the execution of DMN in process orchestrations can be achieved by traditional software systems, such as rule engines, decisions taken during interacting processes have not been investigated. We claim that blockchain technology proves to be useful in such settings due to a set of properties, which are provided in this section. First, we introduce the reader to decision models by means of a running example.

2.1   DMN

Decision Model and Notation (DMN) is a standard of the Object Management Group (OMG) to model operative decisions of enterprises. A decision model considers two layers: the requirements and the logic. The former provides a high level view on the information required for a decision, while the latter provides detailed information on how to take a decision [3].

For the remainder of this paper, we consider the following example: a manufacturer of bearings offers his customers reimbursement for delivered defective units. Therefore, a decision provides the fine as a ratio of the purchase price. In order to perform the decision, both manufacturer and customer must provide data. Figure 1 depicts the respective *Decision Requirements Diagram* (DRD), a graphical representation of the decisions and their dependencies. A node represents an entity, such as data (rounded shapes) and decisions (rectangles). The edges represent the information requirement relation indicating that data or decision results are required by a decision. The example consists of two decisions: *SLA* (service level agreement) and *fine*. SLA requires *years as customer* and *number of units* as data input. The decision *fine* requires the output of SLA and, additionally, the ratio of *defective units*. We marked the defective units in gray because it is contributed by the customer. The other (white) data inputs are provided by the manufacturer.

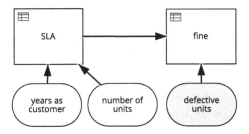

Fig. 1. Decision requirements diagram for the scenario; coloring of data input indicates the provider of the respective data, as detailed in the text.

On a second level, decision models specify the decision logic. DMN offers various notations, but we solely consider decision tables because they are most widely used. The decision tables that we use in this paper represent rules horizontally. A set of conditions on the data inputs is followed by specific decision outputs.

Table 1. Decision table for the SLA

U	Inputs		Output
	years as customer	number of units	SLA
	Number	*Number*	*{1,2 }*
1	<2	<1000	1
2	<2	≥1000	2
3	≥ 2	< 500	1
4	≥ 2	≥ 500	2

Table 2. Decision table for the fine

U	Inputs		Output
	defective units	SLA	Fine
	Number	*{1,2 }*	*Number*
1	<5%	1	0%
2	[5%..10%]	1	2%
3	>10%	1	100%
4	<1%	2	0%
5	[1%..5%]	2	5%
6	>5%	2	105%

Tables 1 and 2 depict the decision tables for our scenario. The table body consists of rules (i.e. four for SLA). In addition, the table header provides meta information. It specifies the names and data types of inputs and outputs: *years as customer* and *number of units* are numbers, the *SLA* is an enumeration with the possible values 1 and 2. The upper left cell of the decision table contains the *hit-policy*. The hit-policy determines in which order the rules of the table are evaluated. The used hit-policy *unique* (U) indicates that no two rules overlap in their conditions. Thus, at most one rule triggers for a specific input, and the evaluation order does not matter. More generally, we distinguish between single and multi-hit policies depending on the number of rules that contribute to the

Table 3. A description of DMN hit-policies

Sign	Name	Description
U	Unique	At most one decision matches a given input
F	First	The rules are evaluated in order. The first rule that matches the input determines the output
P	Priority	Output values must be enumerated and sorted according to priorities. If multiple rules match, the output with the highest priority is chosen
A	Any	If the condition of different rules overlap, the output must be the same. Thus, the evaluation order is irrelevant
C	Collect	All rules are evaluated and the outputs of all matching rules are collected. Afterwards an aggregation function (+, avg, min, max) is applied
R	Rule Order	The results of all matching rules are returned in the order of the rules
O	Output Order	Outputs are enumerated and ordered according to priorities. The outputs of all matching rules are provided in the order of their priority

decision's result (one vs. multiple). Table 3 provides an overview of the different hit policies. In this paper, we map all the elements of DMN decision models and decision tables in particular to Source code for Ethereum. This includes all elements mentioned above.

2.2 Blockchain Technology

Blockchain technology emerged with Bitcoin: a cryptocurrency whose transactions are stored on a blockchain [8]. Ever since, the technology has been adapted for different domains (e.g. BPM and logistics) and for various use-cases. However, all adaptations rely on a set of core properties [9].

A blockchain is a linked list that uses hash pointers. It is shared by a peer-to-peer network. Each element of the list is a block that stores transactions. The blockchain represents a history of transactions that allows us to restore all past and the present states. Some nodes, called *miners*, propose new blocks and find consensus about the set and order of transactions. Once this is achieved, all nodes share the same blockchain [10,11]. Thereby, *relative immutability* is reached: every node can detect inconsistent changes (e.g. changing the past) and reject respective transactions. However, if a majority of nodes agrees on a different version, the chain is adapted accordingly. The probability of a block being immutable increases as the number of succeeding blocks grows.

Another feature of (most) blockchains are cryptocurrencies. They are inherent currencies that are passed between users to reward mining, to pay for transactions, or to perform financial transactions. The so called 2nd generation of blockchains additionally supports complex *smart contracts*: Turing complete scripts, which are stored on the ledger and are executed by all miners. Thereby, immutability, publicity, and cryptocurrencies become available for computer programs. Smart contracts are like classes in an object-oriented language: they encapsulate data and functionality and can be instantiated.

Blockchains' properties are powerful since they enable interactions between participants without the need for trust. In recent research Mendling and others investigated the challenges and opportunities for BPM [12]. For one, process choreographies require trust between participants, but, as we have seen before, a blockchain can be a replacement by offering tamper-proofed monitoring capabilities. Implementations of process choreographies exist for both the Ethereum [6] and the Bitcoin [13] blockchain.

3   Generation of Decision Contracts from DMN

In general, contracts are agreements, about services, products, and money that participants agree to provide to each other. Decisions can be part of these agreements. In order to make a decision, the information has to be provided and logic needs to be executed. In an interactive process, each participant might contribute data to a single decision. Thus, participants need to exchange data to perform a certain decision. The DMN standard suggests to map each decision of a decision model to a decision taken by a single participant. Consequently, each participant will communicate the result of their decision such that others can base their decisions on this result. Figure 2 shows a respective business process collaboration for the running example: the manufacturer determines the SLA, the output of which is used as input by the customer to calculate the fine. In conclusion both decision results are exchanged.

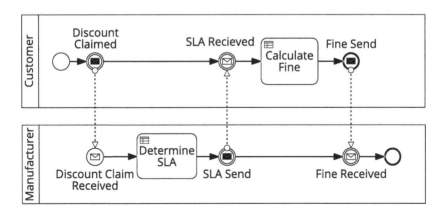

Fig. 2. DMN suggestion for decisions in interactive processes applied to the running example

This method of realizing decisions has a major drawback: due to the lack of a central mediator or state, participants must trust each other to provide the correct information and to take the decisions in the right way. In case of a fraud, conflict resolution is difficult: no accountable source of information exists and an expensive lawsuit can follow.

Weber et al. showed that a blockchain provides accountable audits and can be used as a trusted intermediary for interactive processes [6]. In the same fashion, we use blockchain as an intermediary for decisions. The respective process collaboration via blockchain is depicted in Fig. 3. We use Ethereum and smart contracts to specify the executable decision logic. Further, participants can contribute data to a decision via transactions, and smart contracts execute and log decisions. This leads to a public audit trail, which consists of all information required to reconstruct the decision making process: who provided which inputs, what are the results, and how was the decision made.

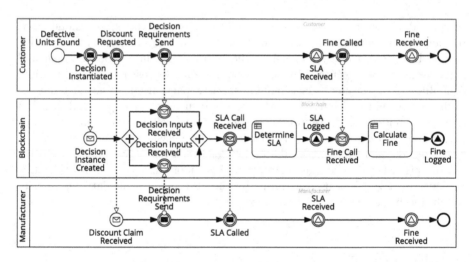

Fig. 3. Blockchain based implementation of a sample decision in interactive processes

Figure 4 depicts the mapping on the model level, where concrete decision tables are defined. As mentioned, a decision table defines the logic of a decision: it specifies inputs, outputs, rules, and a hit-policy. In our approach, we map each DMN decision table to a respective smart contract, which we call *decision contract*. A decision contract encapsulates the logic of the translated decision table and represents it in a function (called *decide*). Further, we provide two auxiliary structures: a *state contract* and a *factory contract*. The state contract encapsulates all inputs and outputs. The decision contract can, therefore, be implemented in a stateless manner following the best practices of DMN. Whenever a decision should be made, we provide a state contract instance with the respective data to the decision contract, which updates the outputs (inside the state contract) accordingly. The factory contract provides auxiliary functions

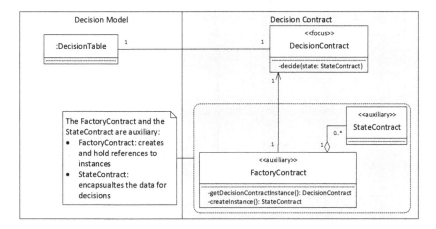

Fig. 4. Mapping of DMN decisions (left) to Ethereum smart contracts (right), smart contracts are deployed on a blockchain

for instantiation: it creates and holds new state contract instances and lazily initializes all decision contracts of a respective decision model.

The DMN standard provides the FEEL expression language to describe execution semantics of decisions. A subset of this language is called S-FEEL and can be used to express decision tables. Such a decision table consists of inputs and outputs, whereby, outputs can be input for other decision tables. A rule of a S-FEEL decision table consists of unary tests for each input (statements that evaluate to true or false for a specific value) and an expression for each output.

Our approach provides a translation of all these elements to Solidity source code. Figure 5 shows the mapping of abstract S-FEEL elements to Solidity code templates. The code is refined as further elements of the decision table are mapped (e.g. for each model, we create one state contract, for each input and output clause we add a respective attribute with a setter method to it). A decision contract additionally has an *event*: a mechanism to broadcast information. It is used to publish the results of a decision. For each decision table, we create one decision contract and for each rule we create/extend an if-statement by adding the unary tests to its condition and the output expression to its consequence. However, different hit policies require different rule generations. We abstracted from this details in Fig. 5 but described them in Table 4.

3.1   Interaction with the Decision Contracts

In order to persist the participant's agreement on a blockchain, it is sufficient to deploy a single instance of the FactoryContract The FactoryContract is capable of instantiating decision contracts and the state contract and, thereby includes, the binary representation of the StateContract and the decision contracts (SLADecisionContract, FineDecisionContract). For the remainder we assume that a respective instance has been deployed.

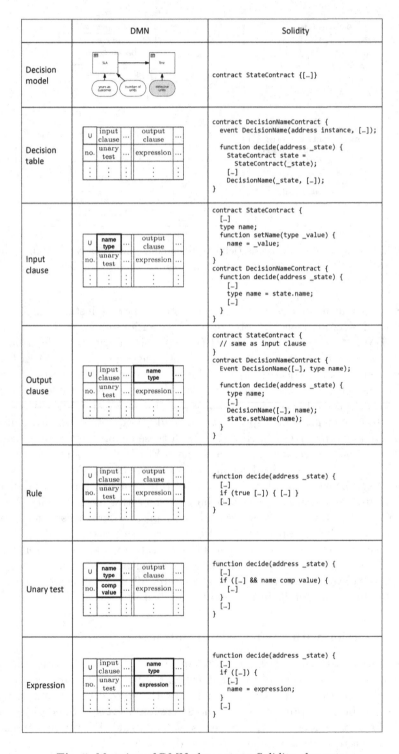

	DMN	Solidity
Decision model		`contract StateContract {[…]}`

Decision table		```contract DecisionNameContract {
		  event DecisionName(address instance, […]);
		
		  function decide(address _state) {
		    StateContract state =
		      StateContract(_state);
		    […]
		    DecisionName(_state, […]);
		}```

Input clause		```contract StateContract {
		  […]
		  type name;
		  function setName(type _value) {
		    name = _value;
		  }
		}
		contract DecisionNameContract {
		  function decide(address _state) {
		    […]
		    type name = state.name;
		    […]
		  }
		}```

Output clause		```contract StateContract {
		  // same as input clause
		}
		contract DecisionNameContract {
		  Event DecisionName([…], type name);
		
		  function decide(address _state) {
		    type name;
		    […]
		    DecisionName([…], name);
		    state.setName(name);
		  }
		}```

Rule		```function decide(address _state) {
		  […]
		  if (true […]) { […] }
		  […]
		}```

Unary test		```function decide(address _state) {
		  […]
		  if ([…] && name comp value) {
		    […]
		  }
		  […]
		}```

Expression		```function decide(address _state) {
		  […]
		  if ([…]) {
		    […]
		    name = expression;
		  }
		  […]
		}```

Fig. 5. Mapping of DMN elements to Solidity elements

Table 4. Handling of different hit-policies in `DecisionContracts`

Hit-policy	Description of mapping
Any, Unique, First	Condition is stored as an *if-else-if*; hence, the evaluation is complete after the first rule hit
Priority	Similar to Any, Unique, and First, but the rules are rearranged according to the output priority
Rule Order	Series of *if-statement* and output values are collected in arrays
Collect	Like Rule Order but an additional aggregation is applied in a postprocessing step
Output Order	Rules are rearranged in respect to the output order. Then, they are handled like Rule Order

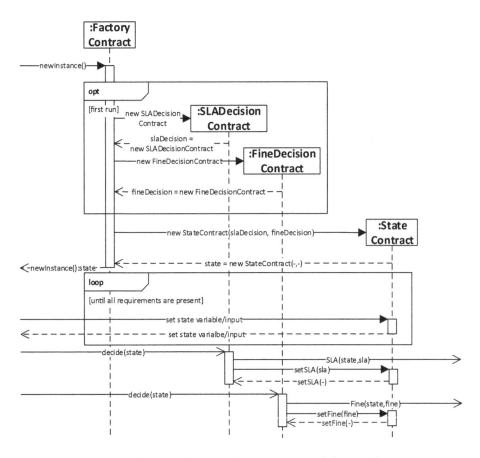

Fig. 6. Sequence diagram showing the interactions of the sample contracts

The sequence diagram in Fig. 6 shows the interaction of participants with smart contracts as well as among smart contracts. A participant (e.g. the manufacturer) calls the `FactoryContract` to create a new instance. If it has not been called before it initializes the decision contracts (i.e. `SLADecisionContract` and `FineDecisionContract`), and then creates a new `StateContract` instance with references to the decisions. The `StateContract`'s address is returned to the caller. Participants can then set input variables (i.e. years as customer, number of units, and defective units) on the `StateContract`, and thereby, provide the requirements for a decision. Once all decision requirements are fulfilled, a participant calls the decision(s) providing a `StateContract` instance. The decision logic is executed, outputs are determined, and eventually the state contract is updated. In this example, the output of SLA together with the defective units fulfills the requirements for the second decision, which can then be called.

4 Evaluation

To evaluate the approach, we investigate its applicability on the basis of a set of real life decision models. In addition, we provide an empirical cost evaluation of the example introduced above.

4.1 Applicability of the Mapping

To study the applicability of the mapping on real life decision models, we implemented the mapping in a proof-of-concept prototype. The prototype consists of a compiler, which translates DMN models (serialized as XML) to Solidity source code and is integrated into the Camunda modeler[1]. In this way, the Solidity source code is automatically derived from the decision model during design time.

Since Solidity is a Turing-complete language, S-FEEL can be translated to Solidity. However, due to Solidity's restricted type system, we encountered some engineering issues when defining the mapping.

S-FEEL's type *Number* is a generic floating-point number, but Solidity supports only integers and fractions. Therefore, floating-point numbers need to be represented with integers, which requires implementing auxiliary functions, for example, to compare two integer-based floating-point numbers. Another limitation comes from the fact that strings are stored as byte-arrays in Solidity, but only one-dimensional arrays can be dynamically sized. Therefore, lists of strings (requiring two-dimensional arrays) must be limited to items of a fixed length. Such lists are, for instance, returned by multi-hit decision tables.

We used the prototype to translate a real life set of decision tables to Solidity and from there to byte code for the Ethereum virtual machine. Our test data set consisted of 51 DMN models that were designed by participants of an online course on BPMN and DMN[2]. The prototype translated 46 of 51 models into

---

[1] https://camunda.org/download/modeler/.

[2] https://open.hpi.de/courses/bpm2016.

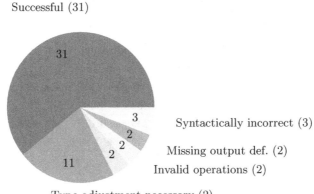

Fig. 7. Results of compiling the decision model test set to Solidity/Ethereum and reasons for failure

Solidity code. Of the remaining five, two models were missing an output definition (optional in DMN but required for our compiler) and three were syntactically incorrect. Afterwards, we ran the Solidity compiler to produce byte code for the Ethereum virtual machine. Out of the 46 remaining models 31 were compiled successfully. Two tables used long strings and compiled after adapting the data types manually. Eleven of the other 13 were not compiled, because identifiers for inputs or outputs used characters (e.g., brackets) or words (e.g., "contract") that are reserved words in Solidity. The remaining two decision models used comparison operators ($<, \leq, =, \geq, >$) on sets, which is forbidden in S-FEEL and in Solidity. These results of the investigation are summarized in Fig. 7.

To conclude, the majority of the decision models could be translated to executable Solidity code. Problems can occur because models are often used for documentation and not defined precisely enough for execution. Furthermore, special guidance during the modeling process could be helpful to prevent the use of reserved words or characters or from omitting necessary information.

4.2   Cost Evaluation

Ethereum is a second generation blockchain that integrates the full capabilities of a 1st generation blockchain—including a cryptocurrency called *Ether*. It is used to pay for transactions, i.e., all operations that store information in the blockchain. In our scenario, creating new contract instances, calling functions, and raising events causes transactions. In Ethereum, transaction costs are calculated in *gas*. Gas is an abstract unit that is mapped to Ether via a dynamic gas-price; in this way, the transaction costs can stay stable even though the value of Ether fluctuates.

The costs to instantiate smart contracts depend mainly on the required storage. Table 5 lists the cost for each contract after manually optimizing the used

Table 5. Instantiation costs for each contract and impact of features (based on avg. gas price and exchange rate on 11/07/2017 from https://etherscan.io/chart/gasprice)

Contract	Stripped	No optimization	Total
FactoryContract	$1,269,518$ gas 5.08€	$153,164$ gas 0.61€	$1,422,682$ gas 5.69€
SLADecision	$320,558$ gas 1.28€	$41,037$ gas 0.16€	$361,595$ gas 1.44€
FineDecision	$494,726$ gas 1.98€	$70,869$ gas 0.28€	$565,595$ gas 2.26€
StateContract	$172,010$ gas 0.69€	$41,233$ gas 0.17€	$213,243$ gas 0.86€
All	$2,256,812$ gas 9.03€	$306,303$ gas 1.23€	$2,563,115$ gas 10.31€

data types and the respective savings. The `FactoryContract` is the most expensive one (5€) because it stores an array of instances, which can grow indefinitely and holds the definition of the other contracts to initialize them. The `FactoryContract` is for convenience: its functionality can be executed off-chain. This would save more than half of the instantiation costs. Moreover, we showed before that DMN data types cannot be mapped perfectly to Solidity types. In general, the latter supports more precise types, for example, integers of different sizes; thus, we can optimize them manually. However, this optimization has a relatively small impact on the costs saving only up to 1.23€.

All but the `StateContract` are initialized only once; if a decision is executed frequently, high instantiation costs can be tolerated. See Table 5 for details. The baseline costs of a single instance is the sum of instantiating the `State-Contract`, calling setters, and executing the decision logic (includes setters for the outcome). Figure 8 depicts the costs for a single instance. From bottom to top we sum the different operations. We show both a version with and a version without optimized types. It should come as no surprise that the instantiation is the most expensive operation. One can see that the costs of a single instance are about 1.70€ and type optimization saves up only 0.10€.

5   Related Work

Within the BPM community increasing interest is spent on blockchain technology. Mendling et al. give an overview of challenges and opportunities that arise from combining BPM and blockchains [12]. So far, existing research in this direction focuses mostly on business process choreographies. These efforts include an Ethereum based solution to enable and monitor choreographies [6]. All messages are logged and state progression is monitored via the blockchain. The current consensus on the blockchain describes the current state of the choreography and conform state progression is enforced. This concept has also been applied to the Bitcoin blockchain [13].

A similar approach is Caterpillar—a blockchain enabled process engine [14]. However, none of these approaches consider decisions and decision models. In Caterpillar XOR gateways are implemented by manually defining Solidity

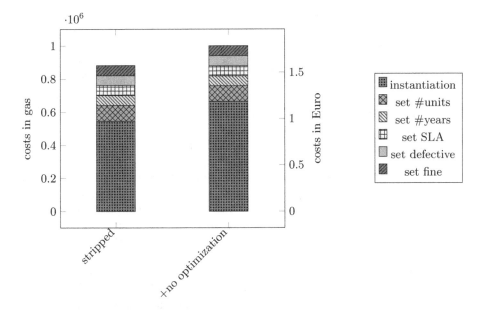

Fig. 8. Cost of an instance in *gas* and Euro—bottom to top: different operations, left to right: additional features

code snippets. Implementing XOR gateways for choreographies is a challenging task since a common understanding of both data and logic is required [2].

We propose smart contracts as a solution for implementing collaborative business decisions. The term smart contract has been formed by Szabo [15]. Kõlvart et al. summarize different smart contract developments and describe some opportunities of smart contracts on blockchains [16]. However, current smart contracts are implemented using procedural languages such as Solidity. In [17] Idelberger et al. discuss using logical programming languages instead because the respective contracts are more comprehensible and less error-prone. We build upon this idea by combining logical programming (decision tables) with a model based approach.

Most blockchains are public. This limits their usage to non-sensitive data [18]. The Ethereum community discusses this limitation and proposes *secret sharing decentralized autonomous* organizations that use data partitioning to sustain privacy [19]. Kosba and others take this idea further and present Hawk: a blockchain, which supports smart contracts and protects privacy through encryption [20]. Privacy protecting blockchains (supporting smart contracts), are not ready for productive usage, yet. Therefore, private and permissioned blockchains can be used to create a peer-to-peer network with a restricted set of nodes.

6   Conclusion and Discussion

Collaborative decisions span across multiple processes and organizations. They apply to interacting business process and influence their behavior. In these cases, it is difficult to establish a trusted exchange of data and a common understanding of both the information and the decision logic [2]. Therefore, we proposed the use of a decentralized blockchain as an intermediary entity. It saves both the data that is required for the decision and the underlying logic. Based on the blockchain technology, trust becomes obsolete due to a technical and reliable mechanism. We also showed a prototypical implementation that automatically translates DMN models into smart contracts. However, current limitations of the type system of Solidity (a language to write smart contracts) requires manual adjustment and workarounds, for example, to handle floating-point numbers.

To make our approach feasible, we extended the mapping with additional features: A push mechanism calls decisions automatically from the state contract. In addition, access rights limit the write access for each attribute to a set of participants or contracts (addresses); hence, unauthorized users cannot temper with the data. Further, only decision contracts can set their respective outputs.

An important issue of modern blockchains, such as Ethereum, is that all data is public. Everything saved to the ledger is visible to every node. This is not an issue for "public" contracts such as the terms of service of a company. However, sensitive or enterprise internal data cannot be stored on the blockchain. Alternatively, one can build a private or permissioned blockchain or use a new kind of blockchain such as Hawk [20] that offers built-in privacy mechanisms such as secret sharing decentralized autonomous organizations [19] and zero-knowledge proofs.

To conclude, DMN decisions can be implemented with blockchain technology making trust and a centralized third party, such as a DMN decision service, obsolete. Therefore, we closed the gap left open by previous work to execute interacting business processes on a blockchain by also taking the decision perspective into account.

Acknowledgements. We thank Alexander Kastius for his valuable contribution to the prototype implementation.

References

1. Weske, M.: Business Process Management - Concepts, Languages, Architectures, 2nd edn. Springer, Heidelberg (2012). https://doi.org/10.1007/978-3-642-28616-2
2. OMG: Business process model and notation, specification 2.0. Version 2 (2011)
3. OMG: Decision model and notation, specification 1.1. Version 1.1 (2016)
4. Batoulis, K., Meyer, A., Bazhenova, E., Decker, G., Weske, M.: Extracting decision logic from process models. In: Zdravkovic, J., Kirikova, M., Johannesson, P. (eds.) CAiSE 2015. LNCS, vol. 9097, pp. 349–366. Springer, Cham (2015). https://doi.org/10.1007/978-3-319-19069-3_22

5. van der Aalst, W.M.P., Weske, M.: The P2P approach to interorganizational work-flows. In: Dittrich, K.R., Geppert, A., Norrie, M.C. (eds.) CAiSE 2001. LNCS, vol. 2068, pp. 140–156. Springer, Heidelberg (2001). https://doi.org/10.1007/3-540-45341-5_10

6. Weber, I., Xu, X., Riveret, R., Governatori, G., Ponomarev, A., Mendling, J.: Untrusted business process monitoring and execution using blockchain. In: La Rosa, M., Loos, P., Pastor, O. (eds.) BPM 2016. LNCS, vol. 9850, pp. 329–347. Springer, Cham (2016). https://doi.org/10.1007/978-3-319-45348-4_19

7. Dannen, C.: Introducing Ethereum and Solidity. Apress, Berkeley (2017). https://doi.org/10.1007/978-1-4842-2535-6

8. Nakamoto, S.: Bitcoin: a peer-to-peer electronic cash system (2008)

9. Tschorsch, F., Scheuermann, B.: Bitcoin and beyond: a technical survey on decentralized digital currencies. IEEE Commun. Surv. Tutor. **18**(3), 2084–2123 (2016)

10. Narayanan, A., Bonneau, J., Felten, E.W., Miller, A., Goldfeder, S.: Bitcoin and Cryptocurrency Technologies - A Comprehensive Introduction. Princeton University Press, Princeton (2016)

11. de Kruijff, J., Weigand, H.: Understanding the blockchain using enterprise ontology. In: Dubois, E., Pohl, K. (eds.) CAiSE 2017. LNCS, vol. 10253, pp. 29–43. Springer, Cham (2017). https://doi.org/10.1007/978-3-319-59536-8_3

12. Mendling, J., Weber, I., et al.: Blockchains for business process management - challenges and opportunities. CoRR abs/1704.03610 (2017)

13. Prybila, C., Schulte, S., Hochreiner, C., Weber, I.: Runtime verification for business processes utilizing the bitcoin blockchain. CoRR abs/1706.04404 (2017)

14. López-Pintado, O., García-Bañuelos, L., Dumas, M., Weber, I.: Caterpillar: a blockchain-based business process management system. In: Proceedings of the BPM Demo Track and BPM Dissertation Award Co-Located with 15th International Conference on Business Process Modeling (BPM 2017), Barcelona, Spain, 13 September 2017 (2017)

15. Szabo, N.: Formalizing and securing relationships on public networks. First Monday **2**(9) (1997)

16. Kõlvart, M., Poola, M., Rull, A.: Smart contracts. In: Kerikmäe, T., Rull, A. (eds.) The Future of Law and eTechnologies, pp. 133–147. Springer, Cham (2016). https://doi.org/10.1007/978-3-319-26896-5_7

17. Idelberger, F., Governatori, G., Riveret, R., Sartor, G.: Evaluation of logic-based smart contracts for blockchain systems. In: Alferes, J.J.J., Bertossi, L., Governatori, G., Fodor, P., Roman, D. (eds.) RuleML 2016. LNCS, vol. 9718, pp. 167–183. Springer, Cham (2016). https://doi.org/10.1007/978-3-319-42019-6_11

18. Atzei, N., Bartoletti, M., Cimoli, T.: A Survey of Attacks on Ethereum Smart Contracts (SoK). In: Maffei, M., Ryan, M. (eds.) POST 2017. LNCS, vol. 10204, pp. 164–186. Springer, Heidelberg (2017). https://doi.org/10.1007/978-3-662-54455-6_8

19. Buterin, V.: Secret sharing DAOs: the other crypto 2.0 (2014)

20. Kosba, A.E., Miller, A., Shi, E., Wen, Z., Papamanthou, C.: Hawk: the blockchain model of cryptography and privacy-preserving smart contracts. In: IEEE Symposium on Security and Privacy, SP 2016, San Jose, CA, USA, 22–26 May 2016, pp. 839–858 (2016)

Shared Ledger Accounting - Implementing the Economic Exchange Pattern in DL Technology

Hans Weigand[1]([⊠]) [iD], Ivars Blums[2] [iD], and Joost de Kruijff[1]

[1] Tilburg University, P.O. Box 90153 5000 LE, Tilburg, The Netherlands
{h.weigand,j.c.dekruijff}@uvt.nl
[2] SIA ODO, Riga, Latvia
ivars@odo.lv

Abstract. DLT suggests a new way to implement the Accounting Information System, but after reviewing the current literature our conclusion is that an ontologically sound consensus-based design is missing to date. Against this research gap, the paper introduces a DLT-based shared ledger solution in a formal way and compliant with Financial Reporting Standards. We build on the COFRIS accounting ontology (grounded in UFO-S) and the blockchain ontology developed by De Kruijff & Weigand that distinguishes between a Datalogical level, an Infological and an Essential (conceptual) level. It is shown how consensual and agent-specific parts of the business exchange transaction can be represented in a concise way, and how this pattern can be implemented using Smart Contracts.

Keywords: Accounting ontology · Smart contracts · UFO

1 Introduction

Blockchain and Smart Contract technology suggests a new way to implement an Accounting Information System (AIS). How exactly this can be done and what are limitations is still very much an open question [8]. A bit more has been said already about the possible benefits. Based on the literature so far, these AIS benefits are the following:

- **Immutability.** The public blockchain as the one underlying Bitcoin claims to provide an immutable tamper-proof storage for transactions that is completely under the control of the technology. This immutability contrasts with the traditional situation where data is under control of the IT center or cloud provider, always with the possibility of management overriding or third-party manipulation. The immutability greatly improves the integrity and verifiability of AIS data and diminishes the need for many of the administrative checks, although the need for a proper design of the control infrastructure is not taken away. Note that the immutability claim is still under discussion [6] and has to be made differently for different blockchain implementations, but we take it as an interface assertion in this paper.

J. Krogstie and H. A. Reijers (Eds.): CAiSE 2018, LNCS 10816, pp. 342–356, 2018.
https://doi.org/10.1007/978-3-319-91563-0_21

- **Actor-independence.** AIS systems are traditionally kept inside an enterprise and represent the company perspective on economic exchanges. Evidence from the environment, e.g. invoices from suppliers, is used by the auditor and considered important, but there is no systematic connection between the invoices sent from company A with the invoices recorded in company B. Triple-entry accounting [9] has been proposed as an independent and secure mechanism to improve the reliability of financial statements based on a neutral intermediary, however, this requires dependence on a third party. A blockchain-based shared ledger (SL) can solve this problem. An actor-independent mechanism may not only drastically reduce the need for multiple copies of the same data, but also contributes to the validity of the transactional data because it is based on consensus.
- **Smart Control.** Smart contracts encoded with accounting and business rules can enable not only efficient control of the recording process ([8]), e.g. authorization checks, and error-detection, but also increase its effectiveness. For traditional internal control measures, auditors must check the design, implementation and operation. Implemented controls could have been switched-off. Building these controls into Smart Contracts that are accessible for auditors (or the parties they represent themselves) makes the design transparent, ensures a 1–1 implementation, and provides a transparent operation (preventive or detective).
- **Tight Integration.** The AIS offers a representation of the (economic) reality of an enterprise, but so far relies on human interfaces with this reality. The "reality" consist of social and physical processes. A purchase order or invoice is such a social process. With SL, the order can be put into the blockchain or be tightly connected to it, so that the relationship between order and the AIS representation of it becomes 1–1. In terms of Grigg [9]: "the entry *is* the transaction". For physical processes, such as the delivery of physical goods, the blockchain combined with IoT infrastructure can achieve a close 1–1 correspondence by setting up the SL as the register of enforceable property rights. We also mention here the integration with and selective disclosure to other parties, such as tax and customs (real-time taxing), regulatory bodies, financial/integrated reporting and assurance services.

Other advantages mentioned in the literature are continuous assurance and real-time reporting, but in our view, these are not specifically bound to the blockchain technology. Given the potential advantages, a few papers have already explored the design of a blockchain-based Distributed Ledger technology (DLT). Dai and Vasarhelyi [8] sketch a system based on triple-entry accounting [9]. In this framework, each company keeps it double-entry bookkeeping system, but the blockchain ledger glues the two together, by (a) having a copy of each account of the local system in the DLT, and (b) adding "obligation" tokens and their transfer from one company account to the other that should match – perhaps enforced by Smart Contract – the Payables or Receivables, and (c) having aggregating accounts of total assets, liabilities and equities whose correspondence with the individual accounts can be monitored by a Smart Contract. Appelbaum and Nehmer [1] discuss the design requirements for a blockchain-based DLT system and its repercussions for auditing tasks, giving special attention to cloud-based DLT solutions. When reviewing the triple-entry solution of [8] we wonder why still so much duplication of accounting entries is needed, given the

DLT robustness. Furthermore, from an accounting ontology point of view the status of "obligation" in this model needs more explanation. Both papers are exploratory in nature. Wang and Kogan [21] introduce a blockchain-based AIS, including a prototype implementation. The main concern addressed in their paper is the tension between protection of private data and the desirable public blockchain transparency. The authors solve the tension using Zero-Knowledge Proof encryption. Apart from the encryption solution, the description of the AIS is sketchy. The paper defines a blockchain-based AIS as "a neutral and independent infrastructure that underpins business event recording" However, whether (or how) such a neutral representation – consensus view – is possible within current accounting standards, is not discussed. Our general conclusion is that an ontologically sound and truly consensus-based design is missing to date.

Against this research gap, the goal of this paper is to introduce a DLT solution in a formal way, grounded in accounting ontology. We build on the blockchain ontology developed in [16] that distinguishes between a Datalogical level (or abstract platform-specific), an Infological (platform-independent) and an Essential (conceptual) level. In the line of [15], we extend the REA ontology [17] used in [16] for the essential layer to the core COFRIS accounting ontology [3, 4] that is based on current Accounting and Financial Reporting Standards (IFRS) [13, 14]. An innovative characteristic of COFRIS is that it does not put the economic event in the center, but the evolving economic relationship in which the economic exchange takes place. Hence events are not viewed in isolation, but as contributing to the development of the exchange. Because of this choice, COFRIS includes an ontological grounding of the obligation concept and provides a good basis for a consensus view. From a Design Research perspective [12] the paper *evaluates* DLT with respect to accounting standards requirements, *builds* a Shared Ledger design, and *demonstrates* its implementability in Smart Contracts.

In the following, we will use the term Shared Ledger rather than Distributed Ledger. The ledger is distributed at the datalogical level, but the key feature at the infological level is that the ledger is shared and provides a consensus view.

Section 2 is a brief overview of the Economic Exchange pattern in COFRIS. Section 3 describes the realization of this pattern in an SL environment.

2   The Economic Exchange

In [3] an economic exchange reference ontology and pattern was introduced in the context of IASB Conceptual Framework for Financial Reporting [13]. This Exchange ontology is grounded on UFO-S – the core Ontology for services [18], which characterizes the service phenomena by considering service *commitments* and *claims* established between service provider and customer along the service life-cycle phases: *offering*, *negotiation/agreement* and *delivery*. UFO-S presents general concepts spanning across several applications domains so that its conceptualization can be reused for the economic exchange activity life-cycle. Economic Resource (Claim) and Transfer/Receipt concepts were added in COFRIS [3, 4] based on the UFO Ontology [10]. The treatment of the *Rights to receive* as Resources, and consequently as material relations make COFRIS different from REA ontology [17], but compliant to existing accounting frameworks [13]. Economic Performance (Revenue), Exchange and Consensus concepts were not enough explicated in the IASB Framework [13] but play a major role in most of the Standards [14]. These concepts are being incorporated in COFRIS in a way described in this paper.

Legal aspects of service contracts were further elaborated in [7] within the UFO-L Legal ontology, that is based on Hohfeld/Alexy's theory of fundamental legal concepts. The legal positions of UFO-L include not only those corresponding to claims and commitments in UFO-S (i.e., right and duty), but also pairs of other elements: permission and no-right, power and subjection, immunity and disability. All these legal relators are from two classes of *entitlement* and *burden* (lack), which we refer further as rights and obligations respectively. The above-mentioned right and obligation pairs form correlative associations [7], [13, para 4.25], which are foundations for a shared ledger view.

2.1 Economic Exchange Life-Cycle

We cannot describe the whole COFRIS ontology but will briefly recall the main economic events and relations of the exchange life-cycle before positioning it within a shared ledger context (see Figs. 1 and 3 for an example).

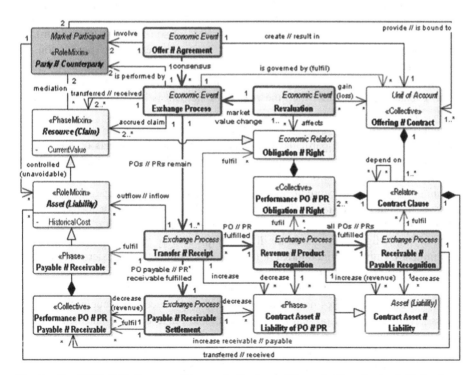

Fig. 1. OntoUML [10] diagram of Economic Exchange life-cycle (Party's view) (Color figure online)

The Economic exchange life-cycle [3] is conceived as an *Offering* of interaction made by *Offer* of one of two *Parties* (played by market participants), followed by its acceptance (*Agreement*) by the *Counterparty* resulting in a *Contract* (of mutual *Obligations and Rights to exchange,* for mutual benefit), that is fulfilled by *Exchange process* – party's *Transfers* of *Resources* (*Claims*) in exchange for accrual of *Claims to receive* against the counterparty, with their subsequent *Settlement* by *Claims to transfer* against the party, accrued for *Receipts* of *Resources* (*Claims*) from the counterparty. An Exchange pattern [3] is a pattern of a party's interaction (or disposition for interaction) with a counterparty. The interaction fulfils party's obligation/right to exchange outflow for inflow, where (with possible reversal): *outflow* is decrease of party's resources and/or increase of claims against party caused by their transfer to the counterparty, and *inflow* is increase of party's resources and/or decrease of claims against party caused by their receipt from the counterparty.

Following [13, 18] we define *Market participant* (or economic agent) as a UFO role-mixin played by social agents – persons and enterprises, contractual groups of people and enterprises, or the society at large. Market participants are capable of self and social committing and fulfilling economic actions, compliant with the market regulations. Market participants are represented by *Actors* that in turn comprise of accounts for economic relationships that mediate a particular party with society and other parties. These relationships are modeled by *Economic relator* - UFO social relator [18] existentially dependent on involved market participants and having two or more pairs of mutually dependent Obligations and/or Rights, valued in monetary terms - *Current Value* (Price), over some *Underlying objects*, at some *Timing* (Condition).

A *Resource* is a Right [13] (claim-right to exchange/receive, permission to use/consume, power and immunity to transfer) that has the disposition to produce economic benefits. A *Claim* against market participant is an obligation to the resource (claim) exchange/transfer to which the market participant is legally or constructively bound.

Underlying object, an UFO mixin, denotes the physical or intellectual object, or their type characterized in particular by: *Quantity* (of collective objects or *Amount* of matter) of underlying objects or object function; *Place* that denotes [fiat] location at [and in] which the object is or will be available for control.

Unit of Account is conceptualized as a bundle of rights and/or obligations which are usually or mandatory transferred (fulfilled, consumed/used, produced, valued) together, such as: a Business; a *Contract clause* of an Economic contract. Unit of Account and its underlying obligations/rights in the life-cycle process go through *Phases,* such as: Commitment/Claim or Obligation/Right to exchange, Contract or Transferred/Received Asset (Liability or Equity), Payable/Receivable.

Party's obligations/rights are often bound together in *Performance Obligation/ Right* (PO/PR) that specifies performance required to produce a revenue/product.

Assets (*Liabilities*) are present rights (obligations) for resources controlled (claims unavoidable) by a party, as a result of past events which form their *Historical cost* [13]. *Receivables/Payables,* are unconditional rights/obligations to receive/transfer.

The blue (darker) lines and boxes of Fig. 1, depict behavioral semantics of the exchange life-cycle. Symmetrical party/counterparty elements here are combined into one by showing party's events and relations before the "*//*" symbol, but counterparty's after, e.g., *Transfer* by a party//*Receipt* from a counterparty, and *Payable* by a party//*Receivable* from a counterparty. Notice though that the diagram represents the party's view.

The Exchange process event, for some Contract [clause], triggers *Transfer* events that *fulfil* Performance obligation - PO, exchanging transferred Assets (Liabilities) valued at *cost*, for *Contract asset of* this *PO*, valued at *price*. A *Receipt* event (e.g. a prepayment from the customer), forms a *Contract liability of PR*.

If some PO is wholly *fulfilled* by the transfers, the *Revenue recognition* event decreases the Contract asset of this PO and increases (by *revenue*) the *Contract asset*.

If all POs are fulfilled, a *Receivable recognition* (or Realization) event takes place that, fulfilling the contract, exchanges the Contract asset for *Performance receivables* (that legally enforce the Exchange rights - PRs).

If all the Exchange rights are fulfilled before the obligations fulfillment, e.g., full prepayment is made, an alternative *Payable recognition* event takes place that, fulfilling the contract, exchanges the *Contract liability* for *Performance payables*, which can be fulfilled by *Transfer* and *Settlement* events. *Payable settlement* event (in accordance with timing) offsets Performance payables against Contract liabilities for each PO.

All events may be actioned by the market participant or its agent or specified in a [smart] contract as automatically executable - triggered by conditions (depicted on the lines that connect event boxes in the diagram), and timing of fulfil.

Due to market conditions the current value of a right//obligation in a contract may change, causing *Revaluation* event and inflow (outflow) called gain (loss). In the exchange process, these changes trigger special transfer events that increase (decrease) contract asset for gain (loss) of rights, and special receipt events that increase (decrease) contract liability for loss (gain) of obligations.

2.2 Towards a Shared Ledger

An advantage of the shared ledger is the actor-independent view that it offers. This does not necessitate that all information is accessible to all parties. Information sharing in a shared ledger must be selective, ranging from global, i.e., among all members of society at large, to particular – among contractual group members, or a party and a counterparty, or participants within an enterprise. The accounting interpretation of the contracts and their fulfillment may be different for each party. The goal should be to obtain more consensus, we call it correlative consensus, for resource/claim interpretation in the contracts.

We assume that conceptually there is a shared contract – a pair of mutual obligations of the parties, and contract fulfillment exchange interactions in consensus. However, the AIS tagging of the entries may be different for different agents, for several syntactic and semantic party specific reasons:

* specific financial period, account name, unit of account granularity, local currency, rounding rules and other qualities;
* specific resource function (purpose) or restriction;
* different accounting classification and valuation requirements.

Therefore, in COFRIS market participants may specialize/generalize at recognition/derecognition the claims and resources in consensus, as their own assets (liabilities) per accounting standards and their own operational purposes and include their specific (de) recognition modules into smart contracts that extend the contract manipulation and transfer events. For example, if a provider sells a product, such as fuel, the customer may classify it as a raw material, as held for sale, or for administrative expense – all these asset types are subtypes of the transferred resource.

The existing accounting often loses the semantics of transfer events, because it recognizes the effects of resource transfer instead of transferred resources. The capturing of interactions that are shared and in consensus should serve as an additional source for (financial) disclosures. An example is services or other resources that are consumed as transferred. The accounts usually recognize only their effects and carrying value increase in e.g. equipment for which installation and testing services were provided. In general, we propose to have the transfer events with the transferred resources (claims) shared and the party specific effects of the interactions on the respective accounts, to be not shared (although this account information can still be part of the smart contract).

To maintain consistency, the phenomena should be correlated in the shared ledger as depicted in Fig. 2. This includes not only relationships like claim-right vs obligation, but also events, e.g. transfer vs receipt. The basic exchange pattern in Fig. 1 remains the same, but a shared ledger reflects not only the party specific view for each participant but also transfer consensus view. The transferor view depicts the events for the contract, transferee shared consensus appears as a correlative view. Specific accounts of the parties – assets (liabilities) are specializations of the resources (claims) affected by the transfer event.

Economic Transfer Events		Affected Contract Economic Relationships					
Transferor view	Transferee correlative view	Transferor view			Transferee correlative view		
		Fulfil	Outflow	Inflow	Fulfil	Inflow	Outflow
Offer [Transfer]	Offer Receipt	Regulations	Commitment to Exchange		Regulations	Claim to Exchange	
Agreement to Exchange [Transfer]	Agreement to Exchange [Receipt]	Commitment to Exchange	Obligation - PO to Exchange	Right - PR to Exchange	Claim to Exchange	Right - PR to Exchange	Obligation - PO to Exchange
Resource (Claim) Transfer	Resource (Claim) Receipt	Exchange Obligation	Transferred Resource (Claim)	Contract Asset of PO	Exchange Right	Received Resource (Claim)	ContractLiability of PR
Revenue Recognition	Product Recognition	PO	ContractAsset of PO	Contract Asset	PR	ContractLiability of PR	ContractLiability
Receivable Recognition	Payable Recognition	Contract	Contract Asset	Receivable - PR	ContractClause	Contract Liability	Payable - PO
[Payable] Settlement	Receivable Settlement	Payable - PO	ContractAsset of PO	Payable - PO	Receivable - PR	ContractLiability of PR	Receivable - PR

Fig. 2. Correlative economic events and relationships in the Shared Ledger. Transferor's obligations - PO correlates with Transferee's rights - PR. (Color figure online)

2.3 Examples

We provide a couple of examples, with particular attention to the question of what should be shared in the shared ledger and what should not. We illustrate the economic Exchange ontology [3] and its extension for shared ledger using an example (Fig. 3), represented in the form of a hierarchical Economic event table.

EID:21 Provider Agreement	29.08.2018			CU:	€	Provider: P		€	Customer: C		€
Fulfil Event	**PO/R**	**Timing**	**Rights**	**Object**	**Qty**	**Price**	**Place**	**Debited**	**Credited**	**Amt**	**Place**	**Debited**	**Credited**	**Amt**
20 Obligation	1	29.08.2018	Ownership	Widget	5	100		Cost	Finished goods	70	1	Raw materials	Contract liability	100
								Contract Asset	Revenue	100				
	2	30.08.2018	Services	Setup	1	10		Cost	Labor	10	1	Raw materials	Contract liability	10
								Contract Asset	Revenue	10				
Consideration		30.09.2018	Ownership	Cash		110	IBAN	Cash in bank	Contract liability	110	IBAN	Contract Asset	Cash in bank	110
EID:22 Provider Transfer	**29.08.2018**			**CU:**	**€**	**Provider: P**		**€**	**Customer: C**		**€**
Fulfil Event	**PO/R**	**Timing**	**Rights**	**Object**	**Qty**	**Price**	**Place**	**Debited**	**Credited**	**Amt**	**Place**	**Debited**	**Credited**	**Amt**
21 Revenue	1	29.08.2018	Ownership	Widget	5	100		Cost	Finished goods	70	1	Raw materials	Contract liability	100
Recognition								Contract Asset	Revenue	100				
EID:23 Provider Transfer	**30.08.2018**			**CU:**	**€**	**Provider: P**		**€**	**Customer: C**		**€**
Fulfil Event	**PO/R**	**Timing**	**Rights**	**Object**	**Qty**	**Price**	**Place**	**Debited**	**Credited**	**Amt**	**Place**	**Debited**	**Credited**	**Amt**
21 Revenue	2	30.08.2018	Services	Setup	1	10		Cost	Labor	10	1	Raw materials	Contract liability	10
Recognition								Contract Asset	Revenue	10				
Realization		30.08.2018	Ownership	Cash		110	IBAN	Receivable	Contract Asset	110	IBAN	Contract liability	Payable	110
EID:24 Customer Transfer	**30.09.2018**			**CU:**	**€**	**Provider: P**		**€**	**Customer: C**		**€**
Fulfil Event	**PO/R**	**Timing**	**Rights**	**Object**	**Qty**	**Price**	**Place**	**Debited**	**Credited**	**Amt**	**Place**	**Debited**	**Credited**	**Amt**
23 Settlement	2	30.09.2018	Ownership	Cash		110	IBAN	Cash in bank	Receivable	110	IBAN	Payable	Cash in bank	110

Fig. 3. Economic event table for Example 1 (Color figure online)

In the header (in dark blue) of an economic event, we have Event identifier (EID), Transferor type - Customer or Provider (or more specialized role), that specifies the context for the correlativity, Event type - Offer, Agreement, Transfer, Revenue or Receivable Recognition, Settlement, or Revaluation (or more specialized subtype), Date or Period and Currency Unit. Provider and Customer identification and their Local currency units with the spot exchange Rates, conclude the event header.

Event detail lines depict events that fulfil the obligations identified by the referenced event, by transferring the promised resource (claim) in exchange of rights to receive entitlement; or obtaining a settlement; or are triggered by the fulfillment. Fulfil events also specify the Phase of transferred resources (claims) since there is a difference e.g. of whether Receivable/Payable is transferred versus settled. The PO/PR, Timing, Rights, Object, Quantity, Price, Place are described in Sect. 2.1. Provider and Customer have their specific, but potentially correlative columns that depict involved Debited/Credited Accounts and Amounts.

Example 1. Revenue. Enterprise E contracts customer C, depicted by EID:21 (that fulfils some offering with EID:20), whereby E obliges to exchange some goods (PO:1) and an accompanying setup (PO:2), by specified dates, for the rights to the cash of 110€ in the P's bank account (IBAN) to be received by 30.09.2018. These rights/obligations are depicted in the agreement details, but the *planned* effect of their fulfillment is specified by underlined provider and customer accounts, and amounts. Event 22 fulfils P's obligation by transferring the goods promised in the Event 21 and acquiring the P's rights to receive - contract asset of PO:1 and recognizing revenue. Event 23 provides setup services, recognizes revenue for PO:2, and completes performance obligation fulfillment that in turn leads to P's realization event that recognizes receivable. Finally, Event 24 represents customer cash transfer and settlement of receivable accrued in Event 23.

Extending [14] we regard *Revenue* as inflow arising in the course of an enterprise ordinary performance, and a fulfillment of performance obligation/payable agreed with a counterparty. It implies that a specific Contract asset of PO and correlative Contract Liability of PR is recognized, as well as a correlative event to Revenue recognition -

Product recognition, by the counterparty. Such an asset (liability) besides revenue recognition may be important to distinguish for legal purposes, in the cases of contract breaches. The example illustrates three important exchange axioms:

- *Transfer duality* – for each Transfer event, the resource (claim) received by one party is the resource (claim) provided by the counterparty: the transferred resource (claim);
- *Reciprocity* – exchange obligations of the two parties balance in Value at contract inception, and the transferred resources (claims) balance in Value with agreed rights to receive (the latter Value may be different from the former);
- *Transfer consensuality* – for each Exchange, a valid Transfer event conforms to a Contract Obligation/Payable.

The third axiom is not commonly found, but inherent to the Exchange perspective. The reverse rule - that each obligation gets fulfilled - is defeasible, and not an axiom.

Example 2. Prepayment in Foreign Currency. If a contract is specified in foreign currency, accounting standards and interpretations [14] require contract asset/liability revaluation into local currency according to the actual exchange rate. Per [14] a *Contract asset* is a party's right to consideration (claim to receive), in exchange for resources transferred to a counterparty, conditioned on something other than the passage of time (for example, the party's future performance), and *Contract liability* is an obligation to transfer resources to a counterparty for which the party has received consideration. These definitions are forward-looking and assign some features of the receivable (payable) resource to these assets (liabilities). Thus IFRIC 21 [14] interprets contract liability, formed from prepayments, as a source for future non-cash assets and thus not subject to revaluation. However, we advocate the present view to these in-process assets and liabilities, meaning that they represent in consensus cash rights/obligations for the transferred resources, to be reimbursed in the case of a breach (for example, a return of a prepayment), thus they need to be constantly revaluated.

Example 3. Cost-Plus Smart Contracts. E, a construction company, enters into a cost-plus smart contract with a customer D to build an object. D reimburses E for all its allowed expenses plus an additional variable payment that allows E to make a profit. E contracts with the subcontractors and vendors Vs and allows these contracts and contract events [complying to IFRS requirements] to serve as inputs to the contract with D, sharing with D [and a global Financial Reporting system] all the required details in consensus with Vs, possibly omitting the names of Vs. Furthermore, in consensus with D, E shares all the required and non-sensitive details of the contract with D with a Financial Reporting system. During the warranty period, D shares all relevant events involving the built object with E. This set-up benefits from having a single source of truth, simplifying administrative and control procedures, and the possibility of automated execution of the smart contract.

It is important that provider and receiver share and have consensus on the asset/liability evaluation/classification, especially in the case of obligations remaining/ongoing, such as in a lease. Unfortunately, existing accounting standards [14] are ambivalent on the correlation and prescribe different (not-correlated) lease accounting for the lessor and lessee [4]. When deciding between services and lease, the decision is not-correlative, while the decision has certain accounting consequences.

3 Shared Ledger

It might seem trivial to realize an AIS on a "Distributed Ledger". However, more is needed than a logistic transfer of money or other resource tokens. To meet the requirements from our ontological analysis – in particular, the distinction between consensus and specific information, and the ability to deal with the whole contract cycle – a basic blockchain does not suffice. However, the contract accounting model can be realized by (translated and extended to) a Smart Contract-based Shared Ledger model. We start by listing the most important principles for this realization:

1. Smart contracts of market participants, containing mutual obligations of resource (claim) transfer, including information sharing specification, and IFRS relevant characteristics are added to a shared ledger by consensus of the parties. Smart contracts comprise a hierarchy of rules and include general principles and regulations, particular rules in consensus, and rules specific to the particular agent for relating its assets (liabilities) to transferred resources (claims).
2. A Digitized resource (claim) or token represents the valued rights of a market participant (for an underlying object) which can be transferred to a counterparty by simply transferring the token. For a referenced resource the token transfer can be a representation of another action of rights transfer or it can effectuate the rights transfer itself (depending on legal context). As explained in the above, we not only have to deal with "traditional" resources such as ownership of goods & (crypto-) money, but also rights/obligations that mark the progress of the Exchange lifecycle. Atomic transfer event happens in point in time or over time when, fulfilling contract obligations, tokens for rights/obligations of resources are conveyed from one market participant to another, with simultaneous conveying the tokens for correlated obligations/rights of resources from the transferee to the transferor.
3. Transfers of digitized resources (claims) are immutably recorded in consensus in a shared ledger, completely, distinctively or partially fulfilling the smart contracts. Transfers together with caused claims accrual or settlement, are accounted within smart contracts, including information sharing and IFRS relevant characteristics.
4. The effects of transfers involving resources (claims) are [de]recognized as assets (liabilities) per IFRS requirements [14] and enterprise policies in the shared or in the individual ledger part, according to information sharing specification.
5. Financial Reporting relevant information gathered in activities 1 through 4 is abstracted to the type level, hiding sensitive instance details and forming an enterprise's multi-dimensional cube within a global Financial Reporting system.
6. The multi-dimensional cube is then aggregated, calculated, viewed, and mined per the IFRS [14] Taxonomy requirements and financial reports are issued.

3.1   Shared Ledger Model

With the advent of smart contracts, also known as coded contracts on the blockchain that automatically move digital assets according to arbitrary pre-specified rules [5], it is possible for two or more parties to exchange digitized resources without the need for a trusted third party [22]. The implementation of smart contracts by enterprise adopters represents a major transformation in business computing, with organizations take advantage of the benefits of perfect replication and high availability, cryptographically-verified transactions, and lucid, robust business logic [19]. Various DLT protocols are being developed around the world that enable the exchange of tokens through smart contracts, Bitcoin and Ethereum being the most affluent today. Hereby, the intent of smart contracts is to both verify and enact the full logic of any given transaction. While Bitcoin scripts are not powerful enough to write full DLT applications to support AIS, Ethereum's smart contract languages (e.g. Solidity, Serpent, Pack) and underlying virtual machines are powerful yet dangerously unconstrained [19].

To separate content and form, we designed an infological representation of an IFRS compliant based smart contract (Fig. 4). Hereby, we have extended the infological blockchain domain ontology as presented in [16] by including COFRIS-related components. In addition, we have designed a platform independent ERC20 compliant representation at the datalogical layer in order to comply to implementation standards and prevent the mentioned implementation constraints. At the infological layer, the notion of transaction has been refined to three levels: transaction – event – posting - which is problematic for blockchain, but very well possible in a Smart Contract based SL.

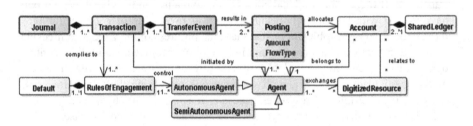

Fig. 4. Infological Shared Ledger model UML diagram.

The classes that are depicted in Fig. 4 are defined in Fig. 5 with references to the corresponding COFRIS concepts.

In Sect. 2, we identified three axioms. To realize the transfer duality rule we assume an equivalent rule on ERC token transfers. The transfer consensuality and reciprocity rules must be built into the Smart Contract.

As uniform standards are essential to (1) ensure interoperability between these different blockchains and (2) increase security of smart contracts in general, we apply the ERC20 common list of token rules to design a smart contract for an AIS. ERC20 is considered to be the standard that facilitates low friction peer-to-peer exchange of so called WRC20 tokens on DLT. The protocol proposed by Ethereum is intended to serve as an open standard and common building block, driving interoperability among

Class	Explanation
Ledger	A shared **ledger** is a set of accounts maintained in a Smart Contract. Shared Ledgers can be said to be part of a (perhaps global) SL environment (not in the model).
Account	An (infologicl) **account** is a container of digital resources corresponding to a COFRIS Unit of account: a group of enforceable/constructive rights and/or obligations, It maps to a datalogical account of tokens.
Agent	A **semi-autonomous agent** is the owner of accounts, registered within the SL environment, corresponding to a market participant in COFRIS. An **autonomous agent** is a registered Smart Contract that initiates a transaction, based on the commitments materialized in the rules of engagement. An agent can initiate a transaction.
Transaction	A **transaction** is a coherent set of transfer events that make bookings on accounts corresponding to a COFRIS transfer (with COFRIS fulfill events)
Journal	A **journal** is a chronological list of transactions controlled by a Smart Contract. At the datalogical level, it corresponds to a continuously growing list of records called blocks. Each block contains a timestamp and a link to a previous block.
Digitized resource	**Digitized resources** represent the resources/claims (COFRIS) or other rights which can be transferred to a counterparty by simply transferring the digital representation
Rules Of Engagement	**Rules of engagement** refer to the explicit rules to which a transaction should behave as specified in a Smart Contract
Default	A **default** is a definitional clause, which defines relevant concepts occurring in the contract as constant variables

Fig. 5. Infological class definitions with mappings

decentralized applications (dApps) that incorporate exchange functionality. In this context, a token hosted on the DLT can be sent, received, checked of its total supply, and checked for the amount that is available on an individual address. This is analogous to sending and receiving Ether or Bitcoin from a wallet, knowing the total amount of coins in circulation, and knowing a particular wallet's balance of a coin. A smart contract that follows this standard is called an ERC20 token. The functions and events are listed in Fig. 6.

AccountBalance	The **AccountBalance** returns the balance of a smart contract account
ClauseFunction	**Clause functions** of a Smart Contract interface implement the Rules of Engagement at the datalogical level
TotalSupply	The **TotalSupply** returns the total supply of a certain token within a smart contract
TransferEvent	A **TransferEvent** is called to transfer tokens between accounts. Infological events are mapped to TransferEvents.
ApprovalEvent	An **ApprovalEvent** is called by the TransferBetweenParties function to make tokens available to the Smart Contract
ClauseEvent	A **ClauseEvent** is an event that is triggered by a ClauseFunction

Fig. 6. ERC20 functions and events

Example. Enterprise E contracts Counterparty C for a simple purchase order contract. We choose not to extend the ERC20 smart contract with business logic between C and E for the sake of simplicity and modularity. Two sets of rules of engagement (smart contracts) are needed: a parent contract, and an ERC20 compliant token contract to represent digitized resources for C and E. The parent contract is initiated through a transaction event that creates the contract with all appropriate clauses (rules of engagement) to exchange goods or services, including the ground rules of the contract (defaults), like payment terms, notice periods etcetera. E and C both become owners of the contract. Once the smart contract is accepted by both owners and stored on the blockchain, it can be effectuated through events that query or update the technical state of the contract on the blockchain. Transactions are stored on the blockchain, optionally anonymized (see below). Either party has the option to reallocate, reclassify or re-evaluate its own resources through accounts, outside of the viewpoint of the counterparty. The main contract instantiates or calls an ERC20 compliant contract by means of a commitment event that transfers obligation tokens between C and E. Once it is confirmed that the physical goods have been exchanged (through a confirmation or directly through IoT), the parent smart contract initiates a transaction to the ERC20 token contract to balance the obligation tokens to zero (as, in more detail, in example Fig. 3).

3.2 Design Choices

Settlements by the smart contract based AIS are executed on the blockchain (on-chain) by imposing consensus over (intermediary) trust. There are two possible scopes for on-chain transaction processing for accounting systems [2]:

- *On-chain.* This model assumes on-chain processing of consensus, settlement with an on-chain order book. Within the context of DLT, this is the most expensive option. Nevertheless, due to the fact that this is a single-entry solution, the individual stakeholders may avoid investments in compute, storage and security products.
- *Hybrid.* This model assumes off-chain order relay with on-chain settlement. In this approach, cryptographically signed orders are broadcast off of the blockchain; an interested counterparty may inject one or more of these orders into a smart contract to execute trades trustlessly, directly on the blockchain.

Our smart contract based AIS assumes that every transaction is triggered and initiated by the smart contract based AIS through the ERC20 transaction functions and events. As a result, no 'standard' DLT transactions are allowed, where the agents transfer tokens directly to each other. It is possible that the ERC20 smart contract is orchestrated by an external smart contract based AIS and/or even a smart contract for zero-knowledge proof (see below). By doing so, the ERC20 token contract is kept original and only serves the purpose of conforming to uniform token standards, making it easier to update and align to best practices.

3.3 Benefits

A smart contract based AIS replaces the necessity of a trusted intermediary between the parties involved as in triple entry accounting, by a DLT-based consensus mechanism. Besides reducing the number of entries, this architecture introduces various side-benefits:

- **Less Coordination.** The smart contract handles the initiation of transfers and the (internal and external) allocation of funds between accounts (e.g. goods payment and transportation). No longer are invoices paid using conventional monies; many transactions are dealt with by internal money transfers through the smart contract and at the edges of the corporation, formal and informal agents work to exchange between internal money and external money.

- **Increased Privacy.** While parties may employ pseudonyms to enhance their privacy, research shows that anyone can de-anonymize DLT transactions by using information in the blockchain, like transaction structure, value, time and date [2]. Zero-knowledge proofs (e.g. Zero-Knowledge Succinct Non-Interactive Argument of Knowledge, zkSNARKS) allow one party (the prover) to prove to another (the verifier) that a statement is true, without revealing any information beyond the validity of the statement itself. By imposing zero-knowledge technology to a smart contract based AIS, participants do not have to share privacy sensitive information, like their private key, resource allocation strategy or transaction size, since the transactions on the blockchain are encrypted [21]. Today, zero knowledge proofs in combination with smart contracts is the future but is still in its early days and has many challenges to overcome to increase practicality, like computational intensity, setup phase and costs.

Similar to triple entry accounting's Shared Transaction Repository, the smart contract based AIS uses a shared environment to store the entry which is the transaction. By applying zero-knowledge proof consensus, the transaction is encrypted on the blockchain, making it impossible for non-participants of the smart contract to de-anonymize these transactions. Due to this encryption, it becomes irrelevant whether or not a public- or a private or blockchain is used, giving participants options with regards to costs and extent of neutrality.

4 Conclusion

Shared ledger systems built on DLT may have a high impact on current AIS, not only because of the claimed immutability of the records but also because of the shift from an internal actor-dependent to an external consensus view. In this paper, we have taken an ontological approach, focusing on the economic exchange pattern. Explicit attention has been given to the question what is to be shared in the shared ledger and what not, and how the two parts can be related in a rigid way. Where there are concerns that triple-entry accounting "may not be advanced enough" [8, p. 18], the paper aims to contribute to a foundation that is both ontologically sound and fully compliant with the accounting standards; not just an add-on to 500 year old double-entry bookkeeping.

We have described how this conceptual model can be realized at the platform-independent (infological) level, by using smart contracts and the ECR20 token standard. The smart contract does not only have the advantage of automated execution (that is, delegated fulfilment of commitments), but also provides an aggregation level close to that of the economic exchange contract. A caveat must be added: our infological model assumes a technical environment to be in place that is still in progress. The model may help to steer its direction.

DLT platforms are evolving rapidly now. For that reason, we have focused on a platform-independent model, and not on the coding, although we are also experimenting with the PIM to PSM level transformation at the moment [20]. We are planning to bring these efforts together. Other research topics include public reporting directly based on the blockchain, the impact of DLT-based AIS on the auditing task, and the further development of a language for commitment-based smart contracts.

References

1. Appelbaum, D., Nehmer, R.: Designing and Auditing Accounting Systems Based on Blockchain and Distributed Ledger Principles. Feliciano School of Business (2017)
2. Ben-Sasson, E., et al.: Decentralized anonymous payments from Bitcoin. In: Proceedings of 2014 IEEE Symposium on Security and Privacy (SP), pp. 459–474 (2014)
3. Blums, I., Weigand, H.: Towards a reference ontology of complex economic exchanges for accounting information systems. In: Proceedings of IEEE EDOC 20, pp. 119–128 (2016)
4. Blums, I., Weigand, H.: Financial reporting by a shared ledger. In: Proceedings of JOWO (2017)
5. Buterin, V.: A Next Generation Smart Contract & Decentralized Application Platform (2014)
6. Coyne, J.G., McMickle, P.: Can blockchains serve an accounting purpose? J. Emerg. Technol. Account. **14**(2), 101–111 (2017)
7. Criffo, C., Almeida, J.P.A., Guizzardi, G.: From an ontology of service contracts to contract modeling in enterprise architecture. In: IEEE EDOC 21 (2017)
8. Dai, J., Vasarhelyi, M.: Toward blockchain-based accounting and assurance. J. Inf. Syst. **31** (3), 5–21 (2017)
9. Grigg, I.: Triple entry Accounting. Systemics Inc. (2005). http://iang.org/papers/triple_entry.html
10. Guerson, J., et al.: OntoUML lightweight editor: a model-based environment to build, evaluate and implement reference ontologies. In: Proceedings of IEEE EDOCW 2015, pp. 144–147 (2015)
11. Guizzardi, G.: Ontological Foundations for Structural Conceptual Models, Telematics Instituut Fundamental Research Series, No. 015, ISSN 1388-1795, The Netherlands (2005)
12. Hevner, A., March, S., Park, J.: Design research in information systems research. MIS Q. **28** (1), 75–105 (2004)
13. IASB Exposure Draft. Conceptual Framework for Financial Reporting, IASB (2015)
14. IASB homepage, IASB (2018). http://www.ifrs.org/issued-standards/list-of-standards
15. ISO/IEC. Information Technology–Business Operational View–Part 4: Business Transactions Scenarios–Accounting and Economic Ontology. ISO/IEC FDIS 15944-4 (2015)
16. de Kruijff, J., Weigand, H.: Understanding the blockchain using enterprise ontology. In: Dubois, E., Pohl, K. (eds.) CAiSE 2017. LNCS, vol. 10253, pp. 29–43. Springer, Cham (2017). https://doi.org/10.1007/978-3-319-59536-8_3
17. McCarthy, W.E.: The REA accounting model: a generalized framework for accounting systems in a shared data environment. Account. Rev. **57**, 544–577 (1982)
18. Nardi, J., et al.: Towards a commitment-based reference ontology for services. In: Proceedings of IEEE EDOC, vol. 17, pp. 175–184 (2013)
19. Popejoy, S.: The Pact Smart-Contract Language, Whitepaper (2017). http://kadena.io
20. Syahputra, H., Weigand, H.: The development of smart contracts for heterogeneous blockchains. In: Enterprise Interoperability VIII, Proc IESA. Springer (2018, to appear)
21. Wang, Y., Kogan, A.: Designing Privacy-Preserving Blockchain Based Accounting Information Systems (2017). https://doi.org/10.2139/ssrn.2978281
22. Warren, W., Bandeali, A.: 0x: An Open Protocol for Decentralized Exchange on the Ethereum Blockchain (2017). https://0xproject.com/pdfs/0x_white_paper.pdf

Process and Multi-level Modelling

Exploring New Directions in Traceability Link Recovery in Models: The Process Models Case

Raúl Lapeña[1]([⊠]), Jaime Font[1], Carlos Cetina[1], and Óscar Pastor[2]

[1] SVIT Research Group, Universidad San Jorge,
Autovía A-23 Zaragoza-Huesca Km. 299, Villanueva de Gállego, Spain
{rlapena,jfont,ccetina}@usj.es
[2] Centro de Investigación en Métodos de Producción de Software,
Universitat Politècnica de València, Valencia, Spain
opastor@pros.upv.es

Abstract. Traceability Links Recovery (TLR) has been a topic of interest for many years. However, TLR in Process Models has not received enough attention yet. Through this work, we study TLR between Natural Language Requirements and Process Models through three different approaches: a Models specific baseline, and two techniques based on Latent Semantic Indexing, used successfully over code. We adapted said code techniques to work for Process Models, and propose them as novel techniques for TLR in Models. The three approaches were evaluated by applying them to an academia set of Process Models, and to a set of Process Models from a real-world industrial case study. Results show that our techniques retrieve better results that the baseline Models technique in both case studies. We also studied why this is the case, and identified Process Models particularities that could potentially lead to improvement opportunities.

Keywords: Traceability Link Recovery · Requirements engineering
Business Process Models

1 Introduction

Traceability Link Recovery (TLR) has been a subject of investigation for many years within the software engineering community [1,2]. Research has shown that affordable Traceability can be critical to the success of a project [3], and leads to increased maintainability and reliability of software systems by making it possible to verify and trace non-reliable parts [4]. Specifically, more complete Traceability decreases the expected defect rate in developed software [5].

In recent years, TLR has been attracting more attention, becoming a subject of both fundamental and applied research [6]. However, most of the works focus on code [7], and the application of Traceability Links Recovery techniques to Process Models is a topic that has not received enough attention yet.

© Springer International Publishing AG, part of Springer Nature 2018
J. Krogstie and H. A. Reijers (Eds.): CAiSE 2018, LNCS 10816, pp. 359–373, 2018.
https://doi.org/10.1007/978-3-319-91563-0_22

Through this work, we study TLR between Natural Language Requirements and Process Models through three different approaches. Given a query Requirement and a Process Model, the three techniques use different means to extract a Model Fragment from the Model, being said Model Fragment relevant to the implementation of the query Requirement. The first technique is a Linguistic technique based on Parts-of-Speech (POS) Tagging and Traceability rules [8]. The technique was designed specifically for TLR in Models, and is used as a baseline against which the proposed techniques are compared. The other two techniques (named 'Aggregation' and 'Mutation Search') are based on Latent Semantic Indexing and Singular Value Decomposition, a well-spread Information Retrieval technique that has been applied previously to TLR in code, obtaining good results in the process [7]. None of the two LSI-based techniques have been applied to extract TLR between Requirements and Process Models previously. Therefore, we adapted them to work for Process Models and propose them as novel techniques in the field.

The three approaches were evaluated through the Camunda BPMN for Research case study (https://github.com/camunda/bpmn-for-research), as well as through a real-world industrial case study, provided by our industrial partner, CAF (Construcciones y Auxiliar de Ferrocarriles, http://www.caf.es/en), a worldwide provider of railway solutions.

Results show that the Mutation Search technique achieves the best results for all the measured performance indicators in both case studies, providing a mean precision value of 63%, a mean recall value of 77%, a combined F-measure of 68%, and an MCC value of 0.60 for the Camunda BPMN for Research case study, and a mean precision value of 79%, a mean recall value of 72%, a combined F-measure of 74%, and an MCC value of 0.69 for the CAF case study. In contrast, the Linguistic baseline and the Aggregation technique present worse results in these same measurements in both case studies.

The overall findings of our paper suggest that adapting techniques that have provided good results in code is beneficial for TLR between Requirements and Process Models, since their results outperform those of a technique created specifically with Models in mind. Moreover, studied why this is the case, and identified Process Models particularities that could potentially lead to improvement opportunities.

The rest of the paper is structured as follows: Sect. 2 describes our Approach, that is, our proposed techniques and how to apply them to TLR between Requirements and Process Model fragments. Section 3 details the baseline technique and the designed Evaluation. Section 4 presents the obtained results. Section 5 discusses the outcomes of the paper. Section 6 presents the Threats to Validity of our work. Section 7 reviews the works related to this one. Finally, Sect. 8 concludes the paper.

2   Approach

Through the following paragraphs, we give an introduction on Latent Semantic Indexing, the technique upon which we base the two novel techniques proposed for TLR between Requirements and Process Models. Afterwards, we describe said techniques, providing insight on their steps, application, and outcomes.

2.1 Latent Semantic Indexing

Latent Semantic Indexing (LSI) [9] is an automatic mathematical/statistical technique that analyzes relationships between *queries* and *documents* (bodies of text). LSI has been successfully used to retrieve Traceability Links between different kinds of software artifacts in different contexts, specially among Requirements and code [7]. This is due to the fact that code often encodes domain knowledge in the form of domain terms, which are also encoded in the Requirements, hence causing LSI to detect similitude between both.

So far, the technique has not been transported to Process Models. We propose two techniques that use LSI for TLR between Requirements and Process Models. In particular, both techniques use LSI to produce a Model Fragment from the Process Model that serves as a candidate for realizing the Requirement. The following sections give more details on the process.

2.2 Aggregation

The first of the two proposed techniques receives a *query* Requirement and a Process Model as input, and generates a ranking of Model Elements through LSI. From the ranking, a Model Fragment is generated. To that extent, the Process Model is firstly split into Model Elements, represented through the text they contain, which is extracted and used as input for LSI. The top part of Fig. 1 shows this process, having the example input Process Model on the left, and the resulting Model Elements on the right, including: (1) lanes 'Inhibition', 'Human', and 'PLC' (ME1, ME2, ME3); (2) the start and end events (ME4, ME10, ME14); (3) the exclusive gateway 'Are the doors open?' (ME8); (4) the 'Push the doors button' and 'Open the doors' tasks (ME6, ME12); and (5) the sequence flows of the diagram (ME5, ME7, ME9, ME11, ME13).

The text of the Requirement and the Model Elements is then treated through Natural Language Processing techniques. To that extent, general phrase styling techniques, Parts-Of-Speech Tagging [10], and Lemmatizing [11] are applied.

Finally, the Requirement and the Model Elements are fed into LSI, which ranks the Model Elements according to their similitude to the Requirement. The bottom left part of Fig. 1 shows an example *term-by-document co-occurrence matrix*, with values associated to our running example. In the following paragraph, an overview of the elements of the matrix is provided.

Each row in the matrix (*term*) stands for each of the words that appear in the processed text of the Requirement and the Model Elements. In Fig. 1, it is possible to notice a subset of said words such as 'Door' or 'Button' as the *terms* of each row. Each column in the matrix (*document*) stands for each of the Model Elements extracted from the input Process Model. In Fig. 1, it is possible to notice identifiers in the columns such as 'ME3' or 'ME12', which stand for the *documents* of those particular Model Elements (namely, the processed text

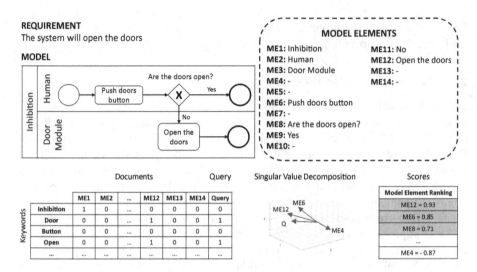

Fig. 1. Aggregation technique example

of 'ME3' and 'ME12'). The final column (*query*), stands for the processed input Requirement. Each cell in the matrix contains the frequency of each *term* in each *document*. For instance, in Fig. 1, the *term* 'Door' appears once in the 'ME12' *document* and once in the *query*.

Vector representations of the *documents* and the *query* are obtained by normalizing and decomposing the *term-by-document co-occurrence matrix* using a matrix factorization technique called *Singular Value Decomposition* (SVD) [9]. SVD is a form of factor analysis, or more properly the mathematical generalization of which factor analysis is a special case. In SVD, a rectangular matrix is decomposed into the product of three other matrices. One component matrix describes the original row entities as vectors of derived orthogonal factor values, another describes the original column entities in the same way, and the third is a diagonal matrix containing scaling values such that when the three components are matrix-multiplied, the original matrix is reconstructed.

In Fig. 1, a three-dimensional graph of the SVD is provided, on which it is possible to notice the vectorial representations of some of the columns. For legibility reasons, only a small set of the columns is represented. To measure the similarity degree between vectors, the cosine between the *query* vector and the *documents* vectors is calculated. Cosine values closer to one denote a high degree of similarity, and cosine values closer to minus one denote a low degree of similarity. Similarity increases as vectors point in the same general direction (as more *terms* are shared between *documents*). Through this measurement, the Model Elements are ordered according to their similarity degree to the Requirement.

The relevancy ranking (which can be seen in Fig. 1) is produced according to the calculated similarity values. In this example, LSI retrieves 'ME12', 'ME6', and 'ME8' in the first, second, and third position of the relevancy ranking due to their *query-documents* cosines being '0.9343', '0.8524' and '0.7112', implying high similarity between the Model Elements and the Requirement. On the opposite, the 'ME4' Model Element is returned in a latter position of the ranking due to its *query-document* cosine being '−0.8736', implying a low similarity degree.

From the ranking, of all the Model Elements, those that have a similarity measure greater than x must be taken into account. The heuristic that we adopted, and that is used in other works, is $x = 0.7$ [12,13]. This value corresponds to a 45° angle between the corresponding vectors. Nevertheless, the selection of this threshold is an issue still under study, and its proper parametrization has not been tackled in Process Models yet.

Following this principle, the Model Elements with a similarity measure equal or superior to $x = 0.7$ are taken to conform a Model Fragment, candidate for realizing the Requirement. Through the example provided in Fig. 1, 'ME12', 'ME6' and 'ME8' are the Model Elements that conform the Model Fragment for the Requirement, due to their cosine values being superior to the 0.7 threshold. The Model Elements below the threshold, except for 'ME4', are not shown in the ranking for space and understandability reasons. The Model Fragment generated in this manner is the final output of the Aggregation technique.

2.3 Mutation Search

The second of the two proposed techniques receives a *query* Requirement and a Process Model as input, generates a population of Model Fragments, and ranks said Model Fragments through LSI. From the ranking, the first Model Fragment is taken as the proposed solution. In order to generate the Model Fragments population, Algorithm 1 is followed. In the algorithm, an empty population and a seed Fragment (chosen randomly from the input Process Model) are created. Then, until the algorithm mets a stop condition (for instance, a certain number of iterations), the Fragment is mutated and each new mutation is added to the population, avoiding the addition of repeated Fragments.

In the algorithm, a mutation in a Fragment can be caused by: (1) adding one new event, gateway, or task that is connected to an already present event, gateway, or task (the flow that causes the connection is also added to the Fragment), (2) removing an Element with only one connection (and the flow that causes said connection), or (3) adding or removing a lane from the Fragment. The performed mutation is chosen randomly on each iteration.

The top part of Fig. 2 shows this process, having the example input Process Model on the left, and some example Model Fragments on the right, generated through the usage of the algorithm. The generated Model Fragments are represented through the text contained in all their elements. The text of both the input Requirement and the generated Model Fragments is then processed through general phrase styling techniques, Parts-Of-Speech Tagging, and Lemmatizing.

Finally, the Requirement and the Model Fragments are fed into LSI, which ranks the Model Fragments according to their similitude to the Requirement. The bottom left part of Fig. 2 shows an example *term-by-document co-occurrence matrix*, with values associated to our running example. The technique works

Algorithm 1. Mutation Search Algorithm

1: $P \leftarrow []$	▷ Initialize the population
2: $F \leftarrow randomFragment(inputModel)$	▷ Create an initial seed Fragment
3: **while** $!(StopCondition)$ **do**	▷ While the stop condition is not met
4: $F \leftarrow mutateFragment(F)$	▷ Mutate the Fragment
5: **if** $!(F \in P)$ **then**	▷ If the new Fragment is not in the population
6: $P \leftarrow P + F$	▷ Add the new mutation to the population
7: **end if**	
8: **end while**	
9: **return** P	▷ Return the population

exactly as it does in the Aggregation technique, except that each column in the matrix (*document*) stands for each of the Model Fragments (MF1 to MFn) generated through the algorithm instead of standing for a single Model Element.

Vector representations of the *documents* and the *query* are obtained by normalizing and decomposing the *term-by-document co-occurrence matrix* using SVD, and the vectorial similarity degrees are calculated through the cosines. The relevancy ranking on Fig. 2 is produced according to the calculated similarity degrees. In this example, LSI retrieves 'MF9' in the first position of the relevancy ranking due to its *query-documents* cosine being '0.9791'. On the opposite, the 'MF6' Model Fragment is returned in the last position of the ranking due to its *query-document* cosine being '−0.9384'.

From the ranking, the first Model Fragment is considered as the candidate solution for the Requirement, and consequently taken as the final output of the Mutation Search technique.

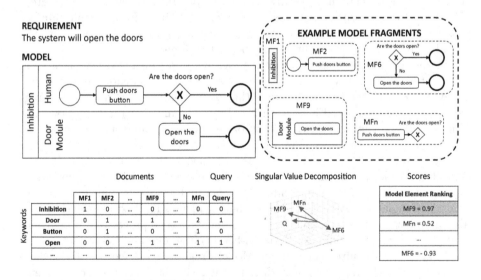

Fig. 2. Mutation search technique example

3   Evaluation

Through the following paragraphs, we introduce the experimental setup and the case studies used to evaluate the baseline and our two proposed approaches, present the oracles used in the evaluation, and detail the design and implementation of said evaluation.

3.1   Experimental Setup

The goal of our work is to perform TLR between Requirements and Process Models through the two proposed techniques, and to compare the results obtained by said techniques against those of a Models specific baseline. Figure 3 shows an overview of the process that was followed to evaluate the Linguistic baseline and our two proposed techniques. The top part shows the inputs, which are extracted from the documentation provided in the case studies: Requirements, Process Models, and approved Traceability between Requirements and Process Models. Each case study comprises a set of Requirements, a Process Model, and an Approved Requirements to Model Fragments Traceability document, which conforms the oracle of our evaluation.

For each case study, the Linguistic baseline and the Aggregation technique take the mentioned inputs, and generate a single Model Fragment for each Requirement. The generated Model Fragments are compared with the oracle Model Fragment. The Mutation Search technique generates a ranking of Model Fragments per Requirement instead. Since the rankings are ordered from best to worst Traceability, the first Model Fragment in each ranking is picked for comparison against its corresponding oracle. Once the comparisons are performed, a confusion matrix is calculated for the baseline and for each technique separately.

A confusion matrix is a table that is often used to describe the performance of a classification Model (in this case the Linguistic baseline and both of our techniques) on a set of test data (the solutions) for which the true values are known (from the oracle). In our case, each solution outputted by the three techniques is a Model Fragment composed of a subset of the Model Elements that are part of the Process Model. Since the granularity is at the level of Model Elements, the presence or absence of each Model Element is considered as a classification. The confusion matrix distinguishes between the predicted values and the real values, classifying them into four categories: (1) True Positive (TP), values that are predicted as true (in the solution), and are true in the real scenario (the oracle); (2) False Positive (FP), values that are predicted as true (in the solution), but are false in the real scenario (the oracle); (3) True Negative (TN), values that are predicted as false (in the solution), and are false in the real scenario (the oracle); and (4) False Negative (FN), values that are predicted as false (in the solution), but are true in the real scenario (the oracle).

Then, some performance measurements are derived from the values in the confusion matrix. In particular, a report including four performance measurements (Recall, Precision, F-measure, and Matthews Correlation Coefficient) is created for the case studies, for each of the three techniques.

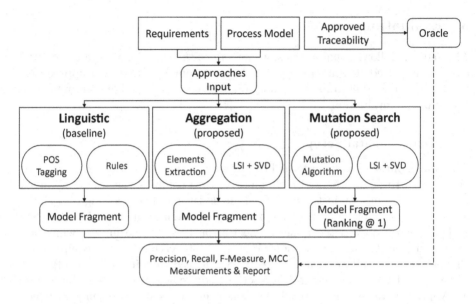

Fig. 3. Experimental setup

Recall measures the number of elements of the solution that are correctly retrieved by the proposed solution. Precision measures the number of elements from the solution that are correct according to the ground truth. F-measure corresponds to the harmonic mean of Precision and Recall [14].

However, none of these previous measures correctly handle negative examples (TN). The **MCC** is a correlation coefficient between the observed and predicted binary classifications that takes into account all the observed values (TP, TN, FP, FN), and is defined as follows:

$$MCC = \frac{TP \cdot TN - FP \cdot FN}{\sqrt{(TP + FP)(TP + FN)(TN + FP)(TN + FN)}}$$

Recall values can range between 0% (which means that no single model element from the realization of the requirement obtained from the oracle is present in the model fragment of the solution) to 100% (which means that all the model elements from the oracle are present in the solution). Precision values can range between 0% (which means that no single model element from the solution is the oracle) to 100% (which means that all the model elements from the solution are present in the oracle). A value of 100% precision and 100% recall implies that both the solution and the requirement realization from the oracle are the same. MCC values can range between −1 (which means that there is no correlation between the prediction and the solution) to 1 (which means that the prediction is perfect). Moreover, a MCC value of 0 corresponds to a random prediction.

3.2   Linguistic Rule-Based Baseline

Spanoudakis et al. [8] present a linguistic rule-based approach to support the automatic generation of Traceability Links between Natural Language Requirements and Models. Specifically, the Traceability Links are generated following two stages: (1) a Parts-of-Speech (POS) tagging technique [15] is applied on the Requirements that are defined using Natural Language, and (2) the Traceability Links between the Requirements and the Models are generated through a set of *Requirement-to-object-Model* (RTOM) rules.

The RTOM rules are specified by investigating grammatical patterns in Requirements. These rules are specified as sequences of terms, and define relations between Requirements and Model Elements. For instance, a rule may attempt to match a *verb-article-noun* pattern that appears in a Requirement with the text that appears in a Model Element. The rules are atomic: the matching succeeds if the Model Element contains the same words in the same pattern.

In [8], the authors propose 26 rules, applied to a Requirement and a Model in order to retrieve a set of Model Elements from the Model that are related to the Requirement. These Model Elements compose the Model Fragment as a result. We worked with a set of rules adapted to work over Process Models.

3.3   Case Study

In order to perform the evaluation of the three approaches, we rely on two different case studies: (1) the Camunda BPMN for Research academic repository, and (2) a set of Process Models provided by CAF, our industrial partner.

Camunda BPMN for Research: The Camunda BPMN for Research case study consists of four Process Modeling exercises. Each exercise contains an associated textual description and the solution Model for the provided description. In order to apply the three approaches to the Camunda case study, a software engineer (with BPMN expertise, and who is not related to the writing of this paper) derived a set of Natural Language Requirements from the problem descriptions. On average, there are around 15 Requirements per problem, with an approximate average of 25 words per requirement. The Models in the case study contain an approximate average of 25 elements per Model.

CAF: For our evaluation, CAF provided us with Natural Language Requirements and Process Models of five railway solutions from Auckland, Bucharest, Cincinnati, Houston, and Kaohsiung. The functionalities are specified through about 100 Natural Language Requirements each, with an approximate average of 50 words per Requirement. Regarding the Process Models, the distinct functionalities are specified through an average 850 total model elements.

3.4   Oracle

In order to obtain the performance results of the three approaches, their outcomes must be compared against the correct solutions of the two case studies.

Camunda BPMN for Research: In the case of the Camunda BPMN for Research case study, each exercise has an associated solution Model for the provided description. The same software engineer who derived the Natural Language Requirements from the problem descriptions also generated a set of Model Fragments from the solution Model, mapping each Fragment to a single Requirement. Thus, we were provided with a set of Requirements, the Model Fragments that implement them, and the TLR mapping between both artifacts.

CAF: Regarding our industrial partner, CAF provided us with their existing documentation on Requirements to Process Models Traceability, where each requirement is also mapped to a single Model Fragment.

In both cases, we use the existing Traceability as the oracle for evaluating the outcomes of each of the three approaches. To do so, we compare the Model Fragments generated for each Requirement by the three of them against the oracle Model Fragment (ground truth Model Fragment) for said Requirements.

3.5 Implementation Details

We have used three libraries to implement the different approaches taken in account through this work: (1) to load and process the Process Models in both case studies, we used the Camunda BPMN Model API [16], (2) to develop the Natural Language Processing operations in our approaches, we have used the OpenNLP Toolkit for the Processing of Natural Language Text [17], and (3) to perform the LSI and SVD carried out in the Aggregation and Mutation Search techniques, the Efficient Java Matrix Library (EJML) was used [18]. For the evaluation, we used a Lenovo E330 laptop, with a processor Intel(R) Core(TM) i5-3210M@2.5 GHz with 16 GB RAM and Windows 10 64-bit.

4   Results

Table 1 outlines the results of the three studied approaches. Each row shows the Precision, Recall, F-measure, and MCC values obtained through each technique.

The Mutation Search technique achieves the best results for all the performance indicators in both case studies, providing a mean precision value of 63%, a mean recall value of 77%, a combined F-measure of 68%, and an MCC value of 0.60 for the Camunda BPMN for Research case study, and a mean precision value of 79%, a mean recall value of 72%, a combined F-measure of 74%, and an MCC value of 0.69 for the CAF case study.

In contrast, both the Linguistic technique and the Aggregation technique present worse results in all the measurements: the Linguistic technique attains a mean precision value of 40%, a mean recall value of 35%, a combined F-measure of 33%, and an MCC value of 0.25 for the Camunda BPMN for Research case study, and a mean precision value of 35%, a mean recall value of 35%, a combined F-measure of 33%, and an MCC value of 0.25 for the CAF case study; and the Aggregation technique attains a mean precision value of 56%, a mean recall value of 72%, a combined F-measure of 61%, and an MCC value of 0.52 for the

Table 1. Mean values and standard deviations for Precision, Recall and F-measure for the three approaches

	Precision	Recall	F-measure	MCC
Linguistic - Camunda	40% ± 25%	35% ± 22%	33% ± 13%	0.25 ± 0.19
Linguistic - CAF	35% ± 28%	35% ± 10%	30% ± 7%	0.18 ± 0.13
Aggregation - Camunda	56% ± 18%	72% ± 22%	61% ± 17%	0.52 ± 0.24
Aggregation - CAF	69% ± 29%	66% ± 17%	64% ± 17%	0.58 ± 0.21
Mutation Search - Camunda	63% ± 21%	77% ± 22%	68% ± 19%	0.60 ± 0.24
Mutation Search - CAF	79% ± 19%	72% ± 19%	74% ± 16%	0.69 ± 0.20

Camunda BPMN for Research case study, and a mean precision value of 69%, a mean recall value of 66%, a combined F-measure of 64%, and an MCC value of 0.58 for the CAF case study.

5   Discussion

The Linguistic technique depends strongly on the language of the Requirements and Models: for a link to be produced between a Requirement a Model Element, exact patterns of words must be atomically matched through the rules. If a single word in a pattern found in a Requirement is different (or missing) in the Model, the rule does not trigger and the link is not produced. On the other hand, in the Aggregation and Mutation Search techniques the atomicity of text patterns is abandoned in favor of the semantic similitude of individual terms. This issue can be illustrated through an example. Consider the Requirement *'The system will open the doors'*, and a Model where the term *'system'* has been swapped for the more technical term *'PLC'*. Due to the vocabulary mismatch, the Linguistic technique would never find the pattern, and thus could never generate the links between Requirement and Model. On the other hand, our techniques would flag the occurrences of the terms *'open'* and *'doors'* in the corresponding Model Elements or Fragments, leading to a potential finding of links.

Moreover, Model Elements with little or no text appear often in Process Models, mainly in the form of flows and sometimes in the form of events. These elements can never be retrieved by the Linguistic technique: since there are no words, there is no pattern that can be matched. They are not retrieved by the Aggregation technique either: they tend to be at the bottom of the ranking produced by LSI since for these elements, all the *term* occurrences are equal to zero and thus, no correlation can be found with the *query* Requirement. However, in the Mutation Search technique, the algorithm does add these Elements to the candidate Fragments. Moreover, the addition of these Elements does not penalize the results technique, since the *term* occurrences are not altered in any way by them. Therefore, the candidate Fragments are more correct and complete, which leads the technique to better Precision and Recall results.

Finally, we also identified certain Process Models particularities that, if leveraged, would improve our Traceability techniques. Some examples of these particularities are: (1) the usage of the term 'if' in a Requirement almost always indicates the presence of an associated gateway in the Process Model, (2) the usage of the terms 'start' or 'end' usually denote events of the same type, (3) questions are often related with gateways in the Models, (4) verbs appear mostly on tasks, or (5) a noun that is often repeated at the start of multiple requirements may be the subject that carries an action (and thus, may appear in the Model as a lane). By studying the patterns of the Process Models language, it could be possible to take in account these particularities in our techniques (by, for instance, weighing the Model Elements accordingly or forcing their appearance), leading them to enhanced Traceability results.

6   Threats to Validity

In this section, we use the classification of threats to validity of [19] to acknowledge the limitations of our approach.

Construct validity: To minimize this risk, our evaluation is performed using four measures: Precision, Recall, F-measure, and MCC. These measures are widely accepted in the software engineering research community.

Internal Validity: The number of requirements and Process Models presented in this work may look small, but they represent a wide scope of different scenarios in an accurate manner.

External Validity: Both Natural Language Descriptions and Business Process Models are frequently leveraged to specify all kinds of different Business Processes. The Camunda Process for Research case study provides different examples from radically different domains. In addition, the real-world CAF Process Models used in our research are a good representative of the railway, automotive, aviation, and general industrial manufacturing domains. Our approach does not rely on the particular conditions of any of those domains. Nevertheless, our results should be replicated with other case studies before assuring their generalization.

Reliability: To reduce this threat, the requirements and Process Models used in our approach were taken from an open-source case study and from an industrial case study. None of the authors of this work was involved in the generation of said data.

7   Related Work

Related works focus on the impact and application of Linguistic techniques to TLR problem resolution at several levels of abstraction. Works like [20,21] or [22], among many others, use Linguistic approaches to tackle specific TLR problems and tasks. In [23], the authors use Linguistic techniques to identify equivalence between Requirements, also defining and using a series of principles for evaluating their performance when identifying equivalent Requirements. The authors of [23] conclude that, in their field, the performance of Linguistic techniques is determined by the properties of the given dataset over which they are performed. They measure the properties as a factor to adjust the Linguistic techniques accordingly, and then apply their principles to an industrial case study. The work presented in [24] uses Linguistic techniques to study how changes in Requirements impact other Requirements in the same specification. Through the pages of their work, the authors analyze TLR between Requirements, and use Linguistic techniques to determine how changes in requirements must propagate.

Our work differs from [20–22,25] since our approach is not based or focused on Linguistic techniques as a means of TLR analysis, but we rather propose novel techniques to perform TLR between Requirements and Process Models, using a Linguistic technique only as a baseline against which our work is compared. Moreover, we do not study how Linguistic techniques must be tweaked for specific problems as [23] does. In addition, differing from [24], we do not tackle changes in Requirements nor TLR between Requirements, but instead focus our work on TLR between Requirements and Process Models.

Finally, other works target the application of LSI to TLR tasks. De Lucia et al. [26] present a tool based on LSI in the context of an artifact management system. [27] takes in consideration the possible configurations of LSI when using the technique for TLR between Requirements artifacts. In their work, the authors state that the configurations of LSI depend on the datasets used, and they look forward to automatically determining an appropriate configuration for LSI for any given dataset. Through our work, we do not study the management of artifacts nor different LSI configurations or how LSI configurations impact the results of TLR, but we rather study TLR between Requirements and Process Models.

8   Conclusions

Traceability Links Recovery (TLR) has been a topic of interest for many years, but its study is an issue that has not received enough attention yet in the field of Process Models. Through this paper, we have studied TLR between Natural Language Requirements and Process Models through three different approaches: a Linguistic approach based on rules, specific from Models (which acts as a baseline for our work), and two techniques (Aggregation and Mutation Search) that we proposed and which we based on Latent Semantic Indexing, a technique that has been used successfully over code. The retrieved TLR results can be utilized by software engineers as a starting point for the development of their solutions.

The three approaches were evaluated by applying them to an academia set of Process Models, and to a set of Process Models from a real-world industrial case study with our industrial partner, CAF, a worldwide manufacturer of railway solutions. Results show that our techniques retrieve better results that the baseline Linguistic technique in both case studies. Through this work, we analyzed why this is the case, and identified some particularities of Process Modeling that could be used in order to improve our techniques in future iterations of our work.

Acknowledgements. This work has been partially supported by the Ministry of Economy and Competitiveness (MINECO) through the Spanish National R+D+i Plan and ERDF funds under the project Model-Driven Variability Extraction for Software Product Line Adoption (TIN2015-64397-R). We also thank ITEA3 15010 REVaMP2 Project.

References

1. Gotel, O.C., Finkelstein, C.: An analysis of the requirements traceability problem. In: Proceedings of the First International Conference on Requirements Engineering, pp. 94–101. IEEE (1994)
2. Spanoudakis, G., Zisman, A.: Software traceability: a roadmap. Handb. Softw. Eng. Knowl. Eng. **3**, 395–428 (2005)
3. Watkins, R., Neal, M.: Why and how of requirements tracing. IEEE Softw. **11**(4), 104–106 (1994)
4. Ghazarian, A.: A research agenda for software reliability. In: IEEE Reliability Society 2009 Annual Technology Report (2010)
5. Rempel, P., Mäder, P.: Preventing defects: the impact of requirements traceability completeness on software quality. IEEE Trans. Softw. Eng. **43**(8), 777–797 (2017)
6. Parizi, R.M., Lee, S.P., Dabbagh, M.: Achievements and challenges in state-of-the-art software traceability between test and code artifacts. IEEE Trans. Reliab. **63**(4), 913–926 (2014)
7. Rubin, J., Chechik, M.: A survey of feature location techniques. In: Domain Engineering, pp. 29–58. Springer, Heidelberg (2013)
8. Spanoudakis, G., Zisman, A., Pérez-Minana, E., Krause, P.: Rule-based generation of requirements traceability relations. J. Syst. Softw. **72**(2), 105–127 (2004)
9. Landauer, T.K., Foltz, P.W., Laham, D.: An introduction to latent semantic analysis. Discourse Processes **25**(2–3), 259–284 (1998)
10. Hulth, A.: Improved automatic keyword extraction given more linguistic knowledge. In: Proceedings of the 2003 Conference on Empirical Methods in Natural Language Processing. Association for Computational Linguistics, pp. 216–223 (2003)
11. Plisson, J., Lavrac, N., Mladenic, D., et al.: A rule based approach to word lemmatization. In: Proceedings of the 7th International Multi-conference Information Society, vol. 1, pp. 83–86. Citeseer (2004)
12. Marcus, A., Sergeyev, A., Rajlich, V., Maletic, J.: An information retrieval approach to concept location in source code. In: Proceedings of the 11th Working Conference on Reverse Engineering, pp. 214–223, November 2004
13. Salman, H.E., Seriai, A., Dony, C.: Feature location in a collection of product variants: combining information retrieval and hierarchical clustering. In: The 26th International Conference on Software Engineering and Knowledge Engineering, pp. 426–430 (2014)
14. Salton, G., McGill, M.J.: Introduction to Modern Information Retrieval. McGraw-Hill Inc., New York (1986)
15. Leech, G., Garside, R., Bryant, M.: CLAWS4: the tagging of the British National Corpus. In: Proceedings of the 15th Conference on Computational Linguistics, vol. 1, pp. 622–628. Association for Computational Linguistics (1994)

16. Camunda: Camunda BPMN Model API (2017). https://github.com/camunda/camunda-bpmn-model. Accessed 3 Nov 2017
17. Apache: OpenNLP Toolkit for the Processing of Natural Language Text (2017). https://opennlp.apache.org/. Accessed 12 Nov 2017
18. Abeles, P.: Efficient Java Matrix Library (2017). http://ejml.org/. Accessed 9 Nov 2017
19. Wohlin, C., Runeson, P., Höst, M., Ohlsson, M.C., Regnell, B., Wesslén, A.: Experimentation in Software Engineering. Springer, Heidelberg (2012)
20. Sultanov, H., Hayes, J.H.: Application of swarm techniques to requirements engineering: requirements tracing. In: 18th IEEE International Requirements Engineering Conference (2010)
21. Sundaram, S.K., Hayes, J.H., Dekhtyar, A., Holbrook, E.A.: Assessing traceability of software engineering artifacts. Requirements Eng. **15**(3), 313–335 (2010)
22. Duan, C., Cleland-Huang, J.: Clustering support for automated tracing. In: Proceedings of the 22nd IEEE/ACM International Conference on Automated Software Engineering (2007)
23. Falessi, D., Cantone, G., Canfora, G.: Empirical principles and an industrial case study in retrieving equivalent requirements via natural language processing techniques. Trans. Softw. Eng. **39**(1), 18–44 (2013)
24. Arora, C., Sabetzadeh, M., Goknil, A., Briand, L.C., Zimmer, F.: Change impact analysis for natural language requirements: an NLP approach. In: IEEE 23rd International Requirements Engineering Conference (2015)
25. Ryan, K.: The role of natural language in requirements engineering. In: Proceedings of IEEE International Symposium on Requirements Engineering (1993)
26. De Lucia, A., Fasano, F., Oliveto, R., Tortora, G.: Enhancing an Artefact management system with traceability recovery features. In: Proceedings of the 20th IEEE International Conference on Software Maintenance, pp. 306–315. IEEE (2004)
27. Eder, S., Femmer, H., Hauptmann, B., Junker, M.: Configuring latent semantic indexing for requirements tracing. In: Proceedings of the 2nd International Workshop on Requirements Engineering and Testing (2015)

Clinical Processes - The Killer Application for Constraint-Based Process Interactions?

Andres Jimenez-Ramirez[1(✉)], Irene Barba[1], Manfred Reichert[2],
Barbara Weber[3,4], and Carmelo Del Valle[1]

[1] Departamento de Lenguajes y Sistemas Informáticos, University of Seville,
Seville, Spain
{ajramriez,irenebr,carmelo}@us.es
[2] Institute of Databases and Information Systems, Ulm University, Ulm, Germany
manfred.reichert@uni-ulm.de
[3] Department of Computer Science, University of Innsbruck, Innsbruck, Austria
Barbara.Weber@uibk.ac.at
[4] Technical University of Denmark, Kongens Lyngby, Denmark
bweb@dtu.dk

Abstract. For more than a decade, the interest in aligning information systems in a process-oriented way has been increasing. To enable operational support for business processes, the latter are usually specified in an imperative way. The resulting process models, however, tend to be too rigid to meet the flexibility demands of the actors involved. Declarative process modeling languages, in turn, provide a promising alternative in scenarios in which a high level of flexibility is demanded. In the scientific literature, declarative languages have been used for modeling rather simple processes or synthetic examples. However, to the best of our knowledge, they have not been used to model complex, real-world scenarios that comprise constraints going beyond control-flow. In this paper, we propose the use of a declarative language for modeling a sophisticated healthcare process scenario from the real world. The scenario is subject to complex temporal constraints and entails the need for coordinating the constraint-based interactions among the processes related to a patient treatment process. As demonstrated in this work, the selected real process scenario can be suitably modeled through a declarative approach.

Keywords: Process flexibility · Declarative process model
Healthcare process · Temporal constraints

1 Introduction

For several years, there has been an increasing interest in aligning information systems in a process-oriented way [6,30]. Thereby, a business process (BP) consists of a set of activities which jointly realize a business goal and whose

© Springer International Publishing AG, part of Springer Nature 2018
J. Krogstie and H. A. Reijers (Eds.): CAiSE 2018, LNCS 10816, pp. 374–390, 2018.
https://doi.org/10.1007/978-3-319-91563-0_23

execution needs to be coordinated in an organizational as well as technical environment [30]. In this context, process-aware information systems (PAISs) offer promising perspectives by enabling enterprises to define their business processes in terms of explicit process models as well as to execute the corresponding process instances in a controlled and efficient manner [24,30].

Declarative approaches are becoming increasingly popular for modeling business processes as they are able to cope with some of the limitations imperative notations are facing [4,9,10,17,20,28,31]. Although declarative modeling languages have been extensively discussed in literature [5,19], even in the context of healthcare [21,25,29], to the best of our knowledge, they have not been used to model complex, real-world scenarios that comprise constraints going beyond control-flow.

In this paper, we propose the use of a declarative language for modeling a sophisticated healthcare process scenario from the real world. Particularly, the latter is subject to complex temporal constraints and entails the need for coordinating the constraint-based interactions among all the processes contributing to the overall patient treatment process. We strongly believe that, if we are able to demonstrate the applicability of declarative modeling languages to clinical process, which can considered as a kind of killer application for PAISs, the respective approach can be applied to many other sophisticated process scenarios as well.

Although the temporal perspective is present in many real-world process scenarios, it has not received sufficient attention yet. In previous work [14], we systematically derived 10 process time patterns (i.e., solutions for representing commonly occurring temporal constraints in PAISs) by analyzing a large collection of non-trivial process models from various domains. In particular, the time patterns (TPs for short) were defined independently of a specific language or paradigm for BP modeling [13]. Despite the needs for enabling process flexibility and dealing with temporal constraints, most existing approaches are unable to manage both. To fill this gap, we proposed the TConDec-R language [2], a declarative process modeling language that enables the specification of temporal constraints related to the TPs. As demonstrated in this work, TConDec-R is suitable for modeling a sophisticated and real-world process scenario from the healthcare domain.

The remainder of the paper is organized as follows. Section 2 presents backgrounds on temporal constraints in declarative process modeling languages and shows why TConDec-R was selected for modeling the considered scenario. Section 3 motivates the need for coordinating the constraint-based interactions among all the processes in a hospital forming the overall patient treatment process. Section 4 details how TConDec-R is used for modeling such a scenario. Section 5 discusses our work. Finally, Sect. 6 concludes the paper with a summary and an outlook.

2 Background

This section discusses work related to the considered time patterns (cf. Sect. 2.1). It then presents a review of proposals that support temporal constraints in

Table 1. Selected process time patterns.

Cat	Time pattern (TP)	Example
I	*TP1 (Time Lags between two Activities)* enables the definition of different kinds of time lags between two activities	*The time lag between registering a Master thesis and submitting it must not exceed 6 months*
	TP2 (Durations) allows specifying the duration of process activities	*Processing 100 requests must not take longer than 1 s*
II	*TP5 (Schedule Restricted Element)* allows restricting the enactment of a particular activity by a schedule	*Comprehensive lab tests in a hospital can only be done from MO-FR between 8 am and 5 pm*
	TP6 (Time-based Restrictions) provides support for restricting the number of times a specific process element may be executed within a given timeframe	*For a specific lab test at least 5 different blood samples have to be taken within 24 h*

contemporary declarative process modeling languages (cf. Sect. 2.2). Finally, it provides an overview of TConDec-R, a declarative process modeling language with extensive support for temporal constraints (cf. Sect. 2.3).

2.1 Process Time Patterns

In [13,14], we systematically identified 10 process time patterns (TPs) by analyzing a large collection of process models from various domains. In this context, 4 of the 10 TPs are considered in the present work as they refer to aspects directly related to the considered scenario (cf. Table 4, column *TP*). Specifically, constraints C8, C17, C18, C19, C21, C22, and C23 correspond to TP1; C11, C12 and C20 correspond to TP5; finally, C3, C14, and C5 correspond to TP6. In addition, the duration of the activities (cf. Table 3) corresponds to TP2. We divided the time patterns into two categories according to pattern semantics (cf. Table 1). Category I (*Durations and Time Lags*) provides support for expressing the durations of different process granularities (i.e., activities, activity sets, processes, or sets of process instances) as well as time lags between activities or process events (e.g., milestones). Category II (*Restricting Execution Times*), in turn, allows constraining execution times of single activities or entire processes (e.g., deadlines). All considered time patterns are highly relevant for the support of patient treatment processes.

Note that there exist numerous variants of the time patterns, also denoted as pattern variants [14]. To cope with this variability and to keep the number of patterns manageable, design choices allow for TP parametrization [14]. For example, whether a time lag represents a minimum value, maximum value, or both (i.e., an interval) constitutes a design choice of pattern TP1 (Time Lags between Activities). Additional variance of this time pattern is caused by the fact that time lags may be specified either between the start of two activities, the start of the first and the end of the second activity, the end of the first and the start of the second activity, or the end of the two activities. Other design choices, in turn, cover the fact that different time granularities (e.g., second or hour) or process granularities (i.e., activity, activity set, process, or set of process instances) may be applied in the context of the time patterns. A complete list of pattern design choices and, hence, pattern variants, is presented in [14].

2.2 Temporal Constraints in Declarative Processes Modeling Languages

This section reviews the works dealing with declarative process modeling languages that support the temporal constraints necessary for modeling the considered scenario.

There exist several proposals for handling the temporal constraints in different domains. We exclude works that are not related to business processes, view temporal constraints solely in the sense of ordering relations, or only mention temporal constraints at an abstract level (i.e., no specific temporal constraints are discussed). In the end, several papers [2,4,9,10,15–18,20,26,32] were identified as relevant works in the context of our research.

In order to asses the support each proposal provides for the time patterns, we analyzed the following temporal constraint features:

F1. Does the proposal allow specifying time lags between activities? (cf. TP1)
F2. Does the proposal allow specifying the duration of process activities? (cf. TP2)
F3. Does the proposal consider constraints that may refer to a calendar or schedule? (cf. TP5)
F4. Does the proposal allow for time-based constraints, restricting the number of times a particular process element may be executed within a pre-specified timeframe? (cf. TP6)

The respective declarative approach allows specifying the temporal constraints related to the considered time patterns (cf. Table 1), if these four features can be checked. Note that we check whether the proposed modeling language *directly* supports the elements needed for assessing the feature. Hence, we do not check for workarounds of the respective approach possible in this context.

Table 2 includes the considered works together with the answers (i.e., yes (✓) or no (✗)) to each feature. It indicates that (1) patterns TP1 and TP2 are well supported, (2) pattern TP4 is only supported by [2,18], and (3) TConDec-R [2] is the only approach supporting TP6. Therefore, to the best of our knowledge, only TConDec-R supports the modeling of business processes enhanced with the temporal constraints related to the time patterns (cf. Table 1). Moreover, for every TConDec-R temporal constraint all relations stated in Allen's interval algebra [1] (i.e., start-start, start-end, end-start, and end-end) can be specified.

Table 2. Consolidated results of the review.

Work	Features			
	F1	F2	F3	F4
Montali et al. [20]	✓	✓	X	X
Hildebrandt et al. [9]	✓	X	X	X
Maggi et al. [15–17]	✓	X	X	X
Jiang et al. [10]	✓	✓	X	X
Burattin et al. [4,26]	✓	X	X	X
Mans et al. [18]	X	✓	✓	X
Barba et al. [2]	✓	✓	✓	✓

✓ = true, X = false

2.3   The TConDec-R Language

As basis of the TConDec-R language [2], we use Declare [23] for specifying activities and their behavioral (i.e., control-flow) constraints. We consider this declarative modeling language as appropriate as it enables the specification of a wide range of process models in a flexible way. Respective process models are denoted as *constraint-based*, i.e., they comprise information about (1) the activities that may be performed during process enactment as well as (2) the constraints to be obeyed in this context. The activities of a constraint-based process model may be executed arbitrarily often unless this is restricted by any constraint. Declare constraints can be categorized as follows [23]:

1. **Existence constraints** are unary relationships constraining the number of times a particular activity may be executed in the context of a process instance. For example, *Exactly(A,n)* specifies that activity A must be executed exactly n times for each process instance.
2. **Relation constraints** are positive binary relationships used to either enforce or allow for the enactment of activities in certain situations. For example, *Precedence(A,B)* specifies that activity B may only be executed if A is executed before.
3. **Negation constraints** are negative binary relationships used to forbid the enactment of certain activities in specific situations. For example, *NotCoexistence(A,B)* expresses mutual exclusion of A and B, i.e., A must not be executed if B is executed, and vice versa.

Definition 1. *(TConDec-R activity). A TConDec-R activity* act = (a, dur) *refers to a process activity* a *with its estimated duration* dur.[1]

---

[1] Respective estimates can be obtained by interviewing subject matter experts or by analyzing the event logs of completed process instances. Moreover, both approaches can be combined to obtain more reliable estimates.

Definition 2. *(TConDec-R process model). A **TConDec-R** process model TCR = (Acts, C_T) corresponds to an extended constraint-based process model, where*

- *Acts is a set of TConDec-R activities and*
- *C_T is a set of constraints that may include any control-flow constraint supported by Declare as well as any temporal constraint related to the time patterns (cf. Table 1).*

TConDec-R constraints are specified according to the graphical notation proposed for Declare constraints [23] and using the graphical notation proposed in [13] for visualizing the temporal constraints.[2]

TConDec-R process model

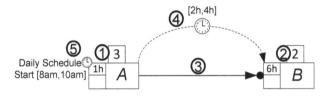

Fig. 1. A simple TConDec-R process model.

Example 1. (**TConDec-R process model**). Figure 1 shows a simple example of a TConDec-R process model:

- *Acts* = {$(A, 1h), (B, 6h)$} consists of two activities A and B; A has an estimated duration of 1 hour. In turn, B has an estimated duration of 6 hours;
- C_T comprises the following constraints:
  (1) *Exactly*$(A, 3)$, expressing that A shall be executed exactly three times,
  (2) *Exactly*$(B, 2)$, expressing that B shall be executed exactly twice,
  (3) *Precedence*(A, B), expressing that activity B may only be executed if A is executed before,
  (4) *TimeLagEndStart*$(A, B, 2h, 4h)$, representing a temporal constraint corresponding to time pattern TP1; it expresses that for each execution A_i of A, there must be at least one execution B_j of B such that there is a time lag of at least 2 hours ($2h$) and at most $4h$ between the end time of A_i and the start time of B_j, and
  (5) *DailyScheduleStart*$(A, 8am, 10am)$, expressing a temporal constraint related to TP5, i.e., each execution of A must start between 8 and 10am.

---

[2] A complete formalization of the TConDec-R constraints is available at http://azarias.lsi.us.es/TCR/Formalization.pdf.

In order to provide support to the TConDec-R language, a web-based tool has been implemented based on previous approaches [2,11].[3] The tool consists of a light-weight web interface that allows for (1) textually modeling declarative business processes using the TConDec-R language or loading already created models, and (2) providing operational support for the models (e.g., generating valid execution traces or checking the conformance of models). For example, Fig. 2 shows the tool where a TConDec-R model has been loaded and a valid trace is generated. The client uses a REST API layer to connect to a server being in charge of all the heavy-duty tasks requested by the user interface.

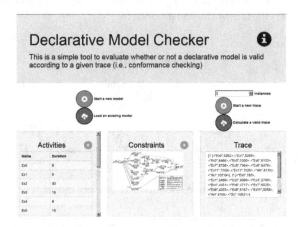

Fig. 2. TConDec-R client interface.

3 Clinical Process Scenarios

3.1 General Insights into Clinical Practice

In the following, an impression of the constraints driving the interactions between the clinical workflows of a patient treatment process is given to emphasize under which conditions constraint-based solutions need to operate in a healthcare environment.

In the context of a patient treatment process, numerous clinical procedures have to be planned, ordered and prepared, appointments be made, and results be obtained and evaluated. Moreover, many procedures need preparatory measures of various complexity. Before a surgery may take place, a patient has to undergo numerous preliminary examinations (i.e., smaller processes), each of them requiring additional preparations. While some of them are known in advance, others may have to be scheduled dynamically, depending on the individual patient and her state of health.

---

[3] It is available at http://azarias.lsi.us.es/TCR/ModelChecker.

In general, the various clinical workflows to be performed in a patient treatment process as well as their tasks may have to consider complex temporal constraints. After an injection with contrast medium was given to a patient, for example, some other tests cannot be performed within a certain period of time. In contemporary healthcare environments, physicians still have to coordinate the tasks related to their patients manually, taking into account all the constraints existing in this context. Therefore, changing a schedule is not trivial and requires time-consuming communication.

For other procedures, medical staff from various departments have to collaborate. Thus, the process is subdivided into organization-oriented views leading to several problems. First, patients have to wait, because resources (e.g., physicians, rooms or technical equipment) are not available due to insufficient coordination. Second, medical procedures cannot be performed as planned if information is missing, preparations are omitted, or a preceding procedure is postponed, canceled or requires latency time. Depending procedures might then have to be re-scheduled resulting in time-consuming phone calls. Third, if urgently needed results are missing, clinical procedures may have to be performed repeatedly causing unnecessary costs and burdening patients.

For all these reasons, from both the patient and the hospital perspective undesired effects occur: hospital stays can take longer than required and costs or even invasiveness of patient treatment increase. Therefore, healthcare process support, in particular regarding the many constraint-based interactions between processes, would be highly welcome by medical staff.

3.2   A Concrete Treatment Scenario

This section deals with the considered clinical scenario, which we derived by interviewing subject matter experts as well as by analyzing process-relevant information (e.g., patient records, process handbooks) [22,27].

This clinical scenario deals with surgeries (including their preparations) in the context of ovarian carcinoma [22,27]. Usually, patients with ovarian carcinoma are treated in the women's hospital (WH); on average 2 patients with this diagnosis emerge to the WH per day. After admitting a patient to the gynecological ward, one of the two ward physicians visits and examines the patient. Afterwards, the physician orders and schedules a number of medical examinations, which need to be performed before the surgery may take place. Additionally, the patient needs to be examined by an anesthetist who may then request additional medical examinations before the surgery.

Some of the examinations are not carried out in the WH itself, but are provided by other clinical departments, which may be either internal or external (i.e., placed at a different location) to the WH. In the given scenario, five external departments (i.e., Endoscopy Department (ED), Radiology Department (RD), Comprehensive Cancer Centre (CCC), Otolaryngology Department (OD), and Neurology Department (ND)) are involved as well as an internal one comprising two units (i.e., Ultrasound Unit (UU) and Laparoscopy Unit (LU)). Each department has limited resources that provide services to many other departments and clinics respectively (including WH). In order to ensure a fair use of the shared resources, a particular department may only order a maximum number of a specific examination from the respective provider per day.

Table 3. Processes relevant in the context of the considered clinical scenario.

ID	Description	Dur	Unit/Dep	%Req
Ex0	First visit and examination of the patient	30 m	WH	100
Ex1	Pelvic Ultra-sound Imaging	30 m	UU	100
Ex2	Cystoscopy & Rectoscopy	2 h 30 m	ED	100
Ex3	Uretero Pyelography	1 h 30 m	RD	100
Ex4	CT scanning	45 m	RD	60
Ex5	Magnetic Resonance Imaging	1 h 15 m	RD	40
Ex6	Colonoscopy	2 h 15 m	CCC	100
Ex7	Colon Contrast Imaging	3 h 30 m	CCC	40
Ex8	X-ray of the gastrointestinal tract	1 h 15 m	RD	35
Ex9	Chest X-ray	30 m	RD	85
Ex10	Blood test	10 m	WH	70
Ex11	Laparoscopy	1 h	LU	100
Ex12	Doppler examinations	30 m	UU	20
Ex13	Medical council with the Otolaryngology Department	60 m	OD	20
Ex14	Medical council with the Neurology Department	30 m	ND	10
AN	The patient is examined and interviewed by an anesthetist	1 h	WH	100
SU	The surgery is performed	2–6 h	WH	75

h = hours, m = minutes

All relevant processes and activities of the considered clinical scenario (i.e., appointments to be made with the physician and anesthetist, medical examinations to be performed, and the surgery to be carried out) are summarized in Table 3: Column ID contains the identifier of the activity, Dur its average duration, and $Unit/Dep$ the unit/department the activity is performed at. Finally, $\%Req$ indicates the frequency with which the respective examination is requested for a patient (e.g., $\%Req = 100$ expresses that the examination is requested once for every patient, i.e., in 100% of all cases). Estimates regarding the average duration of the activities were obtained by interviewing subject matter experts. Considering such indications, an example of the different processes involving 14 patients for the considered scenario can be generated using the provided tool (cf. Sect. 2.3).[4]

---

[4] A graphical example is depicted in http://azarias.lsi.us.es/TCR/PlanEx1.pdf.

Concerning the surgery of a particular patient and required preparations (i.e., medical examinations), several constraints (including temporal ones) need to be obeyed. Examples include chronological orders of activities and time lags between them. Corresponding constraints are informally summarized in Table 4.

4   Modeling the Clinical Scenario with TConDec-R

This section shows how the considered scenario (cf. Sect. 3.2) can be properly described with TConDec-R. Note that all constraints related to the clinical scenario can be modeled using TConDec-R constraints. The resulting TConDec-R constraints are depicted in the right column of Table 4. When capturing the scenario with TConDec-R, we obtain the model depicted in Fig. 3.

On one hand, each process activity is characterized by its estimated duration according to Definition 2. In Fig. 3, the department/hospital performing the activities is depicted for the sake of clarity (e.g., $Ex0$ is performed in the WH). Note that this information is useful, for example, to check whether patient transportation is needed between two directly succeeding activities. In such cases, a time lag of 20 min is added to properly cover the time needed for transportation.

On the other, each constraint is labeled with the related constraint ID (cf. Table 4). In addition, for the sake of clarity, in Fig. 3 some related constraints are depicted together. For example, constraints Precedence(A,B) and TimeLagEnd-Start(A,B,LB,UB) are depicted as a Precedence constraint with a clock and the related time lag [LB,UB] on top of it (e.g., constraint $C8$, Precedence(Ex6,Ex7), and TimeLagEndStart(Ex6,Ex7,6d,-)). For the sake of readability, when the same constraint applies to a set S of activities in Table 4, it is depicted as only one constraint over S, e.g., in constraint $C1$, constraints Exactly(Ex0,1), Exactly(Ex1,1), Exactly(Ex2,1), Exactly(Ex3,1), Exactly(Ex6,1), and Exactly(AN,1) are represented by Exactly({Ex0,Ex1,Ex2,Ex3,Ex6,AN},1). Further, in the TConDec-R model from Fig. 3, the TimeBasedExclusive1of2Daily constraints (e.g., constraint $C3$) are not depicted as each of them relates one activity with (almost) all other activities and, hence, their inclusion would affect readability of the figure.

Finally, constraints requiring that no activity may be performed on Saturday or Sunday (i.e., constraint $C23$) are added by stating corresponding schedule relations over all activities.

We consider the case as a proper clinical scenario for illustrating the proposed approach. It represents a real-world process, includes process-relevant perspectives (i.e., control flow and time), and requires dealing with many activities and numerous constraints of diverse nature. Altogether, the modeling of such scenario entails a rather high complexity, hence it may be considered as killer scenario for constraint-based languages. With the proposed approach, unlike with previous proposals, it becomes possible to model the scenario without the need for any workarounds or extensions.

The considered scenario has been modeled using the proposed web-based tool to generate valid execution traces. Such traces have then be used as a first

Table 4. Constraints related to the clinical scenario.

ID	Constraints informally described	TP	Related TConDec-R constraints
C1	Activities Ex0, Ex1, Ex2, Ex3, Ex6, and AN are always performed once for all patients	-	$Exactly(\{Ex0,Ex1,Ex2,Ex3,Ex6,AN\},1)$
C2	Ex0 is always the first and Ex1 the second activity to be performed	-	$Succession(Ex0,Ex1)$ and $Precedence(Ex1,\{Ex2,Ex3,Ex8,Ex10,Ex11,Ex13,Ex14\})$
C3	No other examination may be scheduled for the day at which Ex2 is performed	TP6	$TimeBasedExclusive1of2Daily(Ex2,\{Ex0,Ex1,Ex3,Ex4,...,Ex14\},1d)$
C4	Ex3 has to be performed in the morning; i.e., it takes place between 8 am and 12 am	TP5	$DailyScheduleStart(Ex3,8h,12h)$ and $DailyScheduleEnd(Ex3,8h,12h)$
C5	Either Ex4 or Ex5 shall be performed for a particular patient, but not both examinations.	-	$Exclusive1of2(Ex4,Ex5)$
C6	Ex3 needs to be completed before Ex4 or Ex5 may be started	-	$Precedence(Ex3,Ex4)$ and $Precedence(Ex3,Ex5)$
C7	Ex4 or Ex5 needs to be finished before starting Ex6	-	$Response(Ex4,Ex6)$ and $Response(Ex5,Ex6)$
C8	If Ex7 is ordered, it must be performed after Ex6 and there must be a time lag between the completion of Ex6 and the start of Ex7 of at least 6 days (i.e., minimum time lag)	TP1	$Precedence(Ex6,Ex7)$ and $TimeLagEndStart(Ex6,Ex7,6d,-)$
C9	The day before Ex6 is performed, no other examination may be scheduled after 12 am. The day Ex6 is performed, no other examination may be scheduled before Ex6 is started	TP6	$TimeBasedNotPrecedenceDaily(\{Ex0,Ex1,Ex2,...Ex14\},Ex6,1d:\ 12,-)$
C10	The day Ex6 is performed, no other examination, except Ex10, may be scheduled	TP6	$TimedBasedExclusive1of2(Ex6,\{Ex0,Ex1,Ex3,Ex4,...,Ex9,Ex11,...,Ex14\},1d)$
C11	Ex6 shall take place in the morning, i.e., between 8 am and 12 am	TP5	$DailyScheduleStart(Ex6,8h,12h)$ and $DailyScheduleEnd(Ex6,8h,12h)$

(continued)

Table 4. (*continued*)

ID	Constraints informally described	TP	Related TConDec-R constraints
C12	Ex10 is performed early in the morning, i.e., between 6 am and 7 am	TP5	$DailyScheduleStart(Ex10,6h,7h)$ and $DailyScheduleEnd(Ex10,6h,7h)$
C13	Ex11, Ex13 and Ex14 are performed before the appointment with the anesthetist (AN) takes place	-	$Response(Ex11,AN)$, $Response(Ex13,AN)$, and $Response(Ex14,AN)$
C14	Ex13 must not be performed more than once per day	TP6	$TimedBasedCrossInstanceAbsence(2,Ex13,1d)$
C15	Ex14 must not be performed more than once per day	TP6	$TimedBasedCrossInstanceAbsence(2,Ex14,1d)$
C16	Ex9 and Ex12 are performed after the appointment with the anesthetist (AN) took place	-	$Precedence(AN,Ex9)$ and $Precedence(AN,Ex12)$
C17	AN is going to be performed at least 1 day after completing Ex8	TP1	$Response(Ex8,AN)$ and $TimeLagEndStart(Ex8,AN,1d,-)$
C18	All examinations need to be completed at least 3 days before the surgery may take place	TP1	$Response(Ex2,SU)$, $TimeLagEndStart(Ex2,SU,3d,-)$, $Response(Ex7,SU)$, $TimeLagEndStart(Ex7,SU,3d,-)$, $Response(Ex9,SU)$, $TimeLagEndStart(Ex9,SU,3d,-)$, $Response(Ex10,SU)$, $TimeLagEndStart(Ex10,SU,3d,-)$, $Response(Ex12,SU)$, and $TimeLagEndStart(Ex12,SU,3d,-)$
C19	No other examination must take place at the day Ex13 or Ex14 is performed	TP1	$TimeBasedExclusive1of2Daily(Ex13,\{Ex0,Ex1,Ex3,Ex4,...,Ex14\},1d)$ and $TimeBasedExclusive1of2Daily(Ex14,\{Ex0,Ex1,Ex3,Ex4,...,Ex13\},1d)$
C20	SU needs to be started in the morning (between 8 am and 12 am)	TP5	$Schedule(SU,\{W,T\})$, $DailyScheduleStart(SU,8h,12h)$, and $DailyScheduleEnd(SU,8h,12h)$
C21	Each department allows each other department to perform a maximum of 2 examinations per day	TP1	$TimedBasedCrossInstanceAbsence(3,\{Ex2,Ex3,Ex4,Ex5,Ex6, Ex7,Ex8,Ex9\},1d)$
C22	Examinations may only be performed between 8am and 6pm (except Ex10, see C12)	TP1	$DailyScheduleStart(\{Ex0,Ex1,Ex2,Ex3,Ex4,Ex5,Ex7,Ex8,Ex9,Ex11,Ex12, Ex13,Ex14,AN\}, 8h, 18h)$ and $DailyScheduleEnd(\{Ex0,Ex1,Ex2, Ex4,Ex5,Ex7,Ex8,Ex9,Ex11,Ex12,Ex13, Ex14, AN\}, 8h, 18h)$
C23	No activity must be performed on Saturday or Sunday	TP1	$Schedule(\{Ex0,Ex1,Ex2,Ex3,Ex4,Ex5,Ex6,Ex7,Ex8,Ex9,Ex10,Ex11, Ex12,Ex13,Ex14, AN\}, \{M,Tu,W,T,F\})$

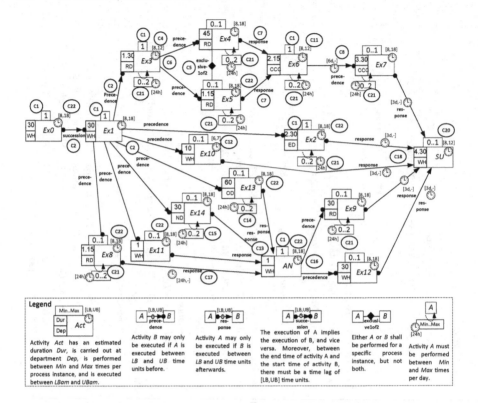

Fig. 3. Simplified TConDec-R model for the scenario.

validation of the model by checking that each constraint of the scenario (cf. Table 4) is fulfilled.[5]

5   Discussion and Lessons Learned

In the current work, we build upon a constraint-based business process modeling language that allows specifying sophisticated temporal constraints, i.e., the TConDec-R language [2]. On one hand, the fact that TConDec-R is framed in the declarative modeling paradigm allows, unlike imperative proposals, for the specification of flexible scenarios. On the other, there are many scenarios from various domains (e.g., healthcare, logistics, and engineering) requiring support for temporal constraints, particularly in the context of long-running processes. More precisely, this work deals with a sophisticated healthcare process scenario from the real world. We derived the scenario by interviewing subject matter experts as well as by analyzing process-relevant information in this context (e.g., patient records, process handbooks) [22,27]. We consider the clinical processes

---

[5] A set of traces generated by the web-based tool can be accessed at http://azarias. lsi.us.es/TCR/instances.zip.

as a proper case for illustrating the proposed approach as they were derived from a real hospital environment, include all process-relevant perspectives, and require dealing with numerous activities and constraints of diverse nature.

Although there exists related work on declarative BP modeling [4,9,10,17, 20,28,31], only few approaches pay attention to the temporal perspective from a wider point of view, i.e., beyond viewing temporal constraints solely just in the sense of ordering relations [4,9,10,15–17,20,26,32]. To assess to what extent these works may support the temporal constraints required by the considered scenario, we performed a review which revealed that there exists other works that partially addresses some of the requirements related to the time patterns (i.e., activity durations and time lags between activities). Unlike TConDec-R, existing works do not consider other requirements that may refer to a calendar or schedule, and repetitions of process elements within a pre-specified timeframe. Therefore, TConDec-R opens a new opportunity for (1) further scenarios that were not supported by the other declarative BP languages, and (2) existing models that can be refined by including the TConDec-R constraints in order to improve the accuracy of the model, i.e., the similarity between the real and the modeled behavior.

Although there are other promising and interesting scenarios where declarative paradigm plays an important role in their effective management [8], we choose a clinical process scenario as it is on the one hand so challenging with respect to the constraints to be supported that the current state of the art was unable to effectively deal with it, and on the other hand not so "exotic" that the tool support needed would not be useful for other application domains as well, making clinical processes the 'killer application' for most declarative approaches currently available.

To demonstrate the applicability and expressiveness of the presented approach, the healthcare scenario is modeled with TConDec-R. The modeling of such scenario entailed a rather high complexity. Despite this complexity, the TConDec-R language could be successfully used for modeling the considered scenario without need for any workaround or extension. Nonetheless, the proposed approach is not only suitable for the illustrated clinical scenario, but for a class of processes with (1) flexible nature, i.e., scenarios for which declarative specifications fit better than imperative ones when designing the process, and (2) temporal constraints playing an important role, i.e., the enactment of process activities must obey complex temporal relations. Hence, we strongly believe that the proposed approach can be successfully applied to many real-world scenarios for enabling flexible process support.

Additionally, to provide a validation of the language, a web-based tool has been implemented to support the TConDec-R.[6] Such a tool allows both (1) modeling the clinical scenario of this work as well as other scenarios which require constraints related to the TPs [13], and (2) providing support for generating valid execution traces according to the modeled scenario. The latter has been conducted based on previous work [2,11].

---

[6] It is available at http://azarias.lsi.us.es/TCR/ModelChecker.

Altogether, we can ensure both the relevance and the novelty of the proposed BP modeling language, i.e., TConDec-R is a declarative approach that is able to cope with all temporal constraints related to the time patterns.

Note that the presented approach has revealed limitations as well. First, designers must deal with a new language for the constraint-based specification of processes. Thus, training is required to make them familiar with TConDec-R. Moreover, although declarative approaches enable a high degree of flexibility, problems in understanding and maintaining the respective process models often impede their adoption [7,33]. Several works related to the understanding of declarative process models have been conducted (e.g., [7,33]). However, respective research should also incorporate temporal constraints to make the results of such studies applicable to the current proposal.

6 Summary and Outlook

This paper analyzed a real-world scenario from healthcare being subject to complex temporal constraints. Thereafter, a review of proposals that may support the time patterns of the scenario was conducted. Unlike existing proposals, TConDec-R allows specifying all constraints related to process time patterns in a comprehensive way. Additionally, we demonstrated that all requirements of the healthcare scenario can be modeled with TConDec-R. Finally, a tool was presented for managing TConDec-R models.

In future work, we will study additional real-world process scenarios in other domains to demonstrate broad applicability of the approach. Furthermore, we will investigate the use and validation of algorithms to improve the support of TConDec-R in several respects, e.g., to provide personal schedules or generate predictions. Related to that, an empirical evaluation for analyzing the efficiency of the proposed approach will be performed. Moreover, the proposed language is intended to be evaluated in terms of usability and scalability. In addition, we will extend the proposed approach by enhancing it with the data perspective as well. Finally, we plan to improve the presented tool in order to (1) enable features like optimization [11], recommendations [3], and predictions [12], and (2) make it stable enough to share it as an open source project.

Acknowledgments. This research has been supported by the Pololas project (TIN2016-76956-C3-2-R) and by the SoftPLM Network (TIN2015-71938-REDT) of the Spanish Ministerio de Economía y Competitividad.

References

1. Allen, J.F.: Maintaining knowledge about temporal intervals. Commun. ACM **26**(11), 832–843 (1983)
2. Barba, I., Lanz, A., Weber, B., Reichert, M., Del Valle, C.: Optimized time management for declarative workflows. In: Bider, I., Halpin, T., Krogstie, J., Nurcan, S., Proper, E., Schmidt, R., Soffer, P., Wrycza, S. (eds.) BPMDS/EMMSAD-2012. LNBIP, vol. 113, pp. 195–210. Springer, Heidelberg (2012). https://doi.org/10.1007/978-3-642-31072-0_14

3. Barba, I., Weber, B., del Valle, C., Jimenez-Ramirez, A.: User recommendations for the optimized execution of business processes. Data Knowl. Eng. **86**, 61–84 (2013)
4. Burattin, A., Maggi, F.M., Sperduti, A.: Conformance checking based on multi-perspective declarative process models. Expert Syst. Appl. **65**, 194–211 (2016)
5. de Leoni, M., Maggi, F.M., van der Aalst, W.M.P.: Aligning event logs and declarative process models for conformance checking. In: Barros, A., Gal, A., Kindler, E. (eds.) BPM 2012. LNCS, vol. 7481, pp. 82–97. Springer, Heidelberg (2012). https://doi.org/10.1007/978-3-642-32885-5_6
6. Dumas, M., van der Aalst, W.M.P., ter Hofstede, A.H. (eds.): Process-Aware Information Systems: Bridging People and Software Through Process Technology. Wiley-Interscience, Hoboken (2005)
7. Haisjackl, C., Barba, I., Zugal, S., Soffer, P., Hadar, I., Reichert, M., Pinggera, J., Weber, B.: Understanding declare models: strategies, pitfalls, empirical results. Softw. Syst. Model. **15**(2), 325–352 (2016)
8. Heuck, E., Hildebrandt, T., Kiærulff Lerche, R., Marquard, M., Normann, H., Iven Strømsted, R., Weber, B.: Digitalising the general data protection regulation with dynamic condition response graphs. In: Proceedings of BPM, pp. 124–134 (2017)
9. Hildebrandt, T., Mukkamala, R.R., Slaats, T., Zanitti, F.: Contracts for cross-organizational workflows as timed dynamic condition response graphs. J. Logic Algebraic Program. **82**(5), 164–185 (2013)
10. Jiang, Y., Xiao, N., Zhang, Y., Zhang, L.: A novel flexible activity refinement approach for improving workflow process flexibility. Comput. Ind. **80**, 1–15 (2016)
11. Jiménez-Ramírez, A., Barba, I., del Valle, C., Weber, B.: Generating multi-objective optimized business process enactment plans. In: Salinesi, C., Norrie, M.C., Pastor, Ó. (eds.) CAiSE 2013. LNCS, vol. 7908, pp. 99–115. Springer, Heidelberg (2013). https://doi.org/10.1007/978-3-642-38709-8_7
12. Jimenez-Ramirez, A., Barba, I., Fernandez-Olivares, J., del Valle, C., Weber, B.: Time prediction on multi-perspective declarative business processes. Knowledge and Information Systems, pp. 1–31 (in press)
13. Lanz, A., Reichert, M., Weber, B.: Process time patterns: a formal foundation. Inf. Syst. **57**, 38–68 (2016)
14. Lanz, A., Weber, B., Reichert, M.: Time patterns for process-aware information systems. Requirements Eng. **19**(2), 113–141 (2014)
15. Maggi, F.M.: Discovering metric temporal business constraints from event logs. In: Johansson, B., Andersson, B., Holmberg, N. (eds.) BIR 2014. LNBIP, vol. 194, pp. 261–275. Springer, Cham (2014). https://doi.org/10.1007/978-3-319-11370-8_19
16. Maggi, F.M., Dumas, M., García-Bañuelos, L., Montali, M.: Discovering data-aware declarative process models from event logs. In: Daniel, F., Wang, J., Weber, B. (eds.) BPM 2013. LNCS, vol. 8094, pp. 81–96. Springer, Heidelberg (2013). https://doi.org/10.1007/978-3-642-40176-3_8
17. Maggi, F.M., Westergaard, M.: Using timed automata for a Priori warnings and planning for timed declarative process models. Int. J. Coop. Inf. Syst. **23**(1), 1440003 (2014)
18. Mans, R.S., Russell, N.C., van der Aalst, W.M.P., Bakker, P.J.M., Moleman, A.J.: Simulation to analyze the impact of a schedule-aware workflow management system. Simulation **86**(8–9), 519–541 (2010)
19. Mertens, S., Gailly, F., Poels, G.: Enhancing declarative process models with DMN decision logic. In: Gaaloul, K., Schmidt, R., Nurcan, S., Guerreiro, S., Ma, Q. (eds.) CAISE 2015. LNBIP, vol. 214, pp. 151–165. Springer, Cham (2015). https://doi.org/10.1007/978-3-319-19237-6_10

20. Montali, M., Maggi, F.M., Chesani, F., Mello, P., van der Aalst, W.M.P.: Monitoring business constraints with the event calculus. ACM Trans. Intell. Syst. Technol. 5(1), 17 (2013)

21. Mulyar, N., Pesic, M., van der Aalst, W.M.P., Peleg, M.: Declarative and procedural approaches for modelling clinical guidelines: addressing flexibility issues. In: ter Hofstede, A., Benatallah, B., Paik, H.-Y. (eds.) BPM 2007. LNCS, vol. 4928, pp. 335–346. Springer, Heidelberg (2008). https://doi.org/10.1007/978-3-540-78238-4_35

22. Ovarian cancer (CG122) (2011). http://www.nice.org.uk/CG122. Accessed 11 Jan 2017

23. Pesic, M.: Constraint-based workflow management systems: shifting control to users. Ph.D. thesis, Eindhoven University of Technology, Eindhoven (2008)

24. Reichert, M., Weber, B.: Enabling Flexibility in Process-Aware Information Systems. Springer, Heidelberg (2012). https://doi.org/10.1007/978-3-642-30409-5

25. Rovani, M., Maggi, F.M., de Leoni, M., van der Aalst, W.M.P.: Declarative process mining in healthcare. Expert Sys. Appl. 23, 9236–9251 (2015)

26. Schönig, S., Di Ciccio, C., Maggi, F.M., Mendling, J.: Discovery of multi-perspective declarative process models. In: Sheng, Q.Z., Stroulia, E., Tata, S., Bhiri, S. (eds.) ICSOC 2016. LNCS, vol. 9936, pp. 87–103. Springer, Cham (2016). https://doi.org/10.1007/978-3-319-46295-0_6

27. Schultheiß, B., Meyer, J., Mangold, R., Zemmler, T., Reichert, M.: Designing the processes for ovarian cancer surgery. Technical report DBIS-6, University of Ulm (1996)

28. van der Aalst, W.M.P., Pesic, M., Schonenberg, M.H.: Declarative workflows: balancing between flexibility and support. Comput. Sci. Res. Dev. 23(2), 99–113 (2009)

29. van Hee, K., Schonenberg, H., Serebrenik, A., Sidorova, N., van der Werf, J.M.: Adaptive workflows for healthcare information systems. In: ter Hofstede, A., Benatallah, B., Paik, H.-Y. (eds.) BPM 2007. LNCS, vol. 4928, pp. 359–370. Springer, Heidelberg (2008). https://doi.org/10.1007/978-3-540-78238-4_37

30. Weske, M.: Business Process Management: Concepts, Languages, Architectures. Springer, Heidelberg (2007). https://doi.org/10.1007/978-3-642-28616-2

31. Westergaard, M., Maggi, F.M.: Looking into the future. In: Meersman, R., Panetto, H., Dillon, T., Rinderle-Ma, S., Dadam, P., Zhou, X., Pearson, S., Ferscha, A., Bergamaschi, S., Cruz, I.F. (eds.) OTM 2012. LNCS, vol. 7565, pp. 250–267. Springer, Heidelberg (2012). https://doi.org/10.1007/978-3-642-33606-5_16

32. Zeising, M., Schönig, S., Jablonski, S.: Towards a common platform for the support of routine and agile business processes. In: Proceedings of CollaborateCom, pp. 94–103 (2014)

33. Zugal, S., Soffer, P., Haisjackl, C., Pinggera, J., Reichert, M., Weber, B.: Investigating expressiveness and understandability of hierarchy in declarative business process models. Softw. Syst. Model. 14(3), 1081–1103 (2015)

Formal Executable Theory
of Multilevel Modeling

Mira Balaban[1(✉)], Igal Khitron[1], Michael Kifer[2], and Azzam Maraee[1]

[1] Computer Science Department,
Ben-Gurion University of the Negev, Beersheba, Israel
{mira,khitron,mari}@cs.bgu.ac.il
[2] Computer Science Department,
Stony Brook University, Stony Brook, NY, USA
kifer@cs.stonybrook.edu

Abstract. Multi-Level Modeling (MLM) conceptualizes software models as layered architectures of sub-models that are inter-related by the instance-of relation, which breaks monolithic class hierarchies midway between subtyping and interfaces. This paper introduces a formal theory of MLM, rooted in a set-theoretic semantics of class models. The MLM theory is validated by a provably correct translation into the FOML executable logic. We show how FOML accounts for inter-level constraints, rules, and queries. In that sense, FOML is an organic executable extension for MLM that incorporates all MLM services. As much as the page budget permits, the paper illustrates how multilevel models are represented and processed in FOML.

Keywords: Multi-level modeling · Herbrand semantics · Class facet
Object facet · Executable logic

1 Introduction

Multilevel system modeling (MLM) views the enterprise as a layered collection of models that are inter-related by the *instance-of* (or *membership*) relation among objects and classes. The grounds for this approach are both philosophical and pragmatic. On philosophical grounds, researchers have been arguing that faithful modeling of real world domains cannot be restricted by the standard two-layer architecture of the OMG meta-modeling approach. They claim that natural domains have multiple levels of classification, and an artificial restriction to two layers yields models that are too weak [4,15]. On pragmatic grounds, researchers have argued that a multilevel architecture of models simplifies the management and evolution of complex software [23].

An important advantage of multilevel models is that in a monolithic class hierarchy structure every change affects the entire hierarchy. The conventional approach in software systems is to break monolithic hierarchies using interfaces

J. Krogstie and H. A. Reijers (Eds.): CAiSE 2018, LNCS 10816, pp. 391–406, 2018.
https://doi.org/10.1007/978-3-319-91563-0_24

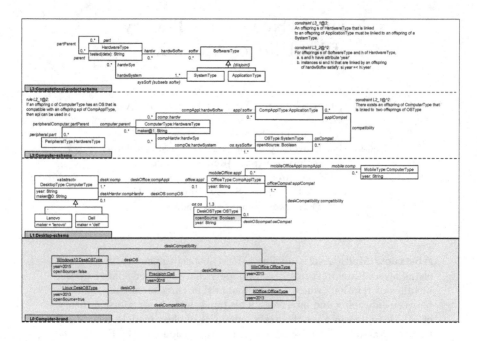

Fig. 1. A multilevel model of computer products

and delegation. In contrast, in MLM, class hierarchies are broken by the instance-of relation — midway between subtyping and interfaces. Thus, multiple modeling layers provide modular representation with controlled limited inheritance.

Figure 1 describes a multilevel model of a computational product. The model is split into disjoint layers L0, L1, L2, L3, where L0, `Computer-brand`, describes desktops brands with their hardware and software components; level L1 is `ComputerKind-schema`, the domain of computer kinds, their hardware and software; level L2 is `Computer-schema`, and level L3 describes the `Computational-Product-schema` for hardware and software components. Level L0 is an *object model* that serves as a partial instance for the *class model* at level L1. (L0 is shaded to indicate that it is not a class model and is unlike L1–L3 in this respect.) In a sense to be detailed in Sect. 2, L1 itself also serves as a partial instance for the class model at level L2. Similarly, L2 serves as an instance for the class model at level L3.

Each layer L_i except the first and the last plays a dual role: as an instance of the class model at level $i+1$ — the *object facet* role, and as a class model at level i — the *class facet* role. In our example, L0 has only the role of an object facet for level 1 (as an instance for L1), L3 has only the class facet role (for level 3). In general, adjacent layers in the multilevel architecture are related by the standard *instance-of* relationship between a class model and its instance object models, while within each class model layer, classes are organized via subclass and general association relationships. Hence, some classes and associations in layer i play the role of objects and links for the class model in layer $i + 1$. This is denoted using the standard notation $o : C$. Such classes are called *clabjects* [4,23], i.e., **cla**sses that are also **object**s, and such associations are called *assoclinks*, i.e., **assoc**iations that are also **link**s. For example, in level L1, the clabjects are DesktopType, OfficeType, CompOSType and MobileType, and the assoclinks are deskOffice, deskOS, mobileOfficeAppl and deskCompatibility. The clabjects and assoclinks of *ComputerKind-schema* in level L1 form a partial legal instance of the class model *Computer-schema* in level L2. Properties (association-ends) of assoclinks are *renamed* when moving up a layer, denoted with the object instantiation notation: $level_i prop : level_{i+1} prop$, as in $desk : comp$ in level L1.

The tradition of multilevel modeling enables assignment of a *potency* numerical attribute to elements of the model [24,28]. The intuition is that the potency of an element specifies the maximum number of allowed consecutive instantiations, which can be smaller than the level of the element. Potency 0 for a class (or association) means that it is not instantiated by a clabject (assoclink) in the immediate lower level. An attribute with potency 0 is not inherited by instance clabjects of the class of the attribute. The default potency of an element is its level, and is not noted explicitly in the diagram. In Fig. 1, potency is denoted with the "@n" sign. Attribute maker of class ComputerType in level L2 has potency 1 and, indeed, its instantiation ends in level L1. Gray background for an attributes indicates that they are inherited (cf. *openSource* in L1).

The schemas in a multilevel model can be constrained and extended by inter-layer *constraints* and *inference rules* (so called "deep" constraints and rules). These constraints often involve the notion of an *offspring*. An offspring of class C is any class or object that is connected to C by a chain of *instance-of* and *subclass* relationships. Figure 1 includes one *deep rule* and three *deep constraints*; they are specified at some levels and affect several layers below them. *Computational-product-schema* includes the constraint *L3_1* of potency 3. This means that it affects the schemas at levels L2, L1, and L0. It states that for an offspring h of HardwareType, if h is linked to an offspring of ApplicationType, then it must also be linked to some offspring of SystemType. Indeed, the MobileType clabject at L1 does not satisfy this constraint, and *ComputerKind-schema* is a partial instance of *Computer-schema*, that can be complemented into a legal instance that satisfies this constraint.

Single-potency [28] of degree n is a different kind of constraint, denoted with "@^n" sign: it is specified at some level i but constrains the level $i - n$. Typically this is done for convenience, as specifying a constraint at a higher level can be more succinct. For instance, in Fig. 1, the constraint *L3_2@^2* is specified at level L3, but applies to L1-classes *DesktopType*, *MobileType*, *OfficeType*, and *DeskOSType* — the offsprings of the L3-classes *HardwareType* and *SoftwareType* that are connected to these L3-classes by chains of *instance-of* relationships of length 2. One could have specified the constraint *L2_2@^2* directly at level L1, but then it would have to be repeated for each of the above four offsprings.

Inference rules provide a powerful representation mechanism that can derive intensional information that is not explicit. Rule *L2_1* has potency 2 and states that if a computer object c is an offspring of `ComputerType`, and has an operating system, which is compatible with an application a, then a is an application of c. With this rule, we can *derive* an L0-level link of the L1-level association `deskOffice` that links the L0-objects `KOffice` and `Precision`.

The overall architecture of a multilevel model organizes schemas in a layered architecture called an *ontological dimension*. Within each schema, the subclass relation is used, while the instance-of relation is used between adjacent schemas. A multilevel model can include several *overlapping* ontological dimensions. Due to space limitation, our formalization deals only with a single ontological dimension. Traditionally, MLM refers to the *linguistic modeling dimension* as a syntactic definition of models. This aspect is covered in Sect. 2.

The main contribution of this paper is a novel formal executable theory for MLM based on a model-theoretic semantics of class models [8,9]. Class models form multilevel models with the help of *Herbrand instances*, in which the syntactic symbols comprise the semantic domain. Our formal theory relies on *direct semantics* for multilevel models and accounts for complex model interactions.

The second contribution is a provably correct translation of the aforesaid MLM theory into the executable logic language FOML [7], which is based on an underlying theory of unrestricted chains of instance-of and subtyping relationships, combined with path expressions.[1] FOML enables direct, seamless encoding of the MLM theory, and is inspired by the industrial multilevel *Ink* project [1].

The third contribution has to do with the correctness of multilevel models and their intra-level and inter-level interactions. We define *consistency* and *finite-satisfiability* as natural extensions of their class model analogies, and show that intra-layer correctness can be checked for each schema independently. Moreover, FOML is capable of validating deep inter-level MLM dependencies.

Paper organization: Sect. 2 introduces a formal abstract syntax and model-theoretic semantics for multilevel models. Section 3 deals with correctness analysis Sect. 4 describes the encoding in FOML, reasoning and proving correctness, Sect. 5 discusses related work and Sect. 6 concludes the paper.

2   Multilevel Models–A Set-Theoretic Formalization

A multilevel model is a finite collection of *ontological dimensions*, each being a sequence of *schemas*. A schema consists of a *Herbrand instance* and a *class model*. The schemas in a dimension are defined over a global, sorted, infinite vocabulary $\mathcal{V} = \langle \mathcal{O}, \mathcal{C}, \mathcal{P}, \mathcal{A}, \rangle$ of *object* ($\mathcal{O}$), *class* ($\mathcal{C}$), *property* ($\mathcal{P}$), and *association* ($\mathcal{A}$) symbols. (Due to page limitation, we omit attributes, qualifiers, datatypes, and some constraints.) The sets $\mathcal{O}$ and $\mathcal{C}$ may overlap, and all other sets are disjoint. We define schemas, and combine them into an overall theory of multilevel models.

---

[1]   Other languages, e.g., Telos [26] and RDF [22], also support these relationships.

2.1   Class Models: Abstract Syntax and Set-Theoretic Semantics

Abstract Syntax:[2] A *class model* over a global sorted vocabulary $\mathcal{V} = \langle \mathcal{O}, \mathcal{C}, \mathcal{P}, \mathcal{A} \rangle$ is a tuple $CM = \langle \mathcal{C}_{CM}, \mathcal{P}_{CM}, \mathcal{A}_{CM}, Mappings, Constraints \rangle$, where $\mathcal{C}_{CM} \subseteq \mathcal{C}$, $\mathcal{P}_{CM} \subseteq \mathcal{P}, \mathcal{A}_{CM} \subseteq \mathcal{A}$ are finite sets of class, property and association symbols, respectively, and *Mappings* and *Constraints* are defined below.

Mappings:

- *inverse* : $\mathcal{P}_{CM} \to \mathcal{P}_{CM}$ is a bijection that assigns to *every* property p its unique inverse, denoted p^{-1}, such that $inverse(p) \neq p$, and $(p^{-1})^{-1} = p$.
- *props*: $\mathcal{A}_{CM} \to 2^{\mathcal{P}_{CM}}$ is an injection where $props(a) = \{p, p^{-1}\}$ and every property from $\mathcal{P}_{CM}$ appears in exactly one $props(a)$. If $p \in \mathcal{P}_{CM}$, $assoc(p)$ denotes the association such that $p \in props(assoc(p))$. In the *Computational-Product* schema (level L3, Fig. 1), $props(hardwSoftw) = \{hardw, softw\}$, $softw = hardw^{-1}$, and $assoc(softw) = assoc(hardw) = hardwSoftw$.
- *source*: $\mathcal{P}_{CM} \to \mathcal{C}_{CM}$ and *target*: $\mathcal{P}_{CM} \to \mathcal{C}_{CM}$ are mappings such that $target(p) = source(p^{-1})$. They define the source and target classes of a property. For $a \in \mathcal{A}_{CM}$, if $props(a) = \{p_1, p_2\}$ and $target(p_i) = C_i$, then $classes(a) = \{C_1, C_2\}$. In the aforesaid *Computational-Product* schema, $target(softw) = source(hardw) = SoftwareType, source(softw) = target(hardw) = HardwareType$, and $classes(hardwSoftw) = \{HardwareType, SoftwareType\}$.

Constraints: Class model constraints include *property multiplicities, association-classes, aggregation/composition, class hierarchy, generalization-set, property subsetting, redefinition, union, association-hierarchy, association-class hierarchy* and *xor*. We discuss only the following three:

- **Multiplicity mappings**: $min\colon \mathcal{P}_{CM} \to \mathbb{N} \cup \{0\}$ and $max\colon \mathcal{P}_{CM} \to \mathbb{N} \cup \{*\}$ assign minimum and maximum multiplicities to property symbols so that $min(p) \leq max(p)$ ($*$ denotes positive infinity).
- **Class hierarchy**: is an acyclic binary relation on class symbols in $\mathcal{C}_{CM}$: $C_2 \prec C_1$, means that C_2 is a subclass of C_1. The relation $\prec^+$ is a transitive closure of $\prec$ and $C_2 \preceq^* C_1$ means $C_2 = C_1$ or $C_2 \prec^+ C_1$. In the *Desktop* schema in Fig. 1, *Dell* is a subclass of *DesktopType*, i.e., $Dell \prec DesktopType$.
- **Property subsetting (subproperties)**: is an acyclic binary relation $\prec$ on property symbols:[3] $p_1 \prec p_2$ says that p_1 subsets (is a subproperty of) p_2. It is also required that (i) $source(p_1) \prec^* source(p_2)$, (ii) $target(p_1) \prec^* target(p_2)$, and (iii) $max(p_1) \leq max(p_2)$. In the *Computational-Product* schema of Fig. 1, $sysSoft \prec softw$ means that a *SystemType* which is a *sysSoft* of a *Hardware-Type* element, is also a *softw* of this element.

Compact Symbolic Notation for Associations: It is often convenient to have a compact notation that shows an association along with its properties,

---

[2] A full formalization of the UML class model, appears in [9].
[3] $\prec$ is overloaded for subproperties and subclasses.

classes and multiplicities. We write $a(C_1 \xrightarrow[m_1..M_1 \quad m_2..M_2]{p_1 \qquad p_2} C_2)$—or $a(C_1 \xrightarrow{p_1 \qquad p_2} C_2)$, if multiplicities are irrelevant—to denote an association a such that $props(a) = \{p_1, p_2\}$, $target(p_i) = C_i$, $min(p_i) = m_i$ and $max(p_i) = M_i$. For instance, the compact notation for association $\mathtt{hardwSoftw}$ in the *Computational-Product* schema of Fig. 1, is $hardwSoftw(Hardware\,Type \xrightarrow[0..*]{hardw} \xrightarrow[0..*]{softw} Software\,Type)$.

Semantics: The standard set-theoretic semantics of class models associates such models with *instances* I, which consist of a semantic *domain* and a *denotation mapping* "$.^I$" that assigns meaning to syntactic elements. Given a class model $CM = \langle \mathcal{C}_{CM}, \mathcal{P}_{CM}, \mathcal{A}_{CM}, Mappings, Constraints \rangle$: (1) each class symbol $c \in \mathcal{C}_{CM}$ is mapped to a set c^I of objects in the domain, called the *extension of* c; (2) each association symbol $a \in \mathcal{A}_{CM}$ is mapped to a relationship a^I, called the *extension* of a, between the extensions of the classes of a, i.e., *classes*(a); (3) each property symbol $p \in \mathcal{P}_{CM}$ is mapped to a multivalued function $p^I : source(p)^I \to target(p)^I$, as follows: If $assoc(p) = a$ then for each $e \in source(p)^I$, $p^I(e) = \{e' \mid (e, e') \in a^I\}$, i.e., p^I is the projection of a^I on $source(p)^I$.

An *object* of a class model CM with respect to an instance I is an element in the domain of I that belongs to the extension of some class. The set of *objects* of CM with respect to I, denoted $objs_I(CM)$, is the union of all extensions in I of classes of CM, i.e., $\cup_{c \in \mathcal{C}_{CM}} c^I$. A *link* of CM with respect to an instance I is an element of the extension of some association a. A link includes a pair of objects o_1, o_2 in the domain of I, an identifier of the association to which it belongs, and the property roles of the above two objects. We denote links as $a(o_1 \xrightarrow{p_1 \quad p_2} o_2)$, i.e., as labeled edges between nodes o_1, o_2, where $props(a) = \{p_1, p_2\}$, and $o_1 \in p_1^I(o_2), o_2 \in p_2^I(o_1)$. For an association symbol $a \in \mathcal{A}_{CM}$, $links_I(a) \overset{def}{=} \{a(o_1 \xrightarrow{p_1 \quad p_2} o_2) \mid (o_1, o_2) \in a^I, o_1 \in p_1^I(o_2), o_2 \in p_2^I(o_1)\}$. The set $links_I(CM)$ of *all* links in CM with respect to I is the union of all links in I of all associations of CM, $\cup_{a \in \mathcal{A}_{CM}} links_I(a)$.

Legal and Herbrand Instances: An instance I of a class model is *legal*, denoted $I \models CM$, if its denotation mapping satisfies the class model constraints:

- **Multiplicity:** $min(p) \leq |p^I(e)| \leq max(p)$ for each $e \in source(p)^I$.
- **Class-hierarchy:** If $C_1 \prec C_2$ then $C_1^I \subseteq C_2^I$.
- **Property subsetting:** If $p_1 \prec p_2$ and $e \in source(p_1)^I$ then $p_1^I(e) \subseteq p_2^I(e)$.

A *partial instance* is one whose denotation mapping can be extended to yield a legal instance. A *Herbrand instance*[4] of a class model CM over a global vocabulary $\mathcal{V} = \langle \mathcal{O}, \mathcal{C}, \mathcal{P}, \mathcal{A} \rangle$ is an instance of CM over the domain $\mathcal{O}$. Herbrand instances are often written using set notation, that explicitly lists the object members (from $\mathcal{O}$) of classes and the object pairs for associations: $\langle (\{o_1^i, \ldots, o_{n_i}^i\} = C_i)^{C_i \in \mathcal{C}_{CM}}, (\{(o_1^i, u_1^i), \ldots, (o_{n_i}^i, u_{n_i}^i)\} = a_i)^{a_i \in \mathcal{A}_{CM}} \rangle$. This writing saves explicit specification of property mappings and of empty extensions.

---

[4] By analogy with Herbrand interpretations in classical logic.

Example 1. *As explained in the introduction, levels L0–L2 in Fig. 1 play the role of instances for the class models at levels L1–L3. The notion of Herbrand instances formally captures this very idea. Specifically, L0 corresponds to the following Herbrand instance H_1 for the class model at level L1:* $H_1 = \{$ *{Windows10, Linux}* $=$ *DeskOSType,* *{WinOffice, KOffice}* $=$ *OfficeType,* *{Precision}* $=$ *Dell, {(Precision, WinOfffice)}* $=$ *deskOffice, {(Windows10, Precision), (Linux, Precision)}* $=$ *deskOS, {(Windows10, WinOfffice), (Linux, KOfffice)}* $=$ *deskCompatibility }.*
H_1 is a partial (legal) instance since *KOffice* *is an instance of* `OfficeType` *but has no* `deskOffice` *link.*

By analogy, we can construct a legal Herbrand instance H_2 for the class model at level L2. Unlike H_1, it is a complete *legal instance of this class model. However, when constructing a legal instance H_3 for the class model at level L3 out of the L2 level class model, we again get only a* partial *instance, as shown next:*
$H_3 = \{$ *{ComputerType, PeripheralType}* $=$ *HardwareType, {CompApplType}* $=$ *ApplicationType, {OSType}* $=$ *SystemType, {CompApplType, OSType}* $=$ *SoftwareType, {(ComputerType, CompApplType)}* $=$ *hardwSoftw, {(ComputerType, OSType)}* $=$ *hardwSystem}.*
H_3 is a partial *Herbrand instance for level L3 because the subsetting constraint sysSoft $\prec$ softw in level L3 implies that the extension hardwSoftw^{H_3} must include the extension of hardwSystem^{H_3} but, by the above, it does not.* □

Semantic Relationships: An instance I of a class model CM is *empty* if all class extensions are empty and it is *infinite* if some class extension is infinite. A class model constraint *constr* is *entailed* by a class model CM, denoted $CM \models$ *constr*, if it holds in every legal instance of CM. A class model CM_2 is *entailed* by class model CM_1, denoted $CM_1 \models CM_2$, if every legal instance of CM_1 is a legal instance of CM_2. Class models are *equivalent*, denoted $CM_1 \equiv CM_2$, if they have the same set of legal instances.

The following claim says that, for reasoning purposes, it is sufficient to consider legal Herbrand instances only.

Claim. For a class model CM and a class model constraint *constr*, *constr* holds in every legal Herbrand instance of CM, denoted $CM \models_H$ *constr*, if and only if it is *entailed* by CM, i.e., $CM \models$ *constr*.

2.2   Multilevel Models: Abstract Syntax and Model-Theory

First we introduce the notion of *mediation*, which connects a class model CM to an immediately higher class model CM'.

Definition 1. *Given a pair of class models over the same vocabulary:* $CM = \langle \mathcal{C}_{CM}, \mathcal{P}_{CM}, \mathcal{A}_{CM}, \text{Mappings}, \text{Constraints} \rangle$, *and* $CM' = \langle \mathcal{C}_{CM'}, \mathcal{P}_{CM'}, \mathcal{A}_{CM'}, \text{Mappings}', \text{Constraints}' \rangle$. *Let H' be partial Herbrand instance of CM'. Then, H' is a Herbrand mediator of type (CM, CM') if the objects and links of H' are classes and associations of CM, respectively. Formally:*

1. $objs_{H'}(CM') \subseteq \mathcal{C}_{CM}$;
2. *There is a 1:1 mapping link2assoc from links$_{H'}(CM')$ to $\mathcal{A}_{CM}$, that satisfies:*
   $$link2assoc(\ a'(o_1 \xrightarrow{\ p_1' \quad p_2' \ } o_2)\ ) = a, \text{ where } a \in \mathcal{A}_{CM} \text{ and } classes(a) = \{o_1, o_2\}.$$
   The mapping link2assoc extends to properties as follows: $link2assoc(p_i') = p_i$, assuming $props(a) = \{p_1, p_2\}$ and $target(p_i) = o_i$. That is, link2assoc maps links in H' to associations of CM, turning end-objects into end-classes and renaming association and properties.

The partiality of the Herbrand mediator implies that not all elements in a higher level must be instantiated in a lower level. The *potency* assignment, introduced below, is an explicit mechanism for controlling inter-level instantiation.

Abstract Syntax: A multilevel model over a global sorted vocabulary $\mathcal{V} = \langle \mathcal{O}, \mathcal{C}, \mathcal{P}, \mathcal{A} \rangle$ is a finite set $\{\Theta^i\}_{1 \leq k \leq m}$, of *ontological* modeling dimensions. Each modeling dimension is a finite sequence of *schemas* $S_1, ..., S_n$, where each S_i is a pair $\langle H_i, CM_i \rangle$. The CM_i components are *class models* over $\mathcal{V}$; for $1 < i \leq n$, the H_i components are *Herbrand mediators* of type (CM_{i-1}, CM_i); H_1 is a partial legal Herbrand instance of CM_1. The sets of classes in S_is are pairwise disjoint.

Returning to Fig. 1, Example 1 shows the legal partial instances H_1, H_2, H_3 for the class models L1, L2, L3 and illustrates how these instances were constructed out of the lower-level models L0, L1, and L2. The ontological modeling dimension depicted in the figure if formally represented as a sequence of these schemas: $\langle H_1, L1 \rangle$, $\langle H_2, L2 \rangle$, $\langle H_3, L3 \rangle$.

Semantics: A *legal instance* of a multilevel ontological dimension $(S_i = \langle H_i, CM_i \rangle)_{1 \leq i \leq n}$ is a sequence of Herbrand instances $(H_i')_{1 \leq i \leq n}$, such that each H_i' is a legal Herbrand instance of CM_i and, for all $i \geq 1$, H_i' includes H_i. That is, H_i' extends the partial Herbrand instance H_i into a full legal instance of CM_i.

Potency Assignment: The tradition of multilevel modeling enables assignment of a *potency* numerical attribute to elements of the model [24,28]. The intuition is that the potency of an element specifies the maximum number of allowed consecutive instantiations, which can be smaller than the level of the element[5]. We define the potency of elements in two steps:

(i) Let $\mathcal{C}_i$, $\mathcal{A}_i$ be the sets of classes and associations at level i. For each i, the partial function *user_defined_potency*: $\mathcal{C}_i \cup \mathcal{A}_i \to \mathbb{N}$ assigns a natural number $\leq i$ to some classes and associations, subject to *consistency requirements* below.

(ii) *user_defined_potency* is extended to a total *potency* function, which is obtained by propagating the values of *user_defined_potency* and using the *level* values as the default. Due to page limitation, we define *potency* for classes only.

1. For each class C' in a class model CM_i such that *user_defined_potency*(C') has a value, set *potency*$(C') =$ *user_defined_potency*(C').

---

[5] Other known forms of potency are constraints on off-springs instead of instantiations.

2. For a class C' in CM_i for which $potency(C') = k$ where $k > 0$ and a class C in CM_{i-1} (i.e., $C \in C'^{H_i}$), set $potency(C) = k - 1$.
3. For every class C in a class model CM_i, $i > 0$, for which $potency(C)$ is undefined, set $potency(C) = level(C)$.

A potency function is **inconsistent** if it assigns more than one value to some class, or if a 0-potency class has direct instances. Formally:

- *Direct instantiation of a 0-potency class*: Let C' be a class in a class model CM_i (level i) with $potency(C') = 0$. The potency function is inconsistent if C' has a direct instance (in level $i - 1$), i.e., if there is a class C in CM_{i-1} in level $i - 1$ ($C \in C'^{H_i}$) with no intermediate class C'' (in level i) between C and C' (i.e., $C'' \prec C' \in CM_i$ and $C \in C''^{H_i}$).
- *Contradiction*: $potency(C)$ has more than one value for some clabject C. This may happen if *user_defined_potency* is over-specified, for example in the presence of multiple inheritance.

3   Analysis of Multilevel Models

Correctness of a multilevel model depends on *intra-level* correctness, which refers to the correctness of each class model and its object facet, and on *inter-level* correctness, which is determined by deep constraints and their interaction with the class models they constrain. The two main correctness problems in class models are *consistency* [10] and *finite satisfiability* [8]. Consistency deals with necessarily empty classes and finite satisfiability with necessarily infinite classes. Detection of consistency and finite satisfiability in full class models are EXPTIME-complete problems. Correctness for multilevel models is defined by extending the class model analogy.

Consistency and Finite Satisfiability of Schemas:

Definition 2. *A schema S_i in a multilevel ontological dimension ($S_i = \langle H_i, CM_i \rangle)_{1 \le i \le n}$ is consistent (or satisfiable) if for every class C of CM_i there is a legal Herbrand instance H'_i of CM_i that extends (includes) H_i, in which the extension of C is not empty, i.e., $C^{H'} \ne \emptyset$.[6] A schema S_i is finitely satisfiable if for every class C of CM_i there is a legal finite Herbrand instance H'_i of CM_i that extends (by inclusion) H_i, in which the extension of C is not empty.*

Proposition 1. *For a schema $S_i = \langle H_i, CM_i \rangle$ in a multilevel ontological dimension: S_i is consistent (respectively, finitely satisfiable) if and only if there is a legal (respectively, finite) Herbrand instance H'_i of CM_i that extends H_i, in which the extension of all classes of CM_i are not empty.*

---

[6] Weaker definitions are possible, following [2].

The proof is based on the closure under disjoint union of legal instances [8]. Hence, can be checked first at the class component, followed by checking the Herbrand instance component for being extendable.

Consistency and Finite Satisfiability of a Multilevel Ontological Dimension: Correctness in a multilevel dimension can be affected by inter-layer constraints. For example, in level L1 of Fig. 1, the association cycle deskOffice, deskCompatibility, deskOS includes *redundant* cardinalities, i.e., cardinalities that cannot be realized in any finite legal instance of the class model. In particular, for the *os* property, only cardinality 1 can be used in finite legal instances, while cardinalities 2 and 3 force instances to be infinite [25]. Therefore, constraint *L2_1*, which affects level L0 and requires existence of an offspring of ComputerType that is linked to two offsprings of OSType, cannot hold in any finite legal instance, thereby violating finite satisfiability.

We identify three possible correctness aspects in a multilevel dimension $\Theta = (S_i = \langle H_i, CM_i \rangle)_{1 \leq i \leq n}$:

1. **Weak local correctness:** Θ is *consistent* (respectively, *finitely satisfiable*) if for every class C in a schema S_i, there is a legal (respectively, finite) instance of Θ in which the extension of C is not empty.
2. **Strong local correctness:** Θ is *consistent* (respectively, *finitely satisfiable*) if for every schema S_i, there is a legal (respectively, finite) instance of Θ in which the extension of all classes of S_i are not empty.
3. **Global correctness:** Θ is *consistent* (respectively, *finitely satisfiable*) if there is a legal (respectively, finite) instance of Θ such that for every schema S_i the extension of all classes of S_i are not empty.

By Proposition 1, weak local correctness implies strong local correctness, so we will deal just with *local correctness*. However, local correctness does not imply global correctness since, as noted above, inter-level constraints may interact.

Validation and analysis of a multilevel model involves checking global and local correctness as well as instance completion, testing, and query answering. Global validation requires correctness checking techniques that operate in a multilevel domain. The FOML implementation described in Sect. 4, checks local and global validation, performs testing, and enables query answering.

4   Multilevel Modeling in FOML

FOML [18] is intended to support model-level activities, such as constraints (extending UML diagrams), dynamic compositional modeling (intensional and transformational), analysis and reasoning about models, model testing, and meta-modeling. In [6] FOML was suggested for multilevel modeling based on its uniform treatment of types and instances and of both abstract syntax and semantics. Moreover, as an executable (i.e., operational) logic language, FOML can express and reason about multiple crosscutting multilevel dimensions, including instantiation constraints. In this paper, we extend this argument by claiming that the FOML encoding is provably correct.

FOML is a semantic layer on top of a compact logic rule language of *guarded path expressions*, called *PathLP* [7,19], an adaptation of a subset of F-logic [21]. Overall, PathLP provides reasoning services over unrestricted instance-of, subtype, and object-link relations, while FOML provides the modeling framework.

4.1   PathLP

The main syntactic constructs of PathLP are *path expressions* for *objects* and *types*, *membership* and *subtyping* relations, *facts, rules, queries*, and *constraints*. **Path expressions** have the general form of `root.link[guard]`, where `root`, `link`, and `guard` are terms that denote semantic entities, and the link applied at `root` evaluates to a set that contains `guard`. For example, referring back to Fig. 1, the `os` values for the object `Precision` can be defined like this:

```
Precision.os[Linux,Windows10];
```

This expression states that `Linux` and `Windows10` belong to the set of OS's for Dell's computer model `Precision`. The target class and multiplicity of property `os` is further constrained at level L1 by a **type path expression**:

```
DesktopType!os[DeskOSType]{1..3};
```

This says that class `DesktopType` has a property `os` that yields objects that must belong to class `DeskOSType` and, for each `DesktopType` object, the property `os` has at least one and at most three objects.

PathLP includes two semantically supported special relations `::` for subtyping and `:` for membership. For example,

```
SystemType::SoftwareType;   OSType:SystemType;   DeskOSType:OSType;
```

are **facts** that declare `SystemType` as a subclass of `SoftwareType`, and `OSType` and `DeskOSType` as members of the classes `SystemType` and `OSType`, respectively. Subtyping is interpreted as a partial order, and PathLP supports the usual properties of subtyping and membership.

Links in path expressions can take arguments. This is used to account for methods. For instance, in Fig. 1, the `HardwareType` class has a method `tested`, which takes a *date*-argument and returns a string.

```
ComputerType:HardwareType;   ComputerType.tested(20170412)[failed];
```

Formally, the link `tested(p)`, when applied to the `ComputerType` object with the argument 20170412 (12th of April, 2017), yields a set that contains the symbol `failed`. This value can be further constrained to be unique using a type path expression at level L3:

```
HardwareType!tested(year)[String]{1..1};
```

PathLP uses the regular Logic Programming nomenclature of **facts, rules, queries**, and **constraints**. Rules represent implications and are recognized via the symbol :-, which separates the *head* (conclusion, on the left) from the *body* (premise, on the right). For example, the rule L2_1 in Fig. 1 is represented by two PathLP rules for the classes at levels L2 and L1, as follows:

$$?\text{cT.appl}[?\text{cApplT}] : -\qquad\qquad ?\text{dkT.appl}[?\text{cApplT}] : -$$
$$\quad ?\text{cT} : \text{ComputerType},\qquad\qquad ?\text{dkT} : \text{DesktopType}, \qquad\qquad (1)$$
$$\quad ?\text{cT.os.applCompat}[?\text{cApplT}];\qquad ?\text{dkT.os.officeCompat}[?\text{officeT}];$$

These rules have *variables*, which are symbols prefixed with "?". Later we show how to represent these rules with a single recursive FOML rule, using the multilevel `offSpring` operation, regardless of how many offspring classes are involved.

Queries are recognized by the prefix ?-. They are used to retrieve information that is derivable from the specification. For example, the following query retrieves pairs `?OS, ?Office` of members of classes `DeskOSType` and `OfficeType` that are related via the property chain `deskHardw.office`:

```
?- ?OS:DeskOSType,?OS.deskHardw.office[?Office];
```

The answer to this query is: ⟨Windows10, WinOffice⟩, ⟨Linux, WinOffice⟩, ⟨Windows10, KOffice⟩, ⟨Linux, KOffice⟩. The last two answers come from rule L2_1, which infers the `deskOffice` link between `Precision` and `KOffice`.

Constraints have the prefix "!-" — they specify forbidden states. The following constraints enforce the disjointness of classes `SystemType`, `ApplicationType` in level L3, and restrict the `parentPart` association to be non-circular, i.e., a hardware type cannot be a direct or indirect part of itself.

```
!-?o:SystemType,?o:ApplicationType   !-?hT:HardwareType,?hT.closure(part)[?hT];
```

4.2   FOML

FOML extends PathLP with basic class modeling facilities, similarly to the OMG MOF. The metalevel categories of FOML are *Class, Property, Association, Attribute, Object*, and *Link*. The FOML meta-modeling tools infer the syntactic structure of the class model components, as shown in the model querying examples below. These tools include a rich, extendable library of meta-level operations. Here are some examples of intensional properties:

Property composition: *objects ?o and ?v are related via* `compose(?p1,?mid,?p2)` *if there is a path ?p1.?p2 from ?o to ?v going through ?mid:*

```
?o.compose(?p1,?mid,?p2)[?v]  :- ?o.?p1[?mid].?p2[?v];
```

Transitive closure: `closure(?p)` *is a property that is the transitive closure of ?p:*

```
?o.closure(?p)[?v]  :- ?o.?p[?v];
?o.closure(?p)[?v]  :- ?o.?p.closure(?p)[?v];
```

Property circularity: ?p is circular if its closure is a self link for some object ?o:

```
?p.circular[true]  :- ?o.closure(?p)[?o];
```

Using these meta-modeling facilities, FOML can inspect the syntactic structure of class models and define derived relationships.

Find pairs of properties of the same association:
```
?- ?assoc._props[?prop1,?prop2];
```
Find properties that connect a class to itself:
```
?- ?Class!?prop[?Class];
```

Correctness Summary:

Claim (Correctness of FOML encoding for class models). Let CM be a UML class model, H be a Herbrand instance of CM, and let CM^{FOML}, H^{FOML} be their respective FOML encodings. Then, $H^{FOML} \models CM^{FOML}$ iff $H \models CM$, i.e., H^{FOML} is valid in CM^{FOML} iff H is a legal Herbrand instance of CM. □

The proof is based on a correspondence between Herbrand models of the FOML encoding to Herbrand instances of the class model.

4.3 Multi-FOML: Multilevel Modeling in FOML

MLM requires mediation between adjacent schemas. In Multi-FOML this is done by introducing Herbrand mediators between adjacent class models. A mediator specifies a class model as a partial instance of its immediate higher class model, including mapping of names. For example, the specification of level L1 in Fig. 1, as an Herbrand instance of L2:

```
DesktopType.appl[OfficeType];      MobileType.appl[OfficeType];
DesktopType.os[DeskOSType];        OfficeType.osCompat[DeskOSType];
```

All other components of the Herbrand instance are inferred by FOML, using the meta-modeling tools.

The MLM support of multi-FOML includes computation of levels, potency and offsprings, attribute inheritance, inter-layer (deep) constraint and rule computation, local schema validation, global validation, and the regular FOML support for on the fly querying, testing and model analysis. Due to space limitations we just show a few.

Inter-layer Constraints and Rules. We show Multi-FOML representations for rule L2_1@2 and constraint L3_1@3 from Fig. 1.

```
?c.?can_use[?apl]  :-                           %% rule L2_1@2
   ComputerType.offspring(?N)[?c], ?N=<2,
   CompApplType.offspring(?N)[?apl], os.propspring(?N)[?comp_os],
   applCompat.propspring(?N)[?compat], ?c.?comp_os.?compat[?apl];
```

This rule replaces the two rules in (1). In general, one higher-layer rule with attached potency can replace many rules at lower layers, and this is the primary reason for this kind of abstraction in multilevel modeling.

```
!- HardwareType.offspring(?N)[?hw], ?N=<3,         %% constraint L3_1@3
   ApplicationType.offspring(?N)[?sw],
   softw.propspring(?N)[?swprop], ?hw.?swprop.[?sw],
   not (SystemType.offspring(?N)[?sw2],
       sysSoft.propspring(?N)[?sysprop]
       ?hw.?sysprop[?sw2]);
```

Here not is read as "not exists ?sysprop such that ..." The FOML code is the negative form of the English description of constraint L3_1@3 in Fig. 1.

Claim (Correctness of multi-FOML encoding for MLM). Let Θ be a multilevel dimension and Θ^{FOML} be its FOML encoding. Then Θ^{FOML} is valid iff Θ is globally correct.   □

The proof is based on a correspondence of the Herbrand-based theories.

5   Related Work

Most MLM tools support modeling activities, but do not provide analysis, validation, or other reasoning services, and are not built on formal MLM theory. Semantic approaches to multilevel modeling include the graph-based theory of [28], and axiomatic approaches such as [12] (FOL) or [14] (OCL). The first approach views multilevel models as graphs that simulate the typed-by and conforms-to relations. The levels are implicitly defined by the structure of the graph. Assignment of potency to elements, and graph flattening rules define the semantics of inheritance. In the axiomatic approach, the full multilevel theory is encoded with the help of large sets of axioms. Both approaches rely on indirect translational semantics, and. it is unclear how well they integrate with existing class model analysis tools.

Telos [26], a rich knowledge representation framework implemented in ConceptBase [17], is based on first-order logic. It supports unrestricted instance-of and subtyping chains, but its reasoning capabilities and expressive power are incomparable to those of FOML (neither subsumes the other). RDF [22] is yet another language that supports unrestricted instance-of and subtyping, but its expressive power is too limited for our purposes. Nivel [3] is another logic-programming based language designed for multilevel metamodeling. In comparison, FOML focuses on software modeling, has simpler path-expression based syntax, and greater expressivity. The expressivity gap widens further when FOML is extended with Transaction Logic [11] and HiLog [13], which enable logic reasoning about statecharts and support higher-order introspection. Neither of these languages support the desiderata for multilevel modeling languages listed in [6].

Recently a number of works [16,27] have proposed to use F-logic [20] as a language for MLM. Indeed, FOML is a derivative of F-logic, but without sacrificing expressiveness, FOML is a much smaller language and its constructs are designed specifically for modeling. The aforesaid works do not provide declarative semantics for MLM, which is a main contributions of the present paper.

A comparison of FOML with Alloy and OCL appears in [5]. FOML appears to subsume and extend the *representational* capabilities of both languages. As an analysis tool, FOML provides services not supported by OCL and Alloy tools, including querying, inference, meta-level reasoning, and MLM.

6 Conclusion and Future Work

We presented a formal theory of multilevel modeling, including analysis and correctness. The theory is validated by (1) encoding it in the executable FOML language; (2) showing that it can account for all MLM features; (3) on proving the correctness of the encoding. Specifically, we showed how FOML is used for formal logic specification of class models, their instances, and constraints; querying and reasoning about multilevel models; and correctness analysis. The expressivity of FOML goes beyond UML's class models and has the necessary wherewithals to be a compact, yet expressive underlying framework for MLM. In future work, we will study the pragmatics of multilevel modeling, including methodologies for breaking monolithic class-hierarchies into multilevel structures.

References

1. Acherkan, E., Hen-Tov, A., Lorenz, D., Schachter, L.: The ink language meta-metamodel for adaptive object-model frameworks. In: OOPSLA 2011 (2011)
2. Artale, A., Calvanese, D., Ibáñez-García, A.: Full satisfiability of UML class diagrams. In: Parsons, J., Saeki, M., Shoval, P., Woo, C., Wand, Y. (eds.) ER 2010. LNCS, vol. 6412, pp. 317–331. Springer, Heidelberg (2010). https://doi.org/10.1007/978-3-642-16373-9_23
3. Asikainen, T., Mannisto, T.: Nivel: a metamodelling language with a formal semantics. Softw. Syst. Model. (SoSyM) 8(4), 521–549 (2009)
4. Atkinson, C., Kühne, T.: Rearchitecting the uml infrastructure. ACM TOMACS 12(4), 290–321 (2002)
5. Balaban, M., Bennett, P., Doan, K.H., Georg, G., Gogolla, M., Khitron, I., Kifer, M.: A comparison of textual modeling languages: OCL, Alloy, FOML. In: 16th International Workshop on OCL and Textual Modeling, Models (2016)
6. Balaban, M., Khitron, I., Kifer, M.: Multilevel modeling and reasoning with FOML. In: IEEE CS International Conference on SwSTE 2016 (2016)
7. Balaban, M., Kifer, M.: Logic-based model-level software development with F-OML. In: Whittle, J., Clark, T., Kühne, T. (eds.) MODELS 2011. LNCS, vol. 6981, pp. 517–532. Springer, Heidelberg (2011). https://doi.org/10.1007/978-3-642-24485-8_38
8. Balaban, M., Maraee, A.: Finite satisfiability of UML class diagrams with constrained class hierarchy. ACM TOSEM 22(3), 24:1–24:42 (2013)
9. Balaban, M., Maraee, A.: UML Class Diagram: Abstract syntax and Semantics (2017). https://goo.gl/UJzsjb
10. Berardi, D., Calvanese, D., Giacomo, D.: Reasoning on UML class diagrams. Artif. Intell. 168, 70–118 (2005)

11. Bonner, A.J., Kifer, M.: A logic for programming database transactions. In: Chomicki, J., Saake, G. (eds.) Logics for Databases and Information Systems. SECS, vol. 436, pp. 117–166. Springer, Boston (1998). https://doi.org/10.1007/978-1-4615-5643-5_5

12. Carvalho, V.A., Almeida, J.P.A.: Toward a well-founded theory for multi-level conceptual modeling. Softw. Syst. Model. **17**, 205–231 (2018)

13. Chen, W., Kifer, M., Warren, D.: HiLog: a foundation for higher-order logic programming. J. Log. Program. **15**(3), 187–230 (1993)

14. Gogolla, M., Sedlmeier, M., Hamann, L., Hilken, F.: On metamodel superstructures employing UML generalization features. In: MULTI 2014 (2014)

15. Henderson-Sellers, B.: On the Mathematics of Modelling, Metamodelling, Ontologies and Modelling Languages. Springer, Berlin (2012). https://doi.org/10.1007/978-3-642-29825-7

16. Igamberdiev, M., Grossmann, G., Selway, M., Stumptner, M.: An integrated multi-level modeling approach for industrial-scale data interoperability. Softw. Syst. Model. **17**(1), 269–294 (2018)

17. Jarke, M., Gallersdörfer, R., Jeusfeld, M., Staudt, M., Eherer, S.: ConceptBase - a deductive object base for meta data management. J. Intell. Inf. Syst. **4**, 167–192 (1995)

18. Khitron, I., Balaban, M., Kifer, M.: The FOML Site (2017). https://goo.gl/AgxmMc

19. Khitron, I., Kifer, M., Balaban, M.: PathLP: a path-oriented logic programming language. The PathLP Web Site (2011). https://goo.gl/877S43

20. Kifer, M., Lausen, G., Wu, J.: Logical foundations of object-oriented and frame-based languages. J. ACM **42**(4), 741–843 (1995)

21. Kifer, M., Lausen, G., Wu, J.: Logical foundations of object-oriented and frame-based languages. J. ACM **42**, 741–843 (1995)

22. Klyne, G., Carroll, J.J.: Resource description framework (RDF): concepts and abstract syntax (2006)

23. de Lara, J., Guerra, E., Cuadrado, J.: When and how to use multilevel modelling. ACM TOSEM **24**(2), 12:1–12:46 (2014)

24. de Lara, J., Guerra, E., Cuadrado, J.S.: Model-driven engineering with domain-specific meta-modelling languages. SoSyM **14**(1), 429–459 (2013)

25. Maraee, A., Balaban, M.: Removing redundancies and deducing equivalences in UML class diagrams. In: Dingel, J., Schulte, W., Ramos, I., Abrahão, S., Insfran, E. (eds.) MODELS 2014. LNCS, vol. 8767, pp. 235–251. Springer, Cham (2014). https://doi.org/10.1007/978-3-319-11653-2_15

26. Mylopoulos, J., Borgida, A., Jarke, M., Koubarakis, M.: Telos: representing knowledge about information systems. ACM TOIS **8**(4), 325–362 (1990)

27. Neumayr, B., Schuetz, C.G., Jeusfeld, M.A., Schrefl, M.: Dual deep modeling: multi-level modeling with dual potencies and its formalization in F-Logic. Softw. Syst. Model. **17**(1), 1–36 (2016)

28. Rossini, A., de Lara, J., Guerra, E., Rutle, A., Lamo, Y.: A graph transformation-based semantics for deep metamodelling. In: Schürr, A., Varró, D., Varró, G. (eds.) AGTIVE 2011. LNCS, vol. 7233, pp. 19–34. Springer, Heidelberg (2012). https://doi.org/10.1007/978-3-642-34176-2_4

Data Management and Visualization

An LSH-Based Model-Words-Driven Product Duplicate Detection Method

Aron Hartveld, Max van Keulen, Diederik Mathol, Thomas van Noort,
Thomas Plaatsman, Flavius Frasincar$^{(\boxtimes)}$, and Kim Schouten

Erasmus University Rotterdam, PO Box 1738, 3000 DR Rotterdam, The Netherlands
{344544ah,360314mk,361103dm,346877tn,342789tp}@student.eur.nl,
{frasincar,schouten}@ese.eur.nl

Abstract. The online shopping market is growing rapidly in the 21st century, leading to a huge amount of duplicate products being sold online. An important component for aggregating online products is duplicate detection, although this is a time consuming process. In this paper, we focus on reducing the amount of possible duplicates that can be used as an input for the Multi-component Similarity Method (MSM), a state-of-the-art duplicate detection solution. To find the candidate pairs, Locality Sensitive Hashing (LSH) is employed. A previously proposed LSH-based algorithm makes use of binary vectors based on the model words in the product titles. This paper proposes several extensions to this, by performing advanced data cleaning and additionally using information from the key-value pairs. Compared to MSM, the MSMP+ method proposed in this paper leads to a minor reduction by 6% in the F_1-measure whilst reducing the number of needed computations by 95%.

Keywords: Duplicate detection · Min-hashing
Locality sensitive hashing · Web shop products
Multi-component similarity method

1 Introduction

The amount of Web shops has rapidly grown over the last years and, along with the Web shops, the range of products available online increased drastically. Unfortunately, these Web shops might use different descriptions and representations for the same products. It is possible that a Web shop provides additional information on a certain product, while another Web shop does not. For example, *MediaMarkt* can provide information about the weight of a certain laptop, while Bol.com might give information about the operating system, but not the other way around. Furthermore, the price of the same product can differ between Web shops. Therefore, product information often varies across different Web shops. Getting the best price or finding a comprehensive description of the specifications of a product is very time consuming for customers, especially when they have to

© Springer International Publishing AG, part of Springer Nature 2018
J. Krogstie and H. A. Reijers (Eds.): CAiSE 2018, LNCS 10816, pp. 409–423, 2018.
https://doi.org/10.1007/978-3-319-91563-0_25

search among different Web shops in order to find the same product. It is manually even impossible to find a complete overview of the product specifications and the prices among all available Web shops.

Different state-of-the-art duplicate detection methods exist, such as the Multi-component Similarity Method (MSM) used in [2]. However, these methods are very time consuming [2]. Especially when performing duplicate detection on a large amount of data, the running time can easily explode. Therefore, it is convenient to apply a pre-selection that provides candidate pairs and to only perform a duplicate detection method on this set of candidate pairs. One way of finding the candidate pairs is by using Locality-Sensitive Hashing (LSH) [10]. By applying the pre-selection with LSH, the amount of products that will be compared by the complex and time consuming duplicate detection method is reduced. Therefore, the time needed to find duplicates is also decreased.

The research proposed in this paper extends the one from [6]. In [6] the authors only use model words from the product title for duplicate detection. We propose to add information from the key-value pairs in the product descriptions in order to reduce the sparsity of the considered data and to lower the number of false negatives. Furthermore, we use data cleaning by detecting inconsistencies and improving the quality of our data. Lastly, in [6] the signature matrix is built by explicitly performing permutation over the rows. However, in practice, this is a very time-consuming approach. Therefore, in this paper, we use random hash functions to simulate involved permutations. The proposed method is called the Multi-component Similarity Method with Pre-selection+ (MSMP+), as the original method from [6] is called Multi-Component Similarity Method with Pre-selection (MSMP), stressing thus the link to the old method.

The structure of the paper is as follows, in Sect. 2, we describe related work in the fields of data cleaning, key-value pairs, model words, LSH, and duplicate detection methods. Next, in Sect. 3, a description of the duplicate detection method we propose in this paper is given. In Sect. 4, we evaluate our method and last, in Sect. 5 we give our conclusions and suggestions for possible further research.

2   Related Work

The number of Web shops increased enormously over the last years, which is accompanied by a large growth in online data. In order to integrate these data from different heterogeneous sources, it is crucial to use an efficient duplicate detection method. Duplicate detection methods for online data are discussed widely in previous literature. Fetterly et al. proposed an algorithm that detects duplicate Web pages by using Web crawlers, and tracked how clusters of duplicate documents evolve over time [8]. Furthermore, Henzinger did a study on different algorithms that identify duplicate Web pages [9]. She compared Broder et al.'s shingling algorithm, in which the similarity of a subset of shingles is computed based on the Jaccard similarity between two documents [3], and Charikar's random projection algorithm [4], and identified the shortages of these two approaches. She proposed an algorithm that combines the quality of both algorithms and that improved the precision compared to the performances of the algorithms of [3] and [4], individually.

Before applying a duplicate detection method it is crucial to define the input of the algorithm as binary vectors that represent the different products. In academic literature, there are several papers that introduce model words, for example [1,6]. Model words are defined as words that contain both numeric and alphabetic/punctuation tokens e.g. 12″. These types of tokens often give valuable information for the duplicate detection, as they usually represent some unique aspects of a product. The model words are the input for an algorithm that creates a binary vector representation. In [6] only model words from the title are used to create binary vectors representing the products. These binary vectors are employed to find the candidate pairs by the LSH method. [6] does not use any form of data cleaning. However, data cleaning by removing errors and inconsistencies will increase the correctness of the data and avoid wrong conclusions, as argued in [12]. Another limitation of the work proposed in [6] is that this method does not use any information provided by the key-value pairs, while de Bakker et al. [1] show that the key-value pairs contain relevant information that can be used for the duplicate detection.

de Bakker et al. [1] introduce the Hybrid Similarity Method (HSM) for duplicate detection and extract model words from title and model words from the key-value pairs, which leads to superior duplicate detection results. An example of a key-value pair is: ('Weight', '20.5 lbs'). It is suggested that one should only use the product attribute values ('20.5 lbs') and disregards the keys ('Weight'). For model words in the key-value pairs, de Bakker et al. [1] use a broader definition of model words. This definition also includes purely numeric tokens in addition to the mixed numeric/non-numeric tokens, e.g., '41.7' from '41.7 inches'. de Bakker et al. [1] show that model words of the key-value pairs can be useful when applying duplicate detection. Therefore, in our work we extend the binary vectors by also taking the information of the key-value pairs into account, and thus reducing the sparsity of the vectors.

After obtaining the binary vectors, LSH is a useful tool to reduce the dimensionality of the binary vectors [10]. This technique reduces the dimensionality of the data sets by mapping similar items of the signature matrix, constructed by minhashing, into the same buckets with a high probability. The LSH technique maximizes the probability of finding similar items in the same bucket and can be compared to the nearest neighbour search clustering algorithm [6].

LSH uses minhashing, in which the probability of a duplicate detection is maximized given a certain desired Jaccard similarity of two different sets. Cohen et al. [5] propose a minhashing function that defines a signature matrix that is constructed by randomly permuting the rows of the characteristic matrix and selecting for each column the first row index in which the column has a 1. Documents with similar signatures can be considered as similar and therefore become duplicate candidates. A disadvantage of using random permutations is that it is computationally intensive and the space that is required to store the permutations is large [5]. The same approach is used by van Dam et al. [6], in which the same minhashing algorithm is used to define the signature matrix.

By combining the LSH algorithm with minhashing, duplicate detection becomes quick without losing a lot on recall and precision, because it retains the high Jaccard similarity items. Duan et al. [7] propose various LSH methods to improve the scalability of matching, in order to handle both large numbers of instances or match a large number of pairs efficiently. They propose to estimate the item similarity based on a small number of random hash functions and make use of the banding technique to avoid the quadratic complexity when comparing all the pairs. In this paper we plan a similar approach based on hash functions to simulate permutations and thus decrease the computation time and required space.

In order to find the final duplicates, after a pre-selection by LSH, the state-of-the-art product duplicate detection method MSM can be used. This method is described and employed in [2,6]. MSM uses a hierarchical clustering which leads to a relatively high F_1-measure, but simultaneously a large computation time.

3   Method Overview

Figure 1 gives a general overview of the approach used in our paper. We refer to our approach as MSMP+, as the approach of [6] is called the MSMP method. We start with data cleaning followed by extracting the model words from the product titles and key-value pairs. Thereafter, we create a binary product representation for each product. The next phase is to apply Locality Sensitive Hashing (LSH) to get candidate pairs and eventually MSM is used to get the final set of duplicates from the candidate pairs.

3.1   Data Cleaning

To increase the correctness of the data and therefore the results, we use a data cleaning approach. The importance of data cleaning is mentioned for example in [12]. In the Web shop data, used for evaluation in this paper, some inconsistencies exist. These inconsistencies will lead to fewer found candidate pairs and therefore a higher number of false negatives. To efficiently correct for inconsistencies in the data, we used a frequency count of the model words. The most frequently occurring units are transformed into a standardized format. This transformation consists of three steps. First, all different representations of the units are transformed into one. E.g., the "''" sign for inch is normalized to 'inch'. In the second step, all upper-case characters are replaced by lower-case characters. Lastly, all spaces and non-alphanumeric tokens in front of the units are removed. The two main inconsistencies found are the representations of 'hertz' and 'inch'. Frequent variations of hertz and inch are transformed according to the steps described above. The results of these transformations of inch and hz are shown in Table 1. For example '23 Inch' becomes '23 inch'.

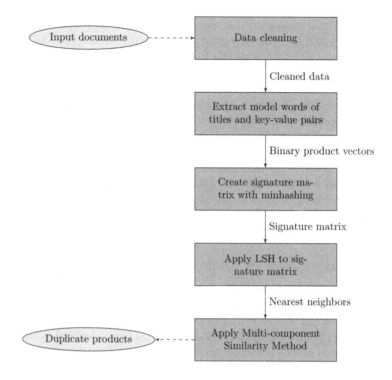

Fig. 1. General overview of MSMP+

Table 1. Transformation of frequent representations

Data value	Normalized value
'Inch', 'inches', '″', '-inch', 'inch', 'inch'	'inch'
'Hertz', 'hertz', 'Hz', 'HZ', 'hz', '-hz', 'hz'	'hz'

3.2 Signature Matrix

In this paper, model words from the title and the key-value pairs as proposed in [2] are used to create binary vectors representing the product. The binary vectors obtained from these model words are used to create a signature matrix with minhashing.

Properties of products are often represented as key-value pairs. The titles in product data are relatively uniformly defined, whereas in the representation of the properties there seems to be more variation across Web shops. The weight of a television, for example, will in one Web shop have value '20.8 lbs' whereas in the other Web shop it will be represented as '20.8 lbs'. Simply using the same definition for model words in the key-value pairs as in the title therefore does not seem logical since only the '20.8 lbs' would be added, creating more dissimilarity between the binary vector representations of the products. To account for this we only add decimal numbers, that either stand alone like '20.8' or that have a numerical part and a qualitative part like, '20.8 lbs' where the 'lbs' part will be deleted so that in both cases the modelwords will only contain a numerical part. Decimal numbers are used since these numbers often show a certain measurement, which contains specific information about the products.

Some product titles, however, contain more information than others. Therefore we use the model words from the title and search for these in the key-value pairs. In the key-value pairs, information from the title is often repeated, thus if one duplicate contains some information in the title and another does not, this information can possibly be extracted from the key-value pairs.

For the model words we use the definitions from [2] and an extended definition. Model words as defined by [2] have the property that they contain both numeric and alphabetic/punctuation tokens. The model words are defined by the following regular expression:

$$ModelWord_{title} = ([a\text{-}zA\text{-}Z0\text{-}9]^*(([0\text{-}9]+[^\wedge 0\text{-}9,]+)|([^\wedge 0\text{-}9,]+[0\text{-}9]+))[a\text{-}zA\text{-}Z0\text{-}9]^*)$$

This regular expression consists of several sub-patterns. The pattern [a-zA-Z0-9] recognises alphanumerical tokens, [0-9] numerical tokens, and [^0-9,] special characters. Model words consist of at least two of these three types.

The extended definition of the model words includes decimal numbers and allow these decimal numbers to have an optional non-numeric part. These model words are defined by regular expression:

$$ModelWord_{key-valuepairs} = (^\wedge\backslash d+(\backslash .\backslash d+)?[a\text{-}zA\text{-}Z]+\$|^\wedge\backslash d+(\backslash .\backslash d+)?\$)$$

This regular expression contains of two parts split by the or sign "|". The part ^\d+(\.\d+)?[a-zA-Z]+$ finds decimal numbers followed by an alphabetic characters. The second pattern ^\d+(\.\d+)?$ finds all the decimal numbers without alphabetic characters.

After identifying these value-based model words, the non-numerical part of these model words is deleted as mentioned earlier. By doing so, we make sure that more similarity is created between products that have the same value for a property, but have a different representation of that value.

We formalize the procedure of obtaining binary vectors in a similar way as [6], but with the additional use of the models words of the key-value pairs. Let P be the set of product descriptions corresponding to the N products we consider in our data. Furthermore, let $title(p)$ denote the title of product $p \in P$ and $values(p)$ denote the set value attributes of the key-value pairs of a product $p \in P$. The procedure of obtaining binary vectors is shown in Algorithm 1. In the first part we initialize MW_{title} as the set containing all model words from titles and MW_{value} as the set containing all the model words from the value attributes of all product descriptions. Secondly, for every product p we define a binary vector b^p by setting element i equal to 1 if the title or a value attribute of product p contains model word $i \in MW_{title}$, or if a value attribute contains a model word $i \in MW_{value}$, and 0 otherwise.

Algorithm 1. Obtaining Binary Vectors

1: $MW = \emptyset$
2: **for all** products $p \in P$ **do**
3:     **for all** model words $mw_{title} \in title(p)$ **do**
4:         $MW = MW \cup \{mw_{title}\}$
5:     **end for**
6:     **for all** model words $mw_{value} \in values(p)$ **do**
7:         $MW = MW \cup \{mw_{value}\}$
8:     **end for**
9: **end for**
10: **for all** products p in P **do**
11:     **for all** model words $mw \in MW$ **do**
12:         **if** $mw_{title} \in title(p) \vee mw_{title} \in values(p) \vee mw_{value} \in values(p)$ **then**
13:             $b_{mw}^{p} = 1$
14:         **else**
15:             $b_{mw}^{p} = 0$
16:         **end if**
17:     **end for**
18: **end for**
19: **return** b^{p} for all $p \in P$

3.3 Defining Duplicate Candidates

After the extraction of the model words and the conversion into binary vectors, we can apply LSH. By applying LSH one can reduce the high dimensionality of the original data set. In addition, the number of computations is reduced. The LSH algorithm maps possible similar items from the original data set into the same bucket. Items within the same bucket can therefore be seen as duplicate candidates. The number of duplicate candidates is much smaller than the original number of input items and therefore the number of computations done by MSM is reduced, which is important because of the previous large running time.

Minhash Signatures. In order to apply the LSH technique efficiently, one can replace the large set of binary codes with a smaller set. These representations are called *signatures*. The final goal of these signatures is to compare them and to give an accurate and fast estimation of the Jaccard similarity of two sets. An effective way to reduce the large set is by applying *minhash signatures*. This technique computes a signature for each set, so that similar documents have similar signatures and dissimilar documents are not likely to have similar signatures. It picks a list of permutations and computes a minhash signature for each set in the data. The permutations are generated randomly using random hash functions. The hash functions are of the following form:

$$h_{a,b}(x) = (a + bx)mod(p) \qquad (1)$$

in which a and b are random integers and p a random prime number $(p > k)$, where k is the dimension of the new vectors [11].

The number of instances in the reduced set (signature matrix) is therefore equal to the number of instances in the characteristic matrix, but the number of rows r is reduced to k. A high number of min-hashes gives more stable results. Since computing similarity with MSMP+ takes up most of the computation time we set the amount of min-hashes to be equal to 50% of the total size of the binary signature vector. We reduce the number of rows by 50% and therefore, k is half of the value of r.

Locality-Sensitive Hashing. In order to find pairs with large similarity efficiently, LSH can be used [13]. LSH divides the signature matrix M into b bands with r rows for each band. The b and r must be chosen in such a way, that the following equation holds:

$$n = r * b, \qquad (2)$$

with n the length of the columns of signature matrix M. In each band, the columns are hashed and divided into buckets. Items are hashed several times, because of the usage of multiple bands. The products that are hashed to the same bucket at least one time are now categorized as candidate pairs, which will be later checked by the MSM algorithm. The LSH algorithm reduces the number of candidate pairs that has to be compared by the MSM algorithm.

The relation between the false positives and the false negatives can be represented by the threshold t. An approximation of this threshold is:

$$t \simeq (1/b)^{1/r}. \qquad (3)$$

A higher threshold reduces the false positives and increase the false negatives, while a lower threshold reduces the false negatives and increases the false positives.

3.4   Multi-component Similarity Method

The final method we use after LSH is the MSM. MSM is a hierarchical adopted single linkage clustering method that makes use of a specific function in order to calculate the similarity of two products. This similarity function consists of three parts. The first part compares matching key-value pairs. The overlapping q-grams are used as a similarity measure. Alternatively, the cosine- and Jaro-Winkler measure could be used, but these are sensitive to misspellings and are token-based. The q-gram uses tokens of q characters, in this case $q = 3$. These are taken from a sliding window from the left to the right of the string. This calculated similarity is added to the final similarity of the two products with a learned weight.

The second part consists of the key-value pairs that were not matched in the first part. For these pairs the HSM method [1] is used. This method uses the model words and calculates the percentage of matches. This calculated similarity will again be added with a learned weight to the final similarity of the two products.

The final part of the similarity function uses the Title Model Words Method (TMWM) [14], which employs the model words from the product titles. This is also added after using a learned weight to the final part of the similarity function of two products.

MSM uses an adapted hierarchical single linkage clustering. This algorithm is performed on the dissimilarity matrix, containing the dissimilarities between products. Some of the dissimilarity values are manually set to infinity. This is true for products of the same Web shop, products that have different brands, using a list of television brands from the Web [15] and for products that are not considered to be candidate pairs by the LSH method. The remaining dissimilarity values are computed by using the same similarity function for each pair of products, as was mentioned before. The adapted hierarchical single linkage clustering is performed on the dissimilarity matrix, where the distances between clusters are defined as the shortest distance between a pair of products from these clusters. A cluster that contains a pair of objects with distance infinity will also have distance infinity. This iterative process will continue to merge the two nearest clusters until the distance exceeds a certain threshold. The clusters obtained can be seen as clusters containing duplicates.

4 Evaluation

In order to evaluate the performance of our method, a data set containing information about televisions sold in four different Web shops is used. The aim of the method is to find duplicate televisions across these Web shops with a high precision and as few comparisons as possible. This is done by using a pre-selection with LSH before we apply the actual duplicate detection method MSM. We compare the results of our method denoted as MSMP+ with the results of the method proposed by [6] denoted as MSMP.

4.1 Data

The four Webshops in our dataset are international Web shops, namely www.amazon.com, www.bestbuy.com, www.thenerds.net and www.newegg.com. The data set contains information of 1629 different televisions in total. All products are represented with a title, as well as additional information stored in key-value pairs. The representation of the title can be seen as a summary of the product as it contains information of several properties of the product. An example of a television product title is: 'Philips 4000 Series 29\" Class 2812\" Diag. LED 720p 60 Hz HDTV 29PFL4508F7 - Best Buy'. This representation gives you information of the brand, television type, resolution refresh rate, and the Web shop that sells the television.

Furthermore, all product descriptions contain other properties in the key-value pairs, such as: 'shop', 'url', and 'ModelID'. If the 'ModelID' for two different products is the same, the products can be considered as duplicates. ModelID's are often not present, which makes the approach of our method useful, but for evaluation purposes we have selected a data set where these ModelID's are present to have a high quality gold standard. We use the comparison made with ModelID's as a benchmark to evaluate our method. The number of properties, represented as keys, and the keys themselves are different across products. For example, one product might contain information on the HDMI-input, while another product might not have this information represented.

4.2   Evaluation Methods

To evaluate the performance of the methods, different metrics are used. In the first part we evaluate the LSH results and in the second part the MSM results. The LSH-method is evaluated with two metrics: Pair Quality (PQ) and Pair Completeness (PC). These are defined as, respectively:

$$PQ = \frac{D_f}{N_c}, \tag{4}$$

where D_f is the amount of duplicates found, and N_c is the number of comparisons made.

$$PC = \frac{D_f}{D_n}, \tag{5}$$

where D_f is the amount of duplicates found and D_n is the total amount of duplicates.

To evaluate the complete method, the F_1-measure is used. The F_1 measure is the harmonic mean of PQ and PC. Its optimal value is 1 and its lowest value is 0. The corresponding formula is:

$$F_1 = \frac{2 * PQ * PC}{PQ + PC}. \tag{6}$$

4.3   LSH Performance

In order to achieve a consistent result we make use of a procedure that is called bootstrapping. Bootstrapping relies on random sampling with replacement. This re-sampling method allows us to evaluate the performance of the method on every bootstrap. In total, 100 bootstraps are performed and around 60% of the products in the data set are used in every bootstrap. This way, each bootstrap contains roughly 1000 products. The final performance of every measure is computed as the average over all bootstraps.

The number of found candidate duplicate-pairs depends on the size n of the signature matrix and the threshold value t as described in Sect. 3. A higher value for t will lead to fewer candidate-duplicate pairs and therefore to fewer comparisons by MSM. However, a higher threshold value t will also lead to a smaller amount of found candidate duplicate-pairs and a potential higher number of false negatives. The aim of this research is to lower the number of comparisons made by MSM, while still finding a large amount of the duplicates. Therefore we run the algorithm for different values of t. A trade-off is made in order to define the best value of t, such that the PQ and the PC are optimized. The fraction of comparisons is defined as the candidate duplicate-pairs found by LSH divided by the total number of possible comparisons. The fraction of comparisons is therefore directly related to the threshold value t. By adding more candidate duplicate-pairs, the PC will increase. However, by doing more pairwise comparisons, the running time of MSM increases as well. The threshold value t, executed for 100 bootstraps, characterizes this trade-off. We vary the value of t from 0 to 1 with a step size of 0.05. All results are averaged over the bootstraps.

In order to compare the performance of the LSH part of our MSMP+ method with the MSMP method, we plot the PQ, PC and F_1 measure against the fraction of comparisons of the methods in Figs. 2, 3 and 4, respectively (after LSH, check only pairs that are in the same bucket). Moreover, these figures report the separate results for the MSMP method with data cleaning (clean), the MSMP method with data cleaning and model words from title and also the model words from the key-value pairs found in the tile (values), and the MSMP+ method this method contains all the elements from above plus the decimal model words as explained in section Method Overview.

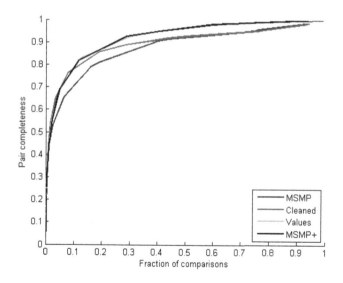

Fig. 2. Pair Completeness for different fractions of comparison

From Fig. 2 it can be shown that the MSMP+ method outperforms the MSMP method. For example by performing 10% of the number of pairwise comparisons a PC of 80% is achieved with the MSMP+ method, while with the MSMP method there is only a PC of 70%. By calculating the area under the

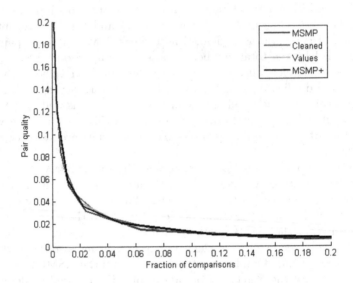

Fig. 3. Pair Quality for different fractions of comparisons

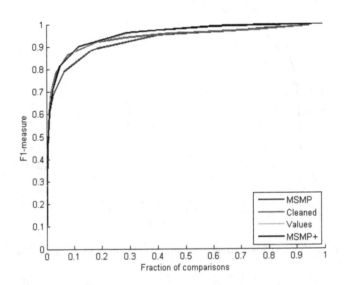

Fig. 4. F_1-measure of MSMP+ compared to the MSMP for different fractions of comparisons

curve using the trapezoid method with the midpoint rule, it can be shown that MSMP+ method has an area under the curve improvement of 12.2% compared to the MSMP method. Furthermore we see that cleaning the data improves MSMP for the lower threshold values and using the key-value pair method does this for the higher threshold values. Finally there is only a slight change adding the decimal numbers.

In Fig. 3 an overview for the PQ against the fraction of comparisons can be found. Although the differences between the different methods of improvement are hard to distinguish, there is a 9.2% improvement of the area under the graph for MSMP+ against MSMP.

Finally we make use of the so-called F_1 measure to compare the results of the LSH part in both methods in Fig. 4. There is a 9.3% improvement of the area under the graph, in this case. Same as in Fig. 2, cleaning the data improves MSMP for the lower threshold values and the key-value pair method does this for the higher threshold values.

We conclude that the LSH part of the MSMP+ method proposed in this paper significantly outperforms the MSMP method on all the evaluation metrics: PC, PQ, and F_1.

4.4   Performance of MSM

In order to evaluate the performance of the MSMP+ method compared to the results of MSMP, the F_1-measure is used again. The results are obtained by running the MSMP+ method for different fixed threshold values t, where every run consists of 10 bootstraps. Figure 5, shows the results. The set of parameters used in the MSMP+ algorithm is optimized over a grid of parameters. For every value t, the best performing parameter is selected.

By applying the LSH pre-selection method before performing the duplicate detection method MSM, we reduced the number of pairwise comparisons with respect to the number of comparisons done by the MSM method. Note that the fraction of pairwise comparisons done by MSM is equal to 1 if all comparisons are performed. As shown in Fig. 5, we compare the F_1-measure obtained by the MSMP+ and MSMP method to the F_1-measure of the MSM method (benchmark). The F_1-measure of the benchmark is equal to 0.525 [2]. As the graph shows, the F_1-measure decreases as the fraction of comparisons decreases. However, a large decrease in faction of comparisons leads to only a small reduction of the F_1-measure. From Fig. 5 we can conclude that the MSMP+ method outperforms the MSMP method. If we look at a reduction of 95% of pairwise comparisons for example, the MSMP method leads to an F-1 measure of 0.46, while the MSMP+ method for this reduction leads to a F-1 measure of 0.49. The MSMP+ method has an improvement of the area under the curve for the F_1-measure of 7.8% compared to the MSMP method.

To summarize, the results show that LSH is an efficient method to significantly reduce the number of pairwise comparisons and therefore the computation time, while still finding an high number of duplicates. Also the MSMP+ method significantly outperforms the MSMP method.

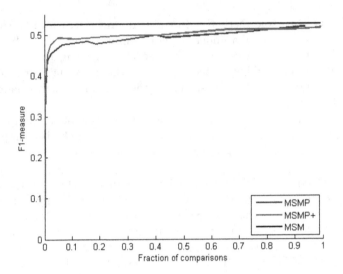

Fig. 5. MSMP+ compared to MSM

5   Conclusions

In this paper we propose a method called MSMP+ in order to reduce the number of calculations involved in product duplicate detection on the Web. This method is an extension of the MSMP method described in paper [6]. The MSMP method uses a LSH pre-selection method before performing the duplicate detection method MSM. Before applying the LSH pre-selection method, it is necessary to define all products as binary vectors. In the MSMP method this is done by only using the model words that appear in the title. In the MSMP+ method, we first start with data cleaning followed by extracting the model words from the product titles and key-value pairs. Then we create a binary product representation for each product, apply LSH, and lastly perform the MSM algorithm.

When we only look at the performance of LSH the MSMP+ method has improved the area under curve compared to MSMP of pair completeness and pair quality with 12.2% and 9.2% respectively. The area under the curve of the F_1-measure is improved by 9.3%. After performing MSM on the candidate duplicates, the MSMP+ method has improved the area under the F_1-measure curve by 7.8%, compared to the MSMP method.

We conclude that the MSMP+ method outperforms the MSM method on all evaluation criteria and therefore the MSMP+ method is a valuable extension of the MSMP method.

References

1. de Bakker, M., Frasincar, F., Vandic, D.: A hybrid model words-driven approach for web product duplicate detection. In: Salinesi, C., Norrie, M.C., Pastor, Ó. (eds.) CAiSE 2013. LNCS, vol. 7908, pp. 149–161. Springer, Heidelberg (2013). https://doi.org/10.1007/978-3-642-38709-8_10
2. van Bezu, R., Borst, S., Rijkse, R., Verhagen, J., Frasincar, F., Vandic, D.: Multi-component similarity method for web product duplicate detection. In: 30th Symposium on Applied Computing (SAC 2015), pp. 761–768. ACM (2015)
3. Broder, A.Z., Glassman, S.C., Manasse, M.S., Zweig, G.: Syntactic clustering of the web. Computer Netw. ISDN Syst. **29**(8), 1157–1166 (1997)
4. Charikar, M.S.: Similarity estimation techniques from rounding algorithms. In: Thirty-Fourth Annual ACM Symposium on Theory of Computing (STOC 2002), pp. 380–388. ACM (2002)
5. Cohen, E., Datar, M., Fujiwara, S., Gionis, A., Indyk, P., Motwani, R., Ullman, J.D., Yang, C.: Finding interesting associations without support pruning. IEEE Trans. Knowl. Data Eng. **13**(1), 64–78 (2001)
6. van Dam, I., van Ginkel, G., Kuipers, W., Nijenhuis, N., Vandic, D., Frasincar, F.: Duplicate detection in web shops using LSH to reduce the number of computations. In: 31th ACM Symposium on of Applied Computing (SAC 2016), pp. 772–779. ACM (2016)
7. Duan, S., Fokoue, A., Hassanzadeh, O., Kementsietsidis, A., Srinivas, K., Ward, M.J.: Instance-based matching of large ontologies using locality-sensitive hashing. In: Cudré-Mauroux, P., Heflin, J., Sirin, E., Tudorache, T., Euzenat, J., Hauswirth, M., Parreira, J.X., Hendler, J., Schreiber, G., Bernstein, A., Blomqvist, E. (eds.) ISWC 2012. LNCS, vol. 7649, pp. 49–64. Springer, Heidelberg (2012). https://doi.org/10.1007/978-3-642-35176-1_4
8. Fetterly, D., Manasse, M., Najork, M.: On the evolution of clusters of near-duplicate web pages. J. Web Eng. **2**(4), 228–246 (2003)
9. Henzinger, M.: Finding near-duplicate web pages: a large-scale evaluation of algorithms. In: 29th Annual International ACM SIGIR Conference on Research and Development in Information Retrieval (SIGIR 2006), pp. 284–291. ACM (2006)
10. Indyk, P., Motwani, R.: Approximate nearest neighbors: towards removing the curse of dimensionality. In: Thirtieth Annual ACM Symposium on Theory of Computing (STOC 1998), pp. 604–613. ACM (1998)
11. Leskovec, J., Rajaraman, A., Ullman, J.D.: Mining of Massive Datasets. Cambridge University Press, Cambridge (2014)
12. Rahm, E., Do, H.H.: Data cleaning: Problems and current approaches. Bull. IEEE Comput. Soc. Tech. Comm. Data Eng. **23**(4), 3–13 (2000)
13. Slaney, M., Casey, M.: Locality-sensitive hashing for finding nearest neighbors. IEEE Sig. Process. Mag. **25**(2), 128–131 (2008)
14. Vandic, D., Van Dam, J.W., Frasincar, F.: Faceted product search powered by the Semantic Web. Decis. Support Syst. **53**(3), 425–437 (2012)
15. Wikipedia: The free encyclopedia. http://wikipedia.org/wiki/List_of_television_manufacturers

A Manageable Model for Experimental Research Data: An Empirical Study in the Materials Sciences

Susanne Putze[(⊠)], Robert Porzel, Gian-Luca Savino, and Rainer Malaka

Digital Media Lab, University of Bremen, Bremen, Germany
{sputze,porzel,gsavino,malaka}@tzi.de

Abstract. As in many other research areas the material sciences produce vast amounts of experimental data. The corresponding findings are then published, albeit the data remains in heterogeneous formats within institutes and is neither shared nor reused by the scientific community. To address this issue we have developed and deployed a scientific data management environment for the material sciences at various test facilities. Unlike other systems this one explicitly models every facet of the experiment and the materials used therein - thereby supporting the initial design of the experiment, its execution and the ensuing results. Consequently, the collection of the structured data becomes an integral part of the research workflow rather than a *post hoc* nuisance. In this paper we report on an empirical study that was performed to test the effects of a paradigm change in the data model to align it better with the actual scientific practice at hand.

Keywords: Information systems for experimental research data
Empirical user study · Experimental results

1 Introduction

In materials science and many other research areas numerous experiments are conducted and large amounts of data are collected. Archiving these experiments is not limited to storing the test results in some digital format [15]. It is equally important to store the underlying experimental parameters which define the varying independent and dependent variables together with the controlled and contextual parameters. Most of the time, these data are collected for the purpose of scientific publications, but the data themselves are not published along with the corresponding results. Consequently, experiments are hardly reproducible by the scientific community, as often, crucial details are missing in the publication. Reproducibility of results is not only a challenge in the case of public publications, but also when data is stored internally within an institute over longer periods of time. Missing experimental details may lead to misinterpretation when re-examining the data or to involuntary duplication of experiments.

© Springer International Publishing AG, part of Springer Nature 2018
J. Krogstie and H. A. Reijers (Eds.): CAiSE 2018, LNCS 10816, pp. 424–439, 2018.
https://doi.org/10.1007/978-3-319-91563-0_26

Many researchers, in theory, support an open research culture with public accessible data [7]. Ideally, according to the "FAIR" principles: Findable, Accessible, Interoperable, Reusable [13]. Data repositories, such as the Open Science Framework (OSF) [3] or Dendro [11], try to embed data storage into the research process, but they do not fully represent the structure of an experiment as part of the digital record.

When it comes to creating digital infrastructures for experimental data, scientist are demanding users:

- On the one hand, they have not much time for data entry, especially when they have to do it more than once.
- On the other hand, they need to store complex and frequently unique experiments with large amounts of parameters.

Therefore, creating data repositories which address only the data storage aspects do not fully address their needs. Researchers avoid additional work and data management is still not seamlessly embedded into their daily research workflow. One of the main challenges when documenting experiments is the need for repeated entry of similar experiment protocols which only vary in a small number of parameters. This process does not only create administrative overhead for the researchers but may also prevent them from discovering structures, similarities, or missing configurations in a series of related experiments.

In this paper, we focus on a method for structured input of scientific experiments which is flexible and not time consuming so that scientists can describe reproducible experiments and store the corresponding data in a way which can be easily understood by others. Therefore, we designed an information system which allows the user to describe scientific experiments and the context in which they are conducted. We present the concept and implementation of an approach for the input of structured *experimental series*. Such a series describe sets of scientific experiments with their controlled parameters, independent and dependent variables. To evaluate this prototype, we designed and conducted a laboratory usability study with naïve users as well as domain experts to compare two data input systems for experiments with or without the ability to define and use an experimental series.

The paper is structured as follows: In Sect. 2, we discuss related work. Section 3 presents our series based experiment management system, while Sect. 4 discusses the study design for our experiment. In Sect. 5, we give an overview over our results; Sect. 6 finally summarizes the paper.

2   Related Work

In recent years a variety of research data repositories have been developed and were made available to the scientific community. These systems represent different approaches to creating digital infrastructures for scientific data. In some cases the approaches differ in their disciplinary scope:

- Some systems are designed as multi-domain systems, e.g. the Open Science Framework [3],
- Others are domain-specific ones focusing on a singular scientific field or type of experiment, e.g. for marine science [8] or systems biology [12].

Another dimension concerns the level of formal explication. Most systems store some meta-data about the experiments using various standards, e.g. using the Dublin Core Meta Data, but they differ in terms of modeling the content data inside explicitly or implicitly.

- One frequent approach is to store data records and their meta data in some human readable format, without having an explicit formal model of the individual entities contained in these records, e.g. PANGEA [4] or OSF [3].
- Others seek to provide structured information of the data themselves along with the published record, e.g. SEEK [14] or InfoSys [15].

Lastly, we see the genesis of the experimental records as an additional dimension to be considered.

- Most of the aforementioned systems constitute *post hoc* data management. That means that after the experiments have been conducted the resulting data is formated and entered into some system, e.g. [3,4,14] or [8].
- Currently only a few systems seek to support concurrent data management. goes a step beyond. Here the question arises of how to address the problem of involving data management as early as possible into the research workflow and to make it an integral part of it, e.g. [11] or [15].

An overview of the dimensions and possible realizations is given in Table 1. Orthogonal, but nonetheless relevant to the data management aspects, there are already some systems that try to help researcher with finding appropriate study designs and also data collection tools, like [9,10] or [5]. A common, but nonetheless important aspect for all those systems is to design user friendly interfaces which support domain experts without specific technical knowledge with their research.

Table 1. Dimension of infrastructures for scientific data management

Dimension	Realization
System scope	Domain-specific vs. Multi-domain
Data model	Implicit vs. Explicit
Workflow	Post-hoc vs. Concurrent

3   System Design

As seen in the related work, different types of systems for experimental scientists to design and store their experimental data and results have recently emerged. In this section, we want to introduce and situate our approach. Our system is:

- domain-specific for research experiment with various kinds of materials in the materials science,
- provides an explicit fine-grained model of the experimental conditions and results, including fixed parameters, independent and dependent variables,
- facilitates the creation and management of experiment series of related experiments,
- is designed to be employed concurrently as part of the actual workflow at hand.

In a nutshell the researcher designs and specifies the experiment using the system, which then serves as a specification for the laboratory technician who, in turn, performs the experiment and files out the specified measurements and results in the same system. Lastly, when finished the results are available to the researcher for further analyses.

3.1   Requirements for Experiment Management

To fulfil our goal of creating a usable information system rather than another burdensome requirement for the researchers involved, it is necessary to get a precise understanding of the research workflow conducted by experimental scientists. Other data infrastructure projects affirm the importance of including domain experts into the design process as well [6].

We analysed the procedure of planning and conducting experiments by running rows of structured and unstructured interviews as well as a number of lab visits where we could observe the scientists while doing experiments. For this purpose, we choose scientists from two materials science labs which work with different classes of materials, namely steel and fiber reinforced materials. While the steel lab already uses an experiment result management system the other lab uses a laboratory information system (LIMS) which covers only the organizational aspects of conducting experiments, e.g. by whom and when the experiment took place, rather then contextual experiment data.

Within the interviews, some researchers reported misgivings being forced to specify each experiment individually. Furthermore, we analysed the existing data stored in the experiment management system. This data showed that experiments are typically conducted with varying only one or two of the whole set of attributes. The other parameters remain fixed, but may be varied in different, but similar experiments. This, consequently led to the inclusion of the concept of *experimental series* which gives the researchers the opportunity to subsume individual experiments, thereby, further integrating the data storage process within their typical research processes.

3.2   The Experiment Series Model

Figure 1 gives an overview of the experiment series model. A series groups multiple individual experiments. It contains the *fixed parameters* common to all grouped experiments of both the examined material, i.e. specimen and settings with their attributes. Every experiment inherits the fixed parameters of the associated experiment series and add values to the *independent variables*. The independent variables in the different experiments instantiate and determine the values of *dependent variables* stored in the test results. Since we want to focus on the experiment designing and conducting part within the research workflow, we do not have a further look on test results in this paper. The specimen describes all attributes of the material and its treatments up to the start of the actual experiment, e.g. a steel with heat treatments. The settings describe the conditions under which the specimen is treated in the experiment, e.g. the test machine and its parameters.

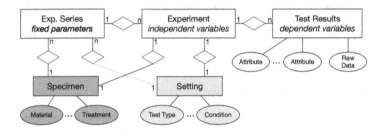

Fig. 1. The experiment series model

Compared to a test result oriented experiment model, the experiment series model promises multiple benefits:

- First, we gain additional information about the structure and relationship between multiple experiments.
- Second, there is much less input effort required for the researcher[1].

The differences become clear on an instance level. In Fig. 2 we show the same example with and without an experiment series. All fixed parameters are captured in a series instance (Fig. 2b) while the experiment only contains the independent variables, i.e. those attributes which are varied across the multiple experiment instances. In the hierarchical structured test result oriented model (Fig. 2a), the settings needs to be redefined for every individual specimen.

---

[1] In those cases where a experimental series consists of only one individual experiment the workload would be the same as before.

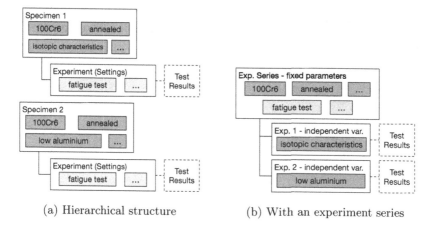

(a) Hierarchical structure   (b) With an experiment series

Fig. 2. An example of two experiment instances with or without an experiment series

Specimen

Material			Treatment		Fixed parameter
Material type	☑ Steel	↕	Heat treatment	☑ Annealed	↕
Material	☑ 100Cr6		Temperature [°C]	☑ 120	
			Duration [min]	☑ 60	
Production					
Modification	☐ independent variable		Independent variable		
Mould	☑ Plate	↕			
Production method	☑ Roled	↕			

Fig. 3. Specimen input step during creation of a new experiment series. The users select all independent variables and enter values for all fixed parameters.

3.3   Interface for Experiment Series

Figure 3 shows the interface for specimen attributes during the creation of a new experiment series. For each attribute the form features three parts: first, the name of the attribute, second a checkbox for specifying whether the attribute is a fixed parameter or an independent variable. Third, for all fixed parameters the user should enter the value. All independent variables need to be explicitly defined as such. To do so, the user has to de-select the selection box that is situated to the left of the input field. After this, the attribute is marked as independent variable and the user cannot enter a value anymore. By asking the user to define independent variables explicitly rather than leaving the value empty, it is possible to differentiate between attributes which do not have a value, e.g. were forgotten during the attribute input and those which are actually independent variables. This may help to avoid mistakes during the data input phase. When adding an experiment instance to an existing experiment series, only the values of all independent variables need to be specified which are typically only a few.

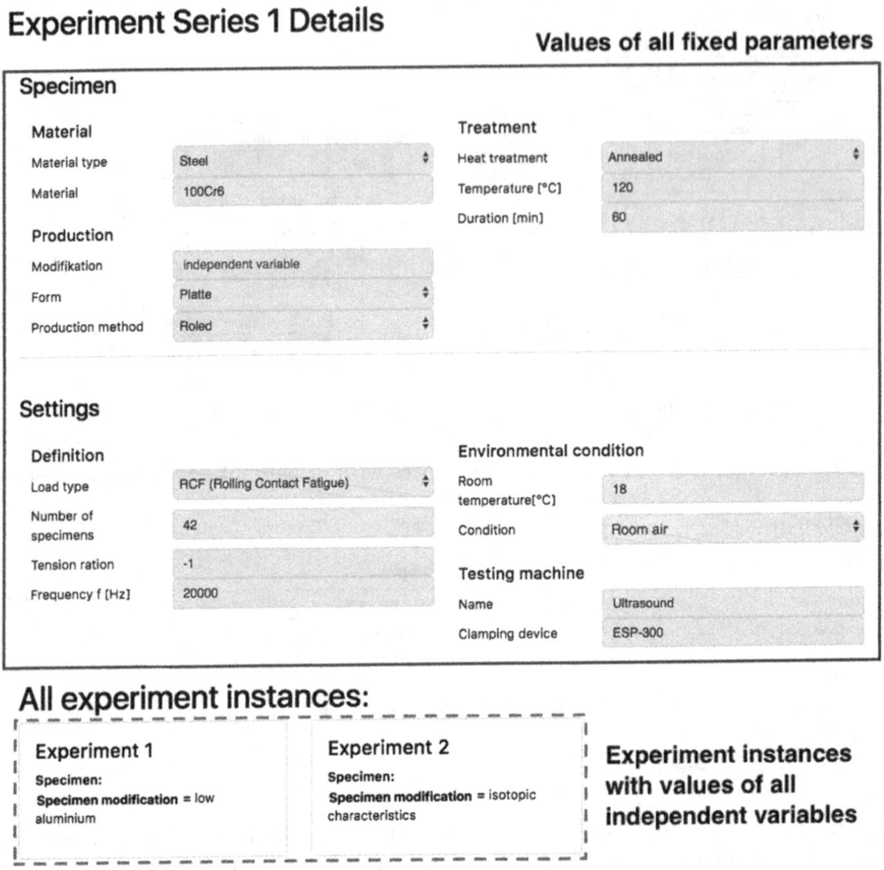

Experiment Series 1 Details

Values of all fixed parameters

Specimen

Material

Material type	Steel	⬍
Material	100Cr6	

Treatment

Heat treatment	Annealed	⬍
Temperature [°C]	120	
Duration [min]	60	

Production

Modifikation	independent variable	
Form	Platte	⬍
Production method	Roled	⬍

Settings

Definition

Load type	RCF (Rolling Contact Fatigue)	⬍
Number of specimens	42	
Tension ration	-1	
Frequency f [Hz]	20000	

Environmental condition

Room temperature[°C]	18	
Condition	Room air	⬍

Testing machine

Name	Ultrasound	
Clamping device	ESP-300	

All experiment instances:

Experiment 1	**Experiment 2**
Specimen:	**Specimen:**
Specimen modification = low aluminium	**Specimen modification** = isotopic characteristics

Experiment instances with values of all independent variables

Fig. 4. Overview of an experiment series. All fixed parameters and their values are listed in the upper part. All experiment instances with the values of the independent variables are listed in the lower part.

Beside the more intuitive interface, the experiment series model allows to neatly arrange multiple experiments on screen, as shown in Fig. 4. The user can get an overview of all attributes, fixed and variable, of a range of experiments without navigating through multiple experiment instances and comparing their individual values. In the upper part of an experiment series detail page, all fixed parameters are listed while in the lower part all experiment instances with the values of the independent variables, as part of the experiment, are catalogued.

4   Study Design

To evaluate the series-oriented input paradigm for the specification of the experimental design, we designed and conducted a laboratory experiment with a total 28 participants. As a benchmark we employed a simplified version the current experiment management system which has been deployed and is being used by the materials science department. This system features a strict hierarchical model, where *Projects* can have multiple albeit unrelated *Experiments* based on individual *Materials*, as depicted in the example shown in Fig. 2a. The existing benchmark system will henceforth be referred as *System B* and compared to the new series-based system called *System S*. Both systems are designed according to the same design basics, as we use the same styles and forms of interaction. Next, we will introduce our hypotheses regarding the series-based experiment management approach and the corresponding study design.

4.1 Hypotheses

With our study we want to examine the following hypotheses regarding the usability of the new approach. In our study we will focus on three usability criteria specifically (1) efficiency, (2) safety (error-avoidance) and (3) user satisfaction. The ensuing corresponding hypotheses, therefore, are:

H1 Data entry with the series-based System S is faster than using the benchmark System B.
H2 Users make less input errors with the series-based System S in comparison to the benchmark System B.
H3 Users are more satisfied with the series based System S than with the benchmark System B.

4.2 Task Design

We use tasks from a real-world scenario as found in the materials science. The participants were asked to enter data from four different sets of experiments. For this we selected attributes and possible values from the existing experiment management system. As our focus lies on the structured input and not on typing speed, we limited the tasks to a small amount of attributes per experiment. The data to be entered was presented to the participants on a sheet of paper which they could consult during data entry.

To examine the input correctness, we do not include any kind of input verification to the study prototype, e.g. no "Not saved" alert or verification of input fields. Doing so, we avoid influences on the correctness of user inputs and expect a better insight in how well they understand the system.

Table 2 summarizes the four tasks and the minimum required effort for input with the two systems, measured in number of operations. An input operation O_I corresponds to the input of a single attribute value, e.g. entering a free text or a selection from a selection list. A structure operation O_S is an operation which manipulates the structure, e.g. adding or saving additional experiments.

4.3 Participants

To collect more information about the system and to be able to perform a quantitative analysis, we use naïve users as well as domain experts for our system

Table 2. Tasks and their Input (O_I) - and Structure (O_S) operations

Task	System S	System B
1 - One experiment Without variation	$4\,O_S + 16\,O_I = 20\,O$	$4\,O_S + 16\,O_I = 20\,O$
2 - Two experiments Vary in specimen	$6\,O_S + 18\,O_I = 24\,O$	$8\,O_S + 32\,O_I = 40\,O$
3 - Three experiments Vary in specimen	$8\,O_S + 19\,O_I = 27\,O$	$12\,O_S + 48\,O_I = 60\,O$
4 - Three experiments Vary in settings	$8\,O_S + 19\,O_I = 27\,O$	$8\,O_S + 32\,O_I = 40\,O$

evaluation. We performed our study with twenty naïve users and eight expert users. The naive users were divided into four groups with five people each. We mixed a Between-Subject and a Within-Subject design, as shown in Table 3. To avoid ordering effects, half of the participants started with the series-based system (System S), the others started with the benchmark system (System B). The tasks were comparable in both runs, using the same structure with changing only the face values of the attributes. In Run 1 each group of users had to interact with only one type of system, either System B or System S. Thereafter, in Run 2, half of the users switched systems for four additional tasks while the other half of each group continued with the system they had been assigned to initially. Each subject had to fill out a standardized usability questionnaire, the System Usability Score (SUS) [2]. Here the *switchers* filled out one questionnaire after the fours tasks in Run1 and another after the four tasks of Run 2, that is once before changing the systems and once at the end, while the *continuers* filled out the questionnaire after eight tasks. At the end, we conducted a structured interview to collect statements about experienced differences. The group of expert users was treated a little bit different, as they have prior knowledge with experiment management system, i.e. System B. All of them started with only one task with System B and continued with all four task with System S.

To get used to the systems, all participants saw an introduction video before they started with the tasks. This video gave a short introduction to experiments in materials science and demonstrated the mechanics of the respective system.

Table 3. Experiment design

	Group 1	Group 2	Group 3	Group 4	Expert group
Run 1 (4 Tasks)	System S		System B		System B (1 Task)
Run 2 (4 Tasks)	System S	System B	System S	System B	System S

5   Results

Based on our study design introduced in Sect. 4, we want examine our hypotheses H1, H2 and H3 respectively. Furthermore, we will summarize pertinent findings from the qualitative data gathered in the interviews.

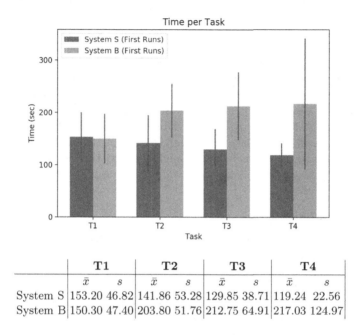

	T1		T2		T3		T4	
	$\bar{x}$	s	$\bar{x}$	s	$\bar{x}$	s	$\bar{x}$	s
System S	153.20	46.82	141.86	53.28	129.85	38.71	119.24	22.56
System B	150.30	47.40	203.80	51.76	212.75	64.91	217.03	124.97

Fig. 5. Mean times needed and standard deviations for fulfilling the tasks of System S (blue) and System B (red) (Color figure online)

5.1   Efficiency

H1 Data entry with the series-based System S is faster than using the benchmark System B.

To examine this, we measured the time a user needed from the start of a task until they finished it by pressing a "Finish Task" button in the user interface. To avoid priming effects we examined only the data obtained in run 1 for all groups of naïve users in this evaluation.

Figure 5 shows the mean values for the task duration together with their standard deviations for all $n_A = n_B = 10$ participants in their first run. As expected, the task duration of the benchmark system (System B, red bars) increases from to Task 1 to Task 4, corresponding to the increasing necessary effort (in number of operations, see Table 2). The task duration therefore increases as the tasks become more extensive. In comparison, task duration for the series-based System S does not increase substantially from Task 1 to Task 4. This is in line with the only moderate increase of required number of operations for these tasks and a progressing familiarization of users with the system. The standard deviation for the series-based system is almost half compared to the benchmark system, which shows that users are consistently more efficient.

An independent-samples t-test was conducted to compare task duration of the two systems. This test revealed a significant difference in the task duration for Task 2 ($t(18) = -2.64, p = 0.017$), Task 3 ($t(18) = -3.47, p = 0.0027$) and Task 4 ($t(18) = -2.44, p = 0.026$). As expected, there is no significant difference between the systems for Task 1 which does not contain an experiment series. Overall, these results support our hypothesis H1.

A special case is Task 4 which users took unexpectedly long to solve with the benchmark system. As shown in Table 2, Task 4 can be solved with fewer operations than Task 3 using the benchmark system when exploiting the hierarchical structure of the data. Some participants did not recognize this and proceeded exactly as for Task 2 and Task 3, although this is not the optimal solution with this system. A reason for this behavior might be that the hierarchical structure of the benchmark system is not easy to understand as expected.

5.2   Error-Avoidance

H2 Users make less input errors with the series-based System S in comparison to the benchmark System B.

To measure input errors the users made, we compared the entered data with the optimal solution and counted every deviation from the sample solution. This includes errors in input operations, for example typing errors or selecting the wrong value from a selection list and errors in structure operations, for example unsaved data.

In both settings the users did not perform many input errors. All users together made absolute in average $\bar{x} = 1.675, s = 1.64$ errors during the task. With $\bar{x}_S = 0.85, s_S = 1.39$ structure errors and $\bar{x}_I = 0.825, s_I = 1.28$ input errors.

The most frequent error was a structural one in task T1 with System S, i.e. an experiment without variation: In this case an empty experiment should be created, since the series is only an abstract data model and would not allow to enter test results. In only nine of twenty runs with System S an experiment was created correctly. In the other eleven runs, the experiments were missing. In a production system with all functions and test results, this error would naturally not occur, as that it is not a key aspect of the system.

Because of the small numbers in errors it is not meaningful to do statistical analysis. This would be different in a production setting with outside influences and should be examined in a different experiment.

5.3   User Satisfaction

H3 Users are more satisfied with the series based System S than with the benchmark System B.

To investigate the usability of the system, we used the System Usability Scale (SUS) [2], a standardized questionnaire for usability evaluation. Continuers (Group 1 & 4), filled all one SUS-questionnaire after both runs, while the switchers filled out two after both runs. Figure 6 shows the mean and the standard deviation for $n_S = 15$ naive users with System S, $n_B = 15$ users with System B and the $n_E = 8$ user of the expert group. With $\bar{x}_S = 77.83, s_S = 17.03$ the series based system is rated as "good" [1]. While the benchmark system ($\bar{x}_B = 59.5, s_B = 20.79$) is only rated as OK. An independent-sample t-test was conducted to compare the SUS-Score for both system with the following result: $t(28) = 2.64, p = 0.013$. There is a significant difference between the two systems and the corresponding null hypothesis can be rejected.

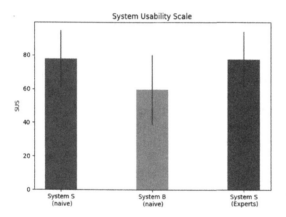

Fig. 6. SUS-Scores of both systems and the expert group

For the two-system groups, Group 2 and Group 4, we evaluated although the differences between ratings of both. For this we ran an dependent-sample t-test. With $t(4) = 4.60, p = 0.010$ (Group 2 - SB) and $t(4) = -2.81, p = 0.048$ (Group 4 - BS) the differences are significant for both groups.

5.4   Qualitative Evaluation

Additionally to the evaluation of the quantitative measures, we interviewed all naive users after finishing both runs. These interviews confirm our findings from the quantitative analysis. 9 out of 10 users from Group 2 and Group 3 (Those groups which handled both systems) find the series based System S more intuitive and even 10 of 10 user from those groups prefer the series based System S over the benchmark System B. Although, we expected the series based system to be more complex, the users viewed the benchmark system to be more complex and unclear because of its hierarchical structure.

For the series based system we received additionally constructive feedback, especially the input of a series using a keyboard only was mentioned very often. Typically, the users did the series input in two steps, first selecting the independent variables using the mouse, and then entering all fixed parameter using the keyboard. It is necessary to skip the "Independent variable"-checkbox by a double tab when switching from one input field to another. This additional effort was criticized very often. Although the users noted the layout of both (checkbox and input-field) next to each other as being very clear.

One of the most common mistakes using the series based system, was the missing additional experiment as task T1. In the interviews, some users mentioned this mistake as self-motivated. It seems that the learning effect by using the system multiple times with different tasks raises the awareness for our concept of experiment series and its abstract characteristics.

5.5   Expert Evaluation

As well as the naïve users we evaluated the system with eight domain experts from the two research labs which we interviewed before. In this expert group were laboratory technicians as well as researchers. All of them are touched with experiments in their daily professional routine.

The expert group did a SUS questionnaire for the series based System S. The corresponding results are displayed in the right bar of in Fig. 6. With $\bar{x} = 77.5$ and $s = 16.48$ their ratings were in the same range as those of the naive user group.

Beside the general questions about the system, for which we received the same answers as from the naive users, we ask them additional questions how their rate the system and how well it fits into their typical research workflow. All domain experts were satisfied with the experiment series based systems and prefer it much over the benchmark system which is comparable to the system some of them are using. Seven out of eight domain experts answered that the system fulfill their requirements for designing experiments. One test user did not answer this question clearly.

6   Discussion and Outlook

This paper investigates a shift in the explicit model for scientific experiments used for research data management. We claim that fitting such models to the research workflow is of paramount importance to systems that aim to provide more than black-box records of experiments. Modelling the data explicitly certainly brings about a range of advantages such as comparability and enhanced data mining and knowledge discovery. Nevertheless, converting data into a structured model in a post-hoc manner requires a lot of effort from the involved researchers, which is why we seek to integrate the data acquisition into the daily workflow of designing, conducting and analysing experiments. In our minds this explains why a system with a better fitting model, even if structurally more complex than the previous one, outperforms the original system in the key usability factors of efficiency, error-avoidance and user satisfaction.

We also want to emphasize that user satisfaction, even when dealing with a professional system not intended for a general consumer market, is critical, as researchers will opt out of using it and revert to their familiar spreadsheets and idiosyncratic formats if they do not like the system and find it too cumbersome to use.

As a next step, we will deploy the new model into the InfoSys system[2], a research data management system from material science. We will further long term evaluations in its daily productive usage. At the same time data mining efforts are underway to discover *hidden* experimental series in the existing data, thereby highlighting the benefits of having explicit fine-grained models of the data.

We are also working on further individualization of the input interfaces so that scientists can individually vary the relevant attributes and the order of the production steps involved in creating the test specimen. These efforts will ultimately lead towards the conversion of the *static* system into an *authoring* system where researchers collaboratively specify the model as they are using the system. In our minds this will facilitate the expansion from a single domain system into a multi domain environment for managing research data from various fields.

Acknowledgments. The research reported in this paper has been partially supported by the German Research Foundation DFG, as part of the project "AimData" MA 1766/3-2 as well as the Collaborative Research Center (Sonderforschungsbereich) 1320 "EASE - Everyday Activity Science and Engineering", University of Bremen (http://www.ease-crc.org/). The research was conducted in subproject H2 "Mining and Explicating Instructions for Everyday Activities".

References

1. Bangor, A., Kortum, P., Miller, J.: Determining what individual SUS scores mean: adding an adjective rating scale. J. Usability Stud. **4**(3), 114–123 (2009)
2. Brooke, J.: SUS-A quick and dirty usability scale. Usability Eval. Ind. **189**(194), 4–7 (1996)
3. Center for Open Science (COS): The Open Science Framework (2018). https://osf.io/
4. Grobe, H., Diepenbroek, M., Dittert, N., Reinke, M., Sieger, R.: Archiving and distributing earth-science data with the PANGAEA information system. In: Fütterer, D.K., Damaske, D., Kleinschmidt, G., Miller, H., Tessensohn, F. (eds.) Antarctica, pp. 403–406. Springer, Heidelberg (2006). https://doi.org/10.1007/3-540-32934-X_51
5. Kim, S., Mankoff, J., Paulos, E.: Sensr: evaluating a flexible framework for authoring mobile data-collection tools for citizen science. In: Proceedings of the 2013 Conference on Computer Supported Cooperative Work, pp. 1453–1462. ACM (2013)

---

[2] http://www.uni-bremen.de/infosys.

6. König-Ries, B., Triebel, D., Huber, R., Glöckler, F., Güntsch, A., Felden, J., Löffler, F., Hoffmann, J.: Setting up an interdisciplinary data infrastructure: why cooperation between domain experts and computer scientists matters - an experience report from the GFBio project. Biodivers. Inf. Sci. Stand. 1, e20198 (2017). https:// doi.org/10.3897/tdwgproceedings.1.20198

7. Nosek, B.A., Alter, G., Banks, G.C., Borsboom, D., Bowman, S.D., Breckler, S.J., Buck, S., Chambers, C.D., Chin, G., Christensen, G., Contestabile, M., Dafoe, A., Eich, E., Freese, J., Glennerster, R., Goroff, D., Green, D.P., Hesse, B., Humphreys, M., Ishiyama, J., Karlan, D., Kraut, A., Lupia, A., Mabry, P., Madon, T., Malhotra, N., Mayo-Wilson, E., McNutt, M., Miguel, E., Paluck, E.L., Simonsohn, U., Soderberg, C., Spellman, B.A., Turitto, J., VandenBos, G., Vazire, S., Wagenmakers, E.J., Wilson, R., Yarkoni, T.: Promoting an open research culture. Science 348(6242), 1422–1425 (2015). https://doi.org/10.1515/9783110494068-038

8. Pfeiffenberger, H.: Data publishing und open access. In: von Söllner, K., Mittermaier, B. (eds.) Praxishandbuch Open Access. De Gruyter Saur, Berlin (2017). https://doi.org/10.1515/9783110494068-038

9. Renaud, G., Azzopardi, L.: SCAMP: a tool for conducting interactive information retrieval experiments. In: IIIX 2012, pp. 286–289. ACM, New York (2012). https:// doi.org/10.1145/2362724.2362776

10. Schobel, J., Pryss, R., Schlee, W., Probst, T., Gebhardt, D., Schickler, M., Reichert, M.: Development of mobile data collection applications by domain experts: experimental results from a usability study. In: Dubois, E., Pohl, K. (eds.) CAiSE 2017. LNCS, vol. 10253, pp. 60–75. Springer, Cham (2017). https://doi.org/10.1007/978-3-319-59536-8_5

11. Rocha da Silva, J., Aguiar Castro, J., Ribeiro, C., Correia Lopes, J.: Dendro: collaborative research data management built on linked open data. In: Presutti, V., Blomqvist, E., Troncy, R., Sack, H., Papadakis, I., Tordai, A. (eds.) ESWC 2014. LNCS, vol. 8798, pp. 483–487. Springer, Cham (2014). https://doi.org/10.1007/978-3-319-11955-7_71

12. Stanford, N.J., Wolstencroft, K., Golebiewski, M., Kania, R., Juty, N., Tomlinson, C., Owen, S., Butcher, S., Hermjakob, H., Le Novère, N., Mueller, W., Snoep, J., Goble, C.: The evolution of standards and data management practices in systems biology. Mol. Syst. Biol. 11(12), 851 (2015)

13. Wilkinson, M.D., Dumontier, M., Aalbersberg, I.J., Appleton, G., Axton, M., Baak, A., Blomberg, N., Boiten, J.W., da Silva Santos, L.B., Bourne, P.E., Bouwman, J., Brookes, A.J., Clark, T., Crosas, M., Dillo, I., Dumon, O., Edmunds, S., Evelo, C.T., Finkers, R., Gonzalez-Beltran, A., Gray, A.J.G., Groth, P., Goble, C., Grethe, J.S., Heringa, J., 't Hoen, P.A.C., Hooft, R., Kuhn, T., Kok, R., Kok, J., Lusher, S.J., Martone, M.E., Mons, A., Packer, A.L., Persson, B., Rocca-Serra, P., Roos, M., van Schaik, R., Sansone, S.A., Schultes, E., Sengstag, T., Slater, T., Strawn, G., Swertz, M.A., Thompson, M., van der Lei, J., van Mulligen, E., Velterop, J., Waagmeester, A., Wittenburg, P., Wolstencroft, K., Zhao, J., Mons, B.: The FAIR Guiding Principles for scientific data management and stewardship. Sci. Data 3, sdata201618 (2016). https://doi.org/10.1038/sdata.2016.18

14. Wolstencroft, K., Owen, S., Krebs, O., Nguyen, Q., Stanford, N.J., Golebiewski, M., Weidemann, A., Bittkowski, M., An, L., Shockley, D., Snoep, J.L., Mueller, W., Goble, C.: SEEK: a systems biology data and model management platform. BMC Syst. Biol. **9**, 33 (2015). https://doi.org/10.1186/s12918-015-0174-y
15. Wuest, T., Tinscher, R., Porzel, R., Thoben, K.D.: Experimental research data quality in materials science. Int. J. Adv. Inf. Technol. **4**(6), 1–18 (2014). https://doi.org/10.5121/ijait.2014.4601

VizDSL: A Visual DSL for Interactive Information Visualization

Rebecca Morgan[1(✉)], Georg Grossmann[1], Michael Schrefl[1],
Markus Stumptner[1], and Timothy Payne[2]

[1] University of South Australia, Adelaide, SA 5000, Australia
rebecca.morgan@mymail.unisa.edu.au
[2] Lockheed Martin STELaRLab, Edinburgh, SA 5111, Australia

Abstract. The development of systems of systems or the replacement of
processes or systems can create unknowns, risks, delays and costs which
are difficult to understand and characterise, and which frequently result
in unforeseen issues resulting in overspend or avoidance. Yet maintaining state of the art processes and systems and utilising best of breed
component systems is essential. Visualization of disparate data, systems,
processes and standards can help end users to understand relationships
such as class hierarchy or communication across system components better. There are many visualization tools and libraries available but they
are either a black box when it comes to specifying possible interactions
between end users and the visualization or require significant programming skills and manual effort to implement. In this paper we propose
a visual language called VizDSL that is based on the Interaction Flow
Modeling Language (IFML) for creating highly interactive visualizations.
VizDSL can be used to model, share and implement interactive visualization based on model-driven engineering principles. The language has
been evaluated based on interaction patterns for visualizations.

Keywords: Model-driven visualization · Domain-specific modelling
Interactive information visualization · IFML

1 Introduction

Modern operational systems are often large, complex and composed of distributed systems of heterogeneous components. Lack of integration and interoperability is costly and often not discovered until late in the development process.
Identification of issues, such as non-matching concepts between specifications,
or just getting an understanding of the scope of an implementation effort is a
significant challenge and can be achieved using modelling over different abstraction levels [1] and domains [2]. An example system discussed in this paper lies
in the digital hand over of design documents to the operation and maintenance

Rebecca Morgan was supported by the Australian Government Research Training
Program Scholarship and by Lockheed Martin Australia.

area in the integrated energy industry using specific standards. The information contained in these standards is very complex, often spanning several parts with hundreds of pages containing several hundred concepts and relationships. Understanding the specifications is extremely challenging and time consuming, even for domain experts like engineers in the energy industry who need to approve what kind of information is passed between information systems and that the information is semantically correct and represented by those standards. Visualization of the standard specifications can be used to reduce the time taken to gain this understanding.

While engineers in this space are reasonably fluent in data modelling and data mapping concerns they are not IT experts or developers who are required to create customised visualizations. Currently, customised visualizations are often hard coded at a low level requiring domain, IT and programming expertise. The solution cannot be reused for other visualizations and is a black box where it is not clear how the visualization and interactivity is specified. If commercial visualization tools are used instead, significant difficulties can be faced in terms of interoperability and extensibility, communication of visualization design between users, product lifespan and support.

The goal of our research is to address these issues by developing a visualization tool based on model-driven engineering principles that creates interactive visualizations quickly for end users to explore and understand complex semantic structures. In this paper, we describe a *platform-independent* and extensible modelling *language*, VizDSL, which allows non-IT experts to describe, model and create interactive visualizations, quickly and easily.

The paper uses a case study from the energy industry based on our ongoing project work which motivated this research and also provides realistic large scale industrial datasets [3]. Lockheed Martin Australia is now supporting this project because of similar requirements in the visualization of complex system architecture frameworks.

The contribution of this paper is threefold: (1) compared to [4] in which we introduce VizDSL, we describe in this paper the detailed metamodel of VizDSL, (2) discuss the implementation and (3) evaluate the implementation based on the visual patterns identified by Heer and Agrawala [5]. In the following, we discuss first the background of the VizDSL approach and related work and then describe the metamodel and the evaluation in more detail.

2   Background

Ensuring interoperability in the face of increasingly large and complex enterprise systems is challenging and expensive. Model driven engineering (MDE) is a way of addressing the difficulties inherent in dealing with complex system architectures. MDE focuses on the model as the primary development driver, rather than the code [6]. Interoperability is assisted by the model being independent of the implementation platform, as well as the provision of customised and reusable software. In MDE, the code is generated directly from the model, as opposed to traditional code-driven development, where the model is referred to separately as an information source during the primary process of writing the code. In MDE, Domain Specific Languages (DSLs) are used to represent domain models in a relatively concise and efficient way [7] leading to greater productivity and reduced costs. In this context, DSLs are used in conjunction with transformation engines/code generators to generate required artifacts such as source code [8]. DSLs with modelling support are used as part of domain specific development, for effective code generation from the model artifacts.

IFML. An example of a DSL with modelling support is the Interaction Flow Modeling Language (IFML)[1]. IFML uses a graphical notation to create visual models of content, user interaction and controls for front end applications. IFML provides a modelling language which sits between a user interface and the code. IFML has been recently accepted as a standard by the OMG for front-end design and it is well established through its predecessor WebML. A commercial implementation of IFML is available through WebRatio[2]. IFML can be extended through its specification written in UML.

Visualization. Visualization amplifies cognition [9] and facilitates knowledge sharing between different user groups with different levels of expertise and experience, without requiring extensive background knowledge or training [10–15]. There is a growing need to support information visualization in a way which is compatible with MDE. Visualization techniques are used as part of MDE to visualise the code, the problem domain and the models used to describe the domain. Models are frequently represented graphically, as a formal specification using graphical syntax, or in an informal sense as a quick means of communicating key features and ideas. Knowledge of underlying semantic structures is of fundamental importance in MDE. This knowledge is collected and shared by visualising semantic information, such as hierarchical, relational or entity-based semantic information, or some combination of these [16].

Although there has been significant research conducted in the areas of visualization tools and techniques, there are no existing visualization tools which can be used to create, share and implement visualizations using model driven techniques and which are based on accepted standards.

To address these gaps, we have developed a graphical DSL which facilitates the creation of interactive visualizations called VizDSL [4]. In [4], we introduced the concept of VizDSL without providing extension details. Since VizDSL is platform-independent and extensible through its UML profile, it is important to provide IFML extension details, to aid implementation in the community. In this paper, we describe the metamodel with the visual syntax of VizDSL and its implementation using the Domain Modelling Environment (DoME) and the

---

[1] http://www.ifml.org.
[2] https://www.webratio.com/site/content/en/home.

Agile Visualization package[3]. Further it includes an industry application and evaluation of the framework.

VizDSL takes a model-based approach rather than a procedural approach to the design process, to meet integration and interoperability requirements in the context of MDE. Since VizDSL has been designed with the aims of MDE in mind, strong support for semantic visualization is provided, including the ability to design interactive visualizations with a hierarchical navigational structure. Before we describe VizDSL in more detail, we discuss its role in the engineering of information systems within an oil & gas industry project.

2.1   Industry Application

To provide context for the application of the proposed DSL, the following section discusses a real life case study which provided the motivation for developing VizDSL. The Open Industrial Interoperability Ecosystem (OIIE)[4] provides a supplier neutral interoperability ecosystem, aimed at asset intensive industries such as the defense industry and the integrated energy industry. The OIIE framework uses system-of-systems interoperability and employs a portfolio of standards for components of representative business processes. These standards are selected according to industry requirements, adoption and community engagement. One instance of the OIIE is found in the Oil and Gas Interoperability (OGI) Pilot[5]. The OGI Pilot is a public collaborative interoperability test bed for the OIIE within the oil and gas industry. Participating organisations in the OGI Pilot include BP, SAP and Yokogawa Electric.

An area of active interest in the OGI Pilot lies in digital information handover, for example, the transfer of operations and management related information from Engineering, Procurement and Construction (EPC) systems to systems that owner/operators are using. In order for safe plant operation and production to occur, operations and management systems must be populated with structural information about the plant, including operating parameters, product data and serialized assets. Businesses in the design domain generally use the ISO 15926 standard for information models, while businesses in the operations and maintenance domain generally use MIMOSA.

For interoperability, the standards must be mapped to each other as part of the digital information handover process. Due to the complexity of the standards information, it is necessary to identify and visualize the hierarchical structure obtained from formal standard specifications, quickly and efficiently, so that this mapping can be established.

2.2   Industry Requirements

From our previous work [4] and from the identified organizational problem as discussed in Sect. 2.1, the following requirements have been identified:

---

[3] http://agilevisualization.com.

[4] http://www.mimosa.org/open-industrial-interoperability-ecosystem-oiie.

[5] http://www.mimosa.org/oil-and-gas-interoperability-ogi-pilot.

R1 Semantic Visualization Using MDE Techniques: Although there are a significant number of visualization solutions available, these solutions tend to rely on procedural programming techniques, with or without graphical user interfaces [4]. Procedural programming techniques are not as well suited to representing information in terms of entities, relationships and hierarchical structure as the model driven approach due to the lower level of abstraction.

Requirement: There is a need for a graphical DSL which can be used to model and implement interactive visualizations. This DSL should be able to visualise complex information such as that contained in the OIIE standards and should be able to visualise semantic structures as well as data content.

R2 Creating Complex Interactive Visualizations Without Programming Experience: With respect to the motivating example given above, the information contained in the standards is currently represented using static visualizations which have been hard coded by software engineers. This approach is problematic. The size and complexity of the information being presented means that static visualizations are not very effective in matching concepts. The structure of the visualization, what it will represent and how it will be represented, has to be inferred directly from the code, which leads to difficulties in communication for non-IT experts, such as design engineers. It would be preferable, in this case, if they are able to create their own visualizations without the need to rely on a third party with necessary programming skills.

Requirement: There is a need for a visualization DSL which can be used by non-IT experts to quickly and easily create highly interactive visualizations.

R3 Support for Standards-Based Interoperability: The problems faced in information handover in the OGI Pilot are further complicated by the fact that the OGI Pilot, as part of the OIIE, employs system-of-systems interoperability. This implies that a visualization solution must remain independent while able to act in concert with other systems used in the OGI Pilot. Standards are employed in interoperable systems to ensure a common grounds for information exchange.

Requirement: There is a need for a visualization DSL which can be employed as part of a heterogeneous system of interoperable components, and which should be based on a standard to increase its adoption.

3   Related Work

There has been significant interest in the field of visualization in recent times. In the following section, we discuss Model Driven Visualization, a framework for the creation of visualizations using concepts from MDE. VizDSL fits into the Model Driven Visualization framework as a DSL which describes software visualizations. We give a brief overview of DaisyViz, another model-based approach to information visualization and briefly discuss CloudMap and RALph, as examples of graphical DSLs which follow a similar structure to VizDSL, although they are not related to visualization as such.

Model Driven Visualization: One proposed approach to visualization in MDE is Model Driven Visualization (MDV) [17–19]. MDV is a model based, platform independent approach to creating visualizations which uses the MDE concepts of metamodels and transformations to describe the structure of visualization tools. One of the open problems identified in [18] was the lack of a suitable transformation language; support was requested to help build a suitable DSL for describing software visualization.

The MDV reference architecture recommends meta-models for the domain and the viewers. Bull [19] noted that there was a need for a DSL for describing software visualization, which can be used for describing translations between domain and viewer meta-models. In [19], it was noted that the work on MDV had focused primarily on the static data structures behind the specification of a view model and that modelling languages could support a broader range of options for view control and behaviour which is addressed in part by VizDSL.

In comparison to presented work, MDV provides a framework for visualization in the context of MDE but cannot be used to create visualizations. VizDSL provides a language to model and execute model-driven visualization.

DaisyViz: DaisyViz [20] is a model-based user interface toolkit for the development of domain-specific information visualization applications without programming. DaisyViz uses three declarative models: data, visualization and control. The data model uses relational database schema to manage data, the visualization model is used to define the visual representation of data and the control model is used to describe the information visualization tasks and techniques used in the views. While DaisyViz has a textual syntax, VizDSL uses a visual modelling language, which improves understandability for non-IT experts.

CloudMap: CloudMap [21] presents a visual notation for representation and management of cloud resources. Notational constructs include entities, links, probes and control actions. These notation constructs were identified and subsequently specified after user surveys. CloudMap implements an interactive mindmap visualization for the navigation of cloud resources, detection and display of events (event management system) and manual and automated actions. The end results were promising, with significantly improved efficiency. CloudMap focuses solely on the visualization of cloud resources whereas VizDSL can be used to visualize any schema or structured data.

RALph: RALph [22] presents a graphical notation for visual modelling of resource selection conditions in process models. The underlying semantic concepts were identified on previous experience and case studies. RALph was integrated into a platform which uses BPMN and provides a graphical editor based on Oryx. Evaluation of RALph is not currently available, however, the authors intend to evaluate the understandability and learnability of RALph in the future, using the Physics of Notation [13]. As with CloudMap, RALph is restricted to a particular domain, in this case, resource assignments in business processes. VizDSL is general purpose in that it is not limited to any particular information domain.

4   VizDSL

In this section, an overview of VizDSL, the graphical notation for VizDSL and discussion of the proposed VizDSL extensions to IFML is given.

VizDSL takes a model driven approach to visualization in the form of a DSL which can be used to model interactive visualizations with a focus on representation of semantic information. VizDSL takes the form of a graphical language since there are cognitive benefits to using graphical languages [23] and graphical languages are more effective in conveying information to non-IT experts [13]. To ensure standards-based interoperability, VizDSL makes use of existing standards. A natural candidate to form the basis of VizDSL is the Information Flow Modeling Language (IFML) released by the OMG: (1) it is a standard, (2) follows MDE principles and (3) can be used to specify user interactions. However, IFML originated from the WebML language aiming at model-driven Web engineering and not the development of interactive information visualizations. Hence, the first step in the development of VizDSL was to identify how to extend IFML with appropriate concepts such as *Data* and *Interaction* for modelling interactive visualizations of data specifications. The goal of this research was to develop a modelling language that is independent from the implementation of a visualization and can be executed by different visualization libraries.

For testing the execution of the language the *Agile Visualization* framework was chosen because it is open source, highly customisable and has strong out of the box support for interactions, animations and web browser integration. It is flexible enough to create completely new visualization techniques. The first step and also challenge in the design of VizDSL was (1) to identify the gap between IFML and Agile Visualization concepts and (2) the integration of both. In some cases the IFML metamodel could be extended with the visualization concepts in a straight forward manner, e.g., place the *ViewContainer* from Agile Visualization as subclass of *ViewElement* in IFML. However, in some cases it was not trivial, e.g., IFML has no or only limited support when it comes to direct representation of data sources including layouts and themes and how they should be associated with other concepts in the metamodel. Below we discuss the extensions of IFML which represent the VizDSL metamodel:

In Fig. 1, the core visual notation for VizDSL is shown. The visual notation is based on the IFML visual notation (shapes in grey), with new notations for the proposed extensions (shapes in white).

In Fig. 2, the class diagram for VizDSL is shown. The VizDSL classes from top left to bottom right which can also be found in IFML are *Parameter*, *ParameterBinding*, *ParameterBindingGroup*, *ViewElement*, *ViewComponentPart*, *Event*, *Action*, *ViewComponent* and *ActionEvent*. These classes have been extended in VizDSL with additional attributes. New classes introduced by VizDSL from top left to bottom right in Fig. 2 are *Data*, *Layout*, *Theme*, *Menu*, *Shape*, *MenuItem*, *Element*, *Edge*, *Interaction*, and *Animation*. Remaining classes are

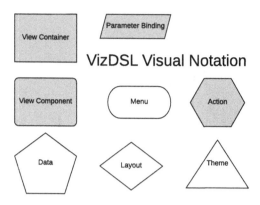

Fig. 1. VizDSL visual notation

VizDSLObject and *ViewContainer* are IFML classes which have been modified for VizDSL[6]. Below some of the classes are described in more detail.

VizDSLObject: This is the VizDSL root class. Like in IFML, which has the root class *InteractionFlowElement*, a root class is required in VizDSL that captures generic attributes that are inherited by subclasses, e.g., *parameter* relationship attribute which maps data to visual objects (see also below about the *Data* concept).

ViewContainer: *ViewContainer* is present in IFML and was extended to represent complex structures by adding new subclasses *Structure, Relationships, Comparison* and *Detail*. Generic concepts *Elements* and *Edges* are rendered on *ViewContainers*, so a relationship was required to model this. Further, *ViewContainers* are associated with *Layouts*, a *Theme, Menus, Interactions* and *Animations*.

Data: One of the significant extensions to IFML is the *Data* class. Although IFML provides means of describing content and data, it does not contain an explicit concept to describe them. This is required to directly represent data source as part of the visualization design process, rather than having to create another data model using a different language. The *Data* class is associated with the IFML *Parameter* class, which provides a way of mapping data to visual objects represented using the *VizDSLObject* class.

Layout and **Theme:** These two classes were introduced to provide users with the capability to use different layout algorithms (*Layout*) and to create a consistent style for multiple views (*Theme*) as there is no way for defining layouts in themes in IFML.

---

[6] In a color-printed version of this paper the IFML classes with extended attributes are in black, new introduced classes are in green and modified classes are in red.

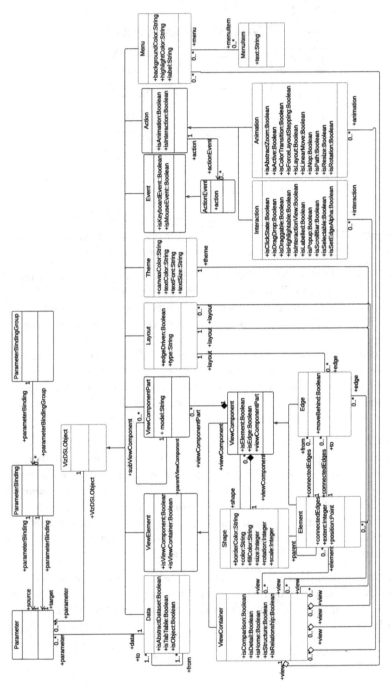

Fig. 2. VizDSL class diagram

Menu and **MenuItem:** Interactive menus were not directly supported by IFML but are required for interactive visualizations. As with *Themes, Menus* are reusable and can be associated with multiple *ViewContainers*.

Shape: The *Shape* class was introduced as a means of defining the visual attributes of the primary visual components for the proposed VizDSL *View-Component* classes: *Elements* and *Edges*. Each *Element* and *Edge* is associated with a *Shape*, which is used to map visual attributes to the final display.

Element: The *Element* class was introduced to provide visual representations of entities, which was identified as a requirement for the VizDSL language in Sect. 2.2. *Elements* can be associated with multiple *ViewContainers, Edges* and with other *Elements*, by nesting.

Edge: Together with *Elements, Edges* make up the fundamental visual components of a VizDSL visualization. *Edges* represent relations between entities, which is one of the requirements for VizDSL as identified in Sect. 2.2. Each *Edge* is from one *Element* to another *Element* and is associated with a *Shape*, which provides the visual representation of the *Edge*. As with *Elements, Edges* can be associated with multiple *ViewContainers*. Each *Edge* is associated with a *to* and *from* element.

Interaction: The *Interaction* class is a child of the IFML *Action* class. *Interactions* are triggered by IFML *Events*. *Events* are occurrences which affect the state of the application, such as clicking on the mouse. When an *Event* occurs, an *Action* is triggered. The *Interaction* class was introduced in VizDSL to represent required user interactions, such as selection, drag and drop, dragging, highlighting, labelling, popups, scrollbars and zooming.

Animation: As with the *Interaction* class, the *Animation* class was introduced as a child of the IFML *Action* class. Animations are well supported in Agile Visualization and can provide a greater level of engagement with visualizations. *Animations* are triggered by an *Event* and include abstract zoom, color transitions, layout transitions and stepping, resizing and rotation.

5   Implementation

In the following section, details are given for a prototype implementation of VizDSL using DoME in conjunction with the Roassal visualization engine that implements Agile Visualization. For this particular implementation, the Roassal2 package was used in the VisualWorks development environment. Roassal also works in Pharo, an open source Smalltalk environment; VisualWorks was chosen because DoME is only available in VisualWorks.

Domain Modelling Environment (DoME): We used the DoME for a prototype implementation of VizDSL. DoME is an integrated set of model-editing, metamodelling and analysis tools. We are using DoME to extend IFML and create an editor for our proposed VizDSL. In this editor we can create models that are executable and generate interactive visualizations in Agile Visualization. DoME is available through VisualWorks[7], the commercial implementation of Smalltalk. The VizDSL metamodel was defined using DoME, in order to create interactive model editors which can be used to define VizDSL diagrams.

Agile Visualization: The Agile Visualization package was used to render visualizations. Agile Visualization is written in Smalltalk and is available in Pharo, the open source Smalltalk implementation, or in VisualWorks, the commercial Smalltalk implementation. Agile Visualization was originally developed for the visual analysis of software systems and takes an object-oriented approach to visualization. Agile Visualization was chosen because it supports the identified requirements except for providing a graphical language (for which we can use IFML) and it provides basic building blocks to create the majority of existing visualizations.

6   Evaluation

We have evaluated the framework against the requirements mentioned in Sect. 2.2 and against software design patterns in information visualization. Before we discuss this in more detail, the industry application is demonstrated on the visualization of ISO 15926.

6.1   Industry Application: Visualization of ISO 15926

To illustrate the capabilities of VizDSL, we have used VizDSL to model and implement an interactive visualization of the ISO 15926 standard. As mentioned in Sect. 2.1, the ISO 15926 standard is generally used by the EPC domain. The information contained in ISO 15926 is complex and visualization of the hierarchical structure is necessary for interoperability. The ISO 15926 standard describes 320 classes with associated attributes and relationships. The relatively high number of classes means that static visualizations of the hierarchical structure are cluttered and difficult to understand. In this case, it is more useful to create an interactive visualization with drill-down navigation, since this eliminates unnecessary visual clutter and allows users to focus more effectively on areas of interest.

VizDSL was used to model and implement an interactive visualization of the ISO 15926 hierarchy (shown in Fig. 3). In Fig. 3 the *Data* source refers to the ISO 15926 classes and subclasses, which are assigned to a collection named *classes* through a parameter binding group. There is one *View Container* for this visualization, containing *Elements* on the *classes* collection, colored according to the number of subclasses, with *Edges* from each *Element* to its superclass. Interactions include element selection and details on demand. Menus include node expansion for drilling down, *Theme* and *Layout* selection and element finder.

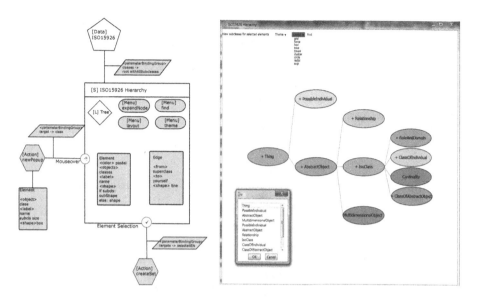

Fig. 3. VizDSL model for an interactive visualization of the ISO 15926 hierarchy

6.2 Evaluation Against Requirements

R1 Semantic Visualization Using MDE Techniques: From the example implementation described above, it is evident that VizDSL meets this requirement, as it was successfully used to model and execute an interactive visualization of the complex information contained in the ISO 15926 standard.

R2 Creating Complex Interactive Visualizations Without Programming Experience: It is important to find a balance between understandability and expressiveness when developing a domain specific modelling language. One of the primary drivers behind the development of VizDSL is the need for non-IT experts to quickly and easily create complex interactive visualizations and as such, understandability is of fundamental importance. IFML, on which VizDSL is based on, was designed for the web development community, which is multidisciplinary and covers a broad spectrum of IT and non-IT skills [24], similar to our target audience. The design principles behind IFML which support model usability and understandability (conciseness, extensibility, implementability, reusability, default modelling patterns and details) is transferred to VizDSL as an extension of IFML.

One of the primary advantages of MDE is the ability to represent complex systems at a relatively high abstraction level, which increases understandability. Again, the example described above highlights that this was possible without writing a single line of code and cannot be achieved with any framework mentioned in the related work section. A comparable visualization to the example above in the popular D3.js JavaScript library is the *d3.layout.tree*[8] which requires

---

[8] http://mbostock.github.io/d3/talk/20111018/tree.html.

170 lines of code. A non-IT expert is not able to write this and that particular D3 example does not even support menus as does the VizDSL example above.

R3 Support for Standards-Based Interoperability: VizDSL, as an extension of the IFML standard, takes advantage of an existing and widely accepted language for modelling user interaction and navigation. As an extension of IFML, a VizDSL model can be connected to other models describing the structural aspects of systems. For example, a VizDSL model can be connected to a UML class diagram to visualize software system structure.

6.3   Software Design Patterns for Information Visualization

Further, we have chosen to evaluate VizDSL using the software design patterns for information visualization as identified by Heer and Agrawala [5], since this is a comprehensive list of patterns. Heer and Agrawala identified a series of twelve design patterns for the domain of information visualization: *Reference Model, Data Column, Cascaded Table, Relational Graph, Proxy Tuple, Expression, Scheduler, Operator, Renderer, Production Rule, Camera, Dynamic Query Binding.* In the following section, we discuss how VizDSL applies to a selection of these interaction patterns.

Reference Model. *Separate data source from visual attributes.*

VizDSL separates the data source from the visual attributes by using the *Data* class to represent the data, the *ViewComponent* subclasses *Element* and *Edge* to represent visual components, the *ViewContainer* class to represent the view and the presentation classes *Layout* and *Theme* for the visual style.

Renderer. *Rendering of visual items performed by dedicated, reusable models which map visual attributes into pixels.*

VizDSL uses the *Shape, Layout* and *Theme* classes to render attributes.

Operator. *Decompose visual data processing into a series of composable operators, enabling flexible and reconfigurable visual mappings.*

VizDSL supports visual data processing using operators by means of the various types of *Layout, Interaction, Animation* and *Shape* classes. For example, the *Layout* class can be used as a basis for composing a custom hybrid layout using a combination of layout types, element groups and edges.

Dynamic Query Binding. *Allow data selection and filtering criteria to be specified dynamically using direct manipulation interface components.*

VizDSL supports the use of dynamic query binding using callbacks in the *Event* and *Action* classes. As an example, users can select elements from the visualization by clicking on them, which will add the element model to a filter group. This filter group can be used as the focus for the next view when navigating through the visualization.

Operator. *Decompose visual data processing into a series of composable operators, enabling flexible and reconfigurable visual mappings.*

VizDSL uses a series of composable builders throughout the visualization pipeline, by the use of classes which are mapped to data structures, visual attributes and control mechanisms such as interactions.

7   Conclusion

In this paper, we have proposed a novel approach for the creation and implementation of interactive visualizations by the development of a visual modelling language (VizDSL). VizDSL extends IFML using its UML profile which facilitates interoperability between models. VizDSL can be used to design and create highly interactive visualizations which improve understanding of both data content and underlying semantic structures.

Evaluation of the quality of modelling languages presents some challenges in the absence of an established modelling language quality evaluation framework [25]. In [26], an empirical framework to evaluate the usability of modelling tools in terms of *satisfaction, efficiency* and *effectiveness* is presented; we will be evaluating usability in this sense by means of user studies with users taken from the OGI Pilot. Future work will also include using the feedback obtained from the usability studies to refine and extend the VizDSL metamodel, with more work on code generation.

References

1. Grossmann, G., Igamberdiev, M., Stumptner, M.: Benefits and challenges of multi-level modelling for ecosystem interoperability. In: Proceedings of BDI4E Workshop at I-ESA (2016)
2. Grossmann, G., Jordan, A., Muruganandha, R., Selway, M., Stumptner, M.: Enabling information interoperability through multi-domain modeling. In: Harmsen, F., Proper, H.A. (eds.) PRET 2013. LNBIP, vol. 151, pp. 16–33. Springer, Heidelberg (2013). https://doi.org/10.1007/978-3-642-38774-6_2
3. Selway, M., Stumptner, M., Mayer, W., Jordan, A., Grossmann, G., Schrefl, M.: A conceptual framework for large-scale ecosystem interoperability and industrial product lifecycles. Data Knowl. Eng. **109**, 85–111 (2017)
4. Morgan, R., Grossmann, G., Stumptner, M.: VizDSL: towards a graphical visualisation language for enterprise systems interoperability. In: Proceedings of Symposium on Big Data Visual Analytics (BDVA). IEEE (2017)
5. Heer, J., Agrawala, M.: Software design patterns for information visualization. IEEE Trans. Visual Comput. Graph. **12**(5), 853–860 (2006)
6. Brambilla, M., Cabot, J., Wimmer, M.: Model-Driven Software Engineering in Practice, 2nd edn. Morgan & Claypool Publishers, San Rafael (2017)
7. Jones, C., Jia, X.: Using a domain specific language for lightweight model-driven development. In: Maciaszek, L.A., Filipe, J. (eds.) ENASE 2014. CCIS, vol. 551, pp. 46–62. Springer, Cham (2015). https://doi.org/10.1007/978-3-319-27218-4_4

8. Schmidt, D.C.: Guest editor's introduction: model-driven engineering. Computer **39**, 25–31 (2006)
9. Fill, H.G.: Visualisation for Semantic Information Systems, 1st edn. Gabler Verlag, Wiesbaden (2009)
10. Howse, J., Stapleton, G., Taylor, K., Chapman, P.: Visualizing ontologies: a case study. In: Aroyo, L., Welty, C., Alani, H., Taylor, J., Bernstein, A., Kagal, L., Noy, N., Blomqvist, E. (eds.) ISWC 2011. LNCS, vol. 7031, pp. 257–272. Springer, Heidelberg (2011). https://doi.org/10.1007/978-3-642-25073-6_17
11. Kocbek, S., Kim, J.D., Perret, J.L., Whetzel, P.L.: Visualizing ontology mappings to help ontology engineers identify relevant ontologies for their reuse. In: Proceedings of 4th International Conference on Biomedical Ontology (2013)
12. Burgstaller, F., Stabauer, M., Morgan, R., Grossmann, G.: Towards customised visualisation of ontologies. In: Proceedings of the Australasian Computer Science Week Multiconference (ACSW), pp. 1–10. ACM Press (2017)
13. Moody, D.: The physics of notations: toward a scientific basis for constructing visual notations in software engineering. IEEE Trans. Softw. Eng. **35**(6), 756–779 (2009)
14. Aranda-Corral, G.A., Borrego-Diaz, J., Chavez-Gonzalez, A.M.: Repairing conceptual relations in ontologies by means of an interactive visual reasoning: cognitive and design principles. In: Proceedings of the 3rd IEEE International Conference on Cognitive Infocommunications (CogInfoCom), pp. 739–744. IEEE (2012)
15. Voigt, M., Pietschmann, S., Meißner, K.: Semantic Models for Adaptive Interactive Systems. Human-Computer Interaction, pp. 1–25 (2013)
16. Nazemi, K., Burkhardt, D., Ginters, E., Kohlhammer, J.: Semantics visualization - definition, approaches and challenges. Procedia Comput. Sci. **75**, 75–83 (2015)
17. Bull, R.I., Favre, J.M.: Visualization in the context of model driven engineering. In: MDDAUI (2005)
18. Bull, R.I., Storey, M.A., Favre, J.M., Litoiu, M.: An architecture to support model driven software visualization. In: Proceedings of the 14th IEEE International Conference on Program Comprehension (ICPC), pp. 100–106. IEEE (2006)
19. Bull, R.I.: Model driven visualization: towards a model driven engineering approach for information visualization. Ph.D. thesis (2008)
20. Ren, L., Tian, F., Zhang, X., Zhang, L.: DaisyViz: a model-based user interface toolkit for interactive information visualization systems. Visual Lang. Comput. **21**(4), 209–229 (2010)
21. Weerasiri, D., Barukh, M.C., Benatallah, B., Jian, C.: *CloudMap*: a visual notation for representing and managing cloud resources. In: Nurcan, S., Soffer, P., Bajec, M., Eder, J. (eds.) CAiSE 2016. LNCS, vol. 9694, pp. 427–443. Springer, Cham (2016). https://doi.org/10.1007/978-3-319-39696-5_26
22. Cabanillas, C., Knuplesch, D., Resinas, M., Reichert, M., Mendling, J., Ruiz-Cortés, A.: RALph: a graphical notation for resource assignments in business processes. In: Zdravkovic, J., Kirikova, M., Johannesson, P. (eds.) CAiSE 2015. LNCS, vol. 9097, pp. 53–68. Springer, Cham (2015). https://doi.org/10.1007/978-3-319-19069-3_4
23. Sendall, S., Kozaczynski, W.: Model transformation: the heart and soul of model-driven software development. IEEE Softw. **20**(5), 42–45 (2003)
24. Brambilla, M., Fraternali, P.: Interaction Flow Modeling Language: Model-Driven UI Engineering of Web and Mobile Apps with IFML, 1st edn. Morgan Kaufmann, San Francisco (2015)

25. Giraldo, F.D., Espana, S., Giraldo, W.J., Pastor, O.: Modelling language quality evaluation in model-driven information systems engineering: a roadmap. In: Proceedings of 9th IEEE Conference on Research Challenges in Information Science (RCIS), pp. 64–69 (2015)
26. Condori-Fernandez, N., Panach, J.I., Baars, A.I., Vos, T., Pastor, O.: An empirical approach for evaluating the usability of model-driven tools. Sci. Comput. Program. **78**(11), 2245–2258 (2013)

Big Data and Intelligence

Evaluating Several Design Patterns and Trends in Big Data Warehousing Systems

Carlos Costa[1,2](✉)⬛ and Maribel Yasmina Santos[2]⬛

[1] CCG - Centre for Computer Graphics, Guimarães, Portugal
carlos.costa@dsi.uminho.pt
[2] ALGORITMI Research Centre, University of Minho, Guimarães, Portugal
maribel@dsi.uminho.pt

Abstract. The Big Data characteristics, namely volume, variety and velocity, currently highlight the severe limitations of traditional Data Warehouses (DWs). Their strict relational model, costly scalability, and, sometimes, inefficient performance open the way for emerging techniques and technologies. Recently, the concept of Big Data Warehousing is gaining attraction, aiming to study and propose new ways of dealing with the Big Data challenges in Data Warehousing contexts. The Big Data Warehouse (BDW) can be seen as a flexible, scalable and highly performant system that uses Big Data techniques and technologies to support mixed and complex analytical workloads (e.g., streaming analysis, ad hoc querying, data visualization, data mining, simulations) in several emerging contexts like Smart Cities and Industries 4.0. However, due to the almost embryonic state of this topic, the ambiguity of the constructs and the lack of common approaches still prevails. In this paper, we discuss and evaluate some design patterns and trends in Big Data Warehousing systems, including data modelling techniques (e.g., star schemas, flat tables, nested structures) and some streaming considerations for BDWs (e.g., Hive vs. NoSQL databases), aiming to foster and align future research, and to help practitioners in this area.

Keywords: Big Data Warehouse · Hive · Presto · NoSQL · SSB+

1 Introduction

Big Data has become a relevant concept for organizations looking to achieve high business value, and it is commonly defined by its unquantifiable characteristics (volume, variety, velocity, value and veracity) [1, 2]. The threshold for which data becomes "big" is subjective, so it remains as an abstract concept [3], being often defined as data "too big, too fast, or too hard for existing tools to process" [4]. The concept gained significant notoriety in many business areas, such as healthcare, retail, manufacturing or modern cities [3, 5].

Big Data is a relatively recent scientific and technical topic, although there are already some efforts of standardizing constructs and logical components of general Big Data Systems (e.g., NIST Big Data Reference Architecture) [6]. Nevertheless, the concept of Big Data Warehousing is even more recent, with even more concerning ambiguity and lack of standard approaches. The BDW can be defined by its characteristics and design

© Springer International Publishing AG, part of Springer Nature 2018
J. Krogstie and H. A. Reijers (Eds.): CAiSE 2018, LNCS 10816, pp. 459–473, 2018.
https://doi.org/10.1007/978-3-319-91563-0_28

changes, including: massively parallel processing; mixed and complex analytical work-loads (e.g., ad hoc querying, data mining, text mining, exploratory analysis and materi-alized views); flexible storage to support data from several sources; real-time operations (stream processing, low latency and high frequency updates); high performance with fast insights and near real-time response; scalability to accommodate growing data, users and analysis; use of commodity hardware to lower costs; and interoperability in a federation of multiple technologies [7–15].

Big Data Warehousing represents a paradigm shift for organizations facing several challenges in their traditional Data Warehousing platforms, namely bottlenecks throughout the collection, storage, processing and analysis of data, due to high demand of input/output efficiency, scalability and elasticity, which can be achieved through parallel storage and processing [3, 16, 17]. The strict modelling approach of traditional DWs is another relevant concerning factor [8], which only highlights the relevance of BDWs. Nevertheless, the state-of-the-art in BDW reflects the young age of the concept, as well as ambiguity and the lack of common approaches to build BDWs according to their characteristics.

Consequently, this work aims to provide guidance when building BDWs, by evaluating several design patterns and trends to avoid potential pitfalls that are inevi-tably the consequence of a never-ending sea of doubts in these emerging contexts, including: are multidimensional models viable in Big Data Warehousing contexts? How does the size of the dimension tables affect query performance? Should we use nested structures in BDWs? For streaming scenarios, are NoSQL databases a more suitable option than Hadoop, and how data volume affects them? All the insights provided by this paper are the result of a laboratory experiment (see Sect. 3) using an in-house developed extension of the Star Schema Benchmark (SSB) [18], which is a Data Warehousing benchmark using the multidimensional modelling strategy [11], whose data model is based on the TPC-H Benchmark [19], an ad hoc querying benchmark based on an operational database. Therefore, this laboratory experiment focuses on an extended version of the SSB, the SSB+ [20], and integrates the eval-uation activity of a broader research process using the Design Science Research Methodology for Information Systems [21] to propose a set of foundations for Big Data Warehousing.

This paper is organized follows: Sect. 2 describes related work; Sect. 3 presents the SSB+ benchmark used in this work; Sects. 4 and 5 evaluate several design patterns and trends in Big Data Warehousing; Sect. 6 concludes with some remarks about the undertaken work.

2 Related Work

Regarding the work related to Big Data Warehousing, some works explore imple-mentations of DWs on top of Not Only SQL (NoSQL) databases, such as document-oriented [22], column-oriented [22] and graph models [23], despite the fact that they were mainly designed to scale Online Transactional Processing (OLTP) applications [24].

Moreover, there are works focusing on the storage technologies and optimizations for BDWs, discussing SQL-on-Hadoop systems like Hive [25] and Impala [26], or improving these technologies through new storage and processing mechanisms, namely using the ORC (Optimized Row Columnar) file format, an efficient columnar format for data analytics, or using Tez, an interactive and optimized execution engine [27]. Some authors propose advancements in analytical and integration mechanisms suitable for BDWs [28–30]. Other works present implementations in specific contexts, which can be related to certain characteristics of a BDW, such as the DW infrastructure at Facebook [31] or DW applications in medicine [32].

Currently, the state-of-the-art shows that the design of BDWs should focus both on the physical layer (infrastructure) and logical layer (data models and interoperability between components) [13], and, in general terms, it can be implemented using two strategies: the "lift and shift" strategy, wherein the capabilities of traditional and relational DWs are augmented with Big Data technologies, such as Hadoop or NoSQL to solve specific use cases, thus a use case driven approach instead of a data modelling approach, which often leads to possible uncoordinated data silos [33]; or the "rip and replace" strategy, wherein a traditional DW is fully replaced by Big Data technologies [13, 14]. However, current non-structured practices and guidelines are not enough. The community needs rigorously evaluated models and methods to design and build BDWs.

In previous works, we have been focusing on different aspects of Big Data Warehousing, in an attempt to provide scientifically supported methods in these contexts. Among these works we can highlight the following: the evaluation of several SQL-on-Hadoop systems [34]; the evaluation of star schemas [11], flat tables and partitions for the modelling and organization of BDWs [35]; and the proposal of architectures to implement BDWs in Smart Cities and Industries 4.0 [36, 37]. This paper continues this line of research, providing a rigorous evaluation obtained through benchmarking (laboratory experiment–Sect. 3) of several practices related to BDWs.

3 Research Method: Laboratory Experiment Using the SSB+

This section presents the SSB+, an extension of the SSB benchmark [18] that we developed to overcome the lack of workloads that combine volume, variety and velocity of data, with adequate customization capabilities and integration with current versions of different Big Data technologies. It can also be used to evaluate different modelling strategies (e.g., flat/denormalized structures, nested structures, star schemas) and different workload considerations (e.g., dimension tables' size, streaming).

3.1 Data Model and Queries

The SSB+ Benchmark data model (Fig. 1) is based on the original SSB Benchmark, so all the original tables remain the same ("lineorder", "part", "supplier", "customer"), with the exception of the "date" dimension, which has been streamlined to remove the several temporal attributes that are not used in the 13 original SSB queries. The 13 queries were only modified to comply with the ANSI SQL joins' syntax, in order to provide an optimal execution plan in the query engines' optimizers, and, obviously,

13 new queries were created to support the new flat *"lineorder"* table. These changes allow us to compare the advantages and disadvantages of star schemas and flat structures.

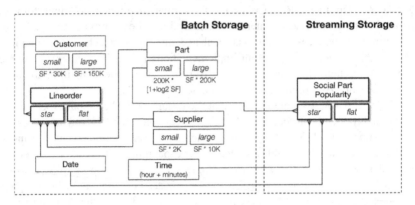

Fig. 1. SSB+ data model. Extended from [18].

Moreover, the SSB+ also considers two different dimensions' size: the original TPC-H sizes [19]; and the original SSB dimensions' size, which are smaller tables that represent typical dimensions in the retail context. This SSB+ feature allow us to understand the impact of the dimension's size in star schema-based BDWs.

Regarding the streaming workloads of the SSB+ Benchmark, a new *"time"* dimension table is included, as the stream has a "minute" granularity. This new dimension can then be joined to the new *"social part popularity"* streaming fact table, as well as other existing dimensions like *"part"* and *"date"*. A flat version of this fact table is also available for performance comparison purposes. The *"social part popularity"* table contains data from a simulated social network, where users express their sentiments regarding the parts sold by a certain organization. Three new queries were developed for both the star schema-based BDW and the flat-based BDW. All the applications, scripts and queries for the SSB+ Benchmark can be found in the respective GitHub repository [20].

3.2 System Architecture and Infrastructure

The SSB+ Benchmark takes into consideration several technologies to accomplish different goals, from data Collection, Preparation and Enrichment (CPE) workloads to querying and OLAP tasks. These technologies are presented in Fig. 2. Starting with the CPE workloads, for batch data, the SSB+ considers a Hive script with several beeline commands that load the data from HDFS to the Hive tables in ORC file format. Regarding streaming data, a Kafka producer generates simulated data at configurable rates, and this data is processed by Spark streaming that finally stores it in Hive and Cassandra. Streaming data is stored both in Hive and Cassandra for benchmarking purposes (see Sect. 5).

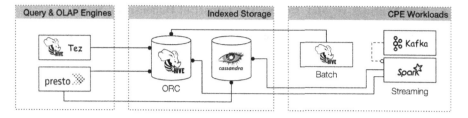

Fig. 2. SSB+ architecture.

For querying and OLAP, this work considers both Hive on Tez and Presto, which are two robust and efficient SQL-on-Hadoop engines [34], used in this work to observe if the conclusions hold true for more than one engine, since one of them may perform better with certain data modelling strategies, for example. However, in the streaming workloads, only Presto is used, since it targets interactive SQL queries over different data sources, including NoSQL databases, which is not a very proclaimed feature in Hive, although it can also be achieved. Besides, despite Tez' tremendous improvements to Hive's performance, Hive on Tez may not be considered a low-latency engine, as it tends to be often outperformed by more interactive SQL-on-Hadoop engines [26, 34]. However, other SQL-on-Hadoop engines can be used, as long as the appropriate scripts are added to the SSB+ Benchmark. All the content of the SSB+ repository [20] is open to the public, in order to facilitate any change or extension.

The infrastructure used in this work is a 5-node Hadoop cluster with 1 HDFS NameNode (YARN ResourceManager) and 4 HDFS DataNodes (YARN NodeManagers). The hardware used in each node includes: 1 Intel core i5, quad core, with a clock speed ranging between 3.1 GHz and 3.3 GHz; 32 GB of 1333 MHz DDR3 Random Access Memory (RAM), with 24 GB available for query processing; 1 Samsung 850 EVO 500 GB Solid State Drive (SSD) with up to 540 MB/s read speed and up to 520 MB/s write speed; 1 gigabit Ethernet card connected through Cat5e Ethernet cables and a gigabit Ethernet switch. The operative system in use is CentOS 7 with an XFS file system, and the Hadoop distribution is the Hortonworks Data Platform (HDP) 2.6.0. Besides Hadoop, a Presto coordinator is also installed on the NameNode, as well as 4 Presto workers on the 4 remaining DataNodes. All configurations are left unchanged, apart from the HDFS replication factor, which is set to 2, as well as Presto's memory configuration, which is set to use 24 GB of the 32 GB available in each worker (identical to the memory available for YARN applications in each DataNode/NodeManager).

4 Batch OLAP for Big Data Warehouses

This section discusses the performance of batch OLAP queries (processing of large amounts of data stored in the BDW's batch storage) for BDWs using two modelling approaches: star schemas and flat tables, considering the impact of dimensions' size in star schemas, and the effects of misusing nested structures in Hive tables.

4.1   Star Schemas vs. Flat Tables: The Impact of Dimensions' Size

Large dimensions can have a considerable impact in star schema-based DWs, as they require more time to compute the join operations between the fact tables and the dimension tables. In previous papers, we used larger dimensions' sizes (TPC-H original tables) [35], and concluded that star schemas can be largely outperformed by flat tables. Although this may not be the usual scenario for many traditional contexts, such as store sales analysis, for example, larger dimensions are typically found in several Big Data contexts, such as Amazon, which has hundreds of millions of customers and parts, or Facebook, which easily surpasses 1 billion users nowadays. Nevertheless, there are also several Big Data contexts wherein dimensions can have a small size, because many organizations can generate millions or billions of transactions only based on a small set of parts, customers and suppliers, for example. For this reason, it becomes interesting to analyze the performance impact caused by dimensions with different sizes. Figure 3 illustrates the results of the Scale Factor (SF) = 300 workload for large and small dimensions (around 1.8 billion sales transactions).

At first glance, looking at Hive's workloads in Fig. 3, the result is surprising. While with large dimensions the flat table is generally the modelling approach with better performance, it is surpassed by the star schema in the small dimensions workload. This

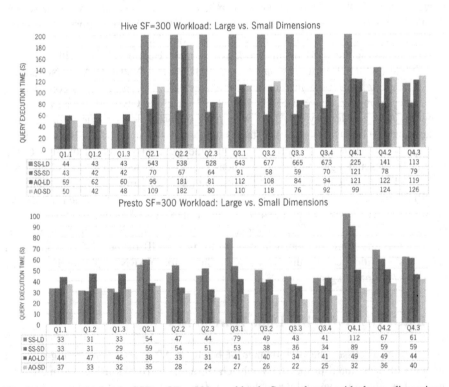

Fig. 3. Large-scale batch SSB+ SF = 300 workload. Star schema with large dimensions (SS-LD); star schema with small dimensions (SS-SD); flat table with large dimensions (FT-LD); flat table with small dimensions (FT-SD). Hive SS-LD/FT-LD based on [35].

shows that with Hive, modelling the BDW using multidimensional structures can not only save a considerably amount of storage size (2.5x in SSB+), but also, as Fig. 3 demonstrates, it can bring significant performance advantages. In this scenario, we conclude that if Hive is able to perform a map/broadcast join, having a larger flat (denormalized) structure may not be beneficial for highly dimensional data like sales data. For the SSB+ flat table, the overhead caused by a storage size that is 2.5x bigger leads to a performance drop, and may become a bottleneck for the Hive query engine. Considering only Hive's results, if the dimensions are small, the star-schema approach would be the most appropriate modelling strategy. Consequently, in certain contexts, it makes sense to model parts of the BDW's data that way.

However, very often Big Data does not adequately fit into the strictures of multi-dimensional and relational approaches (e.g., high volume and velocity sensor data, social media data). Moreover, taking a closer look at Presto's workloads, which are typically much faster than Hive's workloads, it can be observed that, generally, the star schema with smaller dimensions is significantly slower than the corresponding flat table. Even more surprising, the star schema with smaller dimensions is slower than the flat table with the higher attributes' cardinality (corresponding to larger dimensions). Overall, the star schema with smaller dimensions takes 57% more time to complete the workload when compared to the equivalent flat table. The discussion in this subsection is an adequate example to show why we use two SQL-on-Hadoop systems, as the insights retrieved from the workloads may vastly differ depending on the system.

Summarizing the conclusions, there is no hard rule. In certain Big Data Ware-housing contexts, practitioners need to consider their limitations regarding storage size and the characteristics of a particular dataset: is data highly dimensional? Are the dimensions big enough to make inefficient/impossible the use of map/broadcast joins? Furthermore, practitioners may need to perform some preliminary benchmarks with sample data before fully committing to either the extensive use of star schemas or the use of flat tables. Nevertheless, if we only look at query execution times, the results provided in this paper show that the best scenario includes the use of flat tables queried by Presto.

4.2   Are Nested Structures in Tables Always Efficient?

Nested structures like maps, arrays and JSON objects can be significantly helpful in certain contexts. For example, in [36], in which we discuss the implementation of a BDW in a Smart City context, is one of these contexts, as geospatial analytics is a priority, including several complex and nested geometry attributes. Sales analysis is another context where practitioners may find appealing the application of nested structures, namely using a less granular table *"orders"* with the granularity key *"order key"* and using a nested structure to store the data about the products sold in a particular order *(e.g., "product name", "quantity", "revenue")*. Nevertheless, are nested structures the most efficient solution every time? Do the processing of less rows and the smaller storage footprint always create tangible advantages?

Using a SF = 300 workload, the nested table has 95 GB, while the flat table has 139 GB, and the equivalent star schema has only 51 GB. This new nested model is also able to reduce the number of rows from 1.8 billion to just 450 million, since the data regarding the lines of the order is stored in a nested structure, namely an array of Structs. For this test, Q4.1 was chosen, because it involves the need to aggregate and filter data that is stored in the nested attribute *"lines"* across several dimensions. This allows the evaluation of applying different SQL operators to nested attributes, such as lambda expressions, besides the more traditional ones (e.g., Group By, Where).

At first glance, these numbers look promising, but Fig. 4, which presents the results of executing the SF = 300 Q4.1 in all modelling approaches, tells a different story. Despite saving storage space and having much less rows, the nested table is the least performant modelling approach. It can be concluded that storing a large number of dimensions' attributes in a complex structure like an array of Structs may result in a large overhead regarding query processing times. Such data modelling strategy requires the use of lambda expressions (or lateral views) to answer Q4.1, in which Presto wastes the majority of its time. Highly complex nested structures that will be accessed sequentially to answer most of the queries may not be a good design pattern. These results do not mean that processing nested structures are always bad for performance. Nested structures offer great flexibility and can be efficient for certain access patterns, allowing the introduction of new analytical workloads in the BDW, such as intensive geospatial simulations and visualizations.

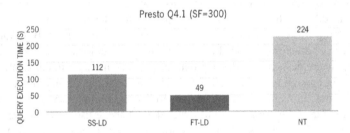

Fig. 4. Performance of a nested table in the SSB+ context. Star schema with large dimensions (SS-LD); flat table with large dimensions (FT-LD); nested table (NT).

5   Streaming OLAP for Big Data Warehouses

Streaming scenarios are common in Big Data Warehousing contexts. The BDW must be able to adequately deal with high velocity and frequencies regarding CPE workloads. Daily or hourly batch CPE workloads may not always be the most effective or efficient solution to solve specific problems, and streaming CPE workloads can be very useful in these cases. This section evaluates the performance of BDWs in streaming scenarios, while discussing several concerns that practitioners must take into consideration regarding stream processing (e.g., Spark streaming, Storm) and storage (e.g., HDFS, Hive, Cassandra, HBase).

Regarding storage technologies, there are two main approaches: using Hive or HDFS, which adequately deal with the sequential access workloads typically found on OLAP queries, but that are not the most adequate for random access (useful for streaming data). In contrast, NoSQL databases like Cassandra are efficient in random access scenarios, but typically fall short in sequential access workloads (useful for OLAP). This section evaluates the performance of Hive and Cassandra as streaming storage systems, using both a star schema and a flat table (see Fig. 1). The data flow is as follows:

1. A Kafka producer generates 10 000 records each 5 s;
2. A Spark streaming application with a 10 s micro batch interval consumes the data for that interval and stores it in Hive and Cassandra;
3. Presto is used to query Hive and Cassandra, every hour, over a period of ten hours.

5.1 How Data Volume Affects the Streaming Storage Component

The streaming storage system of a BDW can only store so much data before its performance starts to degrade, reason why we need to periodically move the data from the streaming storage to the batch storage. Therefore, in this subsection, we analyze how data volume affects the performance of the streaming storage component of the BDW. Figure 5 illustrates the total execution time for all streaming queries (Q5, Q6 and Q7) during a ten-hour workload with roughly constantly increasing data volume. All queries are executed each hour for Hive and Cassandra, and both for the flat table and the star schema.

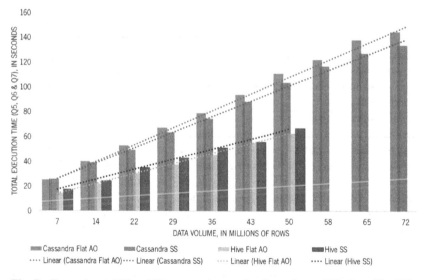

Fig. 5. Cassandra and Hive SSB+ streaming results. Star schema (SS); flat table (FT).

There are several interesting insights that emerge in these tests. The first one focuses on the overall effect of data volume on Hive and Cassandra, in which we conclude that as hours pass by, the increase in the workload's execution time can be modelled as a linear function. As the data volume increases, a significant performance drop is expected in Cassandra, since as previously argued, sequential access over large amounts of data is not one of its strong points. However, this is not expected in Hive, since as we demonstrated in [35], when using an interactive SQL-on-Hadoop engine to query Hive, one is able to achieve much faster execution times than the results obtained in this streaming workload, even with significantly higher SFs (e.g., 30 = 180 000 000 rows).

Despite this observation, detailed afterwards, it can also be concluded that Hive is always much faster than Cassandra until the mark of 58 million rows is reached. At this moment, it becomes clear that the Spark streaming micro batch interval is too short for the demand, and, therefore, the application generated over 9500 small files in HDFS (storage backend for Hive). This causes the streaming micro batches to be consequently delayed, making the results inconclusive, as the number of rows stored in Hive does not match the number of rows stored in Cassandra. Overall, it can be concluded that having small micro batch intervals when using Hive severely deteriorates the performance of the system. This insight corroborates the argument regarding the overhead of having many small files stored in Hadoop [38].

Cassandra also shows some delay in write operations when being queried by Presto, causing the Spark streaming application to queue a few micro batch jobs. However, this phenomenon is significantly less concerning than Hive's phenomenon, as the streaming system is able to control the load without too much delay. Besides, this is not caused by an increase in data volume, but rather by a concurrency issue and resource starvation while Presto queries are running. We can always sacrifice data timeliness by increasing the micro batch interval. However, in order to compare the results between Cassandra and Hive, the write latency and throughput should be identical. In this case, Cassandra adequately handles 20 000 rows each 10 s without significant delays, despite being slower, while Hive fails to do so, despite being faster for all workloads under the 58 million rows mark. This efficiency problem is discussed in more detail in the next subsection, among other relevant considerations regarding streaming for BDWs.

Another interesting analysis focus is the performance of flat tables and star schemas in streaming contexts. In our tests, performance is considerably similar, i.e., the star schema is marginally faster when using Cassandra, while the flat table is marginally faster when using Hive. After monitoring query execution, we concluded that the major difference is the time Presto spends reading from Cassandra, as the flat table is marginally larger. Moreover, in the SSB+ Benchmark, the star schema for the streaming scenario is not very extensive or complex, which in this case favors this modelling approach, since queries do not have to join an extensive set of tables. Despite this, it can be concluded that both modelling strategies are feasible, without any significant performance drawback. When using the star schema, as the dimension tables are stored in Hive, it can also be concluded that using a SQL-on-Hadoop system like Presto, it is also feasible and efficient to combine dimension tables stored in Hive (e.g., *"part"* and *"time"*) with streaming tables stored in Cassandra. These insights motivate practitioners to build BDWs according to the SSB+ architecture (Fig. 2).

5.2   Considerations for Effective and Efficient Streaming OLAP

A successful streaming application can be seen as an adequate balance between data timeliness and resource capacity. To explain these trade-offs, this subsection is divided into three main problems that emerge in our tests, presenting possible solutions to overcome them: high concurrency in multi-tenant clusters can cause severe resource starvation (multiple users and multiple technologies); storage systems oriented towards sequential access (e.g., Hive/HDFS) may present some problems when using small micro batch intervals; operations to move data between streaming and batch storage systems and CPE workloads should be properly planned with adequate resource availability.

Starting with the first problem, in Big Data Warehousing contexts shared-nothing and scale-out infrastructures are promoted [8], being typically capable of multi-tenancy, i.e., handling the storage and processing needs of several Big Data Warehousing technologies and users. Streaming applications, such as the one discussed in the previous subsection, typically require a nearly constant amount of CPU and memory for long periods. Data arrives at the system continuously, thus it needs to assure that the workload has the adequate amount of resources available.

A common setup, used in this work, would be a producer (e.g., Kafka), a consumer (e.g., Spark streaming), a storage system (e.g., Cassandra, Hive), and a query and OLAP engine (e.g., Presto). At first glance, the first three components of this setup may seem to work perfectly fine. However, once we add the query and OLAP engine, resource consumption in the cluster can get significantly high, and the performance of the streaming application may suffer, because we did not choose the adequate trade-off between data timeliness and resource capacity. Take as an example Fig. 6. If observed carefully, in certain periods of time coinciding with the time interval when Presto queries are running, there is a significant increase in the total delay of Spark streaming micro batches, caused by an increase in processing time, which consequently causes a significant increase in the scheduling time of further micro batches.

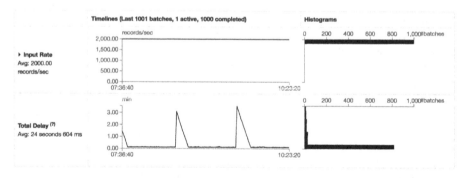

Fig. 6. Spark streaming monitoring GUI showing resource starvation when using Cassandra and Presto simultaneously.

In this case, this happens because there is not enough resource capacity in the current infrastructure to handle the processing demands of Spark streaming, Cassandra and Presto running simultaneously. In these periods, these technologies are mainly racing for CPU usage, and the initial Spark streaming micro batch interval of 10 s is not enough to maintain the demands of the streaming application. Again, these insights bring us back to the trade-off: either resource capacity is increased, in this case more CPUs or CPU cores, or the micro batch time interval is raised, which inevitably affects data timeliness. In this benchmark, queries are only executed each hour, therefore the system is only affected during these periods, but in real-world applications, users are constantly submitting queries, which makes this consideration hard to ignore.

Regarding the use of storage systems like Hive for streaming scenarios, as seen in the previous subsection, it has its advantages, namely faster query execution times than Cassandra. Nevertheless, this performance advantage comes at a cost: as data volume increases, the number of small files stored in HDFS rises considerably, generating a significant load on the system. One small file is created for each RDD partition, in this case each 10 s (micro batch interval). In a matter of hours, the Hive table contains thousands of small files. As the number of files increases, HDFS metadata operations take more time, affecting the time it takes for Spark streaming to save the data in Hive.

A write operation in HDFS includes steps like searching for the existence of the file and checking user permissions [39], which with thousands of files can take longer than usual. Nevertheless, we need to highlight that this problem can be solved by applying an adequate partition scheme to streaming Hive tables, e.g., partitioning by *"date"* and *"hour"*, which creates a folder structure containing fewer files in each folder, and therefore reducing the time to execute metadata operations. With thousands of small files in the same folder, the system is under intensive load and the Spark streaming application starts queuing hundreds of micro batches. Micro batches are queued when the Spark application cannot process them before the defined interval, in this case 10 s. Again, the pre-defined micro batch interval of 10 s is not able to assure that the data is processed before the next batch, and the performance of the streaming application is compromised.

In Hive's case, the small files problem is more severe than the concurrency issue shown by running Cassandra and Presto simultaneously. In Hive's case, even increasing resource capacity is not the best solution, and we should prefer higher micro batch intervals, which will consequently create bigger files. Moreover, it is significantly important to periodically consolidate these into bigger files, or moving them into another table that contains large amounts of historical data. It must be remembered that Hadoop prefers large files, further partitioned, distributed and replicated as blocks.

Finally, and taking into consideration the phenomena discussed above, workloads to move data between the streaming storage and the batch storage, as well as CPE workloads should be carefully planned when streaming applications are using the cluster's resources. These operations can be really heavy on CPU and memory, and can unexpectedly cause resource starvation, as seen with Presto and Cassandra running simultaneously. Practitioners should not take this lightly, and Linux Cgroups, YARN queues and YARN CPU isolation can be extremely useful to assure that the current infrastructure is able to properly assure a rich, complex and multi-tenant environment such as a BDW. These techniques assure that resources are adequately shared by multiple applications, by assigning the resources according to the expected workloads.

Furthermore, practitioners should evaluate their requirements regarding data timeliness, and avoid small micro batch intervals for streaming applications when not needed, as well as plan the execution of intensive background applications. More resource capacity may not always be the answer, since even in commodity hardware environments, buying hardware always come at a cost. In the meanwhile, making some of these changes may increase efficiency without any relevant cost.

6 Conclusion

This paper provided scientific results to support the analysis and understanding of several design patterns and trends in Big Data Warehousing, hoping to foster future research and to support design and implementation choices in real-world applications of BDWs. We discussed the trade-offs between star schemas and flat tables, using large and small dimensions; the usefulness and efficiency of nested structures in BDWs (e.g., array, maps, JSON objects); and several streaming considerations like storage performance, micro batch intervals and multi-tenancy concerns.

The main results provided by this paper are as follows: flat tables tend to outperform star schemas, both with large and small dimensions, but there are contexts wherein star schemas show some advantages; nested structures bring several benefits to BDWs (e.g., geospatial analytics using GeoJSON objects), but are not efficient when we use the attributes in the nested structures to apply heavy filtering or aggregation functions; Hive tends to outperform Cassandra as a streaming storage system, but after a certain period, the number of small files being generated overloads HDFS when performing metadata operations and causes a severe delay in Spark streaming micro batches; interactive SQL-on-Hadoop systems like Presto can efficiently combine a streaming fact table stored in a NoSQL database (e.g., Cassandra) with historical dimensions stored in the batch storage (e.g., Hive), achieving a similar performance to the flat-based fact tables; periodically, we need to move data from the streaming storage to the batch storage of a BDW, in order to maintain the interactivity of the system; in multi-tenant environments, severe attention must be paid to the trade-off between the cluster's resource capacity and the streaming micro batch intervals.

For future work, we aim to assess the overall impact (e.g., redundancy and updates) of flat (denormalized) structures for BDWs, and to extend the SSB+ to support new technologies like Hive LLAP and Kudu, for example.

Acknowledgements. This work is supported by COMPETE: POCI-01-0145- FEDER-007043 and FCT – *Fundação para a Ciência e Tecnologia* within the Project Scope: UID/CEC/00319/2013, by the SusCity project (MITP-TB/CS/0026/2013), and by European Structural and Investment Funds in the FEDER component, through the Operational Competitiveness and Internationalization Programme (COMPETE 2020) [Project no 002814; Funding Reference: POCI-01-0247-FEDER-002814].

References

1. Chandarana, P., Vijayalakshmi, M.: Big Data analytics frameworks. In: 2014 International Conference on Circuits, Systems, Communication and Information Technology Applications (CSCITA), pp. 430–434 (2014)
2. Ward, J.S., Barker, A.: Undefined by data: a survey of big data definitions. arXiv:13095821 CsDB (2013)

3. Chen, M., Mao, S., Liu, Y.: Big data: a survey. Mob. Netw. Appl. **19**, 171–209 (2014). https://doi.org/10.1007/s11036-013-0489-0
4. Madden, S.: From databases to big data. IEEE Internet Comput. **16**, 4–6 (2012). https://doi.org/10.1109/MIC.2012.50
5. Manyika, J., Chui, M., Brown, B., Bughin, J., Dobbs, R., Roxburgh, C., Byers, A.H.: Big Data: The Next Frontier for Innovation, Competition, and Productivity. McKinsey Global Institute, San Francisco (2011)
6. NBD-PWG: NIST Big Data Interoperability Framework, vol. 6, Reference Architecture. National Institute of Standards and Technology (2015)
7. Goss, R.G., Veeramuthu, K.: Heading towards big data building a better data warehouse for more data, more speed, and more users. In: 2013 24th Annual SEMI on Advanced Semiconductor Manufacturing Conference (ASMC), pp. 220–225. IEEE (2013)
8. Krishnan, K.: Data Warehousing in the Age of Big Data. Morgan Kaufmann Publishers Inc., San Francisco (2013)
9. Mohanty, S., Jagadeesh, M., Srivatsa, H.: Big Data Imperatives: Enterprise Big Data Warehouse, BI Implementations and Analytics. Apress, New York City (2013)
10. Kobielus, J.: Hadoop: nucleus of the next-generation big data warehouse (2012). http://www.ibmbigdatahub.com/blog/hadoop-nucleus-next-generation-big-data-warehouse
11. Kimball, R., Ross, M.: The Data Warehouse Toolkit: The Definitive Guide to Dimensional Modeling. Wiley, New York (2013)
12. Golab, L., Johnson, T.: Data stream warehousing. In: 2014 IEEE 30th International Conference on Data Engineering (ICDE), pp. 1290–1293 (2014)
13. Russom, P.: Evolving Data Warehouse Architectures in the Age of Big Data. The Data Warehouse Institute (2014)
14. Russom, P.: Data Warehouse Modernization in the Age of Big Data Analytics. The Data Warehouse Institute (2016)
15. Sun, L., Hu, M., Ren, K., Ren, M.: Present situation and prospect of data warehouse architecture under the Background of Big Data. In: 2013 International Conference on Information Science and Cloud Computing Companion (ISCC-C), pp. 529–535. IEEE (2013)
16. Philip Chen, C.L., Zhang, C.-Y.: Data-intensive applications, challenges, techniques and technologies: a survey on Big Data. Inf. Sci. **275**, 314–347 (2014). https://doi.org/10.1016/j.ins.2014.01.015
17. Hashem, I.A.T., Yaqoob, I., Anuar, N.B., Mokhtar, S., Gani, A., Khan, S.U.: The rise of "big data" on cloud computing: review and open research issues. Inf. Syst. **47**, 98–115 (2015). https://doi.org/10.1016/j.is.2014.07.006
18. O'Neil, P.E., O'Neil, E.J., Chen, X.: The star schema benchmark (SSB) (2009)
19. TPC: TPC-H – Homepage. http://www.tpc.org/tpch/
20. Costa, C.: SSB+ GitHub Repository. https://github.com/epilif1017a/bigdatabenchmarks
21. Peffers, K., Tuunanen, T., Rothenberger, M., Chatterjee, S.: A design science research methodology for information systems research. J. Manage. Inf. Syst. **24**, 45–77 (2007). https://doi.org/10.2753/MIS0742-1222240302
22. Chevalier, M., El Malki, M., Kopliku, A., Teste, O., Tournier, R.: Implementing multidimensional data warehouses into NoSQL. In: International Conference on Enterprise Information Systems (ICEIS 2015), pp. 172–183 (2015)
23. Gröger, C., Schwarz, H., Mitschang, B.: The deep data warehouse: link-based integration and enrichment of warehouse data and unstructured content. In: IEEE 18th International Enterprise Distributed Object Computing Conference (EDOC), pp. 210–217 (2014)
24. Cattell, R.: Scalable SQL and NoSQL data stores. ACM SIGMOD Rec. **39**, 12–27 (2011). https://doi.org/10.1145/1978915.1978919

25. Thusoo, A., Sarma, J.S., Jain, N., Shao, Z., Chakka, P., Zhang, N., Antony, S., Liu, H., Murthy, R.: Hive-a petabyte scale data warehouse using Hadoop. In: IEEE 26th International Conference on Data Engineering (ICDE), pp. 996–1005. IEEE (2010)
26. Floratou, A., Minhas, U.F., Özcan, F.: SQL-on-Hadoop: full circle back to shared-nothing database architectures. Proc. VLDB Endow. **7**, 1295–1306 (2014). https://doi.org/10.14778/2732977.2733002
27. Huai, Y., Chauhan, A., Gates, A., Hagleitner, G., Hanson, E.N., O'Malley, O., Pandey, J., Yuan, Y., Lee, R., Zhang, X.: Major technical advancements in apache hive. In: Proceedings of the 2014 ACM SIGMOD International Conference on Management of Data, pp. 1235–1246. ACM, New York (2014)
28. Song, J., Guo, C., Wang, Z., Zhang, Y., Yu, G., Pierson, J.-M.: HaoLap: A Hadoop based OLAP system for big data. J. Syst. Softw. **102**, 167–181 (2015). https://doi.org/10.1016/j.jss.2014.09.024
29. Wang, H., Qin, X., Zhou, X., Li, F., Qin, Z., Zhu, Q., Wang, S.: Efficient query processing framework for big data warehouse: an almost join-free approach. Front. Comput. Sci. **9**, 224–236 (2015)
30. Li, X., Mao, Y.: Real-time data ETL framework for big real-time data analysis. In: 2015 IEEE International Conference on Information and Automation, pp. 1289–1294 (2015)
31. Thusoo, A., Shao, Z., Anthony, S., Borthakur, D., Jain, N., Sen Sarma, J., Murthy, R., Liu, H.: Data warehousing and analytics infrastructure at Facebook. In: Proceedings of the 2010 ACM SIGMOD International Conference on Management of Data, pp. 1013–1020. ACM, New York (2010)
32. Wang, S., Pandis, I., Wu, C., He, S., Johnson, D., Emam, I., Guitton, F., Guo, Y.: High dimensional biological data retrieval optimization with NoSQL technology. BMC Genom. **15**(Suppl. 8), S3 (2014). https://doi.org/10.1186/1471-2164-15-S8-S3
33. Clegg, D.: Evolving data warehouse and BI architectures: the big data challenge (2015)
34. Santos, M.Y., Costa, C., Galvão, J., Andrade, C., Martinho, B., Lima, F.V., Costa, E.: Evaluating SQL-on-Hadoop for Big Data warehousing on Not-So-Good hardware. In: Proceedings of International Database Engineering & Applications Symposium (IDEAS 2017), Bristol, UK (2017)
35. Costa, E., Costa, C., Santos, M.Y.: Efficient Big Data modelling and organization for Hadoop hive-based data warehouses. In: Presented at the EMCIS 2017, Coimbra, Portugal (2017)
36. Costa, C., Santos, M.Y.: The SusCity Big Data warehousing approach for smart cities. In: Proceedings of International Database Engineering & Applications Symposium, p. 10 (2017)
37. Santos, M.Y., Oliveira e Sá, J., Andrade, C., Vale Lima, F., Costa, E., Costa, C., Martinho, B., Galvão, J.: A Big Data system supporting Bosch Braga Industry 4.0 strategy. Int. J. Inf. Manag. **37**(6), 750–760 (2017). https://doi.org/10.1016/j.ijinfomgt.2017.07.012
38. Mackey, G., Sehrish, S., Wang, J.: Improving metadata management for small files in HDFS. In: 2009 IEEE International Conference on Cluster Computing and Workshops, pp. 1–4 (2009)
39. White, T.: Hadoop: The Definitive Guide. O'Reilly Media, Sebastopol (2015)

KAYAK: A Framework for Just-in-Time Data Preparation in a Data Lake

Antonio Maccioni[1](✉) and Riccardo Torlone[2]

[1] Collective[i], New York City, USA
amaccioni@collectivei.com
[2] Università Roma Tre, Rome, Italy
torlone@dia.uniroma3.it

Abstract. A data lake is a loosely-structured collection of data at large scale that is usually fed with almost no requirement of data quality. This approach aims at eliminating any human effort before the actual exploitation of data, but the problem is only delayed since preparing and querying a data lake is usually a hard task. We address this problem by introducing KAYAK, a framework that helps data scientists in the definition and optimization of pipelines of data preparation. Since in many cases approximations of the results, which can be computed rapidly, are enough informative, KAYAK allows the users to specify their needs in terms of accuracy over performance and produces previews of the outputs satisfying such requirement. In this way, the pipeline is executed much faster and the process of data preparation is shortened. We discuss the design choices of KAYAK including execution strategies, optimization techniques, scheduling of operations, and metadata management. With a set of preliminary experiments, we show that the approach is effective and scales well with the number of datasets in the data lake.

Keywords: Data lake · Data preparation · Big data · Schema-on-read

1 Introduction

In traditional business intelligence, activities such as modeling, extracting, cleaning, and transforming data are necessary but they also make the data analysis an endless process. In response to that, big data-driven organizations are adopting an agile strategy that dismisses any pre-processing before the actual exploitation of data. This is done by maintaining a repository, called "data lake", for storing any kind of raw data in its native format. A dataset in the lake is a file, either collected from internal applications (e.g., logs or user-generated data) or from external sources (e.g., open data), that is directly stored on a (distributed) file system without going through an ETL process.

This work has been supported by the European Commission under the grant agreement number 774571 – Project PANTHEON.

© Springer International Publishing AG, part of Springer Nature 2018
J. Krogstie and H. A. Reijers (Eds.): CAiSE 2018, LNCS 10816, pp. 474–489, 2018.
https://doi.org/10.1007/978-3-319-91563-0_29

Unfortunately, reducing the engineering effort upfront just delays the traditional issues of data management since this approach does not eliminate the need of, e.g., data quality and schema understanding. Therefore, a long process of *data-preparation* (a.k.a. *data wrangling*) is required before any meaningful analysis can be performed [6,13,23]. This process typically consists of pipelines of operations such as: source and feature selection, exploratory analysis, data profiling, data summarization, and data curation. A number of state-of-the-art applications can support these activities, including: (i) data and metadata catalogs, for selecting the appropriate datasets [1,5,10,12]; (ii) tools for full-text indexing, for providing keyword search and other advanced search capabilities [9,10]; (iii) data profilers, for collecting meta-information from datasets [6,9,16]; (iv) processing engines like Spark [24] in conjunction with data science notebooks such as Jupyter[1] or Zeppelin[2], for executing the analysis and visualize the results. In such scenario, data preparation is an involved, fragmented and time-consuming process, thus preventing analysis on-the-fly over the lake.

In this framework, we propose a system, called KAYAK, supporting data scientists in the definition, execution and, most importantly, optimization of data preparation pipelines in a data lake[3]. With KAYAK data scientists can: (i) define pipelines composed by *primitives* implementing common data preparation activities and (ii) specify, for each primitive, their time *tolerance* in waiting for the result. This represents a mechanism to trade-off between performance and accuracy of primitives' results. Indeed, these primitives involve hard-to-scale algorithms that prevent analysis on-the-fly over new datasets [14,16–18], but often an approximate result is informative enough to move forward to the next action in the pipeline, with no need to wait for an exact result. KAYAK takes into account the tolerances by producing quick *previews* of primitive's results, when necessary. In this way, the pipelines are executed much faster and the time for data preparation is shortened.

On the practical side, each primitive in a pipeline is made of a series of tasks implementing built-in, atomic operations of data preparation. Each task can be computed incrementally via a number of steps, each of which can return previews to the user. KAYAK orchestrates the overall execution process by scheduling and computing the various steps of a pipeline according to several optimization strategies that balance the accuracy of results with the given time constraints. Another important feature of KAYAK is its ability to collect automatically, in a metadata catalog, different relationships among datasets, which can be used later to implement advanced analytics. The catalog also keeps the profile of each dataset and provides a high-level view of the content of the data lake.

We have verified the effectiveness of our approach with the first implementation of KAYAK and tested its scalability when the number and size of datasets in the lake increase.

---

[1] http://jupyter.org/.
[2] https://zeppelin.apache.org/.
[3] A demo of KAYAK has been shown in [15].

Currently, the system is used in operation within the PANTHEON project whose aim is supporting precision farming: it is charge of collecting and managing heterogeneous data coming from terrestrial and aerial robots moving in plantations, as well as from ground sensors and weather stations located nearby.

The rest of the paper is organized as follows. In Sect. 2 we provide an overview of our approach. In Sect. 3 we illustrate how KAYAK models data and, in Sect. 4, we describe our strategy for executing primitives. In Sect. 5 we discuss our experimental results and, in Sect. 6, some related works. Finally, in Sect. 7, we draw some conclusions and sketch future works.

2   Overview of the Approach

This section provides an overview, from a high-level perspective, of the main features of the system.

Pipelines, Primitives and Tasks. KAYAK is a framework that lies between users/applications and the file system where data is stored. It exposes a series of *primitives* for data preparation, some of which are reported in Table 1. For example, a data scientist can use primitive P_5 to find interesting ways to access a dataset. Each primitive is composed of a sequence of *tasks* that are reused across primitives (e.g., P_6 is split into T_b, T_c, T_w, while primitive P_7 uses T_c only). A task is atomic and consists of operations that can be executed either directly within KAYAK or involving external tools [16,24], as shown in Table 2.

Table 1. Example of primitives in KAYAK.

Id	Name	Tasks
P_1	Insert dataset	T_a, T_p
P_2	Delete dataset	T_s
P_3	Search dataset	T_o
P_4	Complete profiling	T_a, T_b, T_c, T_d, T_m
P_5	Get recommendation	T_b, T_c, T_d, T_q
P_6	Find related dataset	T_b, T_c, T_w
P_7	Compute joinability	T_c
P_8	Compute k-means	T_g, T_n
P_9	Outlier detection	T_h, T_p, T_r, T_u
...	...	...

Table 2. Example of tasks in KAYAK.

Id	Description
T_a	Basic profiling of a dataset
T_b	Statistical profiling of a dataset
T_c	Compute Joinability of a dataset
T_d	Compute Affinity between two datasets
T_e	Find inclusion dependencies
T_f	Compute joinability between two datasets
...	...

A *pipeline* is a composition of primitives that is representable as a DAG (direct acyclic graph). As an example, Fig. 1 shows a pipeline composed by six primitives: P_1 inserts a new dataset in the lake and P_4 generates a profiles for it; then P_8 processes the dataset with a machine learning algorithm while P_7 identifies possible relationships with another dataset. Eventually, P_5 produces a query recommendation. Users can mark the primitives in the pipeline with a

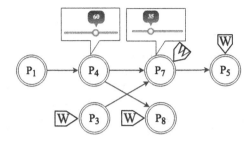

Fig. 1. Example of a data preparation pipeline.

watchpoint to inspect some intermediate result. For example, in Fig. 1 we have defined a watchpoint on P_7, P_5, P_3, and P_8.

Note that, we assume here that the output of a primitive is not directly used as input of the following primitive; they rather communicate indirectly by storing data in the lake or metadata in a catalog. Primitives can be *synchronous* when they do not allow the execution of a subsequent primitive before its completion, or *asynchronous*, when can be invoked and executed concurrently.

Metadata Management. KAYAK extracts metadata from datasets explicitly, with ad-hoc primitives (e.g., P_4), or implicitly, when a primitive needs some metadata and uses the corresponding profiling task (e.g., T_a in P_1). Metadata are organized in a set of predefined attributes and are stored in a catalog so that they can be accessed by any task.

Specifically, KAYAK collects *intra-dataset* and *inter-dataset* metadata. Intra-dataset metadata form the profile associated with each single dataset, which includes descriptive, statistical, structural and usage metadata attributes. Inter-dataset metadata specify relationships between different datasets or between attributes belonging to different datasets. They include integrity constraints (e.g., inclusion dependencies) and other properties proposed by ourselves, such as *joinability* (Ω) and *affinity* (Ψ) between datasets. Inter-dataset metadata are represented graphically, as shown in Fig. 2(a) and (b). Intuitively, joinability measures the mutual percentage of common values between attributes of two datasets, whereas affinity measures the semantic strength of a relationship according to some external knowledge. The *affinity* is an adaptation, to data lakes, of the entity complement proposed by Sarma et al. [21].

Time-to-Action and Tolerance of the User. Let us call *time-to-action* the amount of time elapsing between the submission of a primitive in a pipeline and the instant in which a data scientist is able to take an informed decision on how to move forward to the next step of the pipeline. To shorten primitive computation when unnecessarily long, we let the data scientist specify a *tolerance*. A high tolerance is set by the data scientist who does not want to wait for long and believes that an approximate result is enough informative. On the contrary, a low tolerance is specified when the priority is on accuracy. For instance, in the pipeline of Fig. 1, primitives P_4 and P_7 have been specified with a tolerance of 60% and 35%, respectively.

Incremental Execution for Reducing Time-to-Action. In KAYAK, primitives can be executed incrementally and produce a sequence of *previews* along their computation. A primitive is decomposed into a series of tasks. Each task can be computed as a sequence of steps that returns the previews. A preview is an approximation of the exact result of the task and it is, therefore, computed much faster. Two strategies of incremental execution exist. A *greedy* strategy aims at reducing the time-to-action by producing a quick preview first, and then updating the user with refined previews within her tolerance. Alternatively, a *best-fit* strategy aims at giving the best accuracy according to the given tolerance. It generates only the most accurate preview that fits the tolerance of the user.

Confidence of Previews. Each preview comes with a *confidence* indicating the uncertainty on the correctness of the result with a value between 0 and 1. A confidence is 0 when the result is random and it is 1 when it is exact. A sequence of previews is always produced with an increasing confidence so that the user is always updated with more accurate results and metadata are updated with increasingly valuable information.

Extensibility. KAYAK provides a set of built-in, atomic tasks that can be easily extended for implementing new functionalities. Specifically, tasks implementing common activities of data preparation and therefore can be used by different primitives. For instance, referring to Table 2, task T_b is used by three primitives. In addition, a new task can be defined by the users, who needs to specify also the cost model for the computation of the task and all the possible ways to approximate it, as we will show next.

3   Modeling a Data Lake

In this section, we discuss on how data and metadata are represented and managed in our framework. Let us start with some basic notions.

Definition 1 (Dataset). *A dataset $D(X, C, R)$ has a name D, and is composed by a set X of attributes, a set C of data objects, and a profile R. Each data object in C is a set of attribute-value pairs, with attributes taken from X. The profile R is a set of attribute-value pairs, with attributes taken from a predefined set M of metadata attributes.*

A metadata attribute of a dataset D can refer to either the whole dataset or to an attribute of D. We use the dot notation to distinguish between the two cases. For instance, if D is a dataset involving an attribute *ZipCode*, the profile of D can include the pairs $\langle D.CreationDate : 11/11/2016 \rangle$ and $\langle ZipCode.unique : true \rangle$. For simplicity, we assume that each dataset is stored in a file and therefore we often blur the distinction between dataset and file.

Definition 2 (Data Lake). *A data lake $\mathcal{D}$ is a collection of datasets having distinct names.*

Differently from a profile of a dataset, inter-dataset metadata capture relationships between different datasets and between attributes of different datasets. They are represented as graphs and are introduced next.

Affinity. In the affinity graph of a data lake $\mathcal{D}$, the nodes represent the attributes of the datasets in $\mathcal{D}$ and an edge between two attributes represents the presence of some time-independent relationship between them (e.g., the fact that they refer to the same real-world entity). Edges can have weights that measure the "strength" of the relationship.

Specifically, we consider a *domain-knowledge* affinity $\Omega(D_1.A_i, D_2.A_j)$ that measures the "semantic" affinity between attributes $D_1.A_i$ and $D_2.A_j$, which is computed by taking advantage of some existing external knowledge base (such as a domain ontology). The value assigned by $\Omega(D_1.A_i, D_2.A_j)$ ranges in $[0, 1]$ (i.e. 0 when there is no affinity and 1 when the affinity is maximum). Here, we take inspiration from the notion of *entity complement* proposed by Sarma et al. [21]. However, different kinds of affinity can be used such as those based on text classification.

We can now define the graph representing the affinity of the attributes $\mathcal{A}$ of the datasets in the data lake $\mathcal{D}$.

Definition 3 (Affinity Graph of Attributes). *The affinity graph of attributes in $\mathcal{D}$ is an undirected and weighted graph $G_{\mathcal{A}}^{\Omega} = (N_{\mathcal{A}}, E_{\mathcal{A}}^{\Omega})$ where $N_{\mathcal{A}}$ contains a node for each attribute A in $\mathcal{A}$ and $E_{\mathcal{A}}^{\Omega}$ contains an edge $(A_1, A_2, \Omega(A_1, A_2))$ for each pair of attributes A_1 and A_2 in $\mathcal{A}$ such that $\Omega(A_1, A_2) > 0$.*

The notion of affinity between attributes can be used to define the affinity between two datasets D_1 and D_2.

Definition 4 (Affinity of Datasets). *Let X_1 and X_2 be the set of attributes of the datasets D_1 and D_2, respectively, and let $\hat{X} = X_1 \times X_2$. The affinity between D_1 and D_2, denoted by $\Omega(D_1, D_2)$, is defined as follows:*

$$\Omega(D_1, D_2) = \sum_{(A_j, A_k) \in \hat{X}} \Omega(A_j, A_k)$$

Analogously, we can define an affinity graph of datasets.

Definition 5 (Affinity Graph of Datasets). *The affinity graph of datasets for $\mathcal{D}$ is an undirected and weighted graph $G_{\mathcal{D}}^{\Omega} = (N_{\mathcal{D}}, E_{\mathcal{D}}^{\Omega})$ where $N_{\mathcal{D}}$ contains a node for each dataset D in $\mathcal{D}$ and $E_{\mathcal{D}}$ contains an edge $(D_1, D_2, \Omega(D_1, D_2))$ for each pair of dataset D_1 and D_2 in $\mathcal{D}$ such that $\Omega(D_1, D_2) > 0$.*

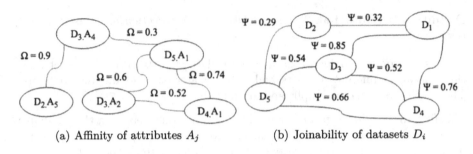

(a) Affinity of attributes A_j (b) Joinability of datasets D_i

Fig. 2. Inter-dataset metadata.

We are clearly interested in highly affine relationships. Therefore, we consider a simplified version of the graph where the edges have weights higher than a threshold τ, defined by the user. This is equivalent to consider irrelevant affinities below τ_Ω. An example of affinity graph of attributes is reported in Fig. 2(a).

Joinability. Another way to relate attributes and dataset is simply based on the existence of common values. We introduce the concept of *joinability* for this purpose.

Definition 6 (Joinability). *Given two attributes A_i and A_j belonging to the datasets D_1 and D_2, respectively, their joinability Ψ is defined as*

$$\Psi(D_1.A_i, D_2.A_j) = \frac{2 \cdot |\pi_{A_i}(D_1 \bowtie_{A_i=A_j} D_2)|}{(|\pi_{A_i}(D_1)| + |\pi_{A_j}(D_2)|)}$$

The joinability measures the mutual percentage of tuples of D_1 that join with tuples of D_2 on $D_1.A_i$ and $D_2.A_j$, and vice versa. This notion enjoys interesting properties, which we can discuss by considering the example in Fig. 3.

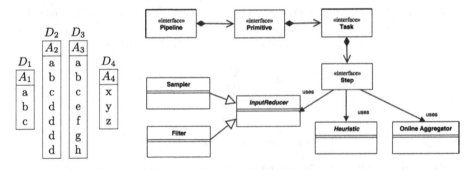

Fig. 3.   Tabular datasets.

Fig. 4. Business logic of the framework.

The maximum joinability, (e.g., $\Psi(D_1.A_1, D_2.A_2) = 1$), is when each value of one attribute matches a value of the other attribute. If the result of the join between the two attributes is empty, the joinability is 0 (e.g., $\Psi(D_2.A_2, D_4.A_4)$ = 0). A joinability in $(0, 1)$ means that there are several matching values. The joinability takes also into account, for both attributes, the number of distinct values that do not match. For the dataset in Fig. 3 we have: $\Psi(D_1.A_1, D_2.A_2) > \Psi(D_1.A_1, D_3.A_3) > \Psi(D_1.A_1, D_4.A_4)$.

Similarly to the property of affinity, we can build a joinability graph of attributes and a joinability graph of datasets, where we represent only those edges whose joinability is higher or equal than a threshold parameter τ_Ψ. An example of a joinability graph is reported in Fig. 2(b).

4   Incremental Execution of Primitives

In this section, we describe the incremental execution of primitives, a mechanism that allows users to obtain previews of a result at an increasing level of accuracy.

Basic Idea. Users submit a primitive over an input I, which is typically a set of datasets. As we can see from Fig. 4, a primitive is composed of one or more tasks of type T. Each task type is associated with one or more *steps*. A step is an operation that is able to return a result for T over I. The result of a step can be either exact or approximate. We use t for indicating the step that computes the exact result r for t for T over I (i.e., $r = t(I)$). We use s_i^T for indicating the i-th approximate step of T, which returns a preview $p_i = s_i^T(I)$. Therefore, a preview p_i is an approximation of r.

We have several types of approximate steps, corresponding to different ways to approximate a task. For instance, some step reduces the input I (e.g., sampling), while other steps apply heuristics over I. In our framework, we have components that support the approximate steps, as shown in Fig. 4. The list of steps S_T for a task type T is declared in the definition of T. For simplicity, the following discussion considers a primitive with a single task type T, but this generalizes easily to primitives with many tasks. The incremental execution of a task type T over I is a sequence of m steps $s_1^T, \ldots, s_m^T$, where possibly the last step is the exact task t (i.e. $t = s_m^T$).

Each preview comes with a *confidence*, indicating the uncertainty on the correctness of the result with a number between 0 and 1. A confidence is 0 when the result is random and it is 1 when the result is exact. KAYAK computes the confidence executing a function embedded in the step definition. This function considers the confidence associated with the input of the step, e.g., the confidence of a metadata attribute. Note that, since previews are produced with an increasing confidence, metadata are stored with increasingly precise information and the user is always updated with more accurate primitive results.

In addition, each step is associated with a cost function that estimates its computational time over an input I, i.e. $cost(s_i^T, I)$. We have defined the cost functions using the Big-Θ time complexity. All other time-based measures like the load of the system and the tolerance have to be comparable with the cost and are therefore expressed in terms of the same virtual time. Moreover, we want to underline that there is no formal relationship between the cost of a step and the confidence of the preview it produces.

Optimization. KAYAK needs to find a suitable incremental execution for a primitive, that is the order of the execution of steps. To this aim, KAYAK takes into account the tolerance of the user and the current workload. The tolerance is fixed by the user for each of the primitives in the pipeline. The workload is given by the sum of the costs of the steps of the primitives to be executed.

In our framework, we devised several incremental execution strategies and further strategies can be defined. For example, we have a so-called *best-fit* strategy that tries to generate only the most accurate preview that fits within the tolerance. This tends to limit the overall delay while still reducing the time-to-action according to the user's tolerance. Another strategy is the so-called *greedy* strategy that aims at minimizing the time-to-action and to update the user with subsequent previews. However, due to lack of space, we do not detail any strategy.

Step Dependency. At the end of the step generation, we set dependencies to enforce a correct execution of primitives composed of many tasks. Since we allow for a parallel execution of steps, a DAG of dependencies is considered. In the DAG, each node T_i is a task type and each edge (T_i, T_j) represents a dependency of a task T_j (the destination node) from another task T_i (the source node). A dependency indicates that T_j can start its execution only after T_i is completed.

In KAYAK, we do not have a centralized representation of the DAG, but dependencies are set within each task. For example, in Fig. 5(a) we have the DAG for the primitive P_5 that is composed of four tasks T_b, T_c, T_d and T_q, as in Table 2. The task T_q is the last task of the primitive and uses metadata provided by T_b, T_c and T_d. For this reason, T_q has a dependency with every other task. In addition, there is a dependency between T_d and T_b. This means that T_b and T_c can execute before the others, possibly in parallel.

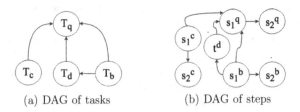

(a) DAG of tasks   (b) DAG of steps

Fig. 5. Dependencies among tasks of a primitive.

When a primitive is executed in incremental mode, the step generation phase produces a DAG that considers every single step. Let us suppose that T_b, T_c and T_q are executed incrementally in two steps each, while T_d is executed in a single step. Figure 5(b) shows the resulting DAG for this primitive execution. We set a dependency between two subsequent steps of the same task, such as (s_1^b, s_2^b), to preserve the increase of the confidence of results. If this dependency is not set, then a preview might overwrite another preview with higher confidence.

We also need to set the inter-task dependencies. In this case, we set a dependency between the first generated steps of the two tasks. Using the example of P_5, the dependency (T_c, T_q) is established by the dependency (s_1^c, t_1^q). No inter-task dependency between following steps is considered. The reason behind this decision is, again, aimed at reducing the time-to-action.

There are two side effects that motivate this decision. The former is when, for example, t_1^c and t_2^c terminate before t_1^q has started. It follows that t_1^q will use metadata produced by t_2^c, resulting in a higher confidence. The latter side effect is when t_1^q is computed between t_1^c and t_2^c. The final result of t_1^q will be less accurate but the time-to-action is minimized. However, the user is notified of the fact that more accurate metadata is present for, possibly, refining the result of the primitive she just launched.

Scheduling of Steps. Dependencies are used for guaranteeing the consistency of primitives' results but they do not suffice to reduce the time-to-action of tasks coming from different primitives. To avoid a random order of execution, we use a step scheduling mechanism where the order of execution is done with respect to a priority assigned to each step. Steps with higher priority are executed first, while low priority steps are treated like processes to execute in background when the system is inactive. The priority function is defined as follows:

$$priority(s) = \frac{1}{cost(s)} + freshness(s) + completeness(s)$$

where:

- The *cost* is used to favor shorter steps, with the aim of reducing the time-to-action. It is given by the same function used in the previous sections.
- The *freshness* is used to avoid starvation. It uses the creation time of steps (with older steps having higher freshness). Let us explain the motivation behind this factor with an example. Let us consider the submission of a heavy task of type T_c followed by the submission of many shorter tasks of type T_a. If we consider only the cost factor, the task T_c will never be executed and it will starve in the queue.
- The *completeness* is used to balance the time-to-action across different primitives. It considers how many steps have already been instantiated for the task type. For instance, the completeness gives an advantage to the first step of task T_c over the second step of another task T_a. In fact, if we use only cost and freshness some step for T_c might not fulfill its time-to-action objective.

Note that our scheduling mechanisms do not conflict with mechanisms of cluster resource managers (e.g., Apache Mesos or Apache Yarn) used by data processing engines. We decide when a step can start its execution, while they schedule jobs of data processing only once their corresponding step has already started.

Use Case: Incremental Computation of Joinability. We now show the incremental execution of the task type T_c that computes the joinability of a dataset against the rest of the datasets in the lake (see Definition 6). Since the data lake $\mathcal{D}$ can have a large number of datasets with many attributes, this task is computationally expensive and does not scale with the size of $\mathcal{D}$. Since this task is used by many primitives, it is often convenient to execute incrementally. Below we list the techniques used to generate previews of T_c.

1. *Joinability over frequent items.* Our least accurate step for T_c is given by a heuristic that takes the most frequent items present in the domain of two attributes along with the number of occurrences. This information is present in our metadata catalog and it is collected by other profiling tasks. We then compute the intersection between the two sets that allows us to determine the joinability of a small portion of the two attribute's domains in a constant time. The confidence is computed by considering the percentage of the coverage that the frequent items have over the entire domain.

2. *Joinability over a subset of attributes and sampled datasets.* This step uses some heuristics that aim at selecting those attributes that are likely to have higher values of joinability against the input dataset. It specifically selects a subset of attributes Z_i of the lake to be used in the computation of the joinability against attributes of D_i. Then, the datasets which the attributes of Z_i belong from are sampled to further reduce the cost of the operation. The approximation of joinability is similar to compute approximate joins [4,11]. The sample rate is chosen dynamically according to the size of the dataset, with lower sample rate for higher dataset size. The selected attributes Z_i are those having: (a) overlapping among the most frequent items of the attributes, (b) an inclusion dependency with D_i (we check from available metadata without computing T_e of Table 2), (c) high affinity with the attributes of D_i as taken from the *affinity graph of attributes*. The confidence of this level is given by the used sample ratios and the number of attributes that have been selected.

3. *Joinability over a subset of attributes.* This step selects the attributes of the previous case but does not apply any sampling over the datasets.

4. *Joinability over a subset of sampled datasets.* This step selects a set Y_i of datasets in $\mathcal{D}$ having high affinity with D_i by checking the *affinity graph of datasets*. Then, it computes the joinability between attributes of D_i and attributes of datasets in Y_i.

5. *Joinability over a subset of datasets.* This step selects the same set Y_i of datasets of the previous case but, differently from it, sampling is not applied.

6. *Joinability over a sampled data lake.* This step selects a sample from each of the datasets in $\mathcal{D}$ and then it applies the joinability between the attributes of D_i and any other attribute in $\mathcal{D}$.

We have described here the steps of the joinability task. The implementation of these steps makes use of other optimizations such as those in presence of inclusion dependencies or those that return zero when data types are different or domain ranges do not overlap. However, we do not discuss them in detail here because they do not deal with the approximation of the exact results.

5   Experimental Results

5.1   Set-Up

The architecture of KAYAK is discussed in [15]. KAYAK is implemented in Java 8 and exploits *java.util.concurrent* library for the concurrent execution of tasks. The current version limits the use to *json* and to *csv* files only. The Metadata Catalog relies on two different database systems, namely MongoDB for the intra-dataset catalog and Neo4j for the inter-dataset catalog. The Queue Manager uses RabbitMQ as a messaging queue system. The User Interface is implemented using JSP pages and servlets on the web application server Tomcat 8. We also rely on external tools such as Spark 2.1 with MLib and SparkSQL add-ons for parallelizing operations on large datasets, and on Metanome[4] for some of the tasks for which the Metadata Collector is in charge.

5.2   Results

This section presents the experimentation that was conducted on a cluster of *m4.4xlarge* machines on Amazon EC2. Each machine is equipped with 16 vCPU, 64 GB and running a 2,3 GHz Intel Xeon with 18 cores. We created a data lake with 200 datasets ranging from hundreds of MBs to few TBs. We have taken datasets from the U.S. open data catalog[5] and from the NYC Taxi trips[6]. We have also generated synthetic datasets to create uses cases that were not covered with downloaded datasets.

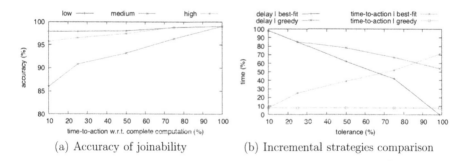

(a) Accuracy of joinability   (b) Incremental strategies comparison

Fig. 6. Incremental step generation at work.

Effectiveness. In this campaign, we have measured the trade-off between accuracy and time-to-action for the joinability task. It is a fundamental task in our framework that is used by many primitives. The results are in Fig. 6(a). We

---

[4] https://github.com/HPI-Information-Systems/Metanome.
[5] https://www.data.gov/.
[6] http://www.nyc.gov/html/tlc/html/about/trip_record_data.shtml.

have divided the tests in three sections according to the range of the joinability value Ψ: *low* $(0.0 \leq \Psi \leq 0.35)$, *medium* $(0.35 < \Psi \leq 0.70)$ and *high* $(0.70 < \Psi \leq 1.0)$. The time-to-action is represented in terms of percentage of the time for the exact step. Each point of the graph represents the average of six joinability computations (we have six different tests for each of the sections). As we can see, the accuracy is constantly high for low values of joinability. This is due to the fact that a reduced input is already able to show that two attributes do not join well. A similar behavior is for medium values, although accuracy slightly degrades for low time-to-action. High values of joinability are more difficult to approximate with shorter time-to-action than previous sections, but we consider this accuracy still good for many practical situations.

Strategies Comparison. In this campaign, we test the differences between two incremental strategies, that we have briefly mentioned above. Again, we consider the joinability task. Let us consider the delay as the extra time spent on the incremental execution with respect to the non-incremental counterpart. We measure how the time-to-action and the delay vary with respect to the tolerance. Both the measures are taken in percentage with respect to the duration of the exact step. As we can see from the results in Fig. 6(b), the time-to-action for the greedy strategy is constant because the same short level is executed independently of the user's tolerance, while for the best-fit strategy the time-to-action increases linearly with the tolerance. However, the time-to-action is always lower than the tolerance due to a fragmentation effect that makes it hard to have the cost of a step that perfectly fits the tolerance. The delay of the greedy strategy is always greater than the delay of the best-fit strategy, because of all the short steps executed at the beginning. The delay for both strategies tends to diminish as the tolerance increases. The delay of the best-fit strategy has an opposite behavior w.r.t. the time-to-action. Indeed, the delay is inversely proportional to the tolerance. This is because as the tolerance increases, the best-fit strategy tends to schedule fewer and fewer steps.

6   Related Work

We divide related work of KAYAK into categories discussed separately.

Data Catalogs. There are several tools that are used for building repositories of datasets [1,3,5,10,12]. Basic catalogs like CKAN [1] do not consider relationships among datasets and metadata are mostly inserted manually. DataHub [5] is a catalog that enables collaborative use and analysis of datasets. It includes features like versioning, merging and branching for datasets, similarly to version control systems in the context of software engineering. GOODS is an enterprise search system for a data lake that is in use at Google [10]. It proposes, among the others, a solution with the semi-automatic realization of a metadata catalog, an annotation service, an efficient tracking of the provenance and advanced search features based on full-text indexing. All above catalogs use basic ways to understand relationships among datasets and give little support to users who are unaware of the content of the datasets, though they are not explicitly designed for data preparation and exploration purposes.

Profiling Tools. In data science, tools such as R^7, IPython [19], Pandas-profiling[8] and notebook technologies[9] are extensively used for data exploration. They mainly compute statistical summaries integrated with some plotting features. More advanced data profiling consists on the discovery of constraints in the data [7,14,16,18]. Metanome, for instance, offers a suite of different algorithms for data profiling [16]. Some of these algorithms run by sharing pieces of computation [7] or by the aid of approximate techniques [14,18]. In KAYAK we have tasks that make use of these algorithms such as for example T_c in Table 2.

Data Wranglers. Schema-on-read data access has opened severe challenges in data wrangling [8,23] and specific tools are aimed at solving this problem [2,3,22]. Data TamR helps in finding insights thanks to novel approaches of data curation and data unification [2,22]. Trifacta is an application for self-service data wrangling providing several tools to the user [3]. All these systems provide features that can be embedded in KAYAK to be executed incrementally for minimizing the time-to-action.

Approximate Querying Systems. Another branch of work specifically focuses on approximating analytical query results [4,11,20]. Hellerstein et al. [11] propose an incremental strategy that aggregates tuples online so that the temporary result of the query is shown to the user, who can decide to interrupt the process anytime. Differently, when computing analytical queries with BlinkDB [4], users are asked the trade-off between time and accuracy in advance, and the system dynamically selects the best sample that allows replying the query under the user's constraints. This is similar to our best-fit strategy but we do not apply only sampling and we do not consider analytical queries. A critical aspect in all these works is the estimation of the error. To overcome these problems, DAQ [20] has recently introduced a deterministic approach to approximating analytic queries, where the user is initially provided with an interval that is guaranteed to contain the query result. Then, the interval shrinks as the query answering proceeds, until the convergence to the final answer. All these techniques work well on OLAP queries but since they require the workload in advance, they cannot be applied in our context where the user has usually not accessed the data yet and sampling cannot be the only technique for reducing the workload.

7 Conclusion and Future Work

In this paper, we have presented KAYAK, a end-to-end framework for data management with a data lake approach. KAYAK addresses data preparation, a crucial aspect for helping data-driven businesses in their analytics processes. KAYAK

---

[7] https://www.r-project.org/.
[8] https://github.com/JosPolfliet/pandas-profiling.
[9] http://zeppelin-project.org/.

provides a series of primitives for data preparation that can be executed by specifying a tolerance when the user prefers a quick result instead of an exact result. The framework also allows to define pipelines of primitives.

We have several future work directions in mind. We want to integrate the framework with components for supporting unstructured data, query expansion, and data visualization. We want to introduce a dynamic scheduling for the tasks and the possibility to set a tolerance for an entire pipeline. Finally, we would like to define a declarative language for designing primitives data preparation.

References

1. CKAN: The open source data portal software. http://ckan.org/. Accessed Nov 2017
2. Tamr. http://www.tamr.com/. Accessed Nov 2017
3. Trifacta. https://www.trifacta.com/. Accessed Nov 2017
4. Agarwal, S., Mozafari, B., Panda, A., Milner, H., Madden, S., Stoica, I.: BlinkDB: queries with bounded errors and bounded response times on very large data. In: EuroSys, pp. 29–42 (2013)
5. Bhardwaj, A.P., Deshpande, A., Elmore, A.J., Karger, D.R., Madden, S., Parameswaran, A.G., Subramanyam, H., Wu, E., Zhang, R.: Collaborative data analytics with DataHub. PVLDB 8(12), 1916–1927 (2015)
6. Deng, D., Fernandez, R.C., Abedjan, Z., Wang, S., Stonebraker, M., Elmagarmid, A.K., Ilyas, I.F., Madden, S., Ouzzani, M., Tang, N.: The data civilizer system. In: CIDR (2017)
7. Ehrlich, J., Roick, M., Schulze, L., Zwiener, J., Papenbrock, T., Naumann, F.: Holistic data profiling: simultaneous discovery of various metadata. In: EDBT, pp. 305–316 (2016)
8. Furche, T., Gottlob, G., Libkin, L., Orsi, G., Paton, N.W.: Data wrangling for big data: challenges and opportunities. In: EDBT, pp. 473–478 (2016)
9. Hai, R., Geisler, S., Quix, C.: Constance: an intelligent data lake system. In: SIGMOD, pp. 2097–2100 (2016)
10. Halevy, A.Y., Korn, F., Noy, N.F., Olston, C., Polyzotis, N., Roy, S., Whang, S.E.: Goods: organizing Google's datasets. In: SIGMOD (2016)
11. Hellerstein, J.M., Haas, P.J., Wang, H.J.: Online aggregation. In: SIGMOD, pp. 171–182 (1997)
12. Hellerstein, J.M., Sreekanti, V., Gonzalez, J.E., Dalton, J., Dey, A., Nag, S., Ramachandran, K., Arora, S., Bhattacharyya, A., Das, S., Donsky, M., Fierro, G., She, C., Steinbach, C., Subramanian, V., Sun, E.: Ground: a data context service. In: CIDR (2017)
13. Heudecker, N., White, A.: The data lake fallacy: all water and little substance. Gartner Report G 264950 (2014)
14. Ilyas, I.F., Markl, V., Haas, P.J., Brown, P., Aboulnaga, A.: CORDS: automatic discovery of correlations and soft functional dependencies. In: SIGMOD, pp. 647–658 (2004)
15. Maccioni, A., Torlone, R.: Crossing the finish line faster when paddling the data lake with KAYAK. PVLDB 10(12), 1853–1856 (2017)
16. Papenbrock, T., Bergmann, T., Finke, M., Zwiener, J., Naumann, F.: Data profiling with metanome. PVLDB 8(12), 1860–1863 (2015)

17. Papenbrock, T., Ehrlich, J., Marten, J., Neubert, T., Rudolph, J., Schönberg, M., Zwiener, J., Naumann, F.: Functional dependency discovery: an experimental evaluation of seven algorithms. PVLDB **8**(10), 1082–1093 (2015)
18. Papenbrock, T., Naumann, F.: A hybrid approach to functional dependency discovery. In: SIGMOD, pp. 821–833 (2016)
19. Pérez, F., Granger, B.E.: IPython: a system for interactive scientific computing. Comput. Sci. Eng. **9**(3), 21–29 (2007)
20. Potti, N., Patel, J.M.: DAQ: a new paradigm for approximate query processing. PVLDB **8**(9), 898–909 (2015)
21. Sarma, A.D., Fang, L., Gupta, N., Halevy, A.Y., Lee, H., Wu, F., Xin, R., Yu, C.: Finding related tables. In: SIGMOD (2012)
22. Stonebraker, M., Bruckner, D., Ilyas, I.F., Beskales, G., Cherniack, M., Zdonik, S.B., Pagan, A., Xu, S.: Data curation at scale: the data tamer system. In: CIDR (2013)
23. Terrizzano, I., Schwarz, P.M., Roth, M., Colino, J.E.: Data wrangling: the challenging journey from the wild to the lake. In: CIDR (2015)
24. Zaharia, M., Xin, R.S., Wendell, P., Das, T., Armbrust, M., Dave, A., Meng, X., Rosen, J., Venkataraman, S., Franklin, M.J., Ghodsi, A., Gonzalez, J., Shenker, S., Stoica, I.: Apache spark: a unified engine for big data processing. Commun. ACM **59**(11), 56–65 (2016)

Defining Interaction Design Patterns to Extract Knowledge from Big Data

Carlos Iñiguez-Jarrín[1,2(✉)], José Ignacio Panach[3],
and Oscar Pastor López[1]

[1] Research Center on Software Production Methods (PROS),
Universitat Politècnica de València, Camino Vera s/n, 46022 Valencia, Spain
{einiguez, opastor}@pros.upv.es
[2] Departamento de Informática y Ciencias de la Computación,
Escuela Politécnica Nacional, Ladrón de Guevara E11-253, Quito, Ecuador
[3] Escola Tècnica Superior d'Enginyeria, Departament d'Informàtica,
Universitat de València, Avenida de la Universidad, s/n, 46100
Burjassot, Valencia, Spain
joigpana@uv.es

Abstract. The Big Data domain offers valuable opportunities to gain valuable knowledge. The User Interface (UI), the place where the user interacts to extract knowledge from data, must be adapted to address the domain complexities. Designing UIs for Big Data becomes a challenge that involves identifying and designing the user-data interaction implicated in the knowledge extraction. To design such an interaction, one widely used approach is design patterns. Design Patterns describe solutions to common interaction design problems. This paper proposes a set of patterns to design UIs aimed at extracting knowledge from the Big Data systems' data conceptual schemas. As a practical example, we apply the patterns to design UI's for the Diagnosis of Genetic Diseases domain since it is a clear case of extracting knowledge from a complex set of genetic data. Our patterns provide valuable design guidelines for Big Data UIs.

Keywords: User Interfaces · Interaction patterns · Big Data

1 Introduction

Extracting knowledge from Big Data is not a trivial task and usually involves interacting with the data by identifying, combining and managing multiple and heterogeneous data sources as well as constructing advanced analysis models to predict outcomes. This task is performed at the User Interface (UI)—the contact point between user and data. To design UIs for Big Data domain, software designers should know the data consumption challenges in this domain, understand the needs of the direct beneficiaries of the information and formulate solutions to design the UI.

The Interaction Design Patterns approach is commonly used to design UIs. An Interaction Design Pattern (IDP) deals with an interaction problem in the UI design and provide a design solution to solve it. The interaction problems in the Big Data analysis are different from the traditional data analysis ones. How to interact with the data when

© Springer International Publishing AG, part of Springer Nature 2018
J. Krogstie and H. A. Reijers (Eds.): CAiSE 2018, LNCS 10816, pp. 490–504, 2018.
https://doi.org/10.1007/978-3-319-91563-0_30

they surpass any of three dimensions volume, velocity, and variety [1] is a problem to address in Big Data and the literature do not fit such Big Data interaction need; therefore, identifying suitable IDPs that help the designers to develop interactive Big Data UIs is a challenge.

We aim to define a set of IDPs for extracting knowledge from Big Data. To do that, we (a) analyze several real Big Data use cases available in the literature, (b) identify the challenges posed by the domain, and (c) derive IDPs as solutions to such challenges. Each IDP describes in detail the user-data interaction, referencing the data schema (conceptual model) that the interactive system supports and highlighting the effect that the dimensions of volume, velocity, and variety produce on the interaction.

We illustrate the IDPs by applying them to the *Diagnosis of Genetic Diseases* (DGD) domain, a concrete Big Data example. In this domain, the researchers extract knowledge by contrasting huge volumes of genetic data with information from data sources of heterogeneous formats [2]. Several tools have been developed to support the researchers in the genetic data analysis [3]. However, the lack of intuitive and interactive-usable mechanisms of such tools converts the analysis activity into a complex and time-consuming one. That is why this domain is the perfect setting to apply the defined IDPs. We illustrate the IDPs through UIs designed for analyzing genetic diseases and highlight the benefits of using such UIs.

This paper is organized as follows: Sect. 2 discusses several endeavors aimed to define IDPs. Section 3 discusses relevant concepts in the DGD, the domain where we will illustrate our patterns. Section 4 describes the Big Data challenges upon which the IDPs have been derived. Section 5 presents the classification of the proposed IDPs and how they have been applied to the illustrative example. Finally, Sect. 6 discusses the conclusions and outlines future work.

2   Related Works

The *pattern* term comes from the design of buildings and architectural planning [4]. Later, the pattern term was adopted by HCI and Software Engineering disciplines. Especially in the UI design context, the patterns emerged to solve interaction design problems, hence its name *"interaction design patterns"*.

There is a wide range of IDPs collections, available in articles, web sites, and books addressing several platforms and domains. Tidwell' book [5] describes a pattern collection to deal with common design problems of the web, desktop, mobile, social media UIs. For example, the "Showing Complex Data" chapter deals with design problems to represent large and complex data sets. Duyne et al. [6] present a pattern collection, organized into 13 groups, to design websites pointed to several domains (e.g., government, e-commerce, educational). The "K. Making Navigation Easy" group, for example, contains 14 patterns to make the web navigation easy to understand and easy to find by applying techniques for organizing and displaying navigational elements. In the Information Retrieval domain, Schmettow's article [7] exposes a pattern language containing 10 patterns for designing UIs in this domain. Other patterns dealing with massive data sets have been presented by the industry such as Expero (an enterprise focused on the user experience for complex domains) that exposes through webinar[1] a set of patterns to solve design interaction problems such as How to configure a complex data table to view the status at a glance? Or When to apply column filtering to display what user wants to see?

In the Big Data domain, the published literature about IDPs is limited, and the little existing literature describes the problems rather than providing solutions. In particular, the problems related to *data consumption* (i.e., the way in which users extract knowledge from data) are documented in Big Data's use cases [8–10], however, there is no formal proposals to solve them.

In relation to how to apply the patterns to design UIs, as far we know, there is no formal method or standard recipe. However, *Situational Method Engineering* (SME) [11] becomes a potential framework for formulating a pattern-based UI design method. SME aims to create a method adapted to a specific situation by harmonizing fragments of existing standard methods. The Big Data UI design is a specific situation where the patterns, which form the UI, behave as fragments that can be selected from situational factors and data characteristics.

At the user interface level, Big Data interaction problems have been little studied. Addressing these problems involves considering the Big Data dimensions and the inherent connection between the UI and what happens behind it. In this work, we will analyze the Big Data's use cases to define the interaction involved in the knowledge extraction and then we will suggest IDPs as solutions to design UIs for extracting knowledge from Big Data.

3   The Diagnosis of Genetic Diseases (DGD)

Differences between humans are registered in the genome under the name of *"genetic variation"*. Genetic variations not only represent different physical traits among humans but can also be the potential cause of genetic diseases (e.g., Alzheimer, Neuroblastoma). The DGD aims to extract knowledge from genetic data to diagnose a certain disease by identifying the disease-causing genetic variations and it represents a complex and time-consuming data comparison process. The researchers use web browsers provided by public biological databases (e.g., PubMed, ClinVar, dbSNP, and other databases maintained by the NCBI[2]) to search for and identify the diseases related to each of the genetic variations of a sample, which are commonly stored in large digital files (e.g., *VCF file*[3]). The main concepts and their relationships surrounding the DGD are described in an earlier work [12] through a conceptual model (CM) upon which an application for DGD, named GenDomus, was designed.

DGD is a clear example of Big Data. Table 1 shows the special data characteristics of this domain in terms of Big Data dimensions.

---

[1] https://www.experoinc.com/post/ui-design-patterns-for-navigating-complex-data-sets-online-seminar.

[2] *National Center for Biotechnology Information.* https://www.ncbi.nlm.nih.gov/guide/all/.

[3] The Variant Call Format (VCF) Version 4.2 Specification. https://samtools.github.io/hts-specs/VCFv4.2.pdf.

Table 1. DGD data characteristics in terms of Big Data dimensions.

VCF files	Volume	The size of a VCF file ranges from gigabytes to terabytes. Frequently, more than one file is required to perform the data analysis
	Variety	The content format of this text file is semi-structured. Each genetic variation in this file is described in one row and several tab-separated text columns; however, several different data can be stored in the same column
	Velocity	VCF files are processed in batch mode
Biological databases	Volume	Especially, PubMed, the database most frequented by researchers, contains more than 28 million citations of biomedical literature, as reported on its website[a]
	Variety	Biomedical literature contained in NCBI biological databases is available as web text, PDF documents, images, videos, or XML format
	Velocity	The data from biological databases can be retrieved on demand by using available APIs or FTP service and processed in batch mode

[a] https://www.ncbi.nlm.nih.gov/pubmed/

A systematic way to perform the DGD is the SILE methodology [13], which consists of four levels: Search (the researchers search for external biological databases and select those that serve the analysis purposes), Identification (for each biological database, the researchers identify the content suitable for the analysis), Load (the VCF files and the biological databases content are loaded into a central repository) and Exploitation (the researchers explore, operate, analyze, and extract knowledge from genetic data by drawing conclusions from them).

In the next sections, we will illustrate how the proposed patterns, embedded into UIs, match with the SILE methodology levels and deal with the interaction difficulties immersed in identifying disease-causing genetic variations.

4 Big Data Challenges in Knowledge Extraction

Our two initial steps in defining IDPs in the Big Data are: to analyze several real Big Data's use cases and to identify clearly the challenges related to the *data consumption* issues (i.e., browsing, exploring, and visualizing the data). We analyze real Big Data's use cases since they become ideas arising from real environments to address a specific need, as explicitly described by IBM [8]. We studied 27 Big Data use cases (5 IBM's Big Data use cases [8], 7 Pentaho's use cases [10], 10 use cases exposed by Laskowski [14], and 5 use cases reported by Datamer [9]) from different domains and industrial sectors (e.g., online stores, financial services, security, real state, pharmaceutical, energy management, IoT and telecommunications service). From the set of Big Data use cases, we identified the common challenges related to the knowledge extraction and summarized them as follow:

- *Challenge 1*: **Enhance Data Discovery.** – Need for novel mechanisms to navigate and explore structured and unstructured data sources. These mechanisms should consider the large amount of data and the diversity of data sources, the provenance (e.g., data within and outside the limits of what you have). This challenge is related to the use cases in [8].
- *Challenge 2*: **Enlarge visualization.** – Need for advanced visualizations and powerful interactive dashboards to present, in a single UI, a complete view about an entity of interest. The visualization should consider different perspectives to represent the data (e.g., data charts) and support static data as well as data in motion (i.e., such as audio, video, social media). This challenge is related to the use cases in [8–10].
- *Challenge 3*: **Perform data analysis operations.** – Need for intuitive mechanisms to operate the data to find differences and similarities between different datasets. Data operations should consider the different data types (e.g., relational data, machine data, social media data) as well as the state of the data (static data or data in motion). This challenge is related to the use cases in [8, 9, 14].
- *Challenge 4*: **Contextualize data by augmenting it.** – Current data analysis relies on the data warehouses content; therefore, the conclusions drawn from the analysis are limited to such data warehouse content. To draw better conclusions, users require enriching the data warehouses with information in context (i.e., place, social space, time) from internal or external multi-structured data sources. This challenge is related to the use cases in [9, 10, 14].

The Big Data challenges are common to all domains dealing with large volumes of heterogeneous data, and the genomic domain is one of them. Therefore, the challenges will serve as a starting point to derive IDPs for extracting knowledge from genetic data.

5   Interaction Design Patterns (IDP)

Based on our context, we define an IDP as *a proven solution to a recurring interaction design problem in the knowledge extraction from Big Data*. Where *"interaction design problem"* refers to the difficulty of designing the user-data dialog when extracting knowledge from data. Therefore, the answer to *"how"* the dialog can be performed becomes an IDP that can be implemented into a UI, allowing the user to complete his/her task. For example, when designing the UI for purchasing an airline ticket, one of the design problems we must face is: *How can the user specify the departure and return dates?* The proven solution to such a design problem is an IDP which has already been documented by other authors (e.g., the Date Selector pattern[4] from the Welie.com catalog).

A UI is an individual piece of presentation containing visual components known as *widgets* (e.g., list box, push button) which can be manipulated by the user to perform an interactive task. Designing UIs is a difficult task since the aesthetic and behavior aspects should be considered as well as the strong dependency with the platform

---

[4] Data Selector pattern. http://www.welie.com/patterns/showPattern.php?patternID=date-selector.

derived from the use of concrete widgets. To avoid such difficulties we adopt a high-level perspective by focusing on the abstract and generalist instead of the concrete and specific. To this end, the concept of **interaction units** (IU) [15] becomes a useful perspective since it defines an individual piece of presentation by abstracting both the presentation and behavior from the UI. The IU represents the *"what"*—the task or set of tasks to be performed by the user—and contains several **interaction patterns** representing the *"how"* such tasks can be performed. Indeed, an IU encloses several interaction patterns aimed to achieve a common goal. The interaction patterns range from the simplest to the most complex interaction. **Elemental patterns** represent the atomic interactions becoming the building blocks of the interaction and can come together to form more complex compound patterns called **composite patterns**. Using the elemental patterns proposed in this paper implies the existence of an underlying CM that represents the data abstractly since such patterns work with the information represented in conceptual schemas. The CM is specific to the problem to solve, in our example, we use the DGD model [12] as illustrative study.

From the implementation point of view, an IU becomes the form or web page with which the user interacts [15]. A standard website for purchasing airline tickets, for example, contains at least three IUs represented by three web forms such as *"Book flights"* to specify departure dates, *"Choose flights"* to specify the available flights and *"Purchasing flights"* to pay for tickets. The user interacts with each IU to achieve his/her goal: *"To buy an air ticket"*.

In the *Book flights* web form, the "how the user can specify the departure and return dates" could be solved through an elemental pattern whereas in the *"Purchasing flights"* web form, the "how to deal with the pay process" could use a composite pattern containing two elemental patterns: *The input data for the credit card information* and *the payment gateway* that informs whether the payment has been approved or denied.

Based on the identified Big Data challenges, we identify the IUs and patterns needed to extract knowledge from the Big Data domain, as shown in Fig. 1. In this section, we will define each IU and interaction pattern and demonstrate how they can be applied through an illustrative example: The Diagnosis of Genetic Diseases.

5.1   Interaction Units

From Fig. 1, the data environment for extracting knowledge is made up of three (3) IUs, defined as follow:

Knowledge extraction. - It acts as an IU container by encapsulating the set of indispensable IUs for extracting knowledge. This IU is related to all Big Data challenges (i.e., challenge 1 to 4) since it provides access to the IUs involved in the knowledge extraction space, as shown in Figs. 2 or 3 for our illustrative example (DGD).

Figure 2, for example, shows the "navigation tabs" to navigate to the Data Management and Exploration UIs. The implicit top-level UI containing the navigation structure becomes the implementation of this IU.

Data Management. – It searches for and identifies the available data sources and highly related to the subject under study. This IU is related to challenge 3.

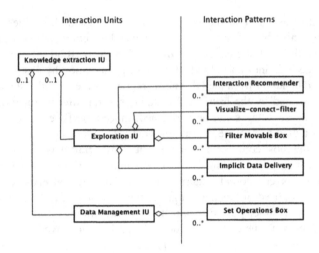

Fig. 1. Overview of the interaction units (IU) and interaction patterns for extracting knowledge from Big Data

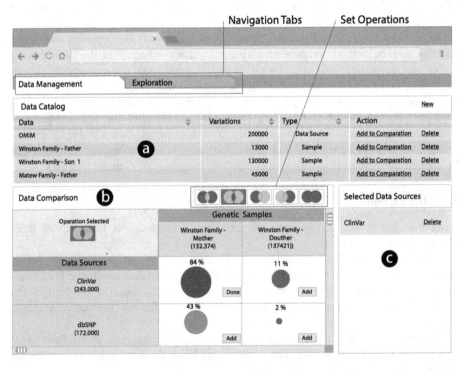

Fig. 2. Design of data management user interface. (Color figure online) Source: The authors.

Figure 2 shows the UI web that implements this UI for our illustrative example (DGD). Because not all the content of the data sources is useful for diagnosis, the content must be verified and selected before being loaded and thus safeguard the performance of the application. Therefore, the user can work with a *comparison matrix* (Fig. 2b) where the data sources and genetic samples available in the "Data Catalog" panel can be added and compared in memory (Fig. 2a). Crossing of genetic samples with the content of the data sources reveals similarities or significant differences that are useful to determine the appropriate content of the sources to be loaded for analysis purposes. The red "Set operation" red box represents an IDP involved in this UI that will be explained in detail in the next section.

Exploration. – It allows the user to explore, visualize, analyze and understand the data, make relevant annotations and share insights to draw conclusions. This IU is related to the challenges 1, 2 and 3.

Figure 3 shows the web UI design that implements this IU for our illustrative example (DGD). Through this UI, the user prioritizes the genetic variations and genes causing a certain disease by exploring and filtering the data, as well as visualizing the data distribution from different perspectives. This UI is made up of three panels located on the left, top and bottom of the UI, respectively. The left panel, hosting two sub-panels, aims to keep the user aware of the data analysis interactions performed and provide contextual information about a certain topic of interest. The top panel ("Navigator" panel) allows to visualize the data from several perspectives and manipulate the data in a direct and intuitive way. The bottom panel displays the

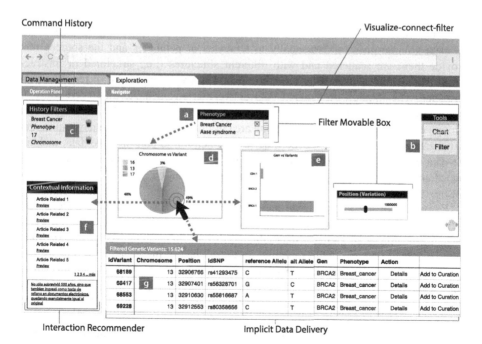

Fig. 3. Design of exploration user interface Source: The authors.

resulting data from the filter conditions applied in the top panel. The IDPs involved in this UI will be explained in the next section.

5.2   Interaction Patterns

In this section, we define the IDPs stated in Fig. 1 by using a five-parts template (*name, context, problem, solution, why,* and *example*) which was defined as a common denominator between other analyzed templates [5, 7]. Especially the "*example*" part of the template describes how the pattern has been applied to the DGD, our illustrative example, by referring to CM [12] of the DGD. What challenge solves each pattern and what level of SILE methodology is affected by each pattern are showed at the end of this section.

Name: Implicit Data Delivery.

Context: User interfaces showing vast volumes of data present performance problems. The user navigates across the dataset by visiting part by part like a person navigates across the pages of a huge book.

Problem: Users experience delays when scrolling through large data volumes. How to improve the user experience when showing and navigating through large data volumes?

Solution: Allow the user to navigate freely across the data set. Loading and delivering the data as the user navigates through a data set of interests. The data is loaded only when the user expresses implicitly her intention to move across them (e.g., navigate freely by an entire data table), rather an explicit intention of the user (e.g., paging control). This solution is applicable to retrieve any data object from the underlying data schema.

Why: Users need to visualize and navigate the data in a fluid manner, even when the data volume grows exponentially to terabytes or petabytes and the sources that provide them are scattered and with different structures (i.e., structured or unstructured). Navigating and viewing the entire data set is very expensive in terms of performance. The implicit on-demand data delivery is an adequate alternative since it motivates the user to discover the unknown data set by visualizing it and navigating it through each of its parts, focusing on one area of interest at a time. The navigation and visualization status is updated according to the intention to consume the data, making the data recovery transparent to the user. This pattern takes advantage of the performance and computational processing by loading the data on demand. Thus, the UI is not overloaded with data.

Example: In the Identification and Exploitation SILE's levels for the DGD, the user scrolls through large sets of genetic variations (represented by the *Variation* class in the CM). Figure 3g shows a data table containing more than 10.000 genetic variations records to be explored (a number easily reached in the genetic test, even when several data filter conditions have been applied), of which only 4 are visible. Instead of using a paging control to navigate the data table, the table implements the "Implicit Data Delivery" pattern allowing the user to fluidly navigate the content. When the user scrolls freely the table, only the section of genetic variations of interest is implicitly requested, loaded, and displayed. Therefore, the user can scroll from the first record to the thousandth record through a single mouse scroll movement without overloading the UI with data.

Name: Filter Movable Box.

Context: The user segregates large volume of data by using predefined filter options which are generally placed in a restrictive area of the UI. To filter and visualize the effects in the data, the user moves to the predefined filter area, then apply the filter and finally, move again to the data chart to visualize the results. The time of response is affected by the required performance to process the large volume of data.

Problem: How can the user define filter conditions and apply them in a fluid way · within the analysis space?

Solution: Allow the user to define data filter conditions from any intrinsic object attribute in the underlying data schema. Of course, the representation aspects and filter operations available in each filter box will depend on the object attribute's data type. A numeric data type, for example, can be well represented by a "single slider" component to select a certain value or a "multiple slider" component to determine a range of values. Likewise, filter operations such as *greater than* (>), *equal to* (=) can be included to express the data filter condition. The user is free to move the filter box across the canvas and to place them wherever he/she need it.

Why: The "pre-defined" filters limit the data filtering to specific criteria of the domain. What if we need to extend the set of filters? This pattern allows the user to create filter options from any attribute from the underlying data schema under analysis. In the front-end, the filter boxes can be moved and placed next to data charts allowing the user to filter the data and to look directly at the effect caused, thus, users avoid moving unnecessarily between the parts of the UI (data charts and the filter options usually predefined and strictly located in a specific area within the UI) to look at the results of the data filtering. In the back-end, performance issues need to be considered to process the large data volume since high performance to index and process the data on the fly is required when filtering the data. Node-based databases engines are suitable to store large volume of structured and structured data; however, the indexation of data is an issue to be considered when filtering the data.

Example: In the Identification and Exploitation SILE's levels for the DGD, a common task is to filter the genetic variations. Figure 3a shows this pattern implemented to assist the user to perform such task. From the CM, the user has selected the *name* and *position* attributes (from *Phenotype* and *Variation* classes respectively) to create the *Phenotype* and *Position* filter boxes. So, when the user selects the "Breast Cancer" as disease name in the Phenotype filter box, the visual data components (i.e., data charts, data table) refresh their state by showing the updated distribution of variations matching the Breast Cancer disease. Likewise, by using the Position filter box, the user filters the genetic variations ranging from 10.000 to 20.000 positions into the chromosome.

Name: Set Operation Box.

Context: Users compare two or more data sets to find similarities or differences between them. Data query languages rely on set operations to compare data sets. Such languages are difficult to use for non-technical users.

Problem: How can the user perform set operations on data in an intuitive way?

Solution: Provide the user a set operations kit (i.e., union, intersection, complement, difference) to compare data sets sharing at least one common attribute of the data schema. The user must select as input, at least two data sets from the data schema, and then select the set operation to be performed. Because of comparison, a new data set is created that can be reused in further comparisons. In the Big Data domain, the set operations must be supported by a sound technological architecture with a high-performance data processing because of a large amount of existent data.

Why: The user identifies valuable information by comparing several data sets. Often. This pattern allows the user to operate the data intuitively avoiding using and understand complex query languages and relying on technical users to extract knowledge from the data. The variety and volume of data are two Big Data dimensions committed in this pattern. The first involves the hardware performance needed to integrate dispersed sources with different formats that participate in the operation. The second involves increasing memory and storage to store the data produced by the operations performed. A hardware architecture with a flexible vertical and horizontal scaling support is appreciated for the implementation of this pattern.

Example: In the Search and Identification SILE's levels for the DGD, one of the key issues is to select the adequate data sources for the analysis. The reliability of the findings depends on large extent on the data used in the analysis. Not all the available data is useful for the data analysis. Selecting the adequate data sources, those containing the most number of variations matching with the variations in the sample under study, ensures valuable and accurate conclusions. Figure 2 shows this pattern applied to identify the adequate data sources for DGD. From the Data Catalog (Fig. 2a), the user adds the data sources (represented by *Data Source* class in the CM) to the comparative matrix (Fig. 2b) in the Data Comparison panel. The panel contains the "set operation" box located in the top-right corner. The intersection, the set operation selected in this case, is applied to each data source located in the first column and to each genetic sample located in the first row. The number of genetic variations matching between the data sources and the samples is shown graphically through a circle of variable area along with its value expressed as a percentage. The user selects the data sources containing the highest degree of concurrence with the samples. Finally, the selected data sources are added to the "Selected Data Sources" panel.

Name: Visualize-connect-filter.

Context: Users use data charts (e.g., bar, lines, maps) to understand the data, even more, if it comes in enormous data volumes. A chart is an isolated view that shows how the data behaves respect to a particular point of view. Working with multiple data charts makes possible to show the behavior of the entire data set. However, working with several charts at the same time is complicated. If a filter condition is applied to the entire data set, each data chart must be manually updated to present the new behavior.

Problem: How can the user visualize the multiple changes in the data behavior caused by a simple filtered data?

Solution: Allow the user to orchestrate all data charts to visualize, from different perspectives, how the data behaves against a certain condition. Provide a canvas where the user is free to place interactive data charts and organize them according to their needs. For each chart, the user sets the data objects, from the underlying data schema, to be displayed. To define the behavior between the charts, the user can define links between the data objects that populate the charts. In this way, an interaction performed (e.g., filter) on a chart' sector will automatically affect the display state (behavior) of the charts linked to it.

Why: This pattern allows users to explore and filter the data visually, intuitively and interactively, focusing on the details while maintaining the overview of the big picture. Combining this pattern with the Command History pattern [5] that shows the list of events performed, the user remains aware of the traceability of the actions. The Big Data dimensions of volume, variety, and velocity are directly related to this pattern. In many cases, visualizing the large volume of data requires an architecture that supports GPU accelerated visualization. Display components must support the variety of content formats (images, text, video, sound) including graphics that support the visualization of multiple variables and various organizational models (e.g., hierarchical, tabular, geographic, linear data, node networks interconnected). The connection between interactive graphics becomes an alternative to integrate dispersed and heterogeneous data sources in format. When graphs are connected through semantically similar data attributes, the underlying data sources that feed each graph are implicitly integrated. Depending on the velocity of data consumption (i.e., real time, or batch mode), visualization components can range from 2D distribution data charts to complex real-time monitoring charts that allow configuring alerts to react to events. Contrary to data processed in batch mode, real-time data imposes high performance on rendering the visualization implying the need for cloud rendering software or an underlying architecture supporting hardware-accelerated interactive data visualization.

Example: Figure 3 shows this pattern implemented for the Exploitation SILE' level for the DGD where the user can filter intuitively and visualize the data. By selecting the *Chromosome* and *Variation* classes from the CM, the user has set the Pie Chart (Fig. 3d) entitled as "Chromosome vs Variant". Similarly, the *Genotype* and *Variation* classes from the CM have been used to set the Bar chart (Fig. 3e) entitled as "Genes vs Variations". The dashed lines indicate the dynamism when filtering and exploring the data. When the user "*click*" the sector corresponding to the chromosome 17 in the "Chromosome vs Variant" chart (arrow pointer), the "Genes vs Variations" chart refreshes its content by presenting the genes contained within the chromosome 17 (i.e., CDH, BRCA2, BRCA1), and how they are distributed according to the number of genetic variations. In addition, each filter interaction is registered into the "History Filters" panel (i.e., an implementation of the Command History pattern).

Name: Interaction Recommender.

Context: Users need to be supported when navigating across the voluminous and heterogeneous set of data. By interacting with the data, the user requires relevant information (e.g., events, people, places, things) relatives to the data of interest.

Problem: Each user-data interaction expresses the user needs in an indirect way. How can the user be provided with knowledge obtained from such interactions?

Solution: Store the interactions performed by the users together with the involved data schema' elements and the related contextual information. For each meaningful interaction performed by the user, the application identifies similar stored interactions, gathers contextual information and defines alternatives exploration ways based on the data schema. The application uses the information to support the user across the data navigation.

Why: The human capacity to figure out all relationships in the large volumes of data is limited. This pattern uses the user-data interaction (e.g., select, filter), compare it with existent ones to obtain contextual information which is made available to user as suggestions to enhance his/her data navigation. The variety and volume Big Data dimensions affect the user-data interaction. The underlying infrastructure must to consider automated and scalable storage to store the large volume of interactions, as well as, the mechanisms to analytically process the interactions (e.g., classification and clustering algorithms), retrieve the interaction-related contextual information (e.g., ETL or microservices), and display the contextual information related to the interactions considering the different content formats (e.g., images, text, video, audio).

Example: This pattern is useful for the Exploitation SILE' level for the DGD. From the interactions performed by the user in the data exploration, the pattern suggests relevant information as illustrated in the Contextual Information panel (Fig. 3f). By selecting a specific genetic variation from the data table (Fig. 3g), the pattern recommends useful resources such as clinical reports, research papers or studies which are related to the selected genetic variation. All resources obtained come from previous interactions similar to the current interaction. In addition, to motivate the exploration of the data, the system identifies the conceptual model elements closely related to the genetic variation (Chromosome, Phenotype, Gen) and suggests alternative contents under the template: *People who searched for "A", searched for "B" too.*

The correspondence between the described patterns with the Big Data challenges and the levels of the SILE methodology mentioned in our illustrative example is shown in Table 2. The Load (L) level of SILE methodology is not covered by the interaction patterns because of this level is pointed to data processing data level instead of front-end level.

Table 2. Interaction patterns vs Big Data challenges and SILE methodology levels.

Interaction pattern	Interaction unit	Big Data challenge				SILE level			
		1	2	3	4	S	I	L	E
Filter movable box	Exploration	X					X		X
Set operation box	Data management	X	X			X	X		
Visualize-connect-filter	Exploration		X						X
Interaction recommender	Exploration		X	X					X
Implicit data delivery	Exploration	X	X				X		X

6   Conclusions and Future Works

In this document, we define interaction patterns for designing user interfaces for extracting knowledge from Big Data. The proposed patterns were derived from data consumption challenges identified from the study of several real use cases of Big Data. The patterns to extract knowledge from Big Data were illustrated through the design of two user interfaces for diagnosing genetic diseases. The first one, to select the right data sources for the data analysis through an intuitive mechanism based on set operations to compare data. The last one, to explore the genetic data by incorporating recommendation mechanisms and interactive data filtering.

Historically, the patterns have been applied in software engineering to increase the performance in the development of applications, as well as the improvement of their quality. In this sense, the application of the proposed interaction patterns aims to accelerate the design of Big Data user interfaces that incorporate quality attributes such as efficiency and usability and that promote productivity in the extraction of knowledge.

From the experience in the design and implementation of GenDomus' UIs for the genetic data analysis, we have identified the interaction needs of the user and the data characteristics involved, as well as, the available solutions to face the user needs and the difficulties to implement them. The proposed patterns gather all those experiences and provide useful recommendations for consuming genetic data considering the dispersion of genetic sources, the type and the volume of data. This experience can be extended perfectly to the Big Data domain.

Once the patterns have been defined, our next step in the research line is to refine, implement, evaluate the proposed patterns and propose a methodology to apply the patterns. Although the defined patterns do not solve all the interaction problems in Big Data, we consider that the patterns presented here become the initiative to create a catalog of interaction patterns to design user interfaces oriented to the extraction of knowledge from Big Data.

Acknowledgments. The authors thank the members of the PROS Center's Genome group for fruitful discussions. In addition, it is also important to highlight that Secretaría Nacional de Educación, Ciencia y Tecnología (SENESCYT) and Escuela Politécnica Nacional from Ecuador have supported this work. This project also has the support of Generalitat Valenciana through project IDEO (PROMETEOII/2014/039) and Spanish Ministry of Science and Innovation through project DataME (ref: TIN2016-80811-P).

References

1. Power, D.J.: 'Big Data' decision making use cases. In: Delibašić, B., Hernández, J.E., Papathanasiou, J., Dargam, F., Zaraté, P., Ribeiro, R., Liu, S., Linden, I. (eds.) ICDSST 2015. LNBIP, vol. 216, pp. 1–9. Springer, Cham (2015). https://doi.org/10.1007/978-3-319-18533-0_1
2. Genetic Alliance: Capítulo 2, Diagnóstico de una enfermedad genética (2009). https://www.ncbi.nlm.nih.gov/books/NBK132200/
3. Pabinger, S., et al.: A survey of tools for variant analysis of next-generation genome sequencing data. Brief Bioinform. **15**(2), 256–278 (2014). https://doi.org/10.1093/bib/bbs086
4. Borchers, J.O.: Pattern approach to interaction design. In: Proceedings of the Conference on Designing Interactive Systems: Processes, Practices, Methods, and Techniques, DIS 2000, pp. 369–378 (2000)
5. Tidwell, J.: Designing Interfaces, vol. XXXIII, no. 2. O'Reilly Media, Sebastopol (2012)
6. Van Duyne, D.K., Landay, J.A., Hong, J.I.: The Design of Sites: Patterns, Principles, and Processes for Crafting a Customer-Centered Web Experience. Addison-Wesley, Boston (2003)

7. Schmettow, M.: User interaction design patterns for information retrieval. In: EuroPLoP 2006, pp. 489–512 (2006)
8. IBM big data use cases – What is a big data use case and how to get started - Exploration. http://www-01.ibm.com/software/data/bigdata/use-cases.html
9. Datamer e-book: Top Five High-Impact Use Cases for Big Data Analytics (2016). https://www.datameer.com/pdf/eBook-Top-Five-High-Impact-UseCases-for-Big-Data-Analytics.pdf
10. Big Data Uses Cases | Pentaho. http://www.pentaho.com/big-data-use-cases
11. Henderson-Sellers, B., Ralyté, J.: Situational method engineering: state-of-the-art review. J. Univers. Comput. Sci. **16**(3), 424–478 (2010)
12. Iñiguez-Jarrin, C., García, A., Reyes, J.F., Pastor, O.: GenDomus: interactive and collaboration mechanisms for diagnosing genetic diseases. In: ENASE 2017 - Proceedings of the 12th International Conference on Evaluation of Novel Approaches to Software Engineering, Porto, Portugal, 28–29 April 2017, pp. 91–102 (2017). https://doi.org/10.5220/0006324000910102
13. Román, J.F.R., López, Ó.P.: Use of GeIS for early diagnosis of alcohol sensitivity. In: Proceedings of the BIOSTEC 2016, pp. 284–289 (2016). https://doi.org/10.5220/0005822902840289
14. Laskowski, N.: Ten big data case studies in a nutshell. http://searchcio.techtarget.com/opinion/Ten-big-data-case-studies-in-a-nutshell
15. Molina, P.J., Meliá, S., Pastor, O.: JUST-UI: a user interface specification model. In: Kolski, C., Vanderdonckt, J. (eds.) Computer-Aided Design of User Interfaces III, pp. 63–74. Springer, Dordrecht (2002). https://doi.org/10.1007/978-94-010-0421-3_5

Continuous Improvement, Business Intelligence and User Experience for Health Care Quality

Annamaria Chiasera[1]([⊠]), Elisa Creazzi[1], Marco Brandi[1],
Ilaria Baldessarini[1], and Cinzia Vispi[2]

[1] CBA Group, Viale Trento 56, 38068 Rovereto, Italy
{annamaria.chiasera,elisa.creazzi,marco.brandi,
ilaria.baldessarini}@cba.it
[2] Auditing, Internal Control and Lean Management, Modena, Italy
cinziavispi@aci-cons.it

Abstract. Long-term health care organizations are facing increased complexity to provide new, high quality services (required by regional laws) keeping costs under control. They have to deal with many internal/external procedures involving outsourced services. We develop a BI solution with a "global" approach to face different issues due to: deep impact on processes, systems, organization's structure and job roles. We propose the combination of different methodologies as Balanced Scorecard, Change management, Lean tools, User Experience with classical Data Warehouse design and development cycle. This new approach can: create "cascading improvement process" for all Departments (medical, administrative); allow timely and easy access to providers' information; bring to governing institutions a considerable saving in time and accurate control of social services' quality. Furthermore, it can develop a new culture towards processes and no value activities to increase overall quality and efficiency. The method has been applied to organizations in north of Italy.

Keywords: Balanced scorecard · User experience · Data warehouse
Continuous improvement · Lean process

1 Context

Italian Nursing homes are a type of residential care (private and public) that provide nursing care for elderly people. Multidisciplinary teams work together in order to manage health care problems but not only. These organizations have administrative, human resources, technical and quality departments with complex issues because of many working people, processes, rules, units and service strategies.

Furthermore, some critical services are outsourced (e.g. meals, laundry service, health care staff, …) by public tenders or by contracts which define prices and quality (Service Level Agreements SLA).

The establishment of an Internal Quality Measurement (QM) [1] becomes an urgent priority because of the lack of control for some critical services.

J. Krogstie and H. A. Reijers (Eds.): CAiSE 2018, LNCS 10816, pp. 505–519, 2018.
https://doi.org/10.1007/978-3-319-91563-0_31

In this scenario various challenges - both organizational and technical – are emerging:

- each service provider and even each internal department has typically its own IT system as a "Silo (or Box)" isolated from the others (no communication);
- there is not a shared infrastructure to coordinate them and integrate data produced by each system in each department;
- even when there is a single IT system (e.g. maintained by a unique service provider) internal procedures are different (from one office to another one), ineffective, and they process data manually not considering IT systems (e.g. using office automation tools like spreadsheets or even paper documents). This practice increases the risk of errors, wrong data, delays and stress because involved offices don't respect internal and external deadlines;
- even if a shared IT system is used and could provide financial, medical, human resource reports, data provided do not support decision makers [2]. They miss the "global vision", inter-functional processes and high data correlations (e.g. Patient Falls-Fractures-Costs related to rehabilitation). The Decision Support System (DSS) has also to include the SLA of the outsourced services which have a critical impact in terms of Quality and Cost.

In the real scenario QM is not useful because data is collected and processed manually, sometimes once a year, often for law requirements. Reports are used for financial needs (budget) and to comply with laws, and data is not considered for planning activities (PDCA) (see [3, 4]). The timing information is very critical to check and anticipate quality service problems and to establish "new relationships" with suppliers in order to improve the global quality to the end users (patients).

Furthermore, those organisations have to collect and manage very critical and sensitive data (e.g. patient's medical data, financial reports for invoices and for public reimbursements) required by public authorities (local Government and Italian Ministry of Health): it is very important and crucial for them to extract data automatically (and not manually as it is currently done). Such reports (required by law) could be internally used to analyse/review **processes** combining **lean** [5] and Business Process Reengineering (**BPR** [6]) tools in order to improve their practices and their quality standards in a Continuous Quality Improvement (CQI).

In some cases, those reports need so many efforts that administrative staff spends at least 1 week/per month for this "no value" activities (NVA).

In this scenario the needs of stakeholders are:

- To make their processes more effective and efficient starting from the possibility to organize the internal and external flow of data (local Authorities and Suppliers)
- To have a data warehouse system capable to collect key data (not only financial but also quality data)
- To have interactive dashboards to support decision makers for long-term and short-term planning activities in different areas: economical, human resource, finance, quality, warehouse, ...
- To employ easily customizable procedures to prepare periodic reports required by external auditors (authorities).

Key data provided by Business Intelligence (BI) can be the first step to plan and set targets, both for quality and for economic goals. A BI/Dashboard is a starting point to go deeper to analyse the critical processes using Lean tools of problems solving as DMAIC [12, 13] for improving quality by eliminating wastes. Particularly, the possibility to access critical information timely and easily will provide new opportunities such as:

1. considerable saving in time by reducing the No Value Activities (NVA)
2. accurate control of internal standard of quality (QM)
3. analysis and design of new processes and new practices both for wards and for administrative area (BPR and Lean tools as PDCA-DMAIC).

We propose an inter-functional process/organizational analysis approach concerning different areas which we wanted to analyze. We combined this holistic approach of analysis with the involvement of the end user in the first step (design and KPI definition) and in the second step (validation, development, interaction). We combined lean tools (DMAIC), BSC and BPR shared with end users of different departments (financial, healthcare, human resources), paying attention to their real needs with classical Data Warehouse design and development cycle.

This method has been applied in real scenarios with medium-large nursing homes (100–600 patients with 80–120 employees located in different regions in the north of Italy: Lombardia, Emilia-Romagna and Trentino) considering the different local requirements and different regional laws.

In Sect. 2 we introduce the techniques adopted, in Sect. 3 we present our approach and in Sect. 4 how it has been conducted in the actual scenario, in Sect. 5 we summarize our results and derive the lessons learned and in Sect. 6 we draw some conclusions and future work.

2 Background and State of the Art

In this section we introduce state of the art methodologies adapted in the project. We adopted a combination of methodologies, that normally are applied to manufacturing industry and we adapted them to health care sector which need new vision and approaches. We considered Lean methodology because its core idea is to create more value for customers with fewer resources increasing quality. It focuses on: i. elimination of wastes (inefficiencies) along entire value streams (instead of isolated points); ii. creation of processes that need less human effort, less space, less capital, and less time. The goal is to make services that are less costly and with much fewer errors, compared with traditional business systems. Lean applies in all industries, services and every process. It is a way of thinking and acting for an entire organization [7].

For Nursing Homes quality, cost control and inter-functional teams are core principles because of their complexity (many services, many law requirements and less resources given by Government).

2.1 LEAN Tools: DMAIC, Root Cause Analysis, Standardization

Our approach starts from the structured method of DMAIC [12, 13] as problem solving approach of Lean Six Sigma because we wanted to check the actual processes and problems in a short time (e.g. we fix 3 meetings in 3 weeks, once a week with the working team) before developing the BI solution.

The DMAIC approach enabled quick follow up of different problems, root causes and solutions/improvements related to the 4 Key areas investigated (Financial, H.R., Medical, Process that are linked to BSC evaluation) (Fig. 1).

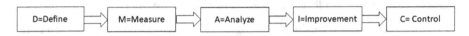

Fig. 1. DMAIC approach.

- **Define:** we define the project, goals with the project sponsor (director of the organization) and the employees of the 4 Key areas (Financial, Human Resources, Medical, Process)
- **Measure:** we collect all the documentation/data necessary for the first analysis of the inter-functional processes in the 4 Key areas
- **Analyze:** we analyze with flow charts processes, wastes and redundancies
- **Improvement:** we identify different steps of improvements (short -3 months- and medium term -9 months)
- **Control:** we validate the project with standard procedures with the identification of the "process owner".

The focus on **process analysis** of each area, within the organizations, showed process redundancies (**Muda**) that is 'waste' which is the primary aim of lean manufacturing. We also try to look deeper into problems with **Root Cause Analysis**. We refer to this Lean Tool because we want to trace a problem to its origins, sharing those information within the group.

In the Control Phase we introduce documented procedures (**Standardized Work**) easy to understand by all the end users involved in the project and useful for future improvement activities.

2.2 Balanced Scorecard Approach

For the analysis of inter-functional processes and identification of Key Performance Indicators (KPIs) we referred to Balanced Scorecard (BSC) [9–11] appropriately adapted to Healthcare Organizations. BSC has been extended from private companies to other sectors. In particular, for the Health and Social Organizations the perspective of BSC should be adapted to the effective needs of "high" quality and economic control required by management and legal regulations. The perspectives listed below identify 4 **key areas** where qualitative information is connected with financial data in a holistic way. We focused on the Financial, Customer, Internal process and Human Capital perspectives.

- Financial area: this perspective views organizational financial performance and the use of financial resources;
- Customer area: this perspective views organizational performance from the point of view of the customer that is the patient of Nursing Homes;
- Internal Process area: views performance of critical processes which have significant impact on quality and efficiency;
- Human Capital area: views performance of human capital, infrastructure, technology, culture.

We decided to adapt these perspectives primary to the needs of our end users and also to the available data sources in order to obtain the right data and KPI's in the following way:

- **Economic** and **Financial** perspective: KPI's as variable, critical costs and staff costs;
- **Residents and Patients** perspective: KPI's as social and healthcare information about patients collected in Electronic Health Records (or external excel files);
- **Process** perspective: KPI's for all processes within the organization which are critical also for legal aspects (e.g. the time/pro patients dedicated by doctors, nurses, therapists, professional careers, the bed rotation, the timing of waiting list, quality problems of external suppliers, …);
- **Personnel** perspective: KPI's for absence/presence, turnover, trainings, salary computation.

Those different perspectives are inter-related: it means that our Business Intelligence (BI) dashboards should provide a standardized nomenclature and support for decision maker using a shared ontology across the different perspectives/key areas (Fig. 2).

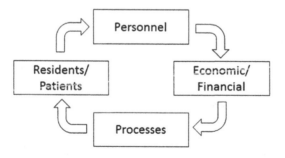

Fig. 2. BSC system.

2.3 Data Warehouse Design and Development

Designing a data warehouse requires a preliminary analysis phase in which the interesting business processes are identified to evaluate the experimental sites' performances. The identified business processes are the foundation to identify facts and dimensions to be translated into a data warehouse (DWH) schema according to the Kimball's

Enterprise Data Warehouse Bus Architecture in [8]. A typical approach to DWH schema design is the star schema approach [8] in which Key Performance Indicators (KPIs) are represented with fact tables containing mainly numeric and quantitative measures. The contextual information allowing to explore the measures along various levels of aggregation and dimensions are represented into separated dimensional tables the fact will refer to. This design approach allows to keep DWH schema clean and flexible in case new KPI or new dimensions will arrive and allows to create queries that are optimized for reporting. Dimensions represents concepts stakeholders can easily understand. This allows to interact more easily with stakeholders to define list of dimensions and level of details along which the KPIs should be explored.

2.4   User Test with Think-Aloud

The Kimball's approach to DWH design is quite effective however if applied as-is it tends to be, in our opinion, a waterfall approach as it lacks the flexibility and agility to deal with new projects in which the interaction mechanisms with users are not yet well defined and should be discovered along the project. For that reason, we combined traditional DWH design techniques with user-centered design (UCD) techniques. Particularly we apply user test technique that is used in user-center interaction to evaluate how users interact with the product.

Different techniques are available in literature to conduct user tests, but we will refer to the *think-aloud* technique. This technique requires users to complete some tasks using the interface and to report aloud thoughts and expectations on the product. Think-aloud is quite simple and cheap to apply in any phase of software development and type of prototype. It may be difficult to participants to be instinctive and to talk alone in front of the screen without thinking too much [14]. The feeling to be "observed" may influence their feedback giving modified comments and consequently biased results. For that reason, it is fundamental the role of the moderator to conduct the test, make the users feel comfortable and keep them focused on the tasks to complete. Another fundamental role is represented by the observer to get as much feedback as possible from users as the moderator is busy with them. Thanks to user tests it is possible to understand if users are capable to complete the tasks autonomously and the time required. Furthermore, it is possible to evaluate the level of satisfaction of users in using the product. [15]

We introduced user experience in the project with the intent to develop an understandable and intuitive BI dashboard. Similar approach was used in [19] to develop a mobile BI dashboard. Their user test was aimed at recreating the use of the product in normal daily routines, adding a heuristic evaluation of the interface to see if the design was in accordance with heuristics. SIBI [20] employed a user-experience approach to evaluate the introduction of BI in the Wright State University, providing dynamic reports on different thematic areas. The goal was to make data accessible to everyone. In the PhD thesis of Micheline Elias [21] user tests have been applied to make dashboards information useful to people who are not experienced in reading graphics. In one user test different types of users, newbies and experts, have been included to highlight several types of approach by capturing audio, video and screen.

Elias approach uses also the think-aloud protocol to better understand the intensions and feelings of users.

In [17] think-aloud technique is used to develop a framework doctors can use to design their own reports. In our case this solution is not suitable because: we are facing several types of users (not only doctors but also administrative staff), we are dealing with a less precise domain (clinical diagnosis and procedures are well defined with ICD-9 codes, but socio-assistive services and procedures are less standardized).

A similar approach is used in [18] with more emphasis on the visualization aspects of the dashboard creation. Thought the idea to let users design autonomously their own dashboards is appealing from our experience it is affordable by few highly skilled and motivated users: our typical user does not have time and motivation to create a dashboard but only to perform little tasks and customizations on existing solutions to "play" with data (e.g. add/remove filters, rearrange graph dimensions and measures).

3   Approach

We applied the approach depicted in Fig. 3 combining the Kimball DWH development lifecycle adapted with Lean-BSC-CM (Lean, Balanced Scorecard and Change Management) and user-tests (see colored boxes).

We created a multidisciplinary and cross functional team (6/7 participants), as follows (Fig. 4).

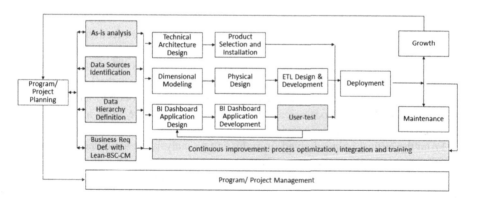

Fig. 3. Kimball DWH development cycle adapted with Lean-BSC-CM and user-tests. (Color figure online)

In addition to the organization director we involved the reference people for financial, HR, medical macro-areas with the coordination of the quality control in supervising the identification of the KPIs and particularly the processes area.

The Six Sigma DMAIC (Define, Measure, Analyse, Improve, Control) was our internal roadmap for problem solving because we had to correct plans, take new actions and initiatives for each issue we met (e.g. incomplete data entry, lack of communication within the 4 areas, fear for transparency and change). DMAIC helped us to follow the project step by step focusing on little daily improvements (Kaizen) with our teams.

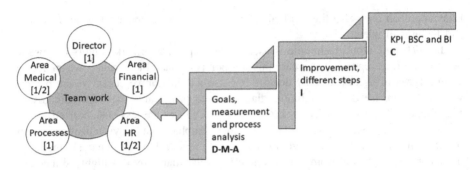

Fig. 4. Multidisciplinary team and flow of steps and techniques applied.

During a first round of interviews (3 meetings in 3 weeks, once a week, in group or single) the working practice of each reference person was depicted and the point of contact and interactions with the IT systems identified. This allowed to perform an *as-is analysis* of the actual business processes, to differentiate primary from support processes and to highlight inefficiencies, waste, costs and weaknesses of the current organizational system. From the outlined information flows we identified some critical points: business procedures are different, inefficient, processing data manually outside IT systems (e.g. using office automation tools like spreadsheets or even paper documents) increasing the risk to introduce errors, incorrect data, delays and overhead on workers; data is replicated across different systems and it is difficult to have a unique and official version of the information especially when it derives from manual manipulation outside the operative IT systems. From this first analysis phase we identified the data sources (*Data Sources Identification*) that can feed the DWH with consistent data, the organization's structure which shape the data hierarchy (*Data hierarchy definition*), the KPIs to be presented and explored. The interconnection among the various stages of business requirements definition was the Lean-BSC-CM approaches trying to cover all the thematic area [10, 11]: Economic and Financial, Residents and Patients, Processes and Personnel. In Table 1 we show an example of KPIs derived by the business requirements definition phase.

From the critical points derived from this initial analysis we identified possibilities to optimize and integrate processes. Furthermore, where current IT systems were used only partially we propose some training and coaching sessions devoted to minimize the manual work and maximize the data processed automatically by the IT systems to make them readily available to the BI platform. This *Continuous improvement: process optimization, integration and training* phase may be triggered also by results in the experimental phase as by the *growth* and *maintenance* steps in which the BI system will evolve to comply changing requirements. The other steps of our approach are the same of the Kimball lifecycle (Technical Architecture Design, Product Selection and Installation, Dimensional Modeling, Physical Design, ETL Design & Development, BI Dashboard Application Design, Deployment) as they are unavoidable steps to develop the DWH. A further extension of the original Kimball lifecycle is the addition of *user-test* phases performed as soon as the first dashboard prototype is available. Based on user-test results the dashboard evolves in a develop-test cycle repeated until an acceptable level of satisfaction is achieved.

Table 1. Example of KPIs.

Area	KPI
Socio-sanitary	*Nr. Falls with fracture* *Nr. Patients with high fall risk* *Nr. Patients with pressure lesions* *Nr. Patients with pain evaluation* *Nr. Patients with malnutrition risk*
Economic	*Monthly Costs of drugs (for advanced medications, antibiotics, Alzheimer,..)* *Monthly cost for meals, cleaning, wardrobe and laundry* *Monthly cost for internal and externalized services* *Monthly cost for internal and external personnel* *Monthly maintenance cost* *Comparison with monthly budget*
Personnel	*Nr. Worked hours by employee (absence type) category* *Nr. Of absence hours by employee (absence type) category*

Since the beginning we were aware that our BI project was related to Change Management which is a style of management which encourage organizations and individuals to deal effectively with the changes taking place in their work.

We had to face some resistances due to the deep change we introduced with our approach with the understanding that it impacted deeply on processes, systems, organizational structure and job roles. Those organizations are hierarchical not used to cross functional work and we had to face some initial "resistances". We involved directors and end users because the change has to start from the bottom and we reinforce our role to facilitate bottom up and top down communication about their processes and critical problems.

With regular meetings and follow up we tried to face the "human" issues which create more difficulties than the technical ones by using all methods and strategies we could manage (Lean philosophy, Balanced scorecard approach, Change Management) in order to overcome different problems and "challenges".

4 Implementation

In this section we present the details of the implementation and the application of the approach.

4.1 The Overall Architecture

We developed a complete Business Intelligence solution entirely based on Pentaho community edition modules (i.e. Pentaho Data Integration[1]) and Apache Superset[2]. We

---

[1] Pentaho Data Integration, http://www.pentaho.com/product/data-integration.
[2] Apache Superset, https://superset.incubator.apache.org/.

realized a monitoring dashboard using Apache Superset sitting on top of a set of data cubes populated with Pentaho Spoon ETL jobs/transformations.

Most of involved pilot sites used some of our IT solutions for Electronic Health Records, Human Resource Management, Warehouse and Financial Management. This allowed us to have full control and knowledge of the majority of source data schema. However, also IT systems owned by different providers are used (e.g. for externalized services). In that case the source schema is not available, and data can be integrated from excel files through ad-hoc ETL.

Furthermore, data is updated even after a long time they are inserted in the IT systems with no automatic way to identify updates (in most of the cases there is no management of history or timestamp of last update). We decided to periodically reload data deleting potentially obsolete data to load the consolidated snapshot (e.g. for economic and personnel data the entire year will be reloaded while for medical data a daily snapshot is enough as medical data are consolidated daily).

We found very useful the capability of Pentaho Kettle in managing several types of sources and data types/format (structured and unstructured) allowing us to prepare a versatile ETL layer (receiving data from input files, databases) enabling the transformation and mapping to the standardized data warehouse schema (Fig. 5).

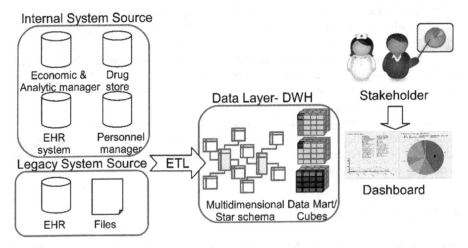

Fig. 5. Business Intelligence system architecture.

4.2   The UX Evaluation

As soon as the structure of the interface has been defined we performed periodic sessions of user tests to highlight strengths and weaknesses of the dashboard to be corrected in the ongoing development process. In the first phase of the user test we asked users to complete some tasks. In the second phase of the test we proposed a questionnaire to evaluate usability of dashboards in terms of: accessibility and clarity of the displayed KPI with different graphical layout (table vs graph views); dashboard interactivity and granularity of exploration; accessibility and exporting capabilities.

We performed the user tests on 4 users that will be the actual users of the system: 2 directors, 1 quality manager, 1 administrative manager. The first time we conducted the user test face to face with the users. The subsequent tests were performed remotely.

The test has been divided into two parts: in the first part we used the think-aloud technique proposing a list of tasks to be completed on the dashboard and asking each user if they manage or not to complete the task; in the second part, we collected through a questionnaire the usability and the dashboard's usage experience. The user test sessions have been registered (video and audio) to perform ex-post some assessments.

Table 2 shows an extract of the evaluation questionnaire provided to users at the end of user test session. For each question (if possible) we asked an evaluation with Likert scale to measure if we managed to improve the prototype. The first question was particularly interesting because it allowed to get the first emotional impression of the user in approaching and using the dashboard for the first time. Answers were quite different going from 'My God! I feel a little lost' to 'Tranquility/Curiosity', 'Very Positive'. This first impression was very useful to decide for which type of users simplification was more valuable that data analysis flexibility and details.

After each user test session, the data analyst identified a list of development activities to be performed to improve the dashboard and to be tested in the next user test.

Table 2. Evaluation questionnaire example.

Group	Question
UX	*What was your experience (emotionally) during use?*
Usability	*How easy is to use the interface?*
	KPIs are shown correctly?
	How much easy is to correlate different KPIs?
	How much easy is to filter data?
	How much easy is to export data?
Adequacy	*How much useful will be the system?*

5 Evaluation

In a previous BI project, we did not involve end-users along all the project phases (process analysis, process re-engineering and BI dashboard design). Our previous approach was aimed at:

- designing a standardize set of KPIs that can satisfy as many customers as possible (30–100 customers) rather than customize indicators to customer's specificities
- providing tools which non-technical users can use autonomously to design their own reports starting from templates, with no a priori knowledge of the underlying data warehouse schema.

That project failed because:

- it is hard to define a common set of KPIs valid for all customers, even if we focus on a very specific set of customers (medium-large residence home) because KPIs should reflect internal organizational procedures, that are only partially standardized by law
- it is hard for non-technical people to design reports on their own, using tools slightly more sophisticated than an excel spreadsheet due to: frequent turnover, lack of time and motivation, reporting tool complexity.

In this project we take a completely different perspective: we asked users which KPIs might help their organization to work more efficiently, instead of which KPIs they should provide by law.

We notice how this project involves both technological and organisational challenges.

On Technological Side

As for any software product, also for a Business Intelligence solution **user tests** are fundamental to improve the solution, avoid errors and correct critical points as soon as possible. In addition to the advantages listed in Sect. 2.4, user tests are cheap and easy to perform: it is not necessary any particular infrastructure and, as we describe through our work, they can be performed even remotely [15]. Furthermore, it is not necessary to test with many people, 5 participants will be enough to discover the majority of problems and 15 participants can discover all usability issues. The number of participants should be enlarged if other key areas are considered [16].

At the end of the experimental phase we asked users an overall evaluation of project's experience with an on-line questionnaire (Table 3). The goals were to evaluate:

1. the overall project approach and especially user's involvement in testing the solution during the first development step and in the experimental session (second step);
2. the final usability of the solution and its level of maturity in satisfying users' expectations;
3. the tangible advantages in terms of time (efficiency) and costs saved.

Table 3. Final project questionnaire.

Section	Topics
Project experience	*Overall evaluation of the experience and approach (time/resources required, analysis, user-tests, on-field experiments)*
Solution's adequacy	*Does the system reach the expected results and gives a tangible advantage (saving in time and costs)?*
Usability	*Is the solution simple and intuitive? How much easy is to explore, correlate and export the analysis?*
Innovative points	*Are the dynamic functionalities (bookmarks, export) useful points of innovations?*

The experimental phase is still on-going, and we have a preliminary evaluation of the prototype. Based on that we completely restructure the dashboard to further improve the user experience. Further feedbacks and complete estimation of the ROI (in terms of costs reduction and time saved) will be collected during an additional 6 months experimental period. The preliminary results are encouraging: the main advantage underlined by stakeholders is the KPI's service quality monitoring. They estimate to save at least 1 day of manual work each month to extract KPIs and at least 3 days each 3 month.

We also expect further time saving and errors reduction when exporting reports, as required by external auditors (3/5 days each month required to make them manually), will no longer be done manually. Furthermore, trend analysis in a simple, intuitive and timely base allows management to perform internal auditing analysis and prevent inefficiencies and quality process issues. In this way, it will be possible to test the effectiveness of operative procedures and the impact of new strategies (e.g. outsourcing decision and ROI evaluation). It will also allow the quality check of outsourced services that are currently hard (or even impossible) to control. During a usability test session one of our end user defined our dashboard "reassuring" because it gives a new experience of interactivity and simplicity by exploring data timely and without errors.

On Organizational Side

From our company side we establish a multidisciplinary team (IT, Sales, Marketing and organizational specialists) that could manage several issues we had to face (e.g. different local regulations and health services, different staff and use of existing data base, different needs and also different culture and attitude to the change).

We had to figure out a way to analyse customer needs, their processes, their hierarchical organization and their motivation because a BI project involves deep internal changes and it can fail because of cultural and mental barriers (definitely not for IT problems).

Technical solutions don't guarantee success, especially for long-term care organizations, because their workers (medical units and staff) are not used to change and to interact as an "inter-functional group". Often those groups focus on medical issues excluding other functions and processes; furthermore, they often obey "unwritten rules", which are difficult to understand and to change.

This BI project made us aware of the need of an **educational** and **training** program, because the different users have to understand what is going to change and how the project will impact their daily work.

Our real internal challenge is to motivate and develop a new "**Vision**", according to Lean Philosophy, which could better help the team to check processes and no value activities in the four BSC key areas. Furthermore, the BI solution can trigger a "cascading process" in all areas for continuous improvement required by the management and also by local rules.

This BI solution linked to Lean Philosophy and Change Management can improve the overall quality of the work within those organizations. It takes into consideration key processes as health care, financial and staff processes, and improves efficiency and effectiveness of each employee, nurse and doctor. Looking at the whole approach this project can reduce waste of time and resources, while it can deeply improve the health service quality.

Usability, KPI's Analysis and the design of the DWH are key "technical" aspects but other "soft" issues can make the difference. This BI project brings to our company a new vision no longer exclusively based on the IT software sales but on a **counselling service** that can really change and improve customer's work. The "BI service" should lead a customer to analyse its organization and processes from different points of view (technological but also organizational) and to find efficient solutions to improve the life quality of their patients.

Even if service quality in healthcare domain is hard to measure we have to include these KPI's (process perspective in the BSC) and consider them as critical for a continuous quality improvement.

6 Conclusions and Future Work

As we have underlined, future challenges will involve technological and especially organizational and educational aspects.

Technically, we have the following open points: (i) some data are still on excel files - outside our systems (frequently data are manually redacted to get a final balance report); (ii) information kept as "free text" by using notes fields instead of categorized fields (frequently for medical/social information); (iii) use of several IT systems owned by different providers.

In order to face these issues, we firstly involved end users as much as possible with training sessions and regular follow up because we had to change not only their process but also their "vision". Secondly, we worked on the technological infrastructure harmonizing data managed by different systems to allow an overall vision of cross-organization processes.

We are not proposing a solution to a novel problem (data warehousing and Business Intelligence systems are now state of the art solutions), but we think that the particular approach, linked to different methodologies as Lean, BSC and Change Management is quite innovative. It was an effort also to involve stakeholders in all the phases of the project, from the analysis to the final solutions, while coordinating our multidisciplinary team.

Furthermore, this BI solution with multidimensional KPIs can provide comparative analysis and graphs in order to observe the dynamics of KPIs. This system can reveal potential areas (medical, financial and human resources areas) where performance is lacking and the management can examine and identify root causes determining performance gaps. BI solutions, combined with Lean Tools and Internal Benchmarking, can make the difference in order to improve health care quality standards and all the processes within those organizations.

Acknowledgement. We thank the teams involved as pilots in the project: "ASP Terre d'Argine" in Carpi and Azienda Speciale Comunale "Cremona Solidale" in Cremona for the valuable support and constant feedback during the project.

References

1. Kukhareva, P., Kawamoto, K., Shields, D., Barfuss, D., Halley, A., Tippetts, T., Warner, P., Bray, B., Staes, C.: Clinical decision support-based quality measurement (CDS-QM) Framework: prototype implementation, evaluation, and future directions. In: AMIA Annual Symposium Proceedings, vol. 2014, pp. 825–834 (2014)
2. Power, D., Sharda, R., Burstein, F.: Decision Support Systems. Wiley Encyclopedia of Management (2015). ISBN 9781118785317
3. Redick, E.L.: Applying FOCUS-PDCA to solve clinical problems. Dimension. Crit. Care Nurs. DCCN 18(6), 30–34 (1999)
4. Berwick, M.: Continuous improvement as an ideal in health care. N. Engl. J. Med. 320(1), 53–56 (1989)
5. Kim, C., Spahlinger, D., Kin, J., Billi, J.: Lean health care: what can hospitals learn from a world-class automaker? Soc. Hosp. Med. (2006). https://doi.org/10.1002/jhm.68
6. Kohn, D.: The role of business process reengineering in health care. Health Inf. Manage. 14(3), 1–6 (1994). PMID: 10131587
7. Womack, J., Jones, D.: Lean Thinking: Banish Waste and Create Wealth in Your Corporation. Simon & Schuster, New York (2003)
8. Kimball, R., Ross, M.: The Data Warehouse Toolkit: The Definitive Guide to Dimensional Modeling, 3rd edn. (2013)
9. McDonald, B.: A Review of the Use of the Balanced Scorecard in Healthcare, BMcDConsulting (2012). www.bmcdconsulting.com
10. Rahman, N.: Measuring performance for data warehouses - a balanced scorecard approach. Int. J. Comput. Inf. Technol. (IJCIT) 4, 1–7 (2013)
11. Wegmann, G.: The balanced scorecard as a knowledge management tool: a French experience in a semi-public insurance company. ICFAI J. Knowl. Manage. 6(3), 22–38 (2008)
12. Sehwail, L., DeYong, C.: Six Sigma in health care. Leadersh. Health Serv. 16(4), 1–5 (2003). https://doi.org/10.1108/13660750310500030
13. Carrigan, M., Kujawa, D.: Six Sigma in health care management and strategy. Health Care Manag. 25(2), 133–141 (2006)
14. Thinking Aloud: The #1 Usability Tool, by Jakob Nielsen, 16 January 2012. https://www.nngroup.com/articles/thinking-aloud-the-1-usability-tool/
15. usability.org: Usability Testing. https://www.usability.gov/how-to-and-tools/methods/usability-testing.html
16. Nielsen, J.: Why You Only Need to Test with 5 Users (2000). https://www.nngroup.com/articles/why-you-only-need-to-test-with-5-users/
17. Chamney, A., Mata, P., Viner, G., Archibald, D., Peyton, L.: Development of a resident practice profile in a business application framework. In: ICTH-2014 (2014)
18. Elias, M., Bezerianos, A.: Exploration views: understanding dashboard creation and customization for visualization novices. In: Campos, P., Graham, N., Jorge, J., Nunes, N., Palanque, P., Winckler, M. (eds.) INTERACT 2011. LNCS, vol. 6949, pp. 274–291. Springer, Heidelberg (2011). https://doi.org/10.1007/978-3-642-23768-3_23
19. Ferreira, G.L.: User-centered design of the interface prototype of a business intelligence mobile application. Master's Degree in Industrial Engineering and Management Dissertation (2013)
20. Prabhala, S.: A User-Centered Approach to Drive Business Intelligence Solution in Higher Education, Strategic Information and Business Intelligence (SIBI), Wright State University (2015)
21. Elias, M.: Enhancing User Interaction with Business Intelligence Dashboards, Ph.D. thesis Ecole Centrale Paris (2012)

Data Modelling and Mining

Embedded Cardinality Constraints

Ziheng Wei and Sebastian Link[(✉)]

Department of Computer Science, University of Auckland, Auckland, New Zealand
{zwei891,s.link}@aucklanduni.ac.nz

Abstract. Cardinality constraints express bounds on the number of data patterns that occur in application domains. They improve the consistency dimension of data quality by enforcing these bounds within database systems. Much research has examined the semantics of integrity constraints over incomplete relations in which null markers can occur. Unfortunately, relying on some fixed interpretation of null markers leads frequently to doubtful results. We introduce the class of embedded cardinality constraints which hold on incomplete relations independently of how null marker occurrences are interpreted. Two major technical contributions are made as well. Firstly, we establish an axiomatic and an algorithmic characterization of the implication problem associated with embedded cardinality constraints. This enables humans and computers to reason efficiently about such business rules. Secondly, we exemplify the occurrence of embedded cardinality constraints in real-world benchmark data sets both qualitatively and quantitatively. That is, we show how frequently they occur, and exemplify their semantics.

Keywords: Cardinality constraint · Data and knowledge intelligence
Data quality · Decision support · Missing information

1 Introduction

Background. Cardinality constraints enforce bounds on the number of data patterns that occur in application domains. Cardinality constraints were introduced in Chen's seminal ER paper [3], and have attracted interest and tool support ever since. A cardinality constraint $card(X) \leq b$ stipulates for an attribute set X and a positive integer b that a relation must not contain more than b different tuples with matching values on all the attributes in X. For example, a social worker may not handle more than five cases at any time. This expressiveness makes cardinality constraints invaluable in applications such as data integration, modeling, and processing [16].

Motivation. Most applications require the efficient handling of missing information. This is particularly true in the big data era where large quantities of data (volume) are integrated from heterogenous sources (variety) with different granularity of completeness (veracity). As such, a major challenge in accommodating missing information in the semantics of classical integrity constraints is

© Springer International Publishing AG, part of Springer Nature 2018
J. Krogstie and H. A. Reijers (Eds.): CAiSE 2018, LNCS 10816, pp. 523–538, 2018.
https://doi.org/10.1007/978-3-319-91563-0_32

Table 1. Snippet of the NCVoter data set

id	v_id	f_name	Lname	gender	address	city	phone	register_date
t_0	480	doris	thompson	f	hwy 119	mebane	5783747	11/05/1940
t_1	612	odessa	teer	f	hwy 119	mebane	$\perp$	5/09/1940
t_2	622	dallie	boswell	f	hwy 119	mebane	$\perp$	5/09/1940
t_3	972	john	smith	m	hwy 119	mebane	$\perp$	10/26/1940
t_4	433	louise	buckner	f	231 s marshall st	graham	2269183	5/04/1940
t_5	577	ruth	albright	f	231 s marshall st	graham	2266060	5/08/1940

the interpretation of null marker occurrences. Indeed, null markers are frequently introduced to integrate data from heterogenous structures in a relation. Previous research has addressed the extension of integrity constraints to incomplete data by uniformly applying one of many possible interpretations to all occurrences of null markers in a relation. As not all null marker occurrences can be interpreted uniformly, in particular in integrated data sets, the results that are derived from such research have only limited applicability. Here, we take a new approach in which the semantics of a cardinality constraint is only dependent on complete fragments that are embedded in a given incomplete relation. Since the definition of our constraints is independent of the interpretation of null markers, they provide a robust approach to describing the semantics of big data, which is fundamental to how such data is processed. We call this new class *embedded cardinality constraints* (eCCs). They consist of a set E of attributes and an ordinary cardinality constraint $card(X) \leq b$ with $X \subseteq E$. The set E specifies the scope r^E of an input relation r on which $card(X) \leq b$ must hold. As such, $card(X) \leq b$ is embedded in r^E. In examples we commonly write $E - X$ instead of E to emphasize on which additional attributes tuples of the scope must be complete. Embedded cardinality constraints provide users with the ability to separate their requirements for the completeness and uniqueness dimensions of data quality. Users specify the set E to declare their completeness requirements, and the cardinality constraint $card(X) \leq b$ with $X \subseteq E$ to declare their uniqueness requirements. As with any constraint, the main target is to improve the consistency of data by enforcing business rules within the database system.

Examples. As an illustration of embedded cardinality constraints we look at the data snippet in Table 1, which is taken from the real-world data set *ncvoter*[1].

An interesting question concerns the number of voters that can live at the same location (address and city). The snippet, and in fact *ncvoter* as a whole, satisfies the embedded cardinality constraint $(\emptyset, card(\{address, city\}) \leq 4)$, since there are at most four different voters that live at the same location and for which address and city have no missing information. However, for marketing campaigns we may only be interested in how many voters can live at the same location that we can reach by telephone, so we are interested in the smallest bound b such that

---

[1] https://hpi.de/fileadmin/user_upload/fachgebiete/naumann/projekte/
repeatability/DataProfiling/FD_Datasets/ncvoter_1001r_19c.csv.

$(\{phone\}, card(\{address, city\})) \leq b)$ holds. For the actual *ncvoter* data set, and therefore for the snippet, this bound is 2, as witnessed by $\{t_4, t_5\}$. Actually, the same tuple block is a showcase that $(\{phone\}, card(\{address, city, gender\})) \leq 2)$ holds. That is, there are up to two voters of the same gender who live at the same location and have a phone number. In contrast, the ordinary cardinality constraint $(\emptyset, card(\{address, city, gender\})) \leq 3)$ says that up to three voters of the same gender live at the same location, for which $\{t_0, t_1, t_2\}$ is a witness.

Contributions. Our contributions can be summarized as follows:

Modeling: We introduce embedded cardinality constraints which hold independently of the interpretation of null marker occurrences in incomplete relations. They subsume previously studied classes of cardinality constraints.

Reasoning: We show that reasoning about embedded cardinality constraints can be done efficiently. Indeed, we characterise the associated implication problem by a finite ground axiomatization and by a linear-time algorithm. Consequently, the gain in expressivity over other classes of constraints does not sacrifice good computational properties. We illustrate how efficient reasoning helps minimise the costs for processing updates, speeds up query evaluation, and prunes the search space when computing the constraints that hold on an incomplete relation.

Case Studies: We illustrate the occurrence of embedded cardinality constraints in actual data sets, previously used as benchmarks for data profiling algorithms. Qualitatively, we present showcases for what embedded cardinality constraints can express in the real world and also provide insight into the lattice structures that these constraints exhibit. Quantitatively, we have implemented a simple heuristics to discover embedded cardinality constraints from an incomplete relation. The heuristics is sound but not complete, so it does not find all embedded cardinality constraints that hold, but those it does find do hold and are minimal. Details of the heuristics are out of scope, but its main purpose is to illustrate that embedded cardinality constraints occur frequently, and that the separation of completeness requirements from uniqueness requirements generates substantial additional patterns that are exhibited by data sets.

Organization. We discuss related work in Sect. 2. Embedded cardinality constraints are introduced in Sect. 3. Some real-world examples and their underlying structure are presented in Sect. 4. Computational problems and their applications are characterized in Sects. 5 and 6. A quantitative analysis of embedded cardinality constraints is presented in Sect. 7. We conclude and mention future work in Sect. 8.

2   Related Work

We demonstrate in this section how embedded cardinality constraints are different from previous work.

Cardinality constraints are an influential contribution of data modeling [16]. They were already present in Chen's seminal paper [3], and are now part of

all major languages for data modeling, including UML, EER, ORM, XSD, and
OWL. Cardinality constraints have been extensively studied [2,4,6–13,15]. Since
the primary goal of cardinality constraints is to improve consistency, they need
to be enforced on actual data sets. Real-world data often exhibits incompleteness
in the form of null marker occurrences. This has necessitated the study of (cardinality) constraints over incomplete relations. According to our best knowledge,
we only know of the approach that ignores tuples with any null marker occurrence on any column over which an integer bound is specified [8]. This approach,
however, is covered by embedded cardinality constraints $(E, card(X) \leq b)$ as
the special case where $E = X$. Hence, embedded cardinality constraints handle
completeness requirements by specifying E, and they handle uniqueness requirements by stipulating $card(X) \leq b$ where $X \subseteq E$. The previous approach, that
is when $E = X$, can only handle both requirements at the same time. Another
special case occurs when $X = \emptyset$: here, b stipulates how many tuples in the given
relation have no null marker occurrences on any of the columns in E. Furthermore, embedded cardinality constraints also subsume the recently introduced
class of contextual keys [17] as the special case where $b = 1$. Embedded cardinality constraints are therefore considerably more expressive than previously
studied classes of cardinality constraints. We also exemplify in this article to
which degree they occur more frequently in the real-world data than contextual
keys. Despite the gain in expressivity, we show that axiomatic and algorithmic
reasoning about embedded cardinality constraints is not much more involved
than that for contextual keys. More precisely, we can establish a finite ground
axiomatization and a linear time algorithm to decide the implication problem
for embedded cardinality constraints. These subsume those established recently
for contextual keys as the special case where $b = 1$ for each given embedded cardinality constraint. Recently, cardinality constraints have also been investigated
for uncertain data models, including possibilistic models [5] and probabilistic
models [14]. These are orthogonal directions of research about cardinality constraints, but possibilistic and probabilistic embedded cardinality constraints can
be investigated in future work.

3   Embedded Cardinality Constraints

We fix concepts from relational databases and introduce the central notion of
embedded cardinality constraints.

A *relation schema* is a finite set R of attributes A. Each attribute A is associated with a domain $dom(A)$ of values. Based on the demand in traditional and
modern applications, data models need to accommodate missing information. In
order to represent the standard approach adapted by relational database technology, we assume that the domain $dom(A)$ of every attribute A contains the
distinguished symbol $\perp$, representing the null marker. We stress that the null
marker is not a domain value, and is only included in the domain of attributes for
convenience and ease of discussion. A tuple t over R is a function that assigns to
each attribute A of R an element $t(A)$ from the domain $dom(A)$. For an attribute

set X, a tuple t is said to be X-total whenever $t(A) \neq \bot$ for all $A \in X$. A *relation* over R is a finite set of tuples over R. An expression $card(X) \leq b$ with some subset $X \subseteq R$ and a positive integer $b \in \mathbb{N}$ is called a *cardinality constraint over* R. A cardinality constraint $card(X) \leq b$ over R is said to *hold* in a relation r of R, denoted by $r \models card(X) \leq b$, if and only if there are not $b+1$ different tuples $t_1, \cdots, t_{b+1} \in r$ such that for all $1 \leq i < j \leq b+1$, $t_i \neq t_j$ and for all $A \in X$, $t_i(A) = t_j(A) \neq \bot$.

Note that this simple model is already expressive enough to address at least three dimensions of big data: volume is represented by the numbers of columns and rows in a relation, veracity is represented by null marker occurrences in relations, and variety is represented by the ability to integrate information from different sources and of different structure by putting domain values into columns where they are known for and null marker occurrences where they are not.

A critical issue in extending integrity constraints to data models with incomplete information is the way in which null marker occurrences are handled. One popular approach is to assign a particular semantics to the occurrences, such as 'value exists but is currently unknown' or 'value does not exist' or 'no information'. In practice, that is in SQL, there is no room to associate different interpretations with different occurrences: Only one universal interpretation is assigned to every occurrence. With this limitation it is difficult to obtain meaningful results when the data is used. This is particularly true for applications that employ integrated data where some occurrences of null markers are bound to have different interpretation. Nevertheless, a plethora of research has been conducted in this area, resulting in different notions of constraints. In contrast, this article follows a complementary approach in which constraints are evaluated independently of any null marker occurrences. That is, the satisfaction of the cardinality constraints is only dependent on the complete fragments in incomplete relations. This has the strong advantage that the results obtained from any use of the constraints are robust under varying interpretations of null marker occurrences. With this motivation in mind, we will now introduce the central notion of embedded cardinality constraints.

Definition 1. *An* embedded cardinality constraint *(eCC) over relation schema* R *is an expression* $(E, card(X) \leq b)$ *where* $X \subseteq E \subseteq R$ *and* $b \in \mathbb{N}$. *We call* E *the* extension *and* $card(X) \leq b$ *the* cardinality constraint associated with the eCC. *For a relation* r *over* R, *the extension* E *defines the* scope r^E *of the eCC as* $r^E = \{t \in r \mid t \text{ is } E - total\}$. *The eCC* $(E, card(X) \leq b)$ *over* R *is* satisfied by, *or said to* hold in, *the relation* r *over* R *if and only if the scope* r^E *satisfies the cardinality constraint* $card(X) \leq b$ *associated with the eCC.*

We sometimes simply write $(E - X, card(X) \leq b)$ instead of $(E, card(X) \leq b)$ in order to save space or emphasize the (non-)existence of additional attributes in the extension. The introduction has already presented several real-world examples of embedded cardinality constraints. The following section provides further insight into the expressivity of this new class of constraints.

4   Real-World Examples with Embedded Lattice View

We illustrate the business rules that embedded cardinality constraints can express, and illustrate the inherent structure that these constraints exhibit. The latter can be exploited as a navigational aid that assists users in understanding, representing and browsing cardinality profiles of their data. As an interesting special case, we present completeness cubes as a navigational aid that users can employ as a model of how many tuples are complete on a given set of attributes. We use the public data set *bridges* as a running example.

The Pittsburgh bridge data set, *bridges*, is a popular reference data set on the UCI machine learning repository[2]. It provides some basic information about 108 different bridges in Pittsburgh. Table 2 shows 14 tuples from the full data set where some columns were removed to focus on the attributes that matter for the embedded cardinality constraints we would like to discuss.

For example, we may want to know the maximum number of bridges that lead over the same river, were built for the same purpose, and are of the same type. Indeed, 14 is the answer, which is the smallest upper bound b with which the eCC $(\emptyset, card(\{river, purpose, type\}) \leq b)$ is satisfied by *bridges*. We may wonder for how many of those the length and the number of lanes are both known. The relevant eCC would be $(\{length, lanes\}, card(\{river, purpose, type\}) \leq 8)$.

For ordinary cardinality constraints the integer bounds are non-increasing with an increasing number of attributes, as illustrated on the left of Fig. 1. For eCCs, additional attributes in the extension E generate an embedded lattice

Table 2. Snippet of the bridges data set

id	river	loccation	erected	purpose	length	lanes	rel-l	type
E17	M	4	1863	RR	1000	2	$\perp$	SIMPLE-T
E21	M	16	1874	RR	$\perp$	2	$\perp$	SIMPLE-T
E25	M	10	1882	RR	$\perp$	2	$\perp$	SIMPLE-T
E26	M	12	1883	RR	1150	2	S	SIMPLE-T
E31	M	8	1887	RR	1161	2	S	SIMPLE-T
E37	M	18	1891	RR	1350	2	S	SIMPLE-T
E45	M	14	1897	RR	2264	$\perp$	F	SIMPLE-T
E47	M	15	1898	RR	2000	2	S	SIMPLE-T
E94	M	13	1901	RR	$\perp$	2	F	SIMPLE-T
E95	M	16	1903	RR	1300	2	S	SIMPLE-T
E51	M	6	1903	RR	1417	2	F	SIMPLE-T
E50	M	21	1903	RR	1154	$\perp$	F	SIMPLE-T
E89	M	4	1904	RR	1200	2	S-F	SIMPLE-T
E92	M	10	1914	RR	2210	$\perp$	F	SIMPLE-T

---

[2] https://archive.ics.uci.edu/ml/index.php.

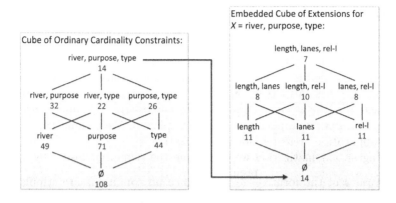

Fig. 1. Embedded lattice of extensions for each ordinary cardinality constraint

structure, for each fixed set X of attributes. Indeed, if E increases, the corresponding bounds cannot increase. This is illustrated on the right of Fig. 1. Table 2 contains those tuples that generate all the integer bounds marked red in the right of Fig. 1.

An interesting special case of these lattices are given by eCCs of the type $(E, card(\emptyset) \leq b)$. These stipulate upper bounds on the numbers of tuples that are E-total. Figure 2 shows these bounds for the data set *bridges*, based on all combinations of the four attributes *length*, *lanes*, *span*, and *rel-l*. For example, there are 90 tuples that are total on *span* and *rel-l*.

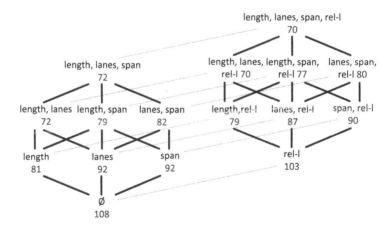

Fig. 2. Lattice of cardinality constraints for completeness dimensions

5 Axiomatic Characterization of the Implication Problem

We establish a finite ground axiomatization for the implication problem of embedded cardinality constraints. This will enable us to effectively enumerate all implied eCCs, that is, to determine the semantic closure $\Sigma^* = \{\sigma \mid \Sigma \models \sigma\}$ of any given eCC set Σ. A finite axiomatization facilitates human understanding of the interaction of the given constraints, and ensures all opportunities for the use of these constraints in applications can be exploited. We comment on a couple of direct application areas for the axiomatization.

In using an axiomatization we determine the semantic closure by applying *inference rules* of the form $\dfrac{\text{premise}}{\text{conclusion}}$. Since no conditions are stipulated for the application of these inference rules, the resulting axiomatization is called a ground axiomatization. For a set $\mathfrak{R}$ of inference rules let $\Sigma \vdash_{\mathfrak{R}} \varphi$ denote the *inference* of φ from Σ by $\mathfrak{R}$. That is, there is some sequence $\sigma_1, \ldots, \sigma_n$ such that $\sigma_n = \varphi$ and every σ_i is an element of Σ or is the conclusion that results from an application of an inference rule in $\mathfrak{R}$ to some premises in $\{\sigma_1, \ldots, \sigma_{i-1}\}$. Let $\Sigma_{\mathfrak{R}}^+ = \{\varphi \mid \Sigma \vdash_{\mathfrak{R}} \varphi\}$ be the *syntactic closure* of Σ under inferences by $\mathfrak{R}$. $\mathfrak{R}$ is *sound* (*complete*) if for every set Σ over every R we have $\Sigma_{\mathfrak{R}}^+ \subseteq \Sigma^*$ ($\Sigma^* \subseteq \Sigma_{\mathfrak{R}}^+$). The (finite) set $\mathfrak{R}$ is a (finite) *axiomatization* if $\mathfrak{R}$ is both sound and complete. Table 3 shows the axiomatization $\mathfrak{C}$ for eCCs that we will prove to be sound and complete. Here, R denotes an arbitrarily given underlying relation schema, $E, E', X, X' \subseteq R$, and b, b' are positive integers.

Theorem 1. *The set $\mathfrak{C} = \{\mathcal{B}, \mathcal{E}, \mathcal{S}, \mathcal{T}\}$ is sound and complete for the implication problem of embedded cardinality constraints.*

We note that the rules

$$\frac{}{card(R) \leq 1} \qquad \frac{card(X) \leq b}{card(XX') \leq b} \qquad \frac{card(X) \leq b}{card(X) \leq b'}$$

are sound and complete for the implication of ordinary cardinality constraints [7,8], and are embedded in our inference rules $\mathcal{T}$, $\mathcal{S}$, and $\mathcal{B}$.

Table 3. The finite ground axiomatization $\mathfrak{C} = \{\mathcal{B}, \mathcal{E}, \mathcal{S}, \mathcal{T}\}$ of eCCs

$\dfrac{}{(R, card(R) \leq 1)}$ (trivially embedded keys, $\mathcal{T}$)	$\dfrac{(E, card(X) \leq b)}{(EE', card(X) \leq b)}$ (super extension, $\mathcal{E}$)
$\dfrac{(E, card(X) \leq b)}{(E, card(XX') \leq b)}$ (super set, $\mathcal{S}$)	$\dfrac{(E, card(X) \leq b)}{(E, card(X) \leq b + b')}$ (weaker bound, $\mathcal{B}$)

Proofs of Soundness and Completeness. Let $\Sigma \cup \{\varphi\}$ denote a set of eCCs over a given relation schema R.

Soundness. We need to show that every eCC φ that can be inferred from a given eCC set Σ by $\mathfrak{C}$ is also implied by Σ. Let r denote a relation of the given relation schema R. It suffices to show the following for every inference rule in $\mathfrak{C}$: If the premise of the rule is satisfied by r, then the conclusion of the rule is also satisfied by r.

For the soundness of $\mathcal{T}$ we observe that the scope r^R contains all tuples of r that are complete on all attributes of R. Since r^R is a set, r^R is also a set and, consequently, there cannot be two different tuples that have matching values on all the attributes of R. Hence, r^R satisfies $card(R) \leq 1$.

For the soundness of $\mathcal{E}$ we assume that $r \models (E, card(X) \leq b)$. By definition, $r^E \models card(X) \leq b$. Consequently, there cannot be $b+1$ different tuples in r^E that all have matching values on all the attributes in X. For every subset $E' \subseteq R$, $r^{EE'}$ is a subset of r^E. Consequently, there cannot be $b + 1$ different tuples in $r^{EE'}$ that all have matching values on all the attributes in X. It follows that $r^{EE'}$ satisfies $card(X) \leq b$. By definition, r satisfies $(EE', card(X) \leq b)$.

For the soundness of $\mathcal{S}$ we assume that r satisfies $(E, card(X) \leq b)$. By definition, r^E satisfies $card(X) \leq b$. Consequently, there cannot be $b+1$ different tuples in r^E that all have matching values on all the attributes in X. For every subset $X' \subseteq R$, X is a subset of XX'. Consequently, there cannot be $b + 1$ different tuples in r^E that all have matching values on all the attributes in XX'. Indeed, otherwise there would have to be $b + 1$ different tuples in r^E that all have matching values on all the attributes in X. It follows that r^E satisfies $card(XX') \leq b$. By definition, r satisfies $(E, card(XX') \leq b)$.

For the soundness of $\mathcal{B}$ we assume that r satisfies $(E, card(X) \leq b)$. By definition, r^E satisfies $card(X) \leq b$. Consequently, there cannot be $b+1$ different tuples in r^E that all have matching values on all the attributes in X. For every non-negative integer b', b is at most $b+b'$. Consequently, there cannot be $(bb')+1$ different tuples in r^E that all have matching values on all the attributes in X. Indeed, otherwise there would already be $b + 1$ different tuples in r^E that all have matching values on all the attributes in X. It follows that r^E satisfies $card(X) \leq b + b'$. By definition, r satisfies $(E, card(X) \leq b + b')$.

Completeness. We need to show that every φ that is implied by Σ can also be inferred from Σ by the use of inference rules in $\mathfrak{C}$ only. We show the contraposition, that is, we assume that φ cannot be inferred from Σ by $\mathfrak{C}$ and construct a relation over R that satisfies Σ but violates φ. Let $\varphi = (E, card(X) \leq b)$ such that $\Sigma \vdash_{\mathfrak{C}} \varphi$ does not hold. We define a relation r over R that consists of $b + 1$ different tuples as follows: For $R - E = \{A_0, \ldots, A_{n-1}\}$ and $j = 0, \ldots, b$, tuple

$$t_j(A) := \begin{cases} 0, & \text{if } A \in X \\ j, & \text{if } A \in E - X \\ j, & \text{if } A = A_j \\ \bot, & \text{if } A \in R - (E \cup \{A_j\}) \end{cases}$$. The relation r may look as follows:

$E-X$	X	$R-E$
$0\cdots 0$	$0\cdots 0$	$0 \perp \quad \perp \quad \perp \cdots \perp \perp$
$1\cdots 1$	$0\cdots 0$	$\perp 1 \quad \perp \quad \perp \cdots \perp \perp$
$\vdots$	$\vdots$	$\vdots$
$b\cdots b$	$0\cdots 0$	$\perp \cdots \quad \perp \quad b \perp \cdots \perp$

The relation is well-defined, that is, contains $b+1$ different tuples for the following reason. If $R - E = \emptyset$ and $E - X = \emptyset$, then $\varphi = (R, card(R) \leq b)$. However, $(R, card(R) \leq 1) \in \Sigma_{\mathcal{C}}^{+}$ by application of $\mathcal{T}$, and this would lead to $\varphi \in \Sigma_{\mathcal{C}}^{+}$ by application of $\mathcal{B}$. Hence, $R - E \neq \emptyset$ or $E - X \neq \emptyset$, and $|r| = b + 1$.

The relation does not satisfy φ since $r^{E} = r$ and r contains $b + 1$ different tuples with matching values on all the attributes in X.

It remains to show that r satisfies every $\sigma = (E', card(X') \leq b') \in \Sigma$. Assume that r violates σ. Then $E' \subseteq E$ (as otherwise $|r^{E'}| \leq 1$ and r would satisfy $(E', card(X') \leq 1)$ and by soundness of $\mathcal{B}$ also σ), $X' \subseteq X$ (as otherwise all tuples would have different projections on X, so r would satisfy $(E', card(X') \leq 1)$ and by soundness of $\mathcal{B}$ also σ), and $b' \leq b$ (as otherwise there couldn't be $b' + 1$ tuples to violate σ). Consequently, we can apply $\mathcal{E}$, $\mathcal{S}$, and $\mathcal{B}$ to $\sigma = (E', card(X') \leq b') \in \Sigma$ to obtain $(E, card(X) \leq b) \in \Sigma_{\mathcal{C}}^{+}$, a contradiction. Consequently, our assumption that r violates σ must have been wrong. Since σ was chosen arbitrarily we have just shown that r satisfies all elements of Σ and violates φ. We conclude that φ is not implied by Σ.

Applications. While axiomatizations facilitate human understanding of how to reason, there are also a number of more tangible applications. This is not surprising, as axiomatizations are usually taken as the first step towards developing automated reasoning tools. Indeed, axiomatizations are commonly employed to develop algorithms that can decide associated implication problems. This, in turn, has numerous applications. The next section deals directly with the development of algorithmic characterizations of the implication problem for eCCs.

Algorithms can decide instances of implication problems efficiently, but they simply return either true or false. People often wonder how an algorithm derived at a particular decision. Here, axiomatizations can provide additional insight. If the answer yes, then a derivation of the candidate constraint from the given constraint set must exist. More intriguingly, if the answer is no, our completeness proof is guaranteed to provide users with an example relation that shows why the candidate constraint is not implied by the given constraint set. The completeness proof can be converted into a tool that constructs such an example relation automatically, whenever the decision algorithm returns false.

As a second application we mention the discovery problem (aka dependency mining or data profiling), in which an algorithm ought to return all those constraints from a given class that hold in a given relation [1]. Quite frequently, the search and solution spaces are massive, which makes it necessary to derive effective pruning strategies that decrease the search space and allow solutions to the discovery problem to be returned efficiently. Here, sound inference rules can directly be translated into pruning strategies. In fact, if an eCC $(E, card(X) \leq b)$

Algorithm 1. Inference

Require: $R, \Sigma, (E, card(X))$ with a set Σ of embedded cardinality constraints
Ensure: $\min\{b : \Sigma \models (E, card(X) \le b)\}$
1: **if** $E = R$ and $X = R$ **then**
2:     **return** 1;
3: **else**
4:     $b \leftarrow \infty$;
5:     **for all** $(E', card(X') \le b') \in \Sigma$ **do**
6:         **if** $E' \subseteq E$ and $X' \subseteq X$ and $b' < b$ **then**
7:             $b \leftarrow b'$;
8: **return** b;

---

has been validated to hold on the input relation, then every check of any eCC $(E', card(X') \le b')$ where $E \subseteq E'$, $X \subseteq X'$, and $b \le b'$ hold is redundant and should be omitted. Having a complete axiomatization ensures that all pruning strategies are known.

6 Algorithmic Characterization

In this section we develop algorithmic tools that decide the implication problem for embedded cardinality constraints in linear time in the input. As outlined before, this complements our axiomatization established in the last section.

Indeed, computing Σ^* and checking whether $\varphi \in \Sigma^*$ is not an efficient approach towards deciding the implication problem. The following theorem allows us to decide the implication problem for embedded cardinality constraints with a single scan of the input. Note that the proof employs the construction from the completeness proof of our axiomatization.

Theorem 2. *Let $\Sigma \cup \{(E, card(X) \le b)\}$ denote a set of eCCs over R. Then Σ implies $(E, card(X) \le b)$ iff (i) $E = R$ and $X = E$ or (ii) there is some $(E', card(X') \le b') \in \Sigma$ such that $E' \subseteq E$, $X' \subseteq X$, and $b' \le b$ hold.*

Proof. If (i) or (ii) hold, then $(E, card(X) \le b)$ can be inferred from Σ by $\mathfrak{C}$. Consequently, the soundness of $\mathfrak{C}$ ensures that $(E, card(X) \le b)$ is implied by Σ.

Vice versa, assume that neither (i) nor (ii) hold. Invalidity of (i) ensures that $R - E \ne \emptyset$ or $E - X \ne \emptyset$ holds. This guarantees that the relation r from the completeness proof has $b + 1$ different tuples. Invalidity of (ii) ensures that r satisfies Σ. Since the relation also violates $(E, card(X) \le b)$ by construction, it follows that Σ does not imply $(E, card(X) \le b)$.

Instead of translating Theorem 2 directly into a decision algorithm, we prefer to establish a linear-time algorithm for the more general computational problem that computes for a given set Σ of eCCs over a given relation schema R, and a given attribute set pair (E, X) with $X \subseteq E \subseteq R$ the minimum positive integer b (or $b = \infty$ if no integer exists) such that $(E, card(X) \le b)$ is still implied by Σ.

Algorithm 1 computes this supremum b as follows: if $E = X = R$, then $b = 1$ is returned according to axiom $\mathcal{T}$ of our axiomatization $\mathfrak{C}$. Otherwise, all input eCCs from Σ are scanned and the current supremum b is revised to b' whenever an eCC is found whose extension E', attribute set X', and bound b' satisfy $E' \subseteq E$, $X' \subseteq X$, and $b' < b$. This is valid due to the remaining inference rules in $\mathfrak{C}$. If no appropriate input eCC is found, then ∞ is returned. The total number of attributes that occur in Σ and R are denoted by $|\Sigma|$ and $|R|$, respectively.

Theorem 3. *On input $(R, \Sigma, (E, card(X)))$, Algorithm 1 returns in $\mathcal{O}(|\Sigma|+|R|)$ time the supremum b with which $(E, card(X) \le b)$ is implied by Σ.*

Proof. The correctness of Algorithm 1 follows directly from Theorem 2. For the time complexity, we only require one scan over all input attributes in Σ plus the input attributes in (E, X). The latter could be provided in the format $(E-X, X)$ ensuring that every attribute in R only occurs once.

Algorithm 1 can directly be used to decide the implication problem of eCCs. Indeed, given an eCC set $\Sigma \cup \{(E, card(X) \le b)\}$ over relation schema R, such a decision algorithm will return yes if and only if $b \ge b'$ where b' is returned by Algorithm 1 on input $(R, \Sigma, (E, card(X)))$.

Corollary 1. *The implication problem of embedded cardinality constraints can be decided in time linear in the input.*   □

Applications. Our algorithm has direct applications in saving update and query costs. When updating a data set, we need to ensure that the resulting data set satisfies all the eCCs that have been established as meaningful business rules of the underlying application domain. Validating the satisfaction of any business rule that is implied by the remaining rules is a waste of time. Being able to detect implied rules enables us to minimal set of business rules in which none is implied by the rest, thereby ensuring a minimal overhead in maintaining the consistency of data sets under updates. For example, if we have already validated that a data set satisfies $\sigma = (\{phone\}, card(address, city) \le 2)$, then there is no need to validate that it satisfies $\sigma' = (\{phone, register\_date\}, card(address, city, gender) \le 3)$. When querying *ncvoter* one may ask to return the voter-id of all voters who live at locations where no more than 3 voters of the same gender live for whom phone numbers and registration dates are known. Being aware that *ncvoter* satisfies the eCC σ and deciding that σ implies σ', the original query can automatically be optimized to the query that returns the voter-id of voters.

7   Quantitative Analysis of Our Real-World Data Sets

This section provides some quantitative insight into the occurrence of embedded and ordinary constraints in five real-world data sets from the UCI machine learning repository. These data sets are frequently used to test the performance of data dependency discovery algorithms [1]. We have implemented a heuristic to discover embedded cardinality constraints from incomplete relations. The heuristic is sound as the eCCs it finds are guaranteed to hold on the given data set and also minimal. The heuristic is not complete, so it is not guaranteed to find all eCCs that hold on the given data set. The point of the heuristic is to show that eCCs occur frequently in real-world data, which we illustrate by the sheer number of their occurrences and also by comparing that to the number of occurrences of ordinary cardinality constraints (oCCs), that is eCCs with an empty extension, embedded uniqueness constraints (eUCs), that is eCCs where $b = 1$, and ordinary uniqueness constraints (oUCs), that is eCCs with an empty extension and where $b = 1$.

Table 4. Characteristics of data sets and numbers of oCCs, pCCs, oUCs, pUCs

	$\#r$	$\#c$	$\#\perp$	$\#ir$	$\#ic$	$\#oCCs$	$\#pCCs$	$\#oUCs$	$\#pUCs$
breast	691	11	16	16	1	557	259	1	1
bridges	108	13	77	38	9	301	1877	0	3
echo	132	13	132	71	12	135	1668	18	27
hepatitis	155	20	167	75	15	312	1262	344	102
ncvoter	1000	19	2863	1000	5	438	976	78	69

7.1 Occurrences of Ordinary and Embedded Constraints

Table 4 shows some characteristics of the five data sets[3] we analyzed: the number of rows ($\#r$), columns ($\#c$), null marker occurrences ($\#\perp$), incomplete rows ($\#ir$), and incomplete columns ($\#ic$).

Our heuristic revealed the number of oCCs ($\#oCCs$), which are eCCs where $E = \emptyset$, and the number of pure *eCCs* ($\#pCCs$), which are eCCs where $E \neq \emptyset$. In previous work we had developed algorithms that determine the total number of oUCs ($\#oUCs$), which are eCCs where $E = \emptyset$ and $b = 1$, and the number of pure eUCs ($\#pUCs$), which are eCCs where $E \neq \emptyset$ and $b = 1$. While $\#pCCs$ and $\#oCCs$ are lower bounds based on our heuristic, $\#pUCs$ and $\#oUCs$ are actual numbers of a sound and complete algorithm from previous work.

With the exception of *breast* our heuristic has found many more pure eCCs than oCCs. Even though the numbers of (ordinary and pure) eCCs are just lower bounds and that of (ordinary and pure) eUCs are exact, the analysis gives an indication of how many more business rules can be expressed by eCCs in comparison to eUCs.

7.2 Cardinality Histograms for All Data Sets

For additional insight we have analyzed the distribution of the integer bounds (cardinalities) in the eCCs we were able to discover. The results are visualized in Fig. 3. The distributions are skewed towards lower cardinalities, which is natural since projections with larger cardinalities are typically less frequent. The

---

[3] https://hpi.de/naumann/projects/repeatability/data-profiling/fds.html#c168191.

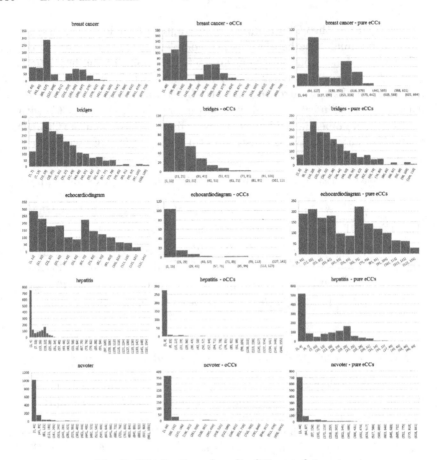

Fig. 3. Distribution of cardinalities on data sets

distributions for pure eCCs are less skewed than the distributions for ordinary eCCs, indicating that the independence of the completeness requirements (as expressed by non-trivial extensions) generates substantial additional constraints with diverse cardinalities.

8 Conclusion and Future Work

We have introduced the new class of embedded cardinality constraints. Their most interesting feature is their independence of any interpretation of missing information, which makes their employment for applications robust in the context of integrated big data sets. Despite the ability of embedded cardinality constraints to express previous classes of constraints as special cases, we showed that embedded cardinality constraints enjoy a finite ground axiomatization and their implication problem can be decided in linear time in the input. This makes their application also effective, as all opportunities of employment can be efficiently checked automatically. In addition, we have exemplified their expressivity on real-world data sets, visualized the interaction they exhibit in the form of embedded lattice structures, and provided quantitative evidence of their frequent occurrence in practice.

There are many more interesting problems associated with embedded cardinality constraints, including their discovery problem and the computation of Armstrong relations. Solutions to these two problems would provide computational support towards the acquisition of embedded cardinality constraints that are meaningful in a given application domain. Other problems include the interaction with other constraints, such as functional dependencies, or the definition of embedded cardinality constraints in models for Web or uncertain data.

References

1. Abedjan, Z., Golab, L., Naumann, F.: Profiling relational data: a survey. VLDB J. **24**(4), 557–581 (2015)
2. Calvanese, D., Lenzerini, M.: On the interaction between ISA and cardinality constraints. In: Proceedings of the Tenth International Conference on Data Engineering, Houston, Texas, USA, 14–18 February 1994, pp. 204–213. IEEE Computer Society (1994)
3. Chen, P.P.: The Entity-Relationship model - toward a unified view of data. ACM Trans. Database Syst. **1**(1), 9–36 (1976)
4. Ferrarotti, F., Hartmann, S., Link, S.: Efficiency frontiers of XML cardinality constraints. Data Knowl. Eng. **87**, 297–319 (2013)
5. Hall, N., Köhler, H., Link, S., Prade, H., Zhou, X.: Cardinality constraints on qualitatively uncertain data. Data Knowl. Eng. **99**, 126–150 (2015)
6. Hartmann, S.: On the implication problem for cardinality constraints and functional dependencies. Ann. Math. Artif. Intell. **33**(2–4), 253–307 (2001)
7. Hartmann, S.: Reasoning about participation constraints and Chen's constraints. In: Schewe, K., Zhou, X. (eds.) Proceedings of the 14th Australasian Database Conference on Database Technologies, ADC 2003, Adelaide, South Australia, February 2003. CRPIT, vol. 17, pp. 105–113. Australian Computer Society (2003)
8. Hartmann, S., Köhler, H., Leck, U., Link, S., Thalheim, B., Wang, J.: Constructing Armstrong tables for general cardinality constraints and not-null constraints. Ann. Math. Artif. Intell. **73**(1–2), 139–165 (2015)
9. Jones, T.H., Song, I.Y.: Analysis of binary/ternary cardinality combinations in Entity-Relationship modeling. Data Knowl. Eng. **19**(1), 39–64 (1996)
10. Liddle, S.W., Embley, D.W., Woodfield, S.N.: Cardinality constraints in semantic data models. Data Knowl. Eng. **11**(3), 235–270 (1993)
11. McAllister, A.J.: Complete rules for n-ary relationship cardinality constraints. Data Knowl. Eng. **27**(3), 255–288 (1998)
12. Queralt, A., Artale, A., Calvanese, D., Teniente, E.: OCL-Lite: finite reasoning on UML/OCL conceptual schemas. Data Knowl. Eng. **73**, 1–22 (2012)
13. Roblot, T.K., Link, S.: Urd: a data summarization tool for the acquisition of meaningful cardinality constraints with probabilistic intervals. In: 33rd IEEE International Conference on Data Engineering, ICDE 2017, San Diego, CA, USA, 19–22 April 2017, pp. 1379–1380. IEEE Computer Society (2017)

14. Roblot, T.K., Link, S.: Cardinality constraints with probabilistic intervals. In: Mayr, H.C., Guizzardi, G., Ma, H., Pastor, O. (eds.) ER 2017. LNCS, vol. 10650, pp. 251–265. Springer, Cham (2017). https://doi.org/10.1007/978-3-319-69904-2_21
15. Thalheim, B.: Fundamentals of cardinality constraints. In: Pernul, G., Tjoa, A.M. (eds.) ER 1992. LNCS, vol. 645, pp. 7–23. Springer, Heidelberg (1992). https://doi.org/10.1007/3-540-56023-8_3
16. Thalheim, B.: Integrity constraints in (conceptual) database models. In: Kaschek, R., Delcambre, L. (eds.) The Evolution of Conceptual Modeling. LNCS, vol. 6520, pp. 42–67. Springer, Heidelberg (2011). https://doi.org/10.1007/978-3-642-17505-3_3
17. Wei, Z., Link, S., Liu, J.: Contextual keys. In: Mayr, H.C., Guizzardi, G., Ma, H., Pastor, O. (eds.) ER 2017. LNCS, vol. 10650, pp. 266–279. Springer, Cham (2017). https://doi.org/10.1007/978-3-319-69904-2_22

Relationship Matching of Data Sources: A Graph-Based Approach

Zaiwen Feng[1,2(✉)], Wolfgang Mayer[1], Markus Stumptner[1],
Georg Grossmann[1], and Wangyu Huang[3]

[1] Advanced Computing Research Centre,
University of South Australia, Adelaide, Australia
{zaiwen.feng,wolfgang.mayer,markus.stumptner,
georg.grossmann}@unisa.edu.au
[2] School of Computer, Wuhan University, Wuhan, China
[3] Data to Decisions CRC, Adelaide, Australia
wangyuh@alumni.cmu.edu

Abstract. Relationship matching is a key procedure during the process of transforming structural data sources, like relational data bases, spreadsheets into the *common data model*. The matching task refers to the automatic identification of correspondences between relationships of source columns and the relationships of the common data model. Numerous techniques have been developed for this purpose. However, the work is missing to recognize relationship types between entities in information obtained from data sources in *instance level* and resolve *ambiguities*. In this paper, we develop a method for resolving ambiguous relationship types between entity instances in structured data. The proposed method can be used as standalone matching techniques or to complement existing relationship matching techniques of data sources. The result of an evaluation on a large real-world data set demonstrated the high accuracy of our approach (>80%).

Keywords: Knowledge graph · Semantic relation · Semantic label
Linked data

1 Introduction

Information pertaining to law enforcement activities is obtained from multiple sources and in a variety of data formats, which must be consolidated into a common data model to facilitate searching and long-term data management. Due to manually mapping data sources to the common data model is a tedious task, a handful of mapping design systems have been developed. These systems include InfoSphere Data Architect (from Clio [15]), BizTalk Mapper [22], Altova MapForce [21], and Stylus Studio [23]. All of these systems are based on the same general methodology that was first proposed in Clio [15]. Several approaches have been proposed to automate this process. Most of these approaches [9–11, 19] focus on *semantic labelling*, annotating data fields, or *source attributes*, with classes and/or properties of common data model. However, a precise mapping needs to describe the semantic relations between the source attributes in addition to their types.

© Springer International Publishing AG, part of Springer Nature 2018
J. Krogstie and H. A. Reijers (Eds.): CAiSE 2018, LNCS 10816, pp. 539–553, 2018.
https://doi.org/10.1007/978-3-319-91563-0_33

In recent years, several works have already addressed *relationship matching* problem. Karma[1] [2–4, 19, 20] is an information integration tool that enables users to quickly and easily integrate data from a variety of data sources including databases, spreadsheets, JSON, and Web APIs. To use Karma, end-users firstly import the domain ontologies they want to use for modeling the data. The system then automatically suggests semantic labels for each columns of source data. Later, they exploit the created semantic labels and the domain ontologies to learn high-quality relationships and finally a *semantic model* for the loaded data source. Karma has been used to model the data from Smithsonian American Art Museum[2] and then publish it into the LD cloud. However, there exist some limitations: Karma is not effective to be applied in disambiguating *multiple* relationship types between two recognized entity *instances* when integrating data sources into knowledge graph based on a semantic model. This requirement, however, is fairly frequent in the Integrated Law Enforcement (ILE) Project of D2D CRC[3] because there might exist multiple relationships between a pair of neighboring classes in the common data models used [17]. For example, there are 54 different kinds of relationship types between the class Person and the class Location, and 119 relationship types between Person and Person.

In this paper, we extend Karma and present a novel approach that disambiguates different types of relations between the data fields in the *data sources* including databases and spreadsheets. The main contribution of our approach is a mechanism to distinguish and then obtain a correct relationship type (e.g. *lives at*) between two recognized entity instances (e.g. John Smith and 5 Long Road) of a knowledge graph, even though there exist multiple kinds of relationships between a pair of classes (e.g. Person and Location) that the above entity instances corresponds to in the common data model. This technique is beneficial to automate tasks of transforming structured data sources into the linked data based on the common data model. To our knowledge, no previous work specially deals with distinguishing relationship types of knowledge graph in the context of data integration.

This paper is structured as follows. Section 2 demonstrates a motivation example of our work, followed by Sect. 3 that describes our approach. Section 4 gives an evaluation of the approach in this paper. Section 5 presents a review of related work. In Sect. 6 we conclude the paper and discuss future work.

2 Motivation Example

In this section, we explain the problem of learning relationship types between recognized entity instances by giving a concrete example that will be used throughout the paper to illustrate our approach. Figure 1 shows a common data model where the ovals represent classes (e.g. Organization, Person, Location, etc.), and rectangles stand for the data attributes of a class (e.g. number, street, postcode and state). We formally

---

[1] http://usc-isi-i2.github.io/karma/.

[2] https://americanart.si.edu/.

[3] http://www.d2dcrc.com.au/rd-programs/integrated-law-enforcement/.

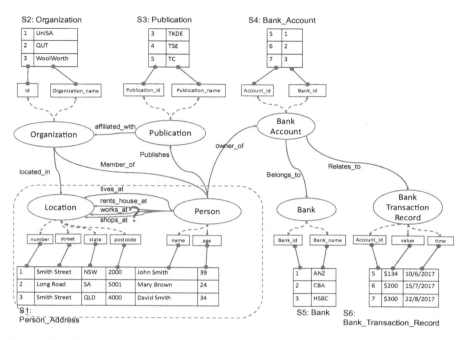

Fig. 1. Sample common data model and the new data source with spreadsheets S1–S6 (Color figure online)

define a *semantic type* to be a pair consisting of a domain class and one of its data properties $\langle class\_uri, property\_uri \rangle$ [19]. The solid lines denote the relationship between different classes (e.g. *located_in* between the class Organization and the class Location), and dashed lines link class and its corresponding data attributes.

As shown above, $S1 - S6$ are spreadsheets of a new data source in this scenario, including *S1: Person_Address, S2: Organization, S3: Publication, S4: Bank_Account, S5: Bank* and *S6: Bank_Transaction_Record*. We want to match all the data values of the new data source to the common data model shown in the top of Fig. 1.

The *first* step in mapping the spreadsheets $S1 - S6$ shown in Fig. 1 to the common data model is to label its attributes with data attributes. For example, the correct semantic types for the fifth column of $S1$ (with data value *John Smith, Mary Brown* and *David Smith*) are $\langle Person, name \rangle$, for the sixth column (with data value *39, 24* and *34*) are $\langle Person, age \rangle$. Various techniques can be employed to automate the labeling task [12, 13, 19]. However, a mapping that only includes the types of the attributes is not enough because it does not reveal how the attributes relate to each other. To build a precise mapping, we need a *second* step that determines how the semantic labels should be connected to capture the intended meaning of the data. In this work, we assume that the labeling step is already done and we focus on distinguishing the *relationship* types.

Assume that we have already obtained a knowledge graph based on the common data model and some other data sources. The initial knowledge graph includes an amount of semantic content. Now we intend to import the new data source $S1 - S6$ into the current knowledge graph. As shown in the red-coloured rounded rectangle at the left bottom of Fig. 1, for the new data source $S1$, all the columns have been annotated by the data properties of the class Person and the class Location in the common data model respectively. The attributes of *name* and *age* in the table are annotated by the data property *Name* and *Age* of the Person entity. Similarly, the attributes of *number of address*, *street name*, *post code*, and *state name* in the table are annotated by the data property *Number*, *Street*, *Postcode*, *State* of the Location entity.

However, the current way of modelling does not always correctly represent the semantics of this new data source at the *instance-level*. The reason is that although there exist multiple relationship types between the class Person and the class Location, i.e. *works at*, *lives at*, *rents house at* and *shops at*, in the common data model (shown at the top of Fig. 1), the relationship between a person instance and a location instance in the knowledge graph is ambiguous unless it is designated manually or captured explicitly in the new data source. For an instance, consider the 3rd tuple of $S1$, in case the relationship type between David Smith and Smith Street is not designated yet, we do not know what the real relationship type between them is if we simply depend on the common data model, leading that the relationship between a person instance and a location instance (e.g. David Smith and Smith Street) in the knowledge graph is ambiguous.

Now the problem is if we leverage the initial linked data as background knowledge to distinguish relationship types between attributes of new data sources? The basic idea of our approach is to exploit the initial linked data as knowledge background to distinguish relationship types between attributes of new data sources. Once we have identified the semantic types of the source attributes, we can search the linked data and slice it into a bundle of fragments of knowledge graph. For example, we will obtain four kinds of smaller linked data graphs, which contains a specific relationship type, including *lives at*, *rents house at*, *works at* and *shops at*, between a Person instance and Location instance. Then, we learn from these knowledge graph fragments as examples to infer relationship types for the new data sources.

3   Our Approach

Our approach to automatically distinguishing multiple relationship types between a pairwise recognized entity instances rests on graph extraction, graph matching, and machine learning for relationship type classification. The inputs to our approach are a repository of (RDF) linked data in a domain, and a data source whose attributes are already labeled with the correct semantic types. The output is an updated knowledge graph expressing how the missed relationship types are disambiguated.

The overall approach includes three steps, which are shown in Fig. 2. Step 1: we slice the knowledge graph into a bundle of graphs (e.g. B1, B2, N1 and N2). Specified relationship types are at the center of a group of these graphs. We then categorize these graphs into groups (e.g. t or n) according to its central relationship. Step 2: we extract frequent subgraphs for each group of graphs. Step 3: we select part of these frequent subgraphs in Step 2 as discriminative feature set (e.g. F1, F2 and F3), code a feature matrix and build an appropriate classifier (e.g. Neural Net or Decision Tree).

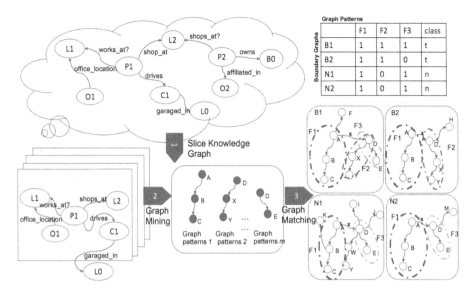

Fig. 2. The overall approach

Now suppose we have new data sources, which contain ambiguous relationships between columns that needs to be clarified. First, we will import all of data values of the new data sources into the existing knowledge graph. Then, similar with Step 1, we slice the knowledge graph into a set of graphs, each of which contains an ambiguous relationship that needs to be clarified. The obtained classifier can be used to classify any of these graphs into a certain group (*t* or *n*) in Step 1, and then we can clarify an ambiguous relationship by identifying the proper relationship type through classifying a graph containing the ambiguous relationship based on a data pool of linked data from and across multiple data sources.

3.1 Building Boundary Graph

The essence of our approach is to analyze the graph structure around a special relationship type (e.g. *lives in*). However, most knowledge graphs are likely to be very large. The knowledge graph needs to be sliced into a set of smaller ones as examples for training. All these sliced boundary graphs are at the center of a special relationship type.

A boundary graph is a directed graph with the relationship *r* between a pair of anchor vertices $x1$ and $x2$, and a given distance from the farthest nodes to the anchor vertices, where:

- $x1, x2$: the *anchor vertices* of the boundary graph;
- *r*: the *central relationship* between $x1$ and $x2$ of the boundary graph;
- *l*: the maximum length of the farthest node of the boundary graph with the start of $x1$ or $x2$.
- *maxDegree*: the maximum degree for each vertex.

We give the procedure for creating a boundary graph from a RDF repository, which is the output of Karma. First, we discover a central relationship r and its corresponding anchor points $x1$ and $x2$ and we construct an initial boundary graph. Subsequently, we extend the initial boundary graph with a process of depth-limited breadth-first search. As such, the size of the extended graph is controlled by the maximum distance from the anchor vertices and maximum degree of each vertex.

Figure 3 shows an example of boundary graph. The anchor points of this boundary graph are an instance of the class Person and an instance of class Location with the relationship typed as *rents_house_at* (green-colored) between them. In this example, the maximum length from the farthest vertex to the anchor points $x1$ or $x2$ (blue-colored) is 3.

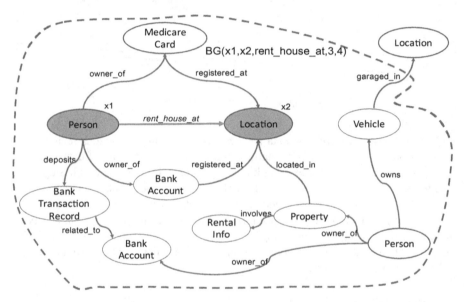

Fig. 3. Example of boundary graph with a relationship *rents_house_at* (Color figure online)

3.2   Extracting Patterns from Boundary Graphs

Although a boundary graph is a sliced fragment from the whole knowledge graph, it contains a lot of graph patterns related to the *anchor points* and the *central relationship*. Given a set of boundary graphs with a specified relationship type, we mine the schema-level patterns connecting the instances of the classes. Each pattern is a graph in which the nodes correspond to classes and the links correspond to relations in the common data model.

Formally, given a Boundary Graph Dataset, $BGD = \{G_0, G_1, \ldots, G_n\}$, each boundary graph $G_i \in BGD$ has the anchor points x_1 and x_2 and the central relationship r, $support(g)$ denotes the number of graphs (in BGD) in which g is a subgraph. The problem of extracting patterns from the set of boundary graphs can be phrased as finding subgraphs g s.t. $support(g) \geq minSup$ (a minimum support threshold). To filter out the subgraphs of which the size is too small, we set a minimal edge number and a minimal node number as the *bound* of size of *g*. We extract the frequent subgraphs in *BGD* using the gSpan algorithm [6]. These extracted frequent subgraphs denote the graph patterns of *BGD*.

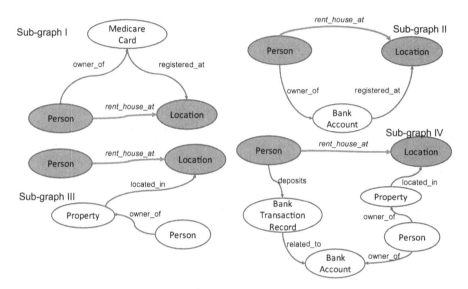

Fig. 4. Part of frequent subgraphs of the boundary graph in Fig. 3

Figure 4 shows four graph patterns extracted from the boundary graphs with the central relationship typed as *rents_house_at* (One of the BG is shown in Fig. 3). The first sub-graph shows that, if a Person rented a house in a certain Location and both were linked to a Medicare Card, then the underlying pattern is that the Person is the owner of this Medicare Card, and the Location is the registered location with the same Medicare Card. The second sub-graph demonstrates that, if a Person rented a house in a certain Location and both were linked to a Bank Account, then the underlying pattern is that the Person is the owner of this Bank Account, and the Location is the registered location with the same Bank Account. The third sub-graph shows a pattern that, if a Person rented a house in a certain Location where a Property was located in, then there might be another Person who is the owner of that Property. The last sub-graph shows that, if a Person rented a house in a certain Location where a property was located in, then if another Person is the owner of that Property, the tenant could be related to a Bank Transaction Record, which is related to a Bank Account that is owned to the property owner.

3.3 Classifying Boundary Graphs

Suppose there are a set of relationship types $R = \{r_1, r_2, \ldots, r_m\}$ and a Boundary Graph Dataset $BGD = \{G_1, G_2, \ldots, G_N\}(m \leq N)$, where each boundary graph $G_i(i \leq N)$ contains a specified relationship type r ($r \in R$). For instance, suppose we have a *BGD* that includes 50 boundary graphs with left anchor vertex Person and right anchor vertex Location. There are 2 different central relations, i.e. $r_1 = rents\_house\_at$, and $r_2 = works\_at$.

We pose the problem of distinguishing relationship types as a boundary graph classification task. Given a set of N training examples of the form $(x_1, y_1), \ldots, (x_N, y_N)$ such that x_i is the feature vector of the ith example (i.e., example boundary graph $G_i, G_i \in BGD$) and y_i is the label (i.e., central relationship type $r, r \in R$), our learning algorithm seeks a function $g : X \to Y$, where X is the input space and Y is the output space.

The features used in our algorithm are graph patterns (see Sect. 3.2) that appear frequently in a set of boundary graphs with a certain relationship type $r(r \in R)$. Let $BG_i(1 \leq i \leq m)$ be a group of boundary graphs with the central relationship type r ($r_i \in R$) (e.g. $works\_at$). We leverage the method described in Sect. 3.2 to find the frequent subgraphs set F_i for BG_i. As such, each $BG_1, BG_2, \ldots, BG_m$ has its corresponding frequent feature set $F_1, F_2, \ldots, F_m$, respectively. Since we are interested in finding the most Discriminative Feature Set (DFS) for the classification work, we ignore all the subgraphs that are common between F_i and $F_1 \cup F_2 \cup \ldots \cup F_{i-1} \cup F_{i+1} \cup \ldots \cup F_m$. Let $F_i' = F_i - (F_1 \cup F_2 \cup \ldots \cup F_{i-1} \cup F_{i+1} \cup \ldots \cup F_m)$ be the DFS for F_i, thus obtaining $DFS = F_1' \cup F_2' \cup \ldots \cup F_m'$, which we use for classifying all of the boundary graphs in BGD. Once we have the feature vectors for all the boundary graph sets, we train a classification algorithm to discriminate between the relationship types we seek to disambiguate.

After obtaining the DFS, we can compute the feature vector of G^r (i.e. G_X^r) using a subgraph matching algorithm [7] to find the exact matching. Let G_X^r be a vector of length $p(p = |DFS|)$, where the ith entry in G_X^r is 1 if $x_i \in DFS$ is a subgraph of G^r.

Table 1. Matrix for training model

	f1	f2	f3	f4	f5	f6	f7	f8	class
G1	1	0	1	1	0	0	0	1	rents_house_at
G2	1	1	1	0	0	0	0	0	rents_house_at
G3	1	1	1	1	0	0	0	0	rents_house_at
G4	0	0	1	0	1	0	1	1	works_at
G5	1	0	0	0	1	1	0	1	works_at
G6	1	0	0	0	1	0	1	1	works_at
...	...	...	...	...	...	...	...	...	...

For an instance, let the frequent sub-graphs of the boundary graphs $G_1 - G_3$ with relationship type $rents\_house\_at$ be $F_1 = \{f_1, f_2, f_3, f_4\}$. Similarly, let the frequent sub-graphs of the boundary graphs $G_4 - G_6$ with relationship type $works\_at$ be $F_2 = \{f_5, f_6, f_7, f_8\}$. There is not any intersection between F_1 and F_2. Thus, we obtain the discriminative feature set $DFS = \{f_1, f_2, f_3, f_4, f_5, f_6, f_7, f_8\}$. We compute the feature vector of each graph using a subgraph matching algorithm and finally code the matrix for training model, as shown in Table 1. A classifier could be built based on this matrix.

Given G^r containing a central relationship r that we seek to distinguish, we compute the feature vector of G^r using a subgraph matching algorithm based on *DFS* and apply the trained classifier to predict the relationship type between the anchors of G^r.

4 Evaluation

A comprehensive performance study has been conducted in our experiments on real world dataset. We applied YAGO (Yet Another Great Ontology) [5] data set, which is a massive semantic knowledge base, derived from Wikipedia, WordNet and GeoNames. Currently, YAGO has knowledge of more than 10 million instances of entities (like persons, organizations, cities, etc.), 99 relationship types, and contains more than 120 million facts about these entities. For an instance, a typical YAGO fact is shown below.

$<$Wouter_Vrancken$>$ $<$playsFor$>$ $<$K.V._Kortrijk$>$

Here, K.V. Kortriik is a Belgian professional football club, which is annotated with the entity *Organization*. Wouter Wrancken, who is annotated with the entity *Person*, is a former Belgian defensive midfielder in association football. The relationship between *Person* and *Organization* is *playsFor* in this case.

Our performance tests show that our method has much better accuracy on the YAGO dataset than Karma. Our method also demonstrates a good scalability on YAGO dataset since it succeeds in completing the match of relationships with 1 K boundary graphs containing over 100 nodes.

All of our experiments are done on a 2.5 GHZ Intel Core i7 PC with 16 GB main memory, running OS X 10.11.6. We used gSpan provided by Yan et al. [6] to get the frequent subgraphs during the process of experiment, and Exact Subgraph Matching algorithm library provided by Liu et al. [7] to verify subgraph-graph isomorphism.

Table 2. Boundary graphs extracted from the YAGO knowledge graph

#	Name of RT	# BG	Anchor point (L)	Anchor point (R)
R1	influences	22820	Person	Person
R2	hasAcademicAdvisor	5253	Person	Person
R3	isMarriedTo	34229	Person	Person
R4	hasChild	42335	Person	Person
R5	wasBornIn	281067	Person	Location
R6	diedIn	94747	Person	Location
R7	isPoliticianOf	32796	Person	Location
R8	livesIn	26483	Person	Location
R9	directed	35603	Person	Movie
R10	wroteMusicFor	24204	Person	Movie
R11	worksAt	10257	Person	Organization
R12	playsFor	525374	Person	Organization

We evaluated our approach using multiple relationship types between different entities in YAGO knowledge graph. Table 2 shows the number of boundary graphs (# BG) we sliced from YAGO with different relationship types. For an instance, there are 4 different relationship types, i.e. influences **(R1)**, hasAcademicAdvisor **(R2)**, isMarriedTo **(R3)** and hasChild **(R4)**, between two person entities. 22820 boundary graphs are sliced and extracted from YAGO with central relation type *influences*.

In our experiment, any group of data set can be described by four parameters: (1) $|N|$, the total number of graphs generated, (2) $\{R_1, R_2, \ldots, R_m\}$, m different central relationship types that the data set involves, (3) $|L|$, the maximum length of the farthest node of the boundary graph with the start of anchor points, and (4) $|I|$, the maximum degree of each node in the graph. We choose these four parameter settings because these determine the characteristics of boundary graph.

Table 3. Comparative experiments on accuracy rate for R1–4

No.	$|N|$	R1	R2	R3	R4	$|L|$	$|I|$	Karma	Our method
1	44	10	10	10	14	2	10	25%	84.09%
2	44	8	20	8	8	2	10	25%	95.45%
3	44	11	11	11	11	2	10	25%	91.11%
4	80	20	20	20	20	2	10	25%	87.34%
5	100	25	25	25	25	2	5	25%	94%
6	120	20	20	20	60	2	5	25%	93.33%
7	140	20	80	20	20	2	5	25%	91.43%
8	400	160	80	80	80	2	5	25%	89.25%
9	400	100	100	100	100	2	5	15%	87.75%
10	1000	250	250	250	250	2	5	10%	91.70%

Table 4. Comparative experiments on accuracy rate for R5–8

No.	$|N|$	R5	R6	R7	R8	$|L|$	$|I|$	Karma	Our method
1	147	43	36	32	36	2	5	25%	55.10%
2	345	85	79	62	119	2	5	25%	54.20%
3	748	174	162	143	269	2	5	25%	54.68%

Table 5. Comparative experiments on accuracy rate for R9–10

No.	$|N|$	R9	R10	$|L|$	$|I|$	Karma	Our method
1	244	90	154	2	5	50%	91.39%
2	1666	906	760	2	5	50%	89.4%

Table 6. Comparative experiments on accuracy rate for R11–12

No.	$|N|$	R11	R12	$|L|$	$|I|$	Karma	Our method
1	103	87	16	2	5	50%	100%
2	998	833	165	2	5	50%	100%

4.1   Accuracy Test

We used *cross-validation* as the strategy for testing the accuracy rate of our method. For testing accuracy of Karma's method, we adopt the strategy described as follows. Suppose that we have a tuple of sets of boundary graphs $(BG_1, BG_2, \ldots, BG_n)$. Each element BG_i represents a set of boundary graphs with the central relationship R_i. We will conduct n rounds of experiments. In the ith round of the experiment, we fetch 5 boundary graphs from BG_i as our testing graphs. We pretend not to know the central relationship types in these 5 graphs and try to predict them. Then, we consolidated all of the rest of boundary graphs from $\{BG_1, \ldots, BG_{i-1}, BG_{i+1}, \ldots, BG_n\}$ into a merged weighted graph as our training graph. According to the scoring formula in [4], each edge is assigned the weigh $1 - x/(n+1)$ where n is the number of known boundary graphs and x is the number of graph identifiers the edge is tagged with. Next, we compare the predicted value with the true value for every testing graph, respectively. For each testing graph, we set the testing result $y_j (j \leq 5)$ as 1 if the predicted value is equivalent to the true one, or 0 if not. We average these 5 testing results, and then obtain the accuracy rate Y_i for the i round of the experiment, i.e. $Y_i = \sum_{j=1}^{5} y_i/5$. We define the final accuracy for the tuple of boundary graphs $\{BG_1, BG_2, \ldots, BG_n\}$ as:

$$Accuracy = \sum_{i=1}^{n} Y_i/n$$

Let us take #1 comparative experiments on accuracy rate for R1-R4 as an example. In the group of experiments, we have 10, 10, 10, 14 boundary graphs with central relationship type influences (R1), hasAcademicAdvisor (R2), isMarriedTo (R3) and hasChild (R4), respectively. We will perform 4 rounds of experiments. In the first round of experiment, 5 boundary graphs are extracted from 10 boundary graphs with central relation type *influences* (R1, the 2nd row and the 3rd column of Table 3). However, we pretend not to know the relationship type of these 5 boundary graphs and try to predict them and them compare them with the true relationship type. The rest of 39 (44 − 5 = 39) boundary graphs are used to be training set. In the second round, we take 5 boundary graph with *hasAcademicAdvisor* (R2) as testing set and so on for the 3rd and 4th round of experiment.

Tables 3, 4, 5 and 6 shows the accuracy comparison between our method and the method used in Karma. The column "No." in Tables 3, 4, 5 and 6 refers to the experiment number. We find that the accuracy rate of our method is 3–4 times better than Karma's method. Karma's method keeps a stable accuracy rate (25%) when we apply it on 44, 80, 100, 120, 140 and 400 boundary graphs. These experimental results match our previous theoretical analysis. The reason is that, no matter how large the generated Steiner tree is, Karma's algorithm always selects the maximum-frequency edge between two anchor points. For each experiment, we take only one boundary graph as the testing graph, which contains one of four special relationship types, and the merged graph is generated by the rest of boundary graphs so that it is constant. The frequency of all the edges between two anchor points is kept unchanged throughout an experiment. That is why the accuracy rate of Karma is 25%.

As we can see in the table, the accuracy in distinguishing RT5–RT8 is lower than other kinds of relationship types. We observed that the graph patterns extracted from RT5–RT8 are very similar, and therefore the available discriminative feature set is smaller than other data sources.

4.2   Scalability Test

We applied our method on 100, 400 and 1000 YAGO boundary graphs ($|L| = 2$, $|I| = 10$) and 1666 graphs ($|L| = 2, |I| = 5$), respectively. Table 7 shows the experimental result of the test. We find that the accuracy rate is over 85% for all the groups of data sets.

Columns T1–T3 in Table 7 show the time required for individual steps of our method. T1 denotes the time taken to obtain the frequent subgraphs. T2 stands for the time consumed for computing discriminative feature set and coding the feature matrix through subgraph matching algorithm. T3 represents the time taken for building a Neural Net based on the feature matrix. T4 represents the time taken for building a merged graph as background knowledge based on known boundary graphs for Karma. During our experiments, slicing the YAGO knowledge graph into boundary graphs was time consuming. For instance, it took around 12 h to slice 1000 boundary graphs. However, considering that the training process is offline, the training time of our method is considered acceptable. The training of the classifier itself is relatively fast. For example, it took 29.7 s for coding a feature matrix for training based on 1000 boundary graphs, and 88 s for training a Neural Net based on the matrix.

Table 7. Scalability test using our method and Karma

#	R	# BG	$|L|$	$|I|$	Precision	Recall	# FS	# DFS	T1 (sec)	T2 (sec)	T3 (sec)	T4 (sec)
1	R1	20	2	10	94.74%	90%	23	51	5.0	1.9	6	0.2
	R2	20			90.91%	100%	19					
	R3	20			94.74%	90%	28					
	R4	20			100%	100%	18					
2	R1	100	2	10	87.00%	87.00%	27	66	45.9	7.0	56	1.5
	R2	100			91.49%	86.00%	24					
	R3	100			92.16%	94.00%	36					
	R4	100			85.58%	89.00%	36					
3	R1	250	2	10	91.67%	88.00%	23	49	21.7	8.0	88	3.2
	R2	250			97.49%	93.20%	19					
	R3	250			88.93%	90.00%	27					
	R4	250			87.31%	93.60%	24					
4	R9	906	2	5	91.72%	94.15%	83	78	5.7	56.0	306	16.3
	R10	760	2	5	92.80%	89.87%	57					

5 Related Work

The work presented in this paper relates to two main streams of research, namely relationship matching and disambiguation in conceptual model.

In recent years, there are some efforts to automatically infer the implicit *relationships* of tables. In Karma [3, 4], given some sample data from the new source, they leverage the knowledge in the domain ontology and the known semantic models to construct a weighed graph that represents the space of plausible semantic model for the new source. They then exploit Steiner Tree algorithm compute the top k semantic models containing the disambiguated relationships. Limaye et al. [12] used YAGO to annotate web tables and generate binary relationships using machine learning approaches. However, this approach is limited to the labels and relations defined in the YAGO ontology. Venetis et al. [13] presented a scalable approach to describe the semantics of tables on the Web. To recover the semantic of tables, they leverage a database of class labels and relationships automatically extracted from the Web. They attach a class label to a column if a sufficient number of the values in the column are identified with that label in the database of class labels, and analogously for binary relationships. Although these approaches are very useful in publishing semantic data from tables, they are limited in learning the semantics of relations. Both of these approaches only infer *binary* relationships between pair of columns via a simple match of source node and target node of the relationship. Some other recent work leverages the Linked Open Data (LOD) cloud to capture the semantics of sources. Schaible et al. [14] extracted schema-level patterns (SLPs) from linked data and generate a ranked list of vocabulary terms for reuse in modelling tasks. SLPs are (sts, ps, ots) triples where sts and ots are sets of RDF types and ps is a set of RDF object properties. For example, the SLP ({*Person, Player*}, {*knows*}, {*Person, Coach*}) indicates that some instances of Person ∩ Player are connected to some instances of Person ∩ Coach via the object property *knows*. Taheriyan et al. [2] mines the small graph patterns occurring in the LOD and combine them to build a graph that will be used to infer semantic relations. Our work differs from these works as our relationship matching method works on distinguishing many relationship types between two entities at the *instance-level*.

The relationship matching in our work is dealing with actually a *disambiguation* problem that disambiguate multiple relationship types between entity instances. There has already existed a lot of work with regard to resolve *ambiguity* for conceptual models. Mens et al. [8] proposed an inconsistency detection approach by using graph transformation rules to detect ambiguity in UML class models and state machine diagrams, and then automatically rework the defects in such models with resolution rules. A prominent instance of ambiguity is the usage of homonymous or synonymous words. Pittke et al. [1] proposed a technique that detects and resolves *terminological ambiguities* in large conceptual model collections. The challenge of *word sense disambiguation* relates to determining the sense of a word in a given context. Supervised machine-learning techniques (e.g. [16]) and clustering approaches (e.g. [18]) are employed to identify *context similar words*. Our idea is analogous to the above work that tries to infer the correct meaning from the context of the ambiguous relationship. However, to the best of our knowledge, there hasn't been work on disambiguating relationship types between entity instances of conceptual model.

6   Conclusion

We proposed a novel method to distinguish relationship types between recognized entity instances as an extension of Karma. How to distinguish multiple relationship types between two recognized entity instances automatically is an essential part of build a precise knowledge graph from huge data sources. The core idea of our work is to exploit the small graph patterns occurring in a bundle of boundary graphs, which are sliced from the existing linked data, with a specific central relationship type, to hypothesize the relationship types between recognized entity instances within a new data source. The experiment result on YAGO demonstrated the high accuracy of the approach (>80%) to distinguish multiple relationship types between recognized entity instances automatically.

There still exist some limitations for our work. On one hand, we observed that our approach is limited in gaining high accuracy rate on distinguishing some of specific relationship types, for example, *wasBornIn* and *diedIn* between two person instances. The reason is that most of the graph patterns related to these two relations are similar so that it is hard to obtain a sufficient discriminative feature set. One direction of our future work is to investigate more efficient features, not only graph structure, to classify boundary graphs more efficiently. One the other hand, our current approach comes with the assumption that the relationship matching problem is a classification problem. The cases where relationship types exist that go beyond the already known ones cannot be addressed in this paper yet.

Acknowledgements. This research was partially funded by the Data to Decisions Cooperative Research Centre (D2D CRC). We appreciate Dave Blockow and Troy Wuttke in D2D CRC for providing technical support as we developed the prototype.

References

1. Pittke, F., Leopold, H., Mendling, J.: Automatic detection and resolution of lexical ambiguity in process models. IEEE Trans. Softw. Eng. **41**(6), 526–544 (2015)
2. Taheriyan, M., Knoblock, C.A., Szekely, P., Ambite, J.L.: Leveraging linked data to discover semantic relations within data sources. In: Groth, P., Simperl, E., Gray, A., Sabou, M., Krötzsch, M., Lecue, F., Flöck, F., Gil, Y. (eds.) ISWC 2016. LNCS, vol. 9981, pp. 549–565. Springer, Cham (2016). https://doi.org/10.1007/978-3-319-46523-4_33
3. Knoblock, C.A., et al.: Semi-automatically mapping structured sources into the semantic web. In: Simperl, E., Cimiano, P., Polleres, A., Corcho, O., Presutti, V. (eds.) ESWC 2012. LNCS, vol. 7295, pp. 375–390. Springer, Heidelberg (2012). https://doi.org/10.1007/978-3-642-30284-8_32
4. Taheriyan, M., Knoblock, C., Szekely, P., Ambite, J.L.: Learning the semantics of structured data sources. Web Sem. Sci. Serv. Agents World Wide Web **37**, 152–169 (2016)
5. YAGO Official Website, 27 July 2017. http://www.mpi-inf.mpg.de/departments/databases-and-information-systems/research/YAGO-naga/YAGO/downloads/
6. Yan, X., Han, J.: gSpan: graph-based substructure pattern mining. In: Proceedings of the 2002 International Conference on Data Mining (ICDM 2002). IEEE (2002)

7. Liu, H., Keselj, V., Blouin, C.: Exploring a subgraph matching approach for extracting biological events from literature. Comput. Intell. **30**(3), 600–635 (2014)
8. Mens, T., Van Der Straeten, R., D'Hondt, M.: Detecting and resolving model inconsistencies using transformation dependency analysis. In: Nierstrasz, O., Whittle, J., Harel, D., Reggio, G. (eds.) MODELS 2006. LNCS, vol. 4199, pp. 200–214. Springer, Heidelberg (2006). https://doi.org/10.1007/11880240_15
9. Ramnandan, S.K., Mittal, A., Knoblock, C.A., Szekely, P.: Assigning semantic labels to data sources. In: Gandon, F., Sabou, M., Sack, H., d'Amato, C., Cudré-Mauroux, P., Zimmermann, A. (eds.) ESWC 2015. LNCS, vol. 9088, pp. 403–417. Springer, Cham (2015). https://doi.org/10.1007/978-3-319-18818-8_25
10. Dhamankar, R., Lee, Y., Doan, A., Halevy, A., Domingos, P.: iMAP: discovering complex semantic matches between database schemas. In: International Conference on Management of Data (SIGMOD), New York, NY, pp. 383–394 (2004)
11. Qian, L., Cafarella, M.J., Jagadish, H.V.: Sample-driven schema mapping. In: SIGMOD 2012 (2012)
12. Limaye, G., Sarawagi, S., Chakrabarti, S.: Annotating and searching web tables using entities, types and relationships. PVLDB **3**(1), 1338–1347 (2010)
13. Venetis, P., Halevy, A., Madhavan, J., Paşca, M., Shen, W., Wu, F., Miao, G., Wu, C.: Recovering semantics of tables on the web. Proc. VLDB Endow. **4**(9), 528–538 (2011)
14. Schaible, J., Gottron, T., Scherp, A.: *TermPicker*: enabling the reuse of vocabulary terms by exploiting data from the linked open data cloud. In: Sack, H., Blomqvist, E., d'Aquin, M., Ghidini, C., Ponzetto, S.P., Lange, C. (eds.) ESWC 2016. LNCS, vol. 9678, pp. 101–117. Springer, Cham (2016). https://doi.org/10.1007/978-3-319-34129-3_7
15. Popa, L., Velegrakis, Y., Hernández, M., Miller, R., Fagin, R.: Translating web data. In: VLDB, pp. 598–609 (2002)
16. Navigli, R., Ponzetto, S.P.: Joining forces pays off: multilingual joint word sense disambiguation. In: EMNLP-CoNLL (2012), pp. 1399–1410. ACL (2012)
17. Grossmann, G., Kashefi, A.K., Feng, Z., Li, W., Kwashie, S., Liu, J., Mayer, W., Stumptner, M.: Integrated law enforcement platform federated data model. Technical report, Data 2 Decision CRC (2017)
18. Pantel, P., Lin, D.: Discovering word senses from text. In: SIGKDD (2002), pp. 613–619. ACM (2002)
19. Pham, M., Alse, S., Knoblock, C.A., Szekely, P.: Semantic labeling: a domain-independent approach. In: Groth, P., Simperl, E., Gray, A., Sabou, M., Krötzsch, M., Lecue, F., Flöck, F., Gil, Y. (eds.) ISWC 2016. LNCS, vol. 9981, pp. 446–462. Springer, Cham (2016). https://doi.org/10.1007/978-3-319-46523-4_27
20. Knoblock, C.A., et al.: Lessons learned in building linked data for the American art collaborative. In: d'Amato, C., Fernandez, M., Tamma, V., Lecue, F., Cudré-Mauroux, P., Sequeda, J., Lange, C., Heflin, J. (eds.) ISWC 2017. LNCS, vol. 10588, pp. 263–279. Springer, Cham (2017). https://doi.org/10.1007/978-3-319-68204-4_26
21. Altova mapforce. http://www.altova.com/mapforce.html
22. Microsoft biztalk server. https://www.microsoft.com/en-au/cloud-platform/biztalk
23. Stylus studio. http://www.stylusstudio.com/

Business Credit Scoring of Estonian Organizations

Jüri Kuusik[1,3(✉)] and Peep Küngas[1,2]

[1] Register OÜ, Tallinn, Estonia
{jyri.kuusik,peep.kungas}@ir.ee
[2] University of Tartu, Tartu, Estonia
[3] STACC, Tartu, Estonia

Abstract. Recent hype in social analytics has modernized personal credit scoring to take advantage of rapidly changing non-financial data. At the same time business credit scoring still relies on financial data and is based on traditional methods. Such approaches, however, have the following limitations. First, financial reports are compiled typically once a year, hence scoring is infrequent. Second, since there is a delay of up to two years in publishing financial reports, scoring is based on outdated data and is not applied to young businesses. Third, quality of manually crafted models, although human-interpretable, is typically inferior to the ones constructed via machine learning.

In this paper we describe an approach for applying extreme gradient boosting with Bayesian hyper-parameter optimization and ensemble learning for business credit scoring with frequently changing/updated data such as debts and network metrics from board membership/ownership networks. We report accuracy of the learned model as high as 99.5%. Additionally we discuss lessons learned and limitations of the approach.

Keywords: Business credit scoring · Machine learning · Boosted decision tree
Hyper-parameter tuning

1 Introduction

Credit scoring is an effective tool to assess credit risks of individuals and businesses. In business credit scoring literature credit risk is often defined as likelihood of default and its numerical representation is called credit score. In this paper, credit score is defined as likelihood of default within the following 12 months such as commonly used in industry [24].

Most often credit scoring is applied in the context of loan applications (in B2C domain) or when setting payment terms of invoices (in B2B domain). Hence the quality of a credit scoring model affects directly the amount of cash companies loose. In the context of business credit scoring, history of credit risk modelling goes back to late 1960s when numerous studies were devoted to model business failure using publicly available data and combining it with statistical classification techniques [27]. Pioneering work and one of the first attempts to perform modern statistical failure analysis was done by Tamari [3]. Since 1990s there has been a steady growth in the

© Springer International Publishing AG, part of Springer Nature 2018
J. Krogstie and H. A. Reijers (Eds.): CAiSE 2018, LNCS 10816, pp. 554–568, 2018.
https://doi.org/10.1007/978-3-319-91563-0_34

number of articles related to credit risk using artificial intelligence and machine learning methods [22, 23, 26]. A few possible reasons can be associated with the increased interest in credit risk modelling - rapid development of some new data mining techniques, the availability of more open credit datasets, the growth of credit products and credit markets [4]. By following the trends, first attempts have been made in application of machine learning methods (support vector machines [28], neural networks [29] and extreme gradient boosting (XGBoost) [1, 26]). XGBoost [16] has proven its superiority in discriminating negative and positive events wrt other learning methods [1, 26] including very popular logistic regression which is extensively used in credit scoring area [2]. Further details on recent advances in the field can be found from a review compiled by Zięba et al. [26].

The business credit scoring models still rely mainly on financial data having the following drawbacks. First, financial reports are compiled typically once a year leading to infrequent scoring. Second, since there is a delay of up to a year in publishing financial reports, scoring is based on outdated data.

Regarding our contribution and goal of paper we advanced the state of art in business credit scoring by demonstrating that combination of rapidly changing datasets and top machine learning method XGBoost will lead to business credit scoring model with utmost accuracy in practical settings. More specifically, we used features extracted from company debts and network metrics from board membership/ownership networks in addition to features used in business credit scoring with traditional approaches. Based on such features we defined multiple models, which in turn we combined with ensemble learning into a model, which will predict with accuracy as high as 99.5% what will be the likelihood of a default event in a company within the following 12 months. Also, by our knowledge we made a first attempt to model forced deletions of companies together with bankruptcies in one model along with using Bayesian hyper-parameter optimization. Wrt model validation credit decisions made by credit specialists were compared with credit scores of fitted model. Wrt model description approach not implemented in [1, 26] has been used (described in Sect. 6.2).

The paper is organized as follows. Problem description is explicitly formulated in Sect. 2. In Sect. 3 dataset and sampling methodology are described. In Sect. 4 modeling and scoring process is described. Section 5 outlines evaluation of results. In Sect. 6 the learned model is described. Finally, conclusions are summarized together with discussion related to threats to validity and future work in Sect. 7. For more detailed documentation refer please to our full report [25] which includes also information of how to access training and validation data set along with used R functions for training and validation of all models.

2 Problem Description and Preliminaries

In its essence, corporate credit risk evaluation is to make a classification of good and bad companies. Importance of the ability to accurately evaluate credit worthiness of customers cannot be overrated as extending credit to insolvent customers directly results in loss of profit and money. Correct determination whether a customer is solvent or there are some indications of default enables the company to minimize costs and maximize earnings by choosing reliable customers. Default occurs when a company cannot meet its contractual financial obligations as they come due. More specifically, a company is technically insolvent if it cannot meet its current obligations as they come due, despite the value of its assets exceeding the value of its liabilities. A company is legally insolvent if the value of its assets is less than the value of its liabilities and a company is bankrupt if it is unable to pay its debts and files a bankruptcy petition [5]. In the model proposed by this paper the bad companies (with positive event of risk) are bankruptcies or forced deletions of companies (informal bankruptcy), good companies (with negative event of risk) are companies non-failed and operating companies (an overview of all research done on credit risk models shows that this is appropriate way of defining bad and good companies for credit risk [6]). To be precise, in Estonia there are certain requirements established for companies that need to be met for legally staying as an operating company. Those requirements are set by Commercial Code of Estonia [7]. These requirements also include regulations for equity and submission of annual reports. So, rather than being just a traditional bankruptcy prediction model, this model is predicting the risk of a company either going bankrupt or being deleted due to their failure of meeting specific requirements set by Commercial Code.

3   Dataset Description

In this section we describe data sample and variables used for modeling.

3.1   Sample Used for Modeling

The sample dataset used in the model training consists of bad companies (in total of 17,953 cases) and good companies that have been active in the same period (in total of 190,804 cases). Information regarding bankrupt and forcefully deleted companies was gathered from Business Register of Estonia and the bankruptcies and forceful deletions from the event period 01.01.2015–31.05.2017 were used. Events from earlier period were not included into analysis due to lack of data availability of explanatory variables (variables used in the models). The date of the bankruptcy or deletion is set to the earliest date of bankruptcy announcement available for us. Companies that have a lower credit risk (negative cases) are the companies that have been active in the event period. Companies registered later than 31.05.2015 were removed from the sample as these companies have little or no historical information that could be used in the training phase. The event dates for negative cases were set so that the event dates of positive and negative cases would be similarly distributed allowing us to reduce the effect that the event period (incl. external macro-economic factors) might have on events if the dates for negative cases would have been selected with a different method.

3.2   Variables Used in Modeling

Variables used in the model were selected by the insights and knowledge about insolvent companies - variables that characterize a typical insolvent company according to domain experts. More precisely, variables formatted as *italic* in the Table 1 below are being used in expert models. Most of the variables have been used in modeling multiple times, i.e. we have used values of these variables at different time points (1–6 months/quarters and 2 years before the event by the results of analysis done in [8]). Set of different time points is used to provide sufficient information for the model to recognize patterns of change in time (e.g. increase or decrease in tax debts, changes in the VAT registration). We used the same 164 variables in each individual model for modeling a positive event happening in each subsequent month of 12-month period and since the computation time for fitting all 12 individual models was acceptable (1 h per model in average) no variable selection was needed.

For fitting individual models following explanatory variables have been used:

Table 1. Variables used for modelling

Category of variable	Granularity	#Variables	Variables
Tax	Month	48	*Total tax debt*, postponed tax debt, disputed tax debt, tax debt interest, postponed tax debt interest, *whether VAT is registered or not*, number of days past due date of VAT registration, number of unsubmitted tax declarations
	Quarter	12	*Paid state taxes, paid personnel taxes*
Annual reports of companies	Year	16	*Turnover*, balance, floating assets, capital assets, current liabilities, long term liabilities, equity, net income
	Month	6	*Number of unsubmitted annual reports*
Network data (board members/co-executives/owners and companies)	Month	42	owner degree*, owner PageRank**, board degree, board PageRank, overall degree, overall PageRank, board degree for negative score***
Company related (as of end of historical period****)	N/A	28	Age of company, initial capital of company, main activity of company, type of company
Debt	Month	6	*Amount of total debts to suppliers*
Derivates	Month	6	*Historical company reputation score*****

* Degree is a feature reflecting the information available on the registry card of the company. Degree shows the number of incoming and outgoing relations of the company.

** PageRank is a link analysis algorithm that assigns a numerical weight to each company. The calculated weight shows the importance of each single company relative to the whole graph. PageRank considers all nodes of the graph and is therefore a global metric. PageRank is a probability of ending up at that node after starting from a random node in the network [8] and therefore it is a numerical value between 0 and 1 for each company.

*** Board degree for negative score shows the number of companies with a bad reputation (reputation of the company is a negative number) related to the board members of the respective company.

**** Historical period is the period before event period, last month/quarter/year in that period belongs to event period.

***** Historical score is an internal derivate of Register OÜ which is computed by different business rules considering different features reflecting the "health" of company.

4   Modeling and Scoring

In this section we first describe sampling and data partitioning approach for individual models and ensemble model (using outputs from all individual models as inputs predicting probability that positive event will be happen within the following 12 months). Then the process of model building is elaborated where in total 13 models are learned. After this scoring mechanism is described and finally we described how we use credit score distribution to define credit classes, which are used in credit management and other applications.

4.1   Sampling

Since the number of positive cases is significantly smaller than the number of negative cases, we need to address the imbalance problem in the training set. In literature various techniques have been documented for both resampling and algorithmic ensemble techniques in addressing the imbalance problem [9, 10]. In treating the imbalance problem of our training set, we applied extreme gradient boosting, avoiding resampling techniques which have many undesirable properties. As a result, no sampling methods are applied to both training and test sample. Therefore, the training and evaluation of the model performance is based on the imbalanced data set which is representative of the actual real-life data.

We actually applied previously under-sampling for balancing training data sets for all individual models and it resulted in too high number of false positives (positive event was predicted by model while no such event actually happened). It was due to the essence of under-sampling – it threw out majority of negative cases from training data and some patterns were lost for distinguishing two classes.

We used 50/50 ratio to split overall sample set of cases into training and test sample via random selection without replacement. Then by using the training cases we created in total, 13 training datasets - 12 datasets for fitting of individual models and one dataset for fitting an ensemble model. All training datasets for individual models contained the same list of companies from training sample. Each training dataset contains therefore 104,378 companies. Each training dataset for certain single model contained the fact of event with event rate of 8.6% and explanatory variables from aligned historical period. The training dataset for ensemble model was constructed from test datasets of individual models. It contains predictions of all individual models on their test datasets and true label (whether a positive event has happened or not).

In total, we created two test datasets from the rest of 50% of the initial sample (from which 50% of data has been used for modeling of all individual models) with random selection without replacement with sample rate of 50%. One test data set was reserved for validation of all individual models while second one was reserved for validation of ensemble model.

4.2   Model Building

In total 13 models were fitted - 12 individual models and an ensemble model. We used state of the art classification method of XGBoost both for individual models and ensemble model. This method has been used in different Kaggle competitions with great success [11] and for creation of credit scoring models both for B2C and B2B domain [1, 26]. For optimizing hyperparameters of the models most promising hyperparameter tuning method was used: model-based (aka Bayesian hyper-parameter) optimization [12]. We compared it also with other optimization methods, but it turned out that above-mentioned method gave the best performance. As implementation, R packages "mlr" and "mlrMBO" were used [12, 13]. In Fig. 1 modeling process of individual models and the ensemble model is summarized. Yellow boxes in Fig. 1 with label "Y" denote months (so-called "target windows" [20], p. 96) where events are found in initial sample and boxes with label "X" denote historical time-periods (so-called "observation windows" [20], p. 96) the values for explanatory variables are taken from.

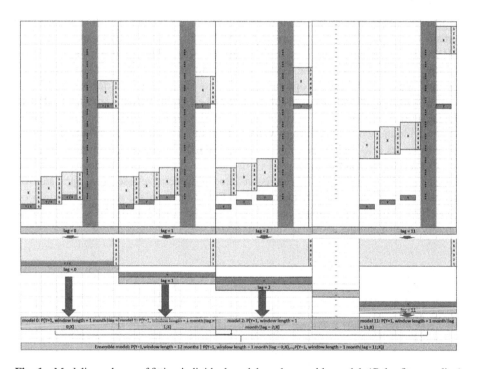

Fig. 1. Modeling scheme of fitting individual models and ensemble model. (Color figure online)

4.3 Scoring

We trained 12 individual models that use data from different time-periods before the event allowing us to predict default of a company at the scoring date and up to 11 months before the potential default event as follows. For "Model 0" we used for training timeseries variables from 0 to 5 months before the event for features with monthly granularity. Using this model with current data allows us to predict if the company becomes insolvent on the current month. In "Model 1" we used for training the model features from 1 to 6 months prior to the event with monthly granularity. Using this model with current data allows us to predict if the company becomes insolvent on the next month from the current date. Analogously other individual models are trained and used until "Model 11" where training data involves data from 11 to 17 months before the event for time series features with monthly granularity. Using this model with current data allows us to predict if the company becomes insolvent on the 11th month from the current date.

Since we defined credit score as likelihood of default within 12 months we combine the preceding 12 individual models via ensemble model to provide a single credit risk score. Additionally, when scoring we by default assign credit score of 1 to all bankrupted companies and companies in liquidation as of scoring date.

4.4   Credit Risk Classes

In practical decision-making often a credit score as a probability is too abstract for humans to interpret. Therefore in practice often (usually, 5 classes by industry standard [24]) credit classes are defined and used instead of credit scores. Hence we also divided credit scores into 5 classes to be usable in the credit management and other processes. We used Jenks natural breaks optimization [14] to calculate boundaries for credit risk classes by using probabilities of event from scored ensemble model as of 28.11.2017 as input. Allocation points for the risk classes are chosen to be the credit score values which divide the whole spectrum of values of credit scores into clusters where density of credit scores is as high as possible - that should indicate that the distinction between the groups is the best at those points [14]. In Fig. 2 credit score distribution (on left side of figure) and the corresponding division into credit score classes (on right side of figure) are depicted. In the left hand side of Fig. 2 one can see usual distribution of scored probabilities of strong classifier i.e. they are distributed like a "bathtub" - majority of probabilities are near values 0 and 1.

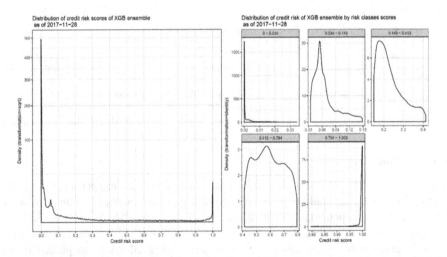

Fig. 2. Distribution of credit scores (left); Distribution of credit scores by risk classes (right).

5   Model Evaluation

In this section we evaluate the learned model from three perspective. First we evaluate model performance wrt test dataset. Then we compare scoring results from model wrt credit specialist decisions for the same cases. Finally, we perform retrospective analysis of the model.

5.1   Performance of the Model on Test Dataset

In Fig. 3 overview of assessment results of ensemble model by test data is compactly presented.

Model	Accuracy	H	Gini	AUC	AUCH	KS	MER	MWL	ER
Ensemble (xgboost)	99.5516	0.9744	0.9982	0.9991	0.9993	0.9826	0.004	0.003	0.004
Maximum (xgboost)	98.6875	0.9605	0.9969	0.9985	0.9986	0.9736	0.008	0.004	0.013
Maximum (rf)	96.2311	0.9253	0.9922	0.9961	0.9962	0.9512	0.015	0.008	0.038

Model	Sensitivity	Specificity	Precision	Recall
Ensemble (xgboost)	0.9722	0.9977	0.9761	0.9722
Maximum (xgboost)	0.9859	0.9870	0.8781	0.9859
Maximum (rf)	0.9861	0.9600	0.7015	0.9861

Model	TP	TN	FP	FN	TPR	FPR	FMES	Youden	NRObs
Ensemble (xgboost)	4,411	47,545	108	126	0.9722	0.0023	0.9742	0.9700	52,190
Maximum (xgboost)	4,473	47,032	621	64	0.9859	0.0130	0.9289	0.9729	52,190
Maximum (rf)	4,474	45,749	1,904	63	0.9861	0.0400	0.8198	0.9462	52,190

Abbreviation	Description
Accuracy	Accuracy
H	H-measure
Gini	Gini coefficient
AUC	Area Under the ROC Curve
AUCH	Area Under the convex Hull of the ROC Curve
KS	Kolmogorov-Smirnov statistic
MER	Minimum Error Rate
MWL	Minimum cost-Weighted Error Rate
ER	Error rate
Sensitivity	Sensitivity
Specificity	Specificity
Precision	Precision
Recall	Recall
TP	True Positives
TN	True Negatives
FP	False Positives
FN	False Negatives
TPR	True Positive Rate / Error II
FPR	False Positive Rate / Error I
F (FMES)	F measure
Youden	Youden index
NRObs	Number of observations

Fig. 3.   Performance of ensemble models and reference models.

For comparison purposes, individual models have been fitted and evaluated with two different methods - extreme gradient boosting ("xgboost") and random forests with under-sampling ("rf") as reference model which was used in production at the time of experimenting with new approach. We used two heuristics for aggregating results of 12 models - "Maximum" means that final score is calculated by taking maximum of scores by individual models, while "Ensemble" means that ensemble learning was used to combine results of individual models. Experimental results indicate that ensemble model based on extreme gradient boosting has best performance wrt accuracy, AUC and several other metrics. Regarding false positives (FP) and false negatives (FN) we can see that extreme gradient boosting based model is producing more false negatives (positive event was not predicted by model while such event has happened actually) while at the same time less false positives at much higher rate. Further details regarding used performance metrics can be found in [15].

5.2   Credit Management Specialist Decisions vs Model Estimates

For understanding in which extent the learned model reflects credit decisions made by humans, we compared scores of companies with the credit decisions made by credit specialists. The decision dataset was provided by company Kreedix OÜ for the period 15.09.2016–14.03.2017. Credit decisions are made in the process of analyzing trade receivables (sales invoices) in the context of Invoice-to-Cash business process [25]. Credit decisions for each debtor may be either "Yes", "No" or "Wait". Credit decision "Yes" means that the invoice receiver (debtor) company is eligible to credit, "No" means that it is not advised to extend credit to the debtor, "Wait" means that further data should be collected before making the credit decision (basically "No"). Credit decisions are made by using in addition to company's public data (plentiful of what we used for model learning) and inside knowledge (e.g. creditor's notes) also invoice data such as the amount of debt and the number of days payment is due date.

Distribution of credit scores along with bootstrapped confidence intervals for the median by credit decisions "NO", "WAIT" and "YES" are depicted in Fig. 4. We can see that credit scores (as logarithm of probabilities of event) by model are quite well in accordance with credit decisions of credit specialists, i.e. distributions of credit scores across different type of credit decisions are quite different and aligned wrt credit decisions (the lower the credit score the better the credit decision). In summary, we can conclude that model reflects credit decisions made by humans.

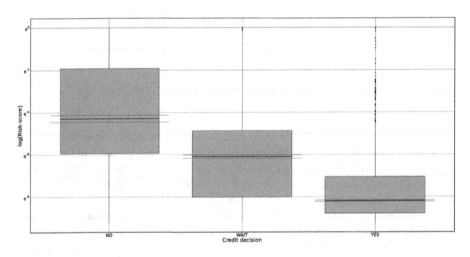

Fig. 4. Comparison of scores of XGboost ensemble model with credit decisions.

5.3   Retrospective Analysis

Here we found retrospective performance of model i.e. knowing actual events during time-period 1.06.2017–31.10.2017 we predicted what would have been the probability of event as of 0, 1, 2, 5, 8 and 11 months before event for each company by scoring them with fitted ensemble model. Sample for validation has been created similarly to the sampling process used during modeling - i.e. including all positive events which occurred in before mentioned time-period and sampling negative events from the same time-period having distribution of dates as similar as possible for both negative and positive events.

As expected, it follows from Fig. 5 that the model is most accurate as of 0 months before the event (bankruptcies and deletions are evaluated with a much higher credit risk compared to normal companies) and less accurate as of 11 months before the event. The prediction accuracy for several months ahead is not very satisfactory wrt prediction accuracy for few months ahead. It might be influenced by used modeling framework in which case for predicting the positive event for several months ahead the observation window and target window are located quite far away from each other and therefore quite outdated data are used for fitting of individual models. We can also see from figure that predicting deletions is more accurate compared to bankruptcies. This can be explained with the presence of annual reports submission information in the model along with larger proportion of deleted companies in training dataset and the fact that not multinomial but binary classification has been used. Also, we can see that variance of predicted credit scores by confidence intervals of median for bankruptcies is very high wrt deletions.

However, in summary, comparing primarily the distribution of credit scores of good companies wrt to credit scores of bad companies we can conclude that model behaves quite acceptable on data set used for retrospective analysis.

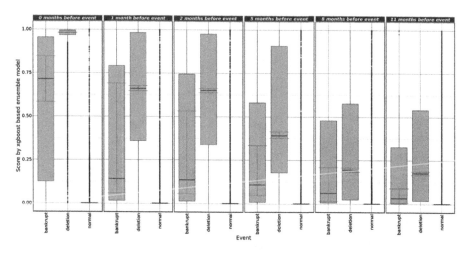

Fig. 5. Credit risk predictions for companies on the data of retrospective analysis.

6 Model Description

In this section description we analyze the key characteristics both of the individual models (taking as an example "Model 0") and ensemble model. Since "black-box" method is used for fitting the classification model, there are by far lesser options to describe the resulting individual models compared to logistic regression with its mature framework [2]. However, there are some tools which could help us to understand the individual models. First, we used relative variable importance (RVI) provided by extreme gradient boosting model [16]. Second, we used so-called partial dependence plots (PDP) [17]. Finally, we described the ensemble model wrt risk classes by comparing distributions of explanatory variables in all of them giving some general patterns for distinction of credit classes and gaining a formal description of credit risk scores classes.

6.1   Relative Variable Importance and Partial Dependence Plots

Here the aim is to understand what kind of explanatory variables are playing supreme role in predicting of positive event in individual models. In the current framework all variables are used for fitting individual models. Therefore, no variable importance evaluation was needed during modeling. However, since extreme gradient boosting is a "black-box" method, then for describing this type of model, RVI can be used. RVI shows how important is the variable in contributing to the ability for differentiate two classes. Also, PDP can be used for the similar purpose. PDP are low-dimensional graphical renderings of the prediction function so that the relationship between the outcome and predictors of interest can be more easily understood [17]. In Fig. 6 RVI (on left) and PDP (on right) partial for the "Model 0" and variable "score" (company reputation score) have been produced.

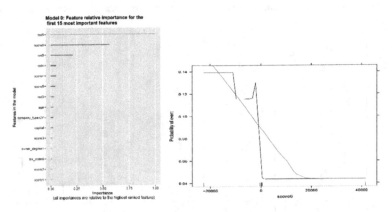

Fig. 6. Plots of relative variable importance (left) and partial dependence (right).

From the RVI we can see that most important variables (considering first 15 most important features in the "Model 0") by extreme gradient boosting model are number of unsubmitted annual reports (the one as of most recent month is the most important variable in "Model 0"), historical score, company age, company type (OÜ/not OÜ), owner degree and state tax. From the PDP we can see that the higher historical reputation score as of most recent month the lower the probability of event.

6.2 Formal Description of Credit Risk Scores Classes

In this section the goal is to describe ensemble model by exploring and differentiating credit risk classes with help of scoring data. Similar approach is currently being used in SAS® Enterprise Miner™ [18].

Columns of matrix ("1"–"5") in Fig. 7 denote credit risk classes, rows of matrix present a selection of explanatory variables used in modelling (the most recent time-period has been chosen). In the figure, next to the name of variable in the brackets there are mean values of that variable by credit risk classes in original scale of variable. Cells of matrix represent result of comparison of distributions in given credit class and population for a given variable (the brighter the cell (lower the value in the cell) the lower the values of variable in risk class than in population in average). For example, if credit risk class is "1" then by population we mean all the rest of other risk classes ("2"–"5"). In the last column "Trend" trend for a given variable is presented wrt moving from best credit risk class ("1") towards worst one ("5"). For instance, we can see that there is a linear trend for variable "Score history" - the lower the values for variable "score history" in average, the worse the credit risk class in terms of credit score.

Risk class	1	2	3	4	5	Scale	Trend
As of 20.11.2017							
Number of companies in risk score class	219,201	17,851	3,310	1,175	8,734		
% of companies in risk score class	88%	7%	1%	0.5%	3.5%		
1. Score history 1(257.9) 2(-91.05) 3(-829.41) 4(-2 879.91) 5(-3 755.81)	6 - Higher than in population	5 - A bit lower than in population	3 - A bit lower than in population	2 - Lower than in population	1 - Much lower than in population	identity	↓
2. Number of unsubmitted annual reports 1(-0.6) 2(-1.35) 3(-0.43) 4(0.98) 5(0.86)	3 - A bit lower than in population	2 - Lower than in population	5 - A bit higher than in population	6 - Higher than in population	4 - No significant difference in class and population	identity	↓↑
3. Number of unsubmitted tax declarations 1(0.14) 2(0.31) 3(1.26) 4(1.8) 5(0.32)	4 - No significant difference in class and population	4 - No significant difference in class and population	6 - A bit higher than in population	5 - A bit higher than in population	4 - No significant difference in class and population	identity	↔↑↓
4. Tax debts 1(178.73) 2(1 339.15) 3(37 0035.17) 4(15 329.02) 5(13 084.72)	2 - Lower than in population	5 - A bit higher than in population	5 - A bit higher than in population	6 - Higher than in population	6 - Higher than in population	log	↑
5. Tax debts interest 1(18.11) 2(149.7) 3(70 139.48) 4(8 485.22) 5(6 680.17)	2 - Lower than in population	5 - A bit higher than in population	6 - Higher than in population	7 - Much higher than in population	7 - Much higher than in population	log	↑
6. Tax debts postponed 1(98.16) 2(304.42) 3(412.991 4(232.24) 5(31.57)	3 - A bit lower than in population	5 - A bit higher than in population	4 - No significant difference in class and population	4 - No significant difference in class and population	4 - No significant difference in class and population	log	↑↔
7. Tax debt disputed 1(1.06) 2(71.96) 3(1 097.57) 4(280.45) 5(369.17)	3 - A bit lower than in population	4 - No significant difference in class and population	5 - A bit higher than in population	5 - A bit higher than in population	5 - A bit higher than in population	log	↑↔
8. Paid taxes (personnel) 1(4 862.43) 2(3 306.72) 3(3007.87) 4(277.17) 5(176.75)	5 - A bit higher than in population	3 - A bit lower than in population	3 - A bit lower than in population	2 - Lower than in population	2 - Lower than in population	log	↓
9. Paid taxes (state) 1(9 443.37) 2(739.93) 3(893.5) 4(1 504.98) 5(391.35)	5 - A bit higher than in population	4 - No significant difference in class and population	4 - No significant difference in class and population	3 - A bit lower than in population	2 - Lower than in population	log	↓
10. Age of company 1(3 310.96) 2(717.36) 3(1 269.98) 4(2 40A.36) 5(4 984.2)	6 - Higher than in population	1 - Much lower than in population	2 - Lower than in population	3 - A bit lower than in population	5 - A bit higher than in population	identity	↓↑
11. Network centrality - degree (score < 0) 1(0.35) 2(1.21) 3(5.59) 4(9.94) 5(2.59)	4 - No significant difference in class and population	4 - No significant difference in class and population	5 - A bit higher than in population	6 - Higher than in population	4 - No significant difference in class and population	log	↓↑↓
12. Network centrality - owner degree 1(0.99) 2(0.17) 3(0.43) 4(0.93) 5(0.4)	5 - A bit higher than in population	4 - No significant difference in class and population	3 - A bit lower than in population	3 - A bit lower than in population	4 - No significant difference in class and population	log	↓
13. Network centrality - owner pagerank 1(0) 2(0) 3(0) 4(0) 5(0)	5 - A bit higher than in population	3 - A bit lower than in population	3 - A bit lower than in population	3 - A bit lower than in population	3 - A bit lower than in population	log	↓↔
14. Network centrality - board degree 1(1.5) 2(1.26) 3(1.28) 4(1.21) 5(0.75)	4 - No significant difference in population	2 - Lower than in population	2 - Lower than in population	2 - Lower than in population	3 - A bit lower than in population	log	↓
15. Network centrality - board pagerank 1(0) 2(0) 3(0) 4(0) 5(0)	4 - No significant difference in class and population	4 - No significant difference in class and population	4 - No significant difference in class and population	3 - A bit lower than in population	3 - A bit lower than in population	log	↔↓
16. Network centrality - degree of all connections 1(3.25) 2(1.62) 3(2.08) 4(2.81) 5(2.04)	6 - Higher than in population	2 - Lower than in population	4 - No significant difference in class and population	5 - A bit higher than in population	2 - Lower than in population	log	↓↑↓
17. Network centrality - pagerank of all connections 1(0) 2(0) 3(0) 4(0) 5(0)	5 - A bit higher than in population	3 - A bit lower than in population	3 - A bit lower than in population	4 - No significant difference in class and population	3 - A bit lower than in population	log	↓↑↓
18. VAT registration (registered or not) 1(31.01) 2(43.74) 3(100.88) 4(252.5) 5(123.47)	3 - A bit lower than in population	2 - Lower than in population	4 - No significant difference in class and population	2 - Lower than in population	1 - Much lower than in population	identity	↓↑↓
19. VAT registration (days past due date) 1(31.01) 2(43.74) 3(100.88) 4(252.5) 5(123.47)	4 - No significant difference in class and population	4 - No significant difference in class and population	3 - A bit lower than in population	3 - A bit lower than in population	5 - A bit higher than in population	log	↔↓↑
20. Claims 1(6.65) 2(74.41) 3(290.94) 4(309.05) 5(963.3)	3 - A bit lower than in population	5 - A bit higher than in population	4 - No significant difference in class and population	5 - A bit higher than in population	5 - A bit higher than in population	log	↑↓↑
21. Balance 1(378.194) 2(32.730) 3(44.034) 4(73.313) 5(113.853)	6 - Much higher than in population	1 - Much lower than in population	1 - Much lower than in population	2 - Lower than in population	1 - Much lower than in population	log (min(s) = 1)	↓↔↑

Fig. 7. Description of credit risk classes.

7   Conclusions, Threats to Validity and Future Work

In this paper we described a novel approach for modelling credit risk of companies. We applied extreme gradient boosting with Bayesian hyper-parameter optimization and ensemble learning with rapidly changing data to learn a model, which predicts likelihood of a default event of a business in 12 months. The learned model has accuracy of 99.5%. For confirming that the high accuracy has impact in practical settings, we performed additionally retrospective analysis and a case study where we analyzed credit scoring results by the model wrt credit management specialist decisions. In retrospective analysis we reviewed the credit scores that the model provided for the insolvent companies 0, 1, 2, 5, 8 and 11 months before event. Retrospective analysis proved that the predictive ability of the model is acceptable even 11 months before the event giving confidence that the model is applicable in predicting events up to 12 months ahead.

Validation of the model wrt credit decisions made by domain experts shows that the model performs quite well in terms of making credit decisions. Furthermore, the analysis shows, that our credit risk model is able to capture some cases where a manual credit decision made by a specialist may be questionable. However, it should be noted that there may be deviations both ways - manual assessments may be incorrect in times and credit risk model also has some deviations and exceptions. By now the model is put into practice and is used by Estonian businesses in everyday decision-making via products of Register OÜ and its partners.

However, there are some threats to validity related to the proposed model wrt to bias in sampling, availability of data, modeling process and usability of model. For remedies please refer to future work activities provided at the end of current section.

Regarding sample used in modeling, we have excluded young companies from training sample since they have only few data points available for training. In the credit model it is reflected in young companies having scores from slightly higher risk class compared to average companies, which seems to make sense if we assume that if not much is known about the company then the potential risk is higher as well. Also, we have excluded from negative cases companies for which tax debts were monotonically increasing since this is an early indicator of a potential default event. Anyway, there should be no need for that explicitly once time series derivates (incl. univariate statistics, linear and non-linear trends, different statistics wrt variations around trends) will be added into modeling process. Positive cases mostly consist of company deletion events due to multiple unsubmitted annual reports. There has been recently significant rise in such events since Estonian Business Register deletes more systematically companies, which have failed to submit their annual reports multiple consecutive years in a row. In the model this means that the number of unsubmitted annual reports is a dominative feature. However, importance of this feature is subject of change if behavior of Estonian Business Register will change.

Regarding modeling process, all missing values for numeric variables were replaced with zeros. This was done primarily due to fact that the same method for handling missing values has been used for reference model. Regarding classification methods "state of the art" algorithm XGBoost has been used alone due to satisfactory results. Currently, in total 13 models needs to be fitted. In case the number of potentially useful explanatory variables and number of observations will grow some more compact and flexible approach for modeling and variable selection is needed. There are also several threats to validity wrt retrospective analysis which are described in according section.

Regarding usage of current model for other than Estonian market there might be present some restrictions - more precisely, some important explanatory variables like tax debts might not be available at all or the collecting process of all data needed for modeling might be too expensive for larger markets.

Currently we use only a limited amount of network features (degree, PageRank [19]) derived from networks of board members/owners and companies. In future we will add support for a wider variety of network types, e.g. networks of companies and locations, and network metrics, e.g. eigenvector centrality, Kleinberg's authority centrality etc. We also see value in adding additional variables derived from time series for improving the model wrt detection of "edge" cases [19–21]. Current model does not consider default events related to disruptions in payments. We have started collecting late payment data and believe that features from this data could be beneficial to the model and everyday credit scoring as well. For improving accuracy of predicting defaults financial ratios have been applied in practice [26]. We are also interested in exploring multinomial discrete hazard survival data mining model which have several good properties wrt current approach [30]. Finally, we have recently initiated a new project, which aims at performing credit scoring by experimenting with plenty of different classification algorithms using freely available web data only. This project represents a shift towards big data scoring, where we intend to explore the enhancements outlined earlier.

References

1. Xia, Y., Liu, C., Li, Y., Liu, N.: A boosted decision tree approach using Bayesian hyper-parameter optimization for credit scoring. Expert Syst. Appl. **78**, 225–241 (2017)
2. Siddiqi, N.: Intelligent Credit Scoring: Building and Implementing Better Credit Risk Scorecards, 2nd edn. Wiley, Hoboken (2017)
3. Tamari, M.: Financial ratios as a means of forecasting bankruptcy. Manag. Int. Rev. **6**(4), 15–21 (1966)
4. Baxter, R.A., Gawler, M., Ang, R.: Predictive model of insolvency risk for Australian corporations. In: Proceedings of the Sixth Australasian Conference on Data Mining and Analytics, vol. 70. Australian Computer Society, Inc. (2007)
5. Investopedia: Bankruptcy risk definition. https://www.investopedia.com/terms/b/bankruptcyrisk.asp
6. Yu, L., Wang, S., Lai, K.K., Zhou, L.: Bio-inspired Credit Risk Analysis: Computational Intelligence with Support Vector Machines. Springer, Heidelberg (2008). https://doi.org/10.1007/978-3-540-77803-5
7. Riigi Teataja: Commercial Code. https://www.riigiteataja.ee/en/eli/504042014002/consolide
8. Ilves, T.: Impact of board dynamics in corporate bankruptcy prediction: application of temporal snapshots of networks of board members and companies. Master thesis. Tartu University (2014)

9. Yap, B.W., Rani, K.A., Rahman, H.A.A., Fong, S., Khairudin, Z., Abdullah, N.N.: An application of oversampling, undersampling, bagging and boosting in handling imbalanced datasets. In: Herawan, T., Deris, M.M., Abawajy, J. (eds.) Proceedings of the First International Conference on Advanced Data and Information Engineering (DaEng-2013). LNEE, vol. 285, pp. 13–22. Springer, Singapore (2014). https://doi.org/10.1007/978-981-4585-18-7_2

10. Analytics Vidhya: How to handle Imbalanced Classification Problems in machine learning? (2017). https://www.analyticsvidhya.com/blog/2017/03/imbalanced-classification-problem/

11. Kaggle Inc.: Kaggle competitions. https://www.kaggle.com/competitions

12. Bischl, B., Richter, J., Bossek, J., Horn, D., Thomas, J., Lang, M.: mlrMBO: a modular framework for model-based optimization of expensive black-box functions. Cornell University Library (2017)

13. Bischl, B., Lang, M., Kotthoff, L., Schiffner, J., Richter, J., Studerus, E., Casalicchio, G., Jones, Z.M.: mlr: machine learning in R. J. Mach. Learn. Res. **17**, 1–5 (2016)

14. Jenks, G.F.: The data model concept in statistical mapping. In: International Yearbook of Cartography, vol. 7, pp. 186–190 (1967)

15. Hand, D.J., Anagnostopoulos, C.: Measuring classification performance. hmeasure.net

16. Chen, T., Guestrin, C.: XGBoost: a scalable tree boosting system. In: KDD 2016 Proceedings of the 22nd ACM SIGKDD International Conference on Knowledge Discovery and Data Mining, San Francisco, pp. 785–794 (2016)

17. Greenwell, B.M.: pdp: an R package for constructing partial dependence plots. R J. **9**(1), 421–436 (2017)

18. SAS Institute Inc.: SAS® Enterprise Miner™ 14.3: Reference Help; Chapter 67 Segment Profile Node (2017)

19. Luke, D.A.: A User's Guide to Network Analysis in R. UR. Springer, Cham (2015). https://doi.org/10.1007/978-3-319-23883-8

20. Svolba, G.: Data preparation for analytics using SAS. SAS Institute Inc. (2006)

21. Афанасьев, В.Н., Юзбашев, М.М.: Анализ временных рядов и прогнозирование. Финансы и статистика (2012)

22. Abdou, H., Pointon, J.: Credit scoring, statistical techniques and evaluation criteria: a review of the literature. Intell. Syst. Account. Financ. Manag. **18**(2–3), 59–88 (2011)

23. Hooman, A., Mohana, O., Marthandan, G., Yusoff, W.F.W., Karamizadeh, S.: Statistical and data mining methods in credit scoring. In: Proceedings of the Asia Pacific Conference on Business and Social Sciences, Kuala Lumpur (2015)

24. Tarver, E.: Business credit score: everything you should know to build business credit. https://fitsmallbusiness.com/how-business-credit-scores-work/

25. Register OÜ: Credit risk prediction for Estonian companies. https://docs.google.com/document/d/1aG9Y6B8J8Q9Ee75tA6X3SJByo9QAJCvugYas7Q9p2VE

26. Zięba, M., Tomczak, S.K., Tomczak, J.M.: Ensemble boosted trees with synthetic features generation in application to bankruptcy prediction. Expert Syst. Appl. **58**, 93–101 (2016)

27. Altman, E.I.: Financial ratios, discriminant analysis and the prediction of corporate bankruptcy. J. Financ. **23**, 589–609 (1968)

28. Shin, K.S., Lee, T.S., Kim, H.J.: An application of support vector machines in bankruptcy prediction model. Expert Syst. Appl. **28**, 127–135 (2005)

29. Geng, R., Bose, I., Chen, X.: Prediction of financial distress: an empirical study of listed Chinese companies using data mining. Eur. J. Oper. Res. **241**, 236–247 (2015)

30. Tutz, G., Schmid, M.: Modeling Discrete Time-to-Event Data. SSS. Springer, Cham (2016). https://doi.org/10.1007/978-3-319-28158-2

Quality Requirements and Software

A Behavior-Based Framework for Assessing Product Line-Ability

Iris Reinhartz-Berger[✉] and Anna Zamansky

Department of Information Systems, University of Haifa, Haifa, Israel
{iris,annazam}@is.haifa.ac.il

Abstract. Systems are typically not developed from scratch, so different kinds of similarities between them exist, challenging their maintenance and future development. Software Product Line Engineering (SPLE) proposes methods and techniques for developing reusable artifacts that can be systematically reused in similar systems. Despite the potential benefits of SPLE to decrease time-to-market and increase product quality, it requires a high up-front investment and hence SPLE techniques are commonly adopted in a bottom-up approach, after individual systems have already been developed. Deciding whether to turn existing systems into a product line – referred to as product line-ability – involves many aspects and requires some tooling for analyzing similarities and differences among systems. In this paper we propose a framework for the identification of "similarly behaving" artifacts and analyzing their potential reuse in the context of product lines. This framework provides metrics for calculating behavior similarity and a method for analyzing the product line-ability of a set of products. The framework has been integrated into a tool named VarMeR – Variability Mechanisms Recommender, whose aim is to systematically guide reuse.

Keywords: Software product line engineering · Variability analysis
Reuse

1 Introduction

The increase in the number and complexity of information systems and software applications sets the ground for the development of methods for increasing reuse. Software Product Line Engineering (SPLE) promotes *systematic reuse* among different, yet similar, systems by creating reusable artifacts, commonly referred to as *core assets* or domain artifacts, and guiding their use in particular systems [7, 12]. It has been shown that SPLE has the potential to decrease time-to-market and increase product quality, yet it requires a high up-front investment in the development of core assets [7]. Hence, SPLE techniques are commonly adopted in a bottom-up approach, namely after several similar variants have been created using ad-hoc reuse techniques [6]. In these cases, the decision whether to set up a Software Product Line (SPL) from the individual systems and adopt SPLE techniques is non-trivial and requires an in-depth assessment. Such assessment should be based on (i) analyzing the similarity and variability of the

© Springer International Publishing AG, part of Springer Nature 2018
J. Krogstie and H. A. Reijers (Eds.): CAiSE 2018, LNCS 10816, pp. 571–586, 2018.
https://doi.org/10.1007/978-3-319-91563-0_35

existing products, and (ii) applying suitable metrics that reflect the effort of creating core assets and generating particular (product) artifacts from these core assets.

While many studies deal with different aspects of similarity analysis of software artifacts, e.g., [1] for variability analysis and [2, 3, 17] for clones detection, only few studies address 'product line-ability' decisions. Berger et al. [4] coined the term 'product line-ability' to refer to the ability of a set of products to form a product line and suggested how to measure it in practice [5]. Several metrics have been proposed for evaluating product line architectures [8, 9, 19]. However, all of these approaches assume a rather high similarity of representations, mainly implementations or architecture models, of the different systems. This makes the reuse of similar components quite straightforward, namely, reused as-is. In reality, however, things are not so black and white; the answer may be degree-based: while some components can be reused as they are, without any adaptations or with small adaptations, others may be quite different in terms of low-level implementations. This is especially relevant in cases where systems were developed by different teams, for different purposes and do not necessarily share similar representations. This leads to the need for defining more robust reusability analysis methods, which take into account the intensions (as specified in behaviors) of artifacts, rather than solely their implementations, and allow for a more refined evaluation of the reuse effort, which reflects the possibilities to adopt specific reuse practices.

In our previous works we presented an approach based on behavioral similarity of object-oriented software artifacts [13–15]. This approach views such artifacts as things exhibiting behaviors, i.e., transformations between *states* due to some *external events*. The approach further supports associating different reuse mechanisms based on the characteristics of related similarity mappings, which intuitively correspond to different degrees of reuse efforts. In this paper we propose a framework for the identification of "similarly behaving" artifacts and analyzing their potential reuse in the context of product lines. The framework is based on three polymorphism-inspired mechanisms: parametric, subtyping, and overloading. It further provides metrics for calculating behavior similarity and a method for analyzing the product line-ability of a set of products, aiming to support developers in making reuse decisions.

The rest of the paper is structured as follows. Section 2 presents related work on product line-ability metrics. Section 3 provides an overview of the suggested approach and introduces similarity relations between a set of systems. Section 4 presents the product line-ability framework that is based on the similarity relations, while Sect. 5 is devoted to implementation aspects and preliminary results. Finally, Sect. 6 summarizes and refers to future directions.

2   Previous Work

Few works propose metrics for evaluating the ability of a set of systems or products to form a product line. We briefly overview some of the most relevant ones below.

In [19], metrics are suggested to assess similarity, variability, reusability, and complexity of product line architectures. These metrics are applicable only on architectures specified in the vADL specification language and thus have a limited scope.

In [4], a more general framework is proposed for deriving a set of metrics for evaluating similar systems, which can be used to estimate the ability of a set of products to form a product line. The framework assumes the availability of a graph-based representation of all products, reflecting dependencies between products' components which can be logical or communicative. In [5], these metrics are applied for a particular case study from the automobile industry for evaluating existing steering systems and estimating the benefit of creating a software product line. In [18], a graph-based representation is also used, incorporating weight values of assets into the metrics. This approach is based on the assumption that similarity of components means at least syntactic similarity, i.e., similar components have identical interfaces. As explained in [4], *"syntactical signature identity* is at least *necessary* but not *sufficient*. Therefore, *semantic signature identity* for two components must additionally be ensured which can for example be evaluated automatically by using the component's test suites in an entangled manner which have to ensure path coverage at least".

Two metrics proposed by Berger et al. [4, 5] are of special interest for assessing product-line ability in general and introducing our work in particular: size of commonality – evaluating the similarity of the existing products (and consequentially the size of potential core assets), and product-related reusability – reflecting the effort of creating particular products from the core assets. Table 1 describes these metrics. In all cases, the set of products is denoted by $p_1, ..., p_n$, and the required and optional components of a product p_i, namely, the components inherently necessary to fulfill the product's basic functionality and those which add further functionality, are denoted by $C_{pi,r}$ and $C_{pi,o}$, respectively.

Table 1. Product line-ability metrics taken from [4]

Name	Formula	Description
Size of commonality (SoC)	$SoC = \left\lvert \bigcap_{i=1..n} C_{p_i,r} \right\rvert + \left\lvert \bigcap_{i=1..n} C_{p_i,o} \right\rvert$	Number of *identical* components among $p_1, ..., p_n$
Product related reusability (PrR)	$PrR_i = \frac{SoC}{\left\lvert C_{p_i,r} \cup C_{p_i,o} \right\rvert}$	Ratio relating the size of commonality for a specific product p_i

It should be noted that the metrics suggested by Berger et al. [4, 5] assume the availability of explicit representation of the required and optional components of each product. Moreover, they take a low-level syntactic view, assuming identity of interfaces of the considered components. Finally, they only consider reuse as is, avoiding other types of reuse that require adaptation of artifacts.

We aim to lift the above strong assumptions, taking a more high-level, behavioral approach. In other words, we are interested also in situations where components may have semantically similar interfaces, and exhibit similar (not necessarily identical) or even different behaviors. We also aim to distinguish between different types of reusability, according to the effort they require for artifacts adaptation. For the sake of completeness, we shortly reproduce definitions of behavioral similarity from our previous publications [13–15] in Sect. 3. Sections 4 and 5 present the new contribution, namely the behavioral product line-ability framework and its application.

3 Behavioral Similarity of Products

For calculating similarity between components (systems or parts of them), we consider their behaviors, each of which is viewed as a transformation between stable states due to some external triggers.

Definition 1 (behavior). A *behavior* is represented by a triple $(S_1, <e>, S*)$, where:

- S_1 is the initial state of the component before the behavior occurs,
- $<e>$ is a sequence of external events triggering the component's behavior,
- $S*$ is the final state of the component after the behavior occurs.

For instance, the behavior of a sorting component can be characterized by the input – unsorted list of items (S_1), the sorting event $(<e>)$, and the output – sorted list $(S*)$.
Behavior can be further represented via two types of descriptors: shallow and deep.

Definition 2 (shallow descriptor). The *shallow behavior descriptor*, commonly known as interface, is a pair (op, params) where:

- op = (bname, btype), bname is the behavior name and btype is the behavior's output (returned type, in the form of a finite or infinite set of possible values);
- params = {(pname, ptype)} denoting the behavior's parameters and types (all ptypes have the form of sets of possible values).

Definition 3 (deep descriptor). The *deep behavior descriptor* is a pair (att_used, att_modified), where:

- att_used = {(aname, atype)}, where aname, atype are respectively the name and type (possible values) of an attribute *involved* in the behavior;
- att_modified = {(aname, atype)}, where aname, atype are respectively the name and type (possible values) of an attribute *being modified* in the behavior.

The aforementioned descriptors enable representation of behavior as a triple $(S_1, <e>, S*)$ where S_1 = deep.att_used $\cup$ shallow.params (namely, all attributes and parameters being used by the behavior before being modified by it), and $S*$ = deep.att_modified $\cup$ shallow.op (i.e., all attributes modified by the behavior and the output). We assume that $<e>$ is derived from the semantics of the behavior name.

Table 2 exemplifies these representations for two sorting algorithms – merge and quick sort[1]. While the implementations of these algorithms are quite different, their

---

[1] Note that the behavior representation considers the impact of procedure calls, such as *doMergeSort* and *mergeParts* for *MergeSort*. Nevertheless, the approach represents only public operations, assuming that private and protected operations may be defined for implementation purposes. Our assumption is that differences in such operations do not necessarily imply differences in intentional behavior.

Table 2. Examples of shallow and deep behavior descriptors

		Merge Sort	Quick Sort
Code		```public class MergeSort { private int[] array; private int[] tempMergeArray; private int length; public void sort(int[] inputArray) { this.array = inputArray; this.length = inputArray.length; this.tempMergeArray = new int[this.length]; doMergeSort(0, this.length - 1); } private void doMergeSort (int lowerIndex, int higherIndex) { if (lowerIndex < higherIndex) { int middle=lowerIndex+(higherIndex- lowerIndex) / 2; doMergeSort(lowerIndex, middle); doMergeSort(middle + 1, higherIndex); mergeParts(lowerIndex, middle, higherIndex); } } private void mergeParts(int lowerIndex, int middle, int higherIndex) { for (int i = lowerIndex; i <= higherIndex; i++) { this.tempMergeArray[i] = this.array[i]; } int i = lowerIndex; int j = middle + 1; int k = lowerIndex; do { if (this.tempMergeArray[i] <= this.tempMergeArray[j]) { this.array[k] = this.tempMergeArray[i]; i++; } else { this.array[k] = this.tempMergeArray[j]; j++; } k++; if (i > middle) { break; } } while (j <= higherIndex); while (i <= middle) { this.array[k] = this.tempMergeArray[i]; k++; i++; } } }```	```public class QuickSort { private int[] array; private int length; public void sort(int[] input) { if ((input == null) || (input.length == 0)) { return; } this.array = input; this.length = input.length; quickSort(0, this.length - 1); } private void quickSort(int lowerIndex, int higherIndex) { int i = lowerIndex; int j = higherIndex; int pivot = this.array[(lowerIndex + (higherIndex - lowerIndex) / 2)]; while (i <= j) { while (this.array[i] < pivot) { i++; } while (this.array[j] > pivot) { j--; } if (i <= j) { exchangeNumbers(i, j); i++; j--; } } if (lowerIndex < j) { quickSort(lowerIndex, j); } if (i < higherIndex) { quickSort(i, higherIndex); } } private void exchangeNumbers(int i, int j) { int temp = this.array[i]; this.array[i] = this.array[j]; this.array[j] = temp; } }```
Shallow	Op	(sort, void)	(sort, void)
	Params	(inputArray, int[])	(input, int[])
Deep	Att-used	(array, int[]) (tempMergeArray, int[]) (length, int)	(array, int[]) (length, int)
	Att-modified	(array, int[]) (tempMergeArray, int[]) (length, int)	(array, int[]) (length, int)

shallow and deep descriptors are quite similar. Yet, we need a formal way to compare them, as the name and types of attributes may be different.

Next, we define relations based on behavioral similarity. For this we assume the availability of a similarity mapping *Sim* between constituents of the shallow and deep descriptors. The similarity mapping can be based on syntactic considerations (e.g., identity). In this case, the parameters of the two implementations of the sorting algorithms, *input* and *inputArray*, will not be considered similar. Alternatively, the similarity mapping can be based on semantic considerations using semantic nets or statistical techniques to measure the distances among words and terms [11]. In this case, the names used in the code need to be meaningful. Another option is using type or schematic similarity, potentially ignoring the semantic roles or essence of the compared elements [10]. This is mainly relevant for comparing systems from different domains.

The choice of a particular similarity mapping *Sim* between the descriptors elements of the compared behaviors b_1 and b_2 induces several types, depending on the following properties of *Sim*:

- Sim is either *covered*, i.e., total and onto, or *not covered*.
- Sim is either *single mapped* (i.e., every element of the descriptors of b_1 has a matching single element of the descriptors of b_2), or *multi-mapped* (there are elements which have more than one match).

Combinations of the above properties of similarity mapping correspond to different similarity (reuse-related) relations which may hold between shallow and deep behavior descriptors.

Definition 4 (similarity relation). Given a similarity mapping *Sim*, the similarity relation between two behaviors b_1 and b_2 can be:

- Use: if *Sim* is covered and single-mapped (intuitively corresponding to the highest degree of similarity between behaviors).
- Refinement: if *Sim* is multi-mapped (intuitively corresponding to splitting a variable of one behavior in the other).
- Extension: if *Sim* is not covered (intuitively corresponding to adding a variable in a behavior).

These similarity relations are summarized in Table 3. Note that both refinement and extension can hold at the same time; we refer to this situation as refined extension. Two behaviors may not be related with a similarity relation at all. The two sorting algorithms depicted in Table 1 can be considered Extension (EXT), as Merge Sort handles (reads and writes) an extra attribute – tempMergeArray of type int[].

Table 3. Similarity relations

	Covered	Not covered
Single-mapped	Use (USE)	Extension (EXT)
Multi-mapped	Refinement (REF)	Refined Extension (REF-EXT)

It should be noted that if we take *Sim* as identity, USE corresponds to the use-as-is notion of reusability considered in [4, 5]. This notion is considerably extended in our framework by (i) allowing similarities other than identity, (ii) considering two different representations of operation: shallow (interface) and deep (transformation), and

(iii) considering other variants of similarity relations besides USE (a 1–1 single-mapped mapping), including REF, EXT, and REF-EXT.

Concerning point (ii) above, the similarity relations defined in Table 3 can indeed hold for both shallow and deep descriptors, leading to 25 possibilities ({USE, REF, EXT, REF-EXT, NONE} for both shallow and deep descriptors). In this paper, we focus on polymorphism-inspired mechanisms, which are characterized by the cases when the USE relationship holds between shallow descriptors. This intuitively corresponds to the case when interfaces of the behaviors are *similar* (but not necessarily identical as usually is the case in polymorphisms; for this reason we term these relations polymorphism-inspired). As listed in Table 4, the differences in the deep descriptors similarity distinguish between three types of potential reuse: parametric (similar transformations), subtyping (refined or extended transformations), and overloading (different transformations). Intuitively, these types require different amount of effort when reusing with parametric requiring the least effort and overloading – the most effort (recall that overloading implies different deep behaviors, namely, different transformations on attributes).

Table 4. Characteristics of polymorphism-inspired mechanisms

Shallow	Deep	Description	Reuse type	Estimated effort
USE	USE	Both interfaces and transformations are similar	Parametric	Low
USE	REF	Interfaces are similar and transformations are refined	Subtyping	Medium
USE	EXT	Interfaces are similar and transformations are extended	Subtyping	Medium
USE	REF-EXT	Interfaces are similar and transformations are both refined and extended	Subtyping	Medium to high
USE	NONE	Interfaces are similar and transformations are different	Overloading	High

Next we introduce a method and metrics for deciding on product line-ability based on the different types of behavioral similarity relations.

4 Behavioral Product Line-Ability Framework

In what follows we assume a set of products $p_1, \ldots, p_N$ ($N \geq 2$), where each p_i is a set of components, namely, $p_i = \{C_{i1}, \ldots, C_{i\,ki}\}$. A component can be a product or its part, e.g., a module or a class. Product line-ability of such a set of products has two directions:

– *bottom-up*, i.e., constructing a core asset out of the products' components
– *top-down*, i.e. creating products out of the core asset.

The simplest case is when all products share a set of identical, mandatory components. Then in the bottom-up direction all these components can be taken to form a core asset. In the top-down direction, these components will be used (as-is) in a new product. However, components may not be identical, but still similar. Then the construction of a core asset (bottom-up direction), while requiring some adaptation effort, may still be worthwhile. Also the creation of new products from the core asset may require adaptation (top-down direction). Product line-ability analysis in this case is intended for providing a way of assessing the required adaptation efforts in both directions. In addition, some products may be less related than the others, and excluding them from the analysis may result in better product line-ability of the remaining products.

To address these challenges, we propose a behavioral product line-ability framework, which is based on the notion of behavior similarity presented in Sect. 3. The framework provides a method for product line-ability analysis, whose inputs are sets of products and the output is a graph highlighting the potential product line-ability of the input product sets. The method is composed of four main sequential steps:

1. Behavior similarity calculation – see Sect. 3 for details.
2. Similarity degree measurement.
3. Product-related variability degree measurement for each product in the set.
4. If there are products in the set whose product-related variability degree is too high (the specific threshold is an analyst's decision), these products can be excluded and the method returns to step 2 for re-measuring the similarity and product-related variability degrees.

Figure 1 depicts the method's steps, as well as its inputs and analyst involvement (upper part) and outputs (lower part). Next, we elaborate on the data structure used in the method – called similarity graph, as well as on the calculations of similarity and product-related variability degrees.

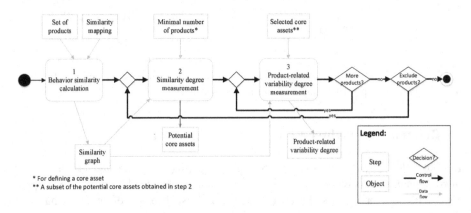

Fig. 1. The suggested method for product line-ability analysis

4.1 Similarity Graphs

The method is based on a graph data structure – called similarity graph – which is a colored undirected graph whose vertices represent components of a set of products (each vertex is colored, where different colors represent different products), and the edges represent reuse types of parametric, subtyping and overloading, defined in Sect. 3.

Definition 5 (similarity graph). Given a set of a set of products $p_1, \ldots, p_N$ ($N \geq 2$), a *similarity graph* is an undirected graph of the form $G = (V, E)$, where V is a set of pairs (C_{ij}, l_i) (C_{ij} is the j-th component of product p_i and l_i is the color associated with p_i) and E is a set of triplets (C_{ij}, C_{kl}, t) which associate components C_{ij} and C_{kl} with a reuse type t ($t \in \{$parametric, subtyping, overloading$\}$).

Returning to our sorting example, assume that we have three products: P_1 with sorting algorithm classes Merge 1 and Quick 1, P_2 with Merge 2 and Quick 2 and P_3 with Merge 3, Quick 3 and Optimized Quick 3 (the optimization extends the behavior of quick sort in order to improve performance). Figure 2 exemplifies a possible similarity graph with seven (class) components. According to this graph, the three quick sort implementations have very similar behaviors, and are therefore connected by parametric edges. As noted, Optimized Quick 3 extends the typical quick sort behavior by writing to additional attributes to increase performance. Thus, the edges between Optimized Quick 3 and the three quick sort algorithms are of type subtyping. Merge 3 behaves differently than Merge 1 and the other sorting algorithms, as it sorts objects rather than simple types (such as integers and strings).

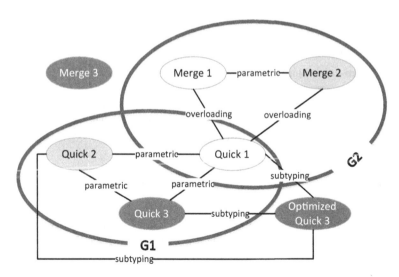

Fig. 2. Example of a similarity graph

4.2 Similarity Degree Measurement

First, we need to assess the degree of similarity of the given set of products: low degree of similarity may not justify transforming the set of products into a product line. This metric relates to the bottom-up direction, and requires analyzing subgraphs of the corresponding similarity graph. Each "good enough" subgraph (i.e., a subgraph with high similarity between its constituent components) will correspond to a potential core asset in case the given set of products will be transformed into a product line. Hence, each such subgraph should include vertices colored in different colors (i.e., belonging to different products). Yet, not all colors must appear in the subgraph, indicating on the existence of optional components.

Intuitively, the most similar subgraphs are cliques in which all edges are parametric (restricting *Sim* to identity mapping corresponds to the case of SoC – Size of Commonality – metric from Table 1). This means that all components are similar to each other and thus potentially can be easily transformed into a core asset. We call such subgraphs m-colored parametric asset, where m is the minimal number of products (colors) that need to appear in the asset (m $\geq$ 2).

Definition 6 (m-colored parametric asset). An *m-colored parametric asset* (m $\geq$ 2) in a similarity graph G = (V, E) is a subgraph G' = (V', E'), where (i) V' $\subseteq$ V, E' $\subseteq$ E, (ii) at least m colors appear in V', and (iii) for each v_1, v_2 $\in$ V' (v_1, v_2, parametric) $\in$ E'.

For instance, G_1 in Fig. 3 is an example of a 3-colored parametric asset. G_2 is not such an asset, but our framework allows also for analyzing subgraphs which are not ("full") parametric assets, yet may yield useful core assets after some adaptation. In these cases, some non-parametric or missing edges may be allowed, indicating different variants, and some of the colors may not appear as well (m < N), identifying optional elements. G_2 is indeed such an example, connecting three algorithms of two different types (merge and quick) in two products, all of which handle sorting of integers.

To support more flexibility of similarity types, we define m-color behavioral similarity degree with respect to how "close" a given similarity subgraph is to being an m-colored parametric asset. Formally expressed:

Definition 7 (m-color behavioral similarity degree). Let G = (V, E) be a similarity graph and G' a subgraph of it with at least m colors (representing different products, m $\geq$ 2). $P_{G'}$, $S_{G'}$, $O_{G'}$ are the numbers of parametric, subtyping and overloading edges in G', respectively; k = |V'| is the number of vertices in G'. The *m-color behavioral similarity degree* of G' is a triplet (PS, SS, OS), where: PS $= \frac{2P_{G'}}{k(k-1)}$, SS $= \frac{2S'_G}{k(k-1)}$, OS $= \frac{2O'_G}{k(k-1)}$.

We further define a total order relation on the behavioral similarity degree as follows:

Definition 8 (total order relation). Let (x_1, x_2, x_3), (y_1, y_2, y_3) be two behavioral similarity degrees. The *total order relation* < is defined as $(x_1, x_2, x_3) < (y_1, y_2, y_3)$ iff $x_1 < y_1$, or $x_1 = y_1$ and $x_2 < y_2$, or $x_1 = y_1$ and $x_2 = y_2$ and $x_3 < y_3$.

As an example, consider G_1 in the similarity graph of Fig. 2. The behavioral similarity degree is (1, 0, 0), indicating on a 3-colored parametric asset. For G_2 the behavioral similarity degree is lower, (0.33, 0, 0.67), indicating on a "less parametric" and "more overloading" 2-colored parametric asset.

Using the order relation defined above, the subgraphs of the similarity graphs can be ordered, so that subgraphs appearing first in the list potentially require less effort to being transformed into core assets. We refer to this list as potential m-color core assets.

Definition 9 (potential m-color core assets). Let G be a similarity graph. The *potential m-color core assets* (m ≥ 2) are a list of all subgraphs of G including at least m colors - {G_i}, such that if G_i precedes G_j then the behavioral similarity degree of G_i is greater than that of G_j (using the < relation defined above).

4.3   Product-Related Variability Degree Measurement

A complementary way of analyzing product line-ability is measuring the differences between each product and the potential m-color core assets, as captured by the m-color behavioral similarity degree. This addresses the top-down direction, i.e., after transforming the set of products into a product line how much effort it will be to derive a specific product p_i. Intuitively, greater coverage of vertices of the form C_{ij} by potential m-color core assets indicates higher product line-ability. Note that there is a tradeoff between covering more components (vertices) and minimizing the number of missing and/or subtyping/overloading edges.

Definition 10 (m-color product-related variability degree). Let G = (V, E) be a similarity graph including product p = {C_j}$_{j=1..k}$ (i.e., C_j are vertices of G), and {G_i} – a set of m-colored subgraphs of G (selected from the potential m-color core assets of G). The m-color product-related variability degree is a quadruplet (PV, SV, OV, PSV), where:

$$PV = \frac{|\{C_j \in p | \exists G_i = (Vi, Ei) \text{ and } C' \in Vi \text{ s.t. } C_j \in Vi \text{ and } (C_j, C', parametric) \in Ei\}|}{k}$$

is the **parametric** variability, namely the percentage of components in p that require **parametric** adaptation (note that k is the number of components in p).

$$SV = \frac{|\{C_j \in p | \exists G_i = (Vi, Ei) \text{ and } C' \in Vi \text{ s.t. } C_j \in Vi \text{ and } (C_j, C', subtyping) \in Ei\}|}{k}$$

is the **subtyping** variability, namely the percentage of components in p that require subtyping adaptation.

$$OV = \frac{|\{C_j \in p | \exists G_i = (Vi, Ei) \text{ and } C' \in Vi \text{ s.t. } C_j \in Vi \text{ and } (C_j, C', overloading) \in Ei\}|}{k}$$

is the **overloading** variability, namely the percentage of components in p that require **overloading** adaptation.

$$PSV = \frac{|\{C_j p | \neg \exists G_i = (Vi, Ei) \text{ s.t. } C_j \in Vi\}|}{k}$$

is the **product-specific** variability, namely the percentage of components in p that require **addition** (are not developed based on existing core assets).

As an example consider the similarity graph in Fig. 2. The 2-color product-related variability degree with respect to $\{G_1, G_2\}$ defined above is:

- For the first (white) product P_1 = {Merge 1, Quick 1}: (1, 0, 1, 0)
- For the second (light grey) product P_2 = {Merge 2, Quick 2}: (1, 0, 0.5, 0)
- For the third (dark grey) product P_3 = {Merge 3, Quick 3, Optimized Quick 3}: (0.33, 0, 0, 0.67)

The analyst may decide to exclude from the analysis the products which vary the most (such as P_3 in our example), after which the bottom-up step of constructing the core assets can be repeated, thereby improving the product line-ability of the remaining products.

5 Implementation Aspects and Preliminary Results

The method defined above has been integrated into VarMeR – Variability Mechanisms Recommender [16]. The inputs of VarMeR are object-oriented code artifacts (in Java) that belong to different products. The outputs include (colored) similarity graphs that highlight the similarity and variability in the behaviors (public operations) of the classes of these products. The edges in these graphs represent different reuse types (parametric, subtyping, and overloading) between classes. Figure 3 depicts a snapshot from VarMeR for implementations of three types of sorting algorithms (Bubble, Quick, and Merge) in four products. To simplify discussion in the sequel, we annotated the vertices with labels of the form Pi-B, Pi-Q, Pi-M, where Pi indicates the product id and B, Q, M – the sorting type. The percentages appearing on the edges are of similar (deep) behaviors of the corresponding classes, as reflected by the public operations of these classes. In other words, classes that have more similar operations of a certain type of relation (USE, REF/EXT/REF-EXT, NONE) will result in higher similarity percentages of the corresponding reuse type (parametric, subtyping, overloading, respectively). The tool supports suppressing low percentages, indicating on low behavioral similarity, by defining a minimal threshold for each type of edges (see the upper bars in Fig. 3). Scalability support of VarMeR, namely, analyzing similarity and variability in different granularity levels (including, product, package, and class), is out of the scope of this paper and is discussed in [16]. On top of the similarity graphs, the analysts can present the different metrics defined in Sect. 4.

We explored the results of our method in the context of product line-ability decisions for a set of four projects consisting of implementations of different sorting algorithms. All projects were downloaded from GitHub repository and developed by different teams (thus low level of syntactic similarity was expected). The four projects involved three sorting algorithms (B-bubble, M-merge, and Q-quick; some projects included multi-classes for each type of sorting; Fig. 3 presents VarMeR outcome for this case).

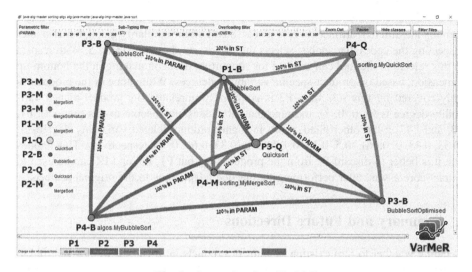

Fig. 3. A snapshot from VarMeR

Due to space limitations, we only exemplify how the tool can be used for supporting decision making in the bottom-up dimension. Table 5 presents two potential core assets for m = 3, namely it uses 3-color behavioral similarity degrees as indication for product line-ability. It can be seen that the first core asset highlights a highly reusable (close to parametric clique) artifact for bubble sort implemented in different projects. The second core asset requires more effort, as some of the artifacts are related via subtyping relations, requiring refinement and/or extension of their deep behaviors. Note that an asset consisting of both is better than the second core asset but worse than the first core asset (similarity degree of (0.33, 0.38, 0)). This means that we do not necessarily desire core assets of maximal size, but of minimal effort, as predicted by the different similarity degrees.

Table 5. Potential core assets and their similarity degrees as predictors for product line-ability

#	Potential core asset (ProjectId-AlgInitial)	Similarity degree	Comments
1	(i) P1-B (ii) P3-B (top left) (iii) P3-B (bottom right) (iv) P4-B	(0.83, 0, 0)	Same algorithm (bubble), implemented by different teams. Upon manual inspection, the implementations indeed turned out to be very similar in terms of behaviors
2	(i) P1-B (ii) P3-Q (iii) P4-M (iv) P4-Q	(0.17, 0.66, 0)	Different algorithms implemented by different teams. Yet P1-B and P3-Q are very similar in terms of behaviors, and P4-M and P4-Q refine them. These were confirmed in manual inspection

Let us assume that after carefully analyzing the specific effort required for transforming the different artifacts to core assets (based on the similarity graph, but also after inspecting the code), the analyst selects the two assets shown in Table 5. With respect to this selection, the tool can be used for supporting decision making in the bottom-up dimension, based on product-specific variability degrees. While none of the products is fully covered by this selection, P2 is completely unrelated (its product-related variability degree is (0, 0, 0, 1), meaning that all classes are specific to this product). P1, P3, and P4, on the other hand, result in higher product-related variability degrees of (0.33, 0.33, 0, 0.66), (0.5, 0.17, 0, 0.5) and (0.33, 0.66, 0, 0), respectively. This implies that it is better to discard P2 from the product line, but P1, P3, and P4 can be used to extract core assets, after performing the needed adaptations to the original artifacts.

6   Summary and Future Directions

Deciding whether to turn existing systems or products into a product line – referred to as product line-ability – is an important practical problem which requires measuring and analyzing similarities and differences among systems. The main challenge here is that detecting similarities in the context of systems developed by different teams and for different purposes is notoriously hard as these software artifacts may differ in their implementation, while still behaving similarly.

In this paper, we propose a framework for the identification of "similarly behaving" artifacts and calculation of their degree of similarity and variability. This framework provides in addition a method for analyzing the product line-ability of a given set of products both in bottom-up and top-down directions. We describe how the proposed framework has been integrated into the VarMeR tool to support product line-ability decisions on Java artifacts. The paper also demonstrates how such decisions can be taken using the framework and its method. While VarMeR is a tool oriented towards analysis of Java artifacts, the presented approach is general, and we plan to extend it to other types of artifacts such as various types of models and textual requirements.

A mandatory future research direction is further validation of the proposed framework by investigation of its potential to support reuse decisions of developers. We intend to empirically evaluate the outcomes of the tool-supported method in the context of academic software engineering courses and professional workshops. Another direction is extending the framework to support more types of reuse relations, including analogy and aggregation, but also cases in which the shallow behaviors are different. Finally, other product line-ability metrics for the bottom-up and top-down dimensions can be explored, drawing further inspiration from metrics proposed in [4].

Acknowledgments. The authors thank Jonathan Liberman and Shai Gutman for their help in the implementation of the VarMeR tool. We also thank Alex Kogan and Asaf Mor for their assistance in the initial steps of the development. The second author was supported by the Israel Science Foundation under grant agreement 817/15.

References

1. Assunção, W.K., Lopez-Herrejon, R.E., Linsbauer, L., Vergilio, S.R., Egyed, A.: Reengineering legacy applications into software product lines: a systematic mapping. Empir. Softw. Eng. **22**, 2972–3016 (2017)
2. Baker, B.S.: Finding clones with dup: analysis of an experiment. IEEE Trans. Softw. Eng. **33** (9), 608–621 (2007)
3. Bellon, S., Koschke, R., Antoniol, G., Krinke, J., Merlo, E.: Comparison and evaluation of clone detection tools. IEEE Trans. Softw. Eng. **33**(9), 577–591 (2007)
4. Berger, C., Rendel, H., Rumpe, B.: Measuring the ability to form a product line from existing products. In: Proceedings of the Fourth International Workshop on Variability Modelling of Software-Intensive Systems (VaMoS) ((2014))
5. Berger, C., Rendel, H., Rumpe, B., Busse, C., Jablonski, T., Wolf, F.: Product line metrics for legacy software in practice. arXiv preprint arXiv:1409.6581 (2014)
6. Berger, T., Rublack, R., Nair, D., Atlee, J.M., Becker, M., Czarnecki, K., Wąsowski, A.: A survey of variability modeling in industrial practice. In: Proceedings of the Seventh International Workshop on Variability Modelling of Software-intensive Systems (VAMOS), p. 7 (2013)
7. Clements, P., Northrop, L.: Software Product Lines. Addison-Wesley, Boston (2002)
8. Dincel, E., Medvidovic, N., van der Hoek, A.: Measuring product line architectures. In: van der Linden, F. (ed.) PFE 2001. LNCS, vol. 2290, pp. 346–352. Springer, Heidelberg (2002). https://doi.org/10.1007/3-540-47833-7_31
9. Gruler, A., Leucker, M., Scheidemann, K.: Calculating and modeling common parts of software product lines. In: Software Product Line Conference (SPLC 2008), pp. 203–212 (2008)
10. Kashyap, V., Sheth, A.: Semantic and schematic similarities between database objects: a context-based approach. VLDB J. Int. J. Very Large Data Bases **5**(4), 276–304 (1996)
11. Mihalcea, R., Corley, C., Strapparava, C.: Corpus-based and knowledge-based measures of text semantic similarity. In: American Association for Artificial Intelligence (AAAI 2006), pp. 775–780 (2006)
12. Pohl, K., Böckle, G., van Der Linden, F.J.: Software Product Line Engineering: Foundations, Principles and Technique. Springer, Heidelberg (2005). https://doi.org/10.1007/3-540-28901-1
13. Reinhartz-Berger, I., Zamansky, A., Kemelman, M.: Analyzing variability of cloned artifacts: formal framework and its application to requirements. In: Gaaloul, K., Schmidt, R., Nurcan, S., Guerreiro, S., Ma, Q. (eds.) CAISE 2015. LNBIP, vol. 214, pp. 311–325. Springer, Cham (2015). https://doi.org/10.1007/978-3-319-19237-6_20
14. Reinhartz-Berger, I., Zamansky, A., Wand, Y.: Taming software variability: ontological foundations of variability mechanisms. In: Johannesson, P., Lee, M.L., Liddle, S.W., Opdahl, A.L., López, Ó.P. (eds.) ER 2015. LNCS, vol. 9381, pp. 399–406. Springer, Cham (2015). https://doi.org/10.1007/978-3-319-25264-3_29
15. Reinhartz-Berger, I., Zamansky, A., Wand, Y.: An ontological approach for identifying software variants: specialization and template instantiation. In: Comyn-Wattiau, I., Tanaka, K., Song, I.-Y., Yamamoto, S., Saeki, M. (eds.) ER 2016. LNCS, vol. 9974, pp. 98–112. Springer, Cham (2016). https://doi.org/10.1007/978-3-319-46397-1_8
16. Reinhartz-Berger, I., Zamansky, A.: VarMeR - A Variability Mechanisms Recommender for Software Artifacts. CAiSE-Forum-DC, pp. 57–64 (2017)

17. Roy, C.K., Cordy, J.R., Koschke, R.: Comparison and evaluation of code clone detection techniques and tools: a qualitative approach. Sci. Comput. Program. **74**(7), 470–495 (2009)
18. Torkamani, M.A.: Metric suite to evaluate reusability of software product line. Int. J. Electr. Comput. Eng. **4**(2), 285 (2014)
19. Zhang, T., Deng, L., Wu, J., Zhou, Q., Ma, C.: Some metrics for accessing quality of product line architecture. In: IEEE International Conference on Computer Science and Software Engineering, vol. 2, pp. 500–503 (2008)

Data-Driven Elicitation, Assessment and Documentation of Quality Requirements in Agile Software Development

Xavier Franch[1(✉)], Cristina Gómez[1], Andreas Jedlitschka[2],
Lidia López[1], Silverio Martínez-Fernández[2], Marc Oriol[1],
and Jari Partanen[3]

[1] Universitat Politècnica de Catalunya (UPC), Barcelona, Spain
{franch, cristina, llopez, moriol}@essi.upc.edu
[2] Fraunhofer IESE, Kaiserslautern, Germany
{andreas.jedlitschka,
silverio.martinez}@iese.fraunhofer.de
[3] Bittium Wireless Ltd., Oulu, Finland
jari.partanen@bittium.com

Abstract. Quality Requirements (QRs) are difficult to manage in agile software development. Given the pressure to deploy fast, quality concerns are often sacrificed for the sake of richer functionality. Besides, artefacts as user stories are not particularly well-suited for representing QRs. In this exploratory paper, we envisage a data-driven method, called Q-Rapids, to QR elicitation, assessment and documentation in agile software development. Q-Rapids proposes: (1) The collection and analysis of design and runtime data in order to raise quality alerts; (2) The suggestion of candidate QRs to address these alerts; (3) A strategic analysis of the impact of such requirements by visualizing their effect on a set of indicators rendered in a dashboard; (4) The documentation of the requirements (if finally accepted) in the backlog. The approach is illustrated with scenarios evaluated through a questionnaire by experts from a telecom company.

Keywords: Quality requirement · NFR · Agile software development

1 Introduction

Software quality is an essential competitive factor for the success of IT companies today [1]. Recent technological breakthroughs as cloud technologies, IoT and 5G, pose new quality challenges in software development. These challenges include quality aspects such as availability, reliability, security, performance, and scalability, which significantly influence the success of current and future software systems. Many well-known cases, which are worth millions of euros in losses, were due to bad quality or defective software (see an example in [2]).

This work is a result of the Q-Rapids project, which has received funding from the European Union's Horizon 2020 research and innovation programme under grant agreement No 732253.

© Springer International Publishing AG, part of Springer Nature 2018
J. Krogstie and H. A. Reijers (Eds.): CAiSE 2018, LNCS 10816, pp. 587–602, 2018.
https://doi.org/10.1007/978-3-319-91563-0_36

Optimal management of software quality demands the appropriate integration of quality requirements into the software life cycle. Quality Requirements (QRs; also known as non-functional requirements) are the artefact that software engineers use to state conditions on, and analyse compliance of, software quality [3]. QRs are as generic as "The software service shall provide optimal response time" (in fact, more of a goal than a requirement) or as detailed as "Once the user completes the request, logging into the system should not take more than 3 s 95% of the time". This level of detail will normally depend on the development stage. However, despite the competitive advantage of ensuring and maintaining high quality levels, software development methodologies still provide limited support to QR management [4].

One example is agile software development (ASD). Quality is essential in rapid software development processes happening in ASD: faster and more frequent release cycles should not compromise software quality. Still, a lack of mechanisms to support continuous quality assurance exists throughout the whole development process in the context of rapid releases [5]. Empirical studies conducted in the last few years in ASD have identified as a major challenge the consideration of QRs [6] whose deficient management was recently qualified as a reason for massive lapses and rework [7].

Given this situation, the present work addresses the following research goal:

> **Research goal : To formulate a tooled method supporting the effective management of QRs in agile software development.**

We decompose this goal into four research questions, see Table 1. As a first step for managing QRs, we need to detect that some quality concern is not well covered in the current release of the system (RQ1). We adopt a data-driven approach in which we gather and analyse data from different sources to detect the need to improve any quality concern currently compromised. This identified need requires to be expressed in the form of one or more QRs (RQ2). We follow here a reuse-based approach in which a catalogue of QR patterns compiles the usual forms in which a QR can be specified. The implications of such QRs over the software system need to be assessed (RQ3) so that decision-makers may decide to accept or discard the QR. With this goal, we use a strategic dashboard that visualizes the effect of accepting the QR over several strategic indicators like product quality, customer satisfaction and team productivity. In case that the QR is finally accepted to be part of the system specification, it needs to be added to the backlog (RQ4). The way in which it is added may change from an epic, a user story or an acceptance condition.

Table 1. Research questions

Id	Question
RQ1	When is it necessary to improve some quality concern of a software system?
RQ2	How can this need be expressed as a quality requirement?
RQ3	How can the consequences of undertaking such quality requirement be assessed?
RQ4	In case that a quality requirement is accepted, how can it be added to the backlog?

As research method, we adopt a design science approach following the engineering cycle as described by Wieringa [8]. The research has been conducted in the context of the Q-Rapids H2020 project (www.q-rapids.eu) which has given us the opportunity to elicit real scenarios and evaluate the results in the context of a number of company-provided pilot cases.

2 State of the Art

Current approaches to manage QRs are usually inadequate in complex scenarios with highly-dynamic environments composed of fast-growing software systems. One the most unresolved challenges concerns their elicitation [9]. Techniques to elicit QRs include structured and unstructured interviews, quality models, checklists, and prioritization questionnaires, among others. These techniques do not exploit runtime data. Recently, data-driven requirements engineering has been advocated as the way to go [10]. In this area, modern approaches like crowdsourcing [11] or mining app stores [12] rely on feedback given by the user. But explicit feedback can be incomplete, biased or ambiguous. This contributes to the conceptual gap between user needs and the interpretation done by the development team of such needs. To overcome such communication problems, some approaches like Shekhovtsov et al. [13] propose a tool that aims at improving the communication in the software process development. In particular, their approach aims at identifying similar information communicated in the past, analyse its properties and, by using machine learning techniques, suggest communication-related decisions that would mitigate communication problems.

A different approach that is not affected by these limitations is implicit feedback (usage data), which has been identified as a promising additional input for requirements elicitation [9]. Existing approaches to implicit feedback [14] neither aim at generating QRs, nor consider the possibility to correlate usage data with data gathered from software repositories or project management tools. Our positioning is that these correlations may be used to uncover relevant observations in the management of QRs.

Another problem with QRs is their refinement in terms of satisfaction criteria. Differently than functional requirements that have clear cut satisfaction criteria, QRs are initially elicited as "soft goals" [15] and need to be elaborated into measurable conditions. Evidence exists that the knowledge needed to determine these values is closer to the development team than to the customer, even to the extreme in which the end-user is not really involved in such process [16]. Some approaches (e.g. QUPER [17]) present techniques for such elicitation, but they are all based in stakeholder explicit involvement before the product is released. These techniques do not try to inject data once the project is in use in order to understand the feasibility and appropriateness of the stated thresholds. We envisage the wide use of different data sources and then complete a fully empirical-based, data-driven approach to QR elicitation.

Another aspect that needs to be emphasised is the use of QRs in decision making. Usually, quality aspects are surveyed by tools used by software developers. Current project management systems deal with scheduling, resource and budget management, etc., but they do not explicitly include support for QR management [18]. Nevertheless, some tools exist to support some activities on quality requirements. For instance, tools to support software architects and developers on decision making that may include quality aspects (e.g. WISED); tools intended to monitor the productivity of developers when implementing such QRs (e.g. JIRA); and tools to evaluate the technical debt based on software quality assessment (e.g. SQALE). Although they may support QRs to a certain extent, their management is not well integrated with the entire life cycle of the software process. Furthermore, they do not provide full support to the assessment of key prioritization criteria for QRs: costs, benefits and risks [19]. For instance, it is reported that they fail on considering risk factors [20]. We envisage the need to provide full support to strategic decision making based on the analysis of the impact of QRs in such key criteria, reconciling: (1) QR-specific universes of discourse with definition of the concepts that matter in their management and assessment [17], (2) reasoning frameworks as the NFR framework [15], (3) a highly informative dashboard that provides full transparency to the decision making process [21].

Last, existing requirements engineering practices fail short regarding QR documentation in ASD. Instead, its reliance on the continuous interaction with customers is thought to minimize the need for specifying QRs [22]. Another reason is that agile developers face challenges while using user stories for documenting NFRs such as security [23]. A recent industrial study reported QR documentation practices ranging from scarce documentation to documentation using agile assets (user stories, DoD,...) and out-of-the-loop artefacts as wiki pages [24]. Other recent works propose quality criteria [25] and traceability [26] to improve QR documentation in ASD.

In summary, we advocate that the current state of the art calls for improved methods for the elicitation, assessment and documentation of QRs especially in ASD.

3 The Q-Rapids Method

Q-Rapids is a data-driven, quality-aware ASD tooled method in which QRs are identified from available data and evaluated with respect to some selected strategic indicators [27]. Q-Rapids aims at increasing software quality through (see Fig. 1):

- Gathering and analyzing data from project management tools, software repositories, quality of service and system usage. Data analysis will permit to systematically and continuously assess quality and eventually suggest QRs.
- Providing decision makers with a highly informative dashboard to help them making data-driven, requirements-related strategic decisions. The dashboard will aggregate the collected data into strategic indicators related to factors as time to market, team productivity, customer satisfaction, and overall product quality.
- Extending the ASD process considering the comprehensive integration of QRs and functional requirements in the product backlog.

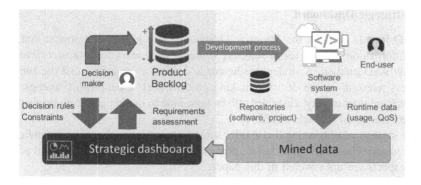

Fig. 1. The Q-Rapids method to QR management in ASD

4   Conceptual Architecture of the Q-Rapids Method

The Q-Rapids method relies on several logical components that together conform a conceptual solution to the data-driven QR management problem. In this section, we present the main components of such solution, put together as shown in Fig. 2.

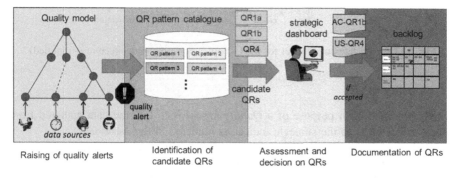

Fig. 2. Main logical components of the Q-Rapids conceptual architecture

These components are rigorously described using a domain reference model expressed as a UML class diagram (see Fig. 3; the diagram also includes concepts presented later on) that is an extension of the decision-making ontology presented in [28]. The core concept of the reference model is the QR concept. We adopt an ontological interpretation of QR based on qualities in foundational ontologies [29] as done in [30]. A quality is defined as a basic perceivable or measurable characteristic that inheres in and existentially depends on its subject [29]. So, a QR is a need expressed by the stakeholders about a quality. Class and attribute names appear in italics when referenced in this section. A definition of all the concepts appearing in the reference model may be found in https://tinyurl.com/QRrefmod.

4.1 Strategic Dashboard

In the Q-Rapids method, the *Strategic Dashboard* is a logical component that visualizes information about a set of selected *Strategic Indicators* for a project developed under an ASD method. A strategic indicator is defined as an aspect that the company considers relevant for the decision-making process [28]. Examples of strategic indicators are: product quality, customer satisfaction, time to market and team productivity. They are measured through a *KPI*.

The strategic dashboard will also offer some techniques for analysing the indicators, e.g. trend and what-if analysis, failure prediction and suggestion of mitigation actions. These aspects are not covered in this paper.

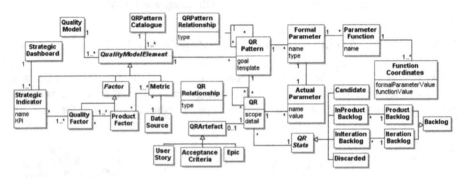

Fig. 3. Domain reference model for the Q-Rapids conceptual architecture (simplified)

4.2 Quality Model

In Q-Rapids, the main purpose of a *Quality Model* is to link the data gathered from some *Data Sources* to the strategic indicators rendered in the dashboard.

We have adopted the Quamoco approach [31] to define Q-Rapids quality models. In Quamoco, *Metrics* gather data from data sources using some software data collectors and are elaborated into *Product Factors* and ultimately aggregated into *Quality Factors*. We distinguish four types of metrics related to categories of data sources:

- Development metrics. Gathering data from software development repositories. E.g., test coverage gathers data from SonarQube using the SQALE plugin.
- Project management metrics. Gathering data from project management tools. For instance, personnel availability from Jira and MS Project.
- Feedback metrics. Gathering data from system users, either directly through given feedback, or indirectly through usage monitoring. For instance, the number of tweets referring to a particular (system-related) hashtag.
- Quality of service metrics. Gathering data from system behaviour through monitoring components. For instance, response time of a given system functionality.

4.3   Requirement Patterns

We envisage the use in Q-Rapids of a catalogue of *QR Patterns* that act as templates for deriving the concrete QRs. This *QR Pattern Catalogue* will be based on the PABRE approach [32]. Its structure complies the following principles:

- Its natural language *template* includes *Formal Parameters* that may have one or more *Parameter Function* defined as a set of *Function Coordinates*.
- QR patterns may present relationships (*QR Pattern Relationship*) of different *types* (e.g., conflict, dependency, synergy, …).
- QR patterns are bound to *Quality Model Elements*. This binding is fundamental in order to select the appropriate QR patterns once a quality alert is issued.

A QR pattern is instantiated into a *QR* by binding *Actual Parameters* to formal parameters. QRs preserve the relationships (*QR Relationship*) established at the level of the patterns they are an instantiation of.

5   The Q-Rapids Process

We present below the four phases of the Q-Rapids process already depicted in Fig. 2.

5.1   Raising of Quality Alerts

In the first phase of the Q-Rapids process, we propose the analysis of data gathered from several sources as to raise quality alerts when some quality factor yields an evaluation outside the acceptable thresholds.

In Fig. 4, we exemplify the hierarchy of a quality model already described Fig. 3. We adopt the Quamoco approach [31], which proposes weighted sums to implement this aggregation. Albeit simple, this makes explicit the knowledge about quality, and allows transparency of quality-related decisions.

In Fig. 4, the *Cyclomatic Complexity* and *Comments Density* metrics are aggregated into the *Analyzability* product factor. This product factor is aggregated together with another one, *Adaptability*, into the ISO 25010 *Maintainability* quality factor. Each product factor should have a specified weight.

Once the quality model is built, the next step is to apply utility functions as the way to embody knowledge on what is an acceptable level of quality. Utility functions model the preferences of decision makers with respect to the values of the factors, so that they can define an alert to be triggered by Q-Rapids when a basic metric or a factor has exceeded a defined threshold. In Fig. 4, we can see examples of utility functions for *Cyclomatic Complexity*, *Comments Density* and *Maintainability*. In a normalized value from 0 to 1, we can see that, in the current state of the system, *Maintainability* has the value of 0.44. If the goal had been set to have maintainability at its best value (this means to be greater than 0.92, as defined in the utility function), the corresponding alert would be triggered to inform that maintainability of the product should be improved. Quality alerts are how Q-Rapids identifies the need for new QRs.

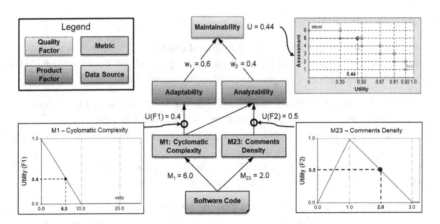

Fig. 4. Example of utility functions to assess the Maintainability quality factor –adapted from Quamoco [31]

5.2   Identification of Candidate Quality Requirements

Quality alerts are bound to elements from the quality model. As the reference model in Fig. 3 shows, QR patterns in the catalogue are bound to such elements, therefore Q-Rapids proposes the use of this binding to select the QR patterns whose instantiation will produce candidate QRs.

Figure 5 shows an excerpt of a QR pattern that will be used as example in Sect. 6. We can see how the link with a factor, the *Availability* product factor (subfactor of the ISO 25010 *Reliability* quality factor), is explicitly written. The template is expressed in natural language and shows two parameters: the *entity* (the whole system, a service, a component, ...) and the required *availabilityLevel* (a percentage). A cost function is attached to one of its parameters, measuring the relative cost of implementing the required availability level. This function will be defined as a set of pairs (parameter value, function value) (see Fig. 3).

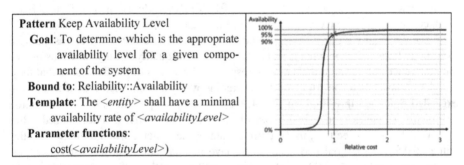

Fig. 5. Requirement pattern Keep Availability Level (left) and cost function of one its parameters (right; graphical representation)

Q-Rapids will select all the patterns bound to the factor for which the alert was triggered and for each, will propose one or more instantiations to be rendered to the decision maker. Having one or more instantiations will depend on the knowledge available to propose different alternative values to the formal parameters: these values should be determined smartly in terms of the utility function of the factor, the effect on the strategic indicators and the available parameter functions.

5.3   Assessment of, and Decision on, Candidate Quality Requirements

Once a candidate QR is derived from the pattern catalogue, it will be assessed by the decision-maker using the dashboard. To assess the impact of the candidate QRs on the indicators, we plan to use Bayesian networks as proposed in the VALUE framework [33]. Models are to be built with a mix of expert knowledge and data, so that they will evolve as the system learns from decisions made. In addition to the indicators, the dashboard will show those other functions bound to QR (coming from the corresponding QR pattern, see Fig. 5), e.g. implementation cost, so that it will be possible to visualize the fluctuation of these functions if the candidate QR were selected.

As an outcome of this assessment, several actions are possible onto the candidate QR (see state diagram for candidate QRs at Fig. 6):

- *Consider in the current iteration*. The QR is added into the current iteration backlog and correctly documented (see next subsection). This action will be applied only in the case of detecting a critical quality concern violation (e.g., resulting in massive negative feedback from users in a short period of time).
- *Consider in future iterations*. Usual action when the QR is accepted but it is moved to the product backlog waiting for its future inclusion in the iteration backlog. As in the previous action, the QR should be correctly documented.
- *Postpone decision*. In this case, the QR is considered plausible but no action is taken yet. The QR is included in the product backlog but with the minimal possible metadata (e.g., if it a user story, its effort is not computed) and it may be removed in the future or even reconsidered using the same casuistic described here.
- *Negotiate*. A QR that is considered necessary may not be possible to just accept, but requires joint analysis with conflicting requirements, eventually calling for a negotiation among involved stakeholders. The QR will not be included in any backlog until the dependencies are managed, and eventually it may end up slightly modified or even just rejected.
- *Detail*. Since the QRs are derived from a pattern catalogue, it may eventually happen that a template misses some detail that is necessary in the current situation or even because Q-Rapids has not access to enough knowledge as to generate values for the formal parameters. In this case, the QR is not ready to be included in the backlog and further refinement is requested.
- *Rephrase*. It may happen that the QR is not representing the quality concern in a valid way. This action asks for an alternative formulation. In this action and the previous one, the pattern catalogue needs to be examined to decide whether it can be improved with the knowledge stemming from this candidate QR.

- *Discard.* The QR analysis using the dashboard concludes that its acceptance would be harmful because some strategic indicators would be seriously damaged.
- *Redefine quality thresholds.* Candidate QRs are proposed considering utility functions and thresholds defined by decision makers for the factors. This action is to be taken when the analysis of the QR points out that the identified quality misbehaviours is in fact not such. Therefore, it is necessary changing the utility functions and thresholds in the QM so that the reason for triggering a QR disappears.

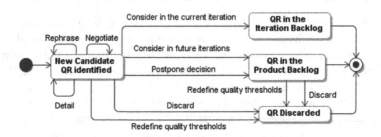

Fig. 6. State diagram of candidate QRs

5.4   Representation of Quality Requirements

In case the candidate QR is accepted as a valid requirement for the system, it needs to be correctly included in the backlog (either iteration or product backlog). As stated in Sect. 2, the documentation of QRs in ASD is a well-known challenge [7, 34].

In [24], we analysed different agile artefacts to represent QRs in ASD: user stories, epics and acceptance criteria. From a practical perspective, the type of artefact impacts on the information that is required to determine when entering the QR into the backlog. For instance, if it is represented as a user story, it will require fixing its priority (according to its business value and effort required), determining its acceptance criteria, etc.

6   Example Scenarios

The following scenarios are inspired on real-based challenges and problems the companies in the Q-Rapids project's consortium face. The scenarios aim at illustrating situations in which a company using ASD would eventually benefit from using Q-Rapids. They are presented with increasing ambition. In the first scenario, the quality alert is not automatically triggered by Q-Rapids but it is the quality expert who raises it through inspection of the dashboard. In the second scenario, the process is automatized. In the third scenario, in order to identify the root cause of the quality alert, two sources of data need to be correlated. Figure 7 shows the indicators, factors and metrics used.

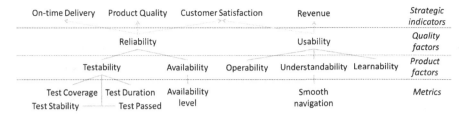

| On-time Delivery | Product Quality | Customer Satisfaction | Revenue | Strategic indicators |

Reliability · Usability — Quality factors

Testability · Availability · Operability · Understandability · Learnability — Product factors

Test Coverage · Test Duration · Availability level · Smooth navigation — Metrics
Test Stability — Test Passed

Fig. 7. Strategic indicators and quality model elements used in the three scenarios

6.1 Scenario 1: Test Coverage

Context. ACME is a software-intensive company continually evolving its software products, with rapid release environment. Currently, they are releasing software every four weeks. To improve time-to-market, the company decides reducing the release period from four to three weeks. They enter the new release time in the Q-Rapids tool.

Raising of quality alerts. All elements of the ACME quality model are regularly monitored, assessed and visualized through the dashboard. Using the Q-Rapids' dashboard, release engineers see the impact of the release decision on the quality model assessment: *On-time delivery* strategic indicator is positively impacted whereas *Product Quality* is negatively impacted, even under the targeted threshold.

Identification of candidate QRs. By browsing patterns in the catalogue related to factors that influence *Product Quality*, Q-Rapids finds *Keep High Coverage Tests*, linked to *Test Coverage*, whose template is: The *<entity>* shall have a test coverage not lower than *<percentage>*. The parameter *<percentage>* is key for assessing the benefit of executing a test. Given his domain knowledge, the engineer predicts a solution to reduce testing time without dramatically decreasing product quality: the prioritization of test cases with greater coverage that cover a greater number of potential bugs [35].

Assessment of, and decision on, candidate QRs. With the support of the platform, the engineer generates the QR from the pattern: The *prioritized tests* shall have a minimal condition coverage rate of 70%. The engineer can visualize the effect of this QR using the drill down feature of the dashboard, allowing to see prioritized tests and their characteristics (i.e., Test Coverage, Test Duration, …). Release engineers can use this information to postpone the execution of tests with lower or redundant test coverage.

Representation of the QR. Three QRs may be inserted into the backlog. First, an issue indicating the reduced set tests to be executed by release engineers because they potentially cover more bugs. Second, the improvement of tests with low quality levels. Third, the test quality product factor also shows the untested code. As Martin Fowler's argue: "It's worth running coverage tools every so often and looking at these bits of untested code" (https://martinfowler.com/bliki/TestCoverage.html). Therefore, another QR should be inserted: manually check the parts of code that are untested.

6.2   Scenario 2: Availability

Context. ACME provides regular over-the-air (OTA) updates of the software installed in their smartphones. However, due to scalability problems, the service that provides such updates, the *update service*, may experience availability issues. This is especially true when there is a critical update and multiple devices try to download such update at the same time. These problems may lead to customer dissatisfaction. The availability rate of the update service is monitored through a data collector that regularly checks if the service is up-and-running. Unfortunately, the current implementation of the system has not considered a QR related to availability specific for this functionality; just a system-wide availability requirement exists, used to design the utility function.

Raising of quality alerts. In this context, ACME releases a critical OTA update, and the data collector detects that the *update service* availability goes down from 95% to 60%. This value is checked against the quality model, which links this gathered data to the quality factors and strategic indicators. To do so, the availability is normalized using the utility function: availability of 60% corresponds to a utility value of 0.35. Such utility value is used to compute the quality factor *Reliability*, and ultimately the strategic indicator *Customer Satisfaction*, whose value falls below the required threshold.

Identification of candidate QRs. Due to the indicator's violation, Q-Rapids browses the pattern catalogue and finds the *Keep Availability Level* pattern (see Fig. 5) linked to the factor whose utility function is failing. In this case, the parameter <*system*> is instantiated as *update service*. For the instantiation of <*availabilityLevel*>, Q-Rapids uses the cost information embodied in the pattern and selects 3 alternatives, 99%, 90% and 80%, which yields to 3 different candidate requirements, one for each value.

Assessment of, and decision on, candidate QRs. The Q-Rapids dashboard offers the three QRs so that the decision-maker can analyse the consequences of each of them by inspecting the fluctuations of the *Customer Satisfaction* strategic indicator value. The decision maker can see that availability of 80% reaches adequate results in terms of *Customer Satisfaction* and cost. Availability of 90% improves significantly the *Customer Satisfaction* with a minimum increase of cost compared to the first option. The last option (availability of 99%) provides a *Customer Satisfaction* only slightly better than the previous option but doubling the cost. After considering these facts, the decision maker selects the QR with 90% availability rate and discards the other two.

Representation of the QR. Q-Rapids considers that the *update service* was implemented in response to a user story identified in some past iteration and marked as "done": "As a user, I want to download OTA updates so that the smartphone is secure". The QR refers to this user story and no other functionality, therefore Q-Rapids adds the QR as new acceptance criteria. By adding a new (unsatisfied) acceptance criteria, such–user story is ready to go back to a backlog. Given the importance to fix quickly this problem, Q-Rapids suggest to include it in the current iteration backlog.

6.3   Scenario 3: Usability

Context. ACME has just released a major update on the user interface of an *e-commerce platform* used to sell their mobile phones and they expect that this update will improve the user experience and hence increase the number of sales.

Raising of quality alerts. In this context, the data collector detects a significant amount of users who get stuck in a particular step of the process and quit without buying the phone. The data collector informs that the *Smooth Navigation* metric (which measures how many users can complete the process of buying the phone smoothly) goes from 90% down to 80%. This metric is normalized using the utility function, yielding to a utility value of 0.45, which impacts negatively on the upper layers in *Understandability*, *Usability*, and finally, *Revenue*. Because of this process, the value of the strategic indicator *Revenue* goes below the required threshold.

Identification of candidate QRs. A drop in the utility value of *Smooth Navigation* causes the violation of the *Revenue* indicator. However, this information does not suffice to identify a candidate QR. This is because a drop in *Smooth Navigation* could be caused by several factors (e.g., is there a problem related to the browser of the user? Or is it related to an issue in the internet connection?). In this scenario, the violation of the indicator triggers Q-Rapids to correlate *Smooth Navigation* with other metrics to detect the cause of the problem. After inspecting different metrics from the multiple data sources under monitoring, Q-Rapids detects a strong correlation between *Smooth Navigation* and the user device type. In particular, it detects that the device of those users who present that problem is a smartphone. Q-Rapids browses the pattern catalogue and finds the *Interface Type* pattern: The *<system>* shall be supported by *<device-types>* devices, which is linked to the *Understandability* factor. The instantiation yields to the candidate QR: The e-commerce platform shall be supported by smartphone devices.

Assessment of, and decision on, candidate QRs. The Q-Rapids dashboard offers the QR so that the decision-maker can analyse the consequences of applying it by inspecting the fluctuations of *Revenue* and other impacted indicators, e.g. *On-Time delivery*. Q-Rapids shows that considering the QR would return the utility value of *Smooth Navigation* close to 1, and hence a very good *Revenue*. *On-Time delivery* would worsen but, by computing different metrics, Q-Rapids demonstrates that not significantly. After considering these implications, the decision-maker decides to apply the QR.

Representation of the QR. Since the QR is asking for a well-defined implementation task, Q-Rapids suggests a new user story: "As a user, I want to be able to use the e-commerce platform using my smartphone in order to ensure user's experience for this kind of device". Finally, Q-Rapids adds the user story to the product backlog, ready to be scheduled for the next iteration.

7   Evaluation

In order to get early feedback from our exploratory ideas, we ran a questionnaire in one of the companies of the Q-Rapids consortium, Bittium Wireless Ltd. We simplified the TAM evaluation questionnaire [36] since the technology is not available; therefore, we asked them to evaluate their vision on the scenarios, using the questions in Table 2, scored from 1 (strongly disagree) to 7 (strongly agree). We got 3 responses.

All the responses, without exceptions, score positively. In addition, we got free text for most responses. Among the feedback, we can mention: (1) acknowledgment that a lot of knowledge needs to be managed inside Q-Rapids ("When thinking about all possible scenario in real life, quite a "machine" is needed!"); (2) statement that even if the system behaves well, decision-makers want to have the opportunity to confirm themselves the recommendations by "digging deeper in the data manually"; (3) Scenario 3 is really valuable ("the third scenario is really genuine, if it works").

Table 2. Questions to practitioners for evaluating Q-Rapids

Criteria	Question	R1	R2	R3
Perceived usefulness	Using Q-Rapids for eliciting, deciding upon and documenting QRs as described in the scenarios above, could be useful for my job	6	5	7
Perceived ease of use	I find easy using Q-Rapids for eliciting, deciding upon and documenting QRs as described in the scenarios above	6	6	7
Output quality	The output given by Q-Rapids as described in the scenarios above, is high	7	6	7
Result demonstrability	It would be easy for me telling others about the results of using Q-Rapids as described in the scenarios above	5	6	7
Feasibility	I find that the scenarios described above can be implemented in my work environment	7	5	6

8   Final Discussion

In this exploratory paper, we have presented a data-driven method, Q-Rapids, for improving QR management in ASD. Q-Rapids argues for the intensive use of data to detect quality misbehaviours, then to use the knowledge embedded in the method to propose candidate QRs, to visualize the consequences of their acceptance in a dashboard and, if the decision-maker accepts them, to document them in a backlog.

We are aware that the main goal of the method is very ambitious and can be the case that not all QRs can be detected as proposed in Q-Rapids; one of the main results of the method evaluation will be to understand its limitations and identify which QR types are the best suited for this data-driven analysis. The Scenario 1 has tried to illustrate this eventual need of putting the human in the loop, which in fact has been one of the points highlighted in the preliminary evaluation presented in Sect. 7. Other threats also exist: e.g., privacy issues are always delicate in data-driven approaches.

Q-Rapids has been conceived to focus on QRs. However, as the project progresses, we see that the main concepts could eventually be used both for suggesting new features and for dealing not only with product factors, but process factors (e.g., factors affecting team performance). This can become a further goal in later stages of our research. Also, the implementation, validation and a complete evaluation of the proposed method are planned for the near future. Last, the adoption of Q-Rapids beyond ASD (e.g., in organizations adopting DevOps) is part of our research agenda.

References

1. Capgemini: World Quality Report 2015-16, 7th edn. https://www.capgemini.com/thoughtleadership/world-quality-report-2015-16
2. Reuters (2015). http://www.reuters.com/article/us-hsbc-it-idUSKBN0UJ0ZB20160105
3. Pohl, K.: Requirements Engineering: Fundamentals, Principles and Techniques (2010)
4. Wagner, S.: Software Product Quality Control. Springer, Heidelberg (2013). https://doi.org/10.1007/978-3-642-38571-1
5. Rodríguez, P., et al.: Continuous deployment of software intensive products and services: a systematic mapping study. J. Syst. Softw. **123**, 263–291 (2017)
6. Ramesh, B., Baskerville, R., Cao, L.: Agile requirements engineering practices and challenges: an empirical study. Inf. Syst. J. **20**(5), 449–480 (2010)
7. Inayat, I., et al.: A systematic literature review on agile requirements engineering practices and challenges. Comput. Hum. Behav. **51**(B), 915–929 (2014)
8. Wieringa, R.J.: Design Science Methodology for Information Systems and Software Engineering. Springer, Heidelberg (2014). https://doi.org/10.1007/978-3-662-43839-8
9. Berntsson-Svensson, R., Host, M., Regnell, B.: Managing quality requirements: a systematic review. In: SEAA (2010)
10. Maalej, M., Nayebi, M., Johann, T., Ruhe, G.: Toward data-driven requirements engineering. IEEE Softw. **33**(1), 48–54 (2016)
11. Groen, E.C., et al.: The crowd in requirements engineering: the landscape and challenges. IEEE Softw. **34**(2), 44–52 (2017)
12. Kurtanovic, Z., Maalej, W.: Mining user rationale from software reviews. In: RE 2017 (2017)
13. Shekhovtsov, V.A., Mayr, H.C., Kucko, M.: Implementing tool support for analyzing stakeholder communications in software development. In: ICSTW 2015 (2015)
14. Liu, X., et al.: Deriving user preferences of mobile apps from their management activities. ACM Trans. Inf. Syst. **35**(4), 39 (2017)
15. Chung, L., Nixon, B.A., Yu, E., Mylopoulos, J.: Non-Functional Requirements in Software Engineering. Springer, New York (2000). https://doi.org/10.1007/978-1-4615-5269-7
16. Ameller, D., Ayala, C.P., Cabot, J., Franch, X.: Non-functional requirements in architectural decision making. IEEE Softw. **30**(2), 61–67 (2013)
17. Berntsson Svensson, R., Regnell, B.: A case study evaluation of the guideline-supported QUPER model for elicitation of quality requirements. In: Fricker, S.A., Schneider, K. (eds.) REFSQ 2015. LNCS, vol. 9013, pp. 230–246. Springer, Cham (2015). https://doi.org/10.1007/978-3-319-16101-3_15
18. Caracciolo, A., Lungu, M.F., Nierstrasz, O.: How do software architects specify and validate quality requirements? In: Avgeriou, P., Zdun, U. (eds.) ECSA 2014. LNCS, vol. 8627, pp. 374–389. Springer, Cham (2014). https://doi.org/10.1007/978-3-319-09970-5_32

19. Daneva, M., Buglione, L., Herrmann, A.: Software architects' experiences of quality requirements: what we know and what we do not know? In: Doerr, J., Opdahl, A.L. (eds.) REFSQ 2013. LNCS, vol. 7830, pp. 1–17. Springer, Heidelberg (2013). https://doi.org/10. 1007/978-3-642-37422-7_1

20. Letier, E., Stefan, D., Barr, D.T.: Uncertainty, risk, and information value in software requirements and architecture. In: ICSE 2014 (2014)

21. Franch, X., Kenett, R., Mancinelli, F., Susi, A., Ameller, D., Annosi, M.C., Ben-Jacob, R., Blumenfeld, Y., Franco, O.H., Gross, D., Lopez, L., Morandini, M., Oriol, M., Siena, A.: The RISCOSS platform for risk management in open source software adoption. In: Damiani, E., Frati, F., Riehle, D., Wasserman, A.I. (eds.) OSS 2015. IAICT, vol. 451, pp. 124–133. Springer, Cham (2015). https://doi.org/10.1007/978-3-319-17837-0_12

22. Sillitti, A., Succi, G.: Requirements engineering for agile methods. In: Aurum, A., Wohlin, C. (eds.) Engineering and Managing Software Requirements. Springer, Heidelberg (2005). https://doi.org/10.1007/3-540-28244-0_14

23. Martakis, A., Daneva, M.: Handling requirements dependencies in agile projects: a focus group with agile software development practitioners. In: RCIS 2013 (2013)

24. Behutiye, W., Karhapää, P., Costal, D., Oivo, M., Franch, X.: Non-functional requirements documentation in agile software development: challenges and solution proposal. In: Felderer, M., Méndez Fernández, D., Turhan, B., Kalinowski, M., Sarro, F., Winkler, D. (eds.) PROFES 2017. LNCS, vol. 10611, pp. 515–522. Springer, Cham (2017). https://doi. org/10.1007/978-3-319-69926-4_41

25. Heck, P., Zaidman, A.: A systematic literature review on quality criteria for agile requirements specifications. Soft. Qual. J. **26**, 127–160 (2018)

26. Furtado, F., Zisman, A.: Trace++: A traceability approach to support transitioning to agile software engineering. In: RE 2016 (2016)

27. Guzmán, L., Oriol, M., Rodríguez, P., Franch, X., Jedlitschka, A., Oivo, M.: How can quality awareness support rapid software development? – A research preview. In: Grünbacher, P., Perini, A. (eds.) REFSQ 2017. LNCS, vol. 10153, pp. 167–173. Springer, Cham (2017). https://doi.org/10.1007/978-3-319-54045-0_12

28. Gómez, C., Ayala, C., Franch, X., López, L., Behutiye, W., Martínez-Fernández, S.: Towards an ontology for strategic decision making: the case of quality in rapid software development projects. In: de Cesare, S., Frank, U. (eds.) ER 2017. LNCS, vol. 10651, pp. 111–121. Springer, Cham (2017). https://doi.org/10.1007/978-3-319-70625-2_11

29. Guizzardi, G.: Ontological foundations for structural conceptual models. CTIT, Centre for Telematics and Information Technology (2005)

30. Li, F.-L., Horkoff, J., Mylopoulos, J., Guizzardi, R.S., Guizzardi, G., Borgida, A., Liu, L.: Non-functional requirements as qualities, with a spice of ontology. In: RE 2014 (2014)

31. Wagner, S., et al.: Operationalised product quality models and assessment: the quamoco approach. Inf. Softw. Technol. **62**, 101–123 (2015)

32. Quer, C., et al.: PABRE: pattern-based requirements elicitation. In: RCIS 2009 (2009)

33. Mendes, E., et al.: Towards improving decision making and estimating the value of decisions in value-based software engineering: the VALUE framework. Softw. Qual. J. (2017)

34. Heikkilä, V.T., Damian, D., Lassenius, C., Paasivaara, M.: A mapping study on requirements engineering in agile software development. In: SEAA (2015)

35. Noor, T.Z., Hemmati, H.: Studying test case failure prediction for test case prioritization. In: PROMISE@ESEM 2017 (2017)

36. Davis, F.F.: Perceived usefulness, perceived ease of use, and user acceptance of information technology. MIS Q. **13**(3), 319–340 (1989)

A Situational Approach for the Definition and Tailoring of a Data-Driven Software Evolution Method

Xavier Franch[1(✉)], Jolita Ralyté[2], Anna Perini[3], Alberto Abelló[1],
David Ameller[1], Jesús Gorroñogoitia[4], Sergi Nadal[1], Marc Oriol[1],
Norbert Seyff[5], Alberto Siena[6], and Angelo Susi[4]

[1] Universitat Politècnica de Catalunya (UPC), Barcelona, Spain
{franch, aabello, dameller, snadal, moriol}@essi.upc.edu
[2] University of Geneva, Geneva, Switzerland
jolita.ralyte@unige.ch
[3] Fondazione Bruno Kessler (FBK), Trento, Italy
perini@fbk.eu
[4] ATOS, Madrid, Spain
jesus.gorronogoitia@atos.net, susi@fbk.eu
[5] University of Applied Sciences Northwestern Switzerland (FHNW),
Windisch, Switzerland
norbert.seyff@fhnw.ch
[6] Delta Informatica SpA, Trento, Italy
alberto.siena@gmail.com

Abstract. Successful software evolution heavily depends on the selection of the right features to be included in the next release. Such selection is difficult, and companies often report bad experiences about user acceptance. To overcome this challenge, there is an increasing number of approaches that propose intensive use of data to drive evolution. This trend has motivated the SUPERSEDE method, which proposes the collection and analysis of user feedback and monitoring data as the baseline to elicit and prioritize requirements, which are then used to plan the next release. However, every company may be interested in tailoring this method depending on factors like project size, scope, etc. In order to provide a systematic approach, we propose the use of Situational Method Engineering to describe SUPERSEDE and guide its tailoring to a particular context.

Keywords: Software evolution · Situational method engineering
Software process

1 Introduction

Software evolution aims at keeping software systems of any kind aligned with users' needs, which are influenced by individual, social, economic, and technological changes. That part of software engineering is receiving more and more attention since software has become a pervasive and key element in modern society [1, 2].

© Springer International Publishing AG, part of Springer Nature 2018
J. Krogstie and H. A. Reijers (Eds.): CAiSE 2018, LNCS 10816, pp. 603–618, 2018.
https://doi.org/10.1007/978-3-319-91563-0_37

Basic principles and processes of software evolution [3] have been revisited in the light of the high availability of data, reflecting the variety of modern software, and of efficient techniques to mine such data. Indeed, key knowledge to foster short and frequent evolution cycles can be extracted from these data, which is both produced explicitly by users (i.e. user feedback), and resulting from monitoring the context in which the software is executed, as well as the software itself at runtime [1].

As in any other software engineering process, the formulation of methods for data-driven software evolution is of paramount importance. These methods should allow clearly identifying the roles involved, the activities undertaken, the resources involved, and the tools utilized. One of such methods is SUPERSEDE, which has been produced in the context of an H2020 EU project with the same name (www.supersede. eu). The SUPERSEDE method drives the data-driven evolution process in a systematic way. It aims at reconciling generality to accommodate different types of organizations and customisability, as to allow the method to be effectively adopted by an organization that may differ in characteristics as size, business domain and other factors.

The research goal addressed in this paper is to provide a systematic definition of the SUPERSEDE method for data-driven software evolution. We aim at SUPERSEDE acting as a reference method that can be tailored to different situations. Therefore, we adopt Situational Method Engineering (SME) [4] as engineering approach to design the method as a composition of reusable components called method chunks. The research questions that we address in relation to this type of evolution are:

RQ1. What are the constituent parts of method chunks for data-driven evolution?
RQ2. What are the criteria whose combination allows expressing the context in which these method chunks apply?
RQ3. What are the most fundamental method chunks for data-driven evolution?
RQ4. How can the different method chunks be combined in order to create organization-specific customizations of the SUPERSEDE data-driven evolution method?

2   Research Method

The formulation of the SUPERSEDE method has followed a 3-stage process:

- **Stage 1:** *Requirements gathering.* We elicited requirements through dedicated workshops with key stakeholders in relation to the topic of software evolution from three companies participating in the SUPERSEDE project (Siemens, SEnerCON and Atos) and created a requirements document [5]. Requirements comprised as-is and to-be goal models, plus the user stories emerging from this to-be model, all of them coming from workshops held at the companies' site (see protocol in [5]).
- **Stage 2:** *Elaboration of SUPERSEDE engineering artefacts (models, techniques and tools).* In the second stage, we analysed the collected requirements as input to determine the elements composing the SUPERSEDE method. The elaboration and evaluation of these engineering artefacts helped to understand the intricacies of the elements to be integrated in the method.

- **Stage 3:** *Formulation of the SUPERSEDE method for evolution.* In this third stage, we adopted SME as scientific approach in order to organize under a holistic view all the artefacts gathered in the previous stage. Regular meetings allowed the incremental synchronization of the method components. Advances were systematically checked against the result of the first stage (requirements compiled in [5]).

3 Background

3.1 Situational Method Engineering

To formalise the SUPERSEDE method, we apply Situational Method Engineering (SME) [4] principles and techniques. In SME, a method is viewed as a collection of autonomous and interoperable method components that can be selected and assembled in a way to satisfy the particular situation of the project at hand. The definition of method components and their assembly techniques vary from one SME approach to another [4]), but the main objective stays the same: to make the method knowledge modular and reusable for the construction and/or adaptation of situation-specific methods. Most of the assembly techniques support method construction from scratch based on the identified situational context and requirements, other deal with incremental organization's method adaptation or even method family construction.

In our work, we adopt and adapt the *assembly-based and method chunk-based SME* approach [6, 7] that supports situation-specific method construction and extension in three steps: method requirements specification, method chunks selection and assembly of the selected chunks (details given in Sect. 6 as needed). Method chunks are reusable method components. A method chunk includes a process (i.e., the guidelines provided by the method chunk) and its related product knowledge (i.e., the formalisation of concepts and artefacts used and produced by the method chunk), and is specified by the situation in which it can be applied (i.e., the required input artefacts) and the intention (i.e. the engineering goal) to be reached. Finally, the reuse context of the method chunk is specified by a set of criteria that can be defined by using a taxonomy like the reuse frame proposed in [8]. We apply an *ad hoc method chunk definition* approach [9] that creates chunks from experts' knowledge, based on a method chunk metamodel.

3.2 The SUPERSEDE Data-Driven Control Loop for Software Evolution

SUPERSEDE's data-driven software evolution process takes inspiration from the autonomic control loop proposed for adaptive systems [10]. In SUPERSEDE, the control loop drives also software evolution, considering runtime and context data, and also explicit user's feedback, which the user might deliver upon having used the software.

The SUPERSEDE process can be characterized by the following steps (see Fig. 1):

Collect. Multi-modal feedback gathering techniques allow users to express their feedback as textual comments, emoticons, rating and pictures. Flexible and configurable monitoring components collect a huge volume of data from the context and

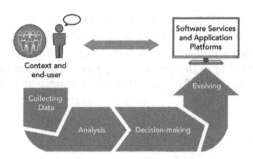

Fig. 1. The SUPERSEDE control loop

system usage [11]. These data, of different types, are stored in a big data storage, which maps data to a semantic model at support of analysis [12].

Analyse. Different types of analysis are tool-supported, for instance, sentiment analysis and extraction of feature requests and bug issues from user textual comments [13], tweet mining to understand perceived quality of experience [14], and combined analysis of end-user feedback and contextual data.

Decide. Focusing on software evolution tasks, automated reasoning techniques support collaborative-decision making concerning, for instance, the identification of new requirements, and their prioritization with respect to multiple criteria [15].

Act. Operationalizations of the decisions made are performed at this step. Selected features are included in a release plan that takes into account available resources, deadlines and organization priorities [16]. In addition, this step can be refined during the actual implementation of the next release using a continuous release planning approach [17].

4   Method Chunks Metamodel for SUPERSEDE

In this section, we address RQ1 (What are the constituent parts of method chunks for data-driven evolution). Following the advices in [9], we implement the concept of method chunk using a process-oriented view. We base our work on existing software process modelling approaches, e.g. from SPEM [18]. We build a metamodel articulating the subset of method elements that are relevant for our purposes, namely activities, artefacts, roles and tools, each one yielding a particular class in the metamodel. Then, we link method chunks to these classes.

4.1   SUPERSEDE Method Elements

Activities, artefacts, roles and tools are declared as specializations of an abstract class MethodElement (see Fig. 2), which links every method element to one or more phases of the SUPERSEDE loop, allowing also referencing to external concepts (e.g., a role not needed by the method but referenced for the sake of completeness). We declare a reflexive association class to express relationships among method elements that will be specialized according to the different types of method elements whenever needed. All method elements may present structural relationships (composition and specialization) and each method element may present other particular types, as we detail below.

Activities involve artefacts, tools and roles. In a nutshell, activities are tasks in which one or more Roles are involved, and are carried out with the support of Tools, which receive one or more Artefacts as input and generate other Artefacts as output. Examples of activities are Feedback Collection (linked to the Collect phase), Domain Ontology Definition (Analyse), Requirements Prioritization (Decide) and Release Planning (Act). Structural relationships will be widely used; for instance, Requirements Prioritization will be specialized into several sub-activities, each adopting a particular prioritization strategy (e.g. Requirements Prioritization with AHP). In addition, activities may present several types of temporal relationships; we adopt the proposal in [19] (richer than SPEM when it comes to temporal relationships) with relationships like start-start, end-start, exclusion, etc. (e.g. Release Planning cannot start before Requirements Prioritization has ended). In all the types of process elements, particular relationships are defined by specializing the MEOtherRel class (see Fig. 2) and including integrity constraints to enforce application to the right type.

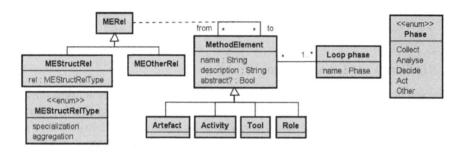

Fig. 2. Elements of the SUPERSEDE method

Artefacts represent informational resources that are produced, consumed or just used as a working asset by an activity. Examples of artefacts are Feedback Document (produced as output by the Feedback Collection activity), Prioritized List of Requirements (consumed as input by Release Planning) and AHP Matrix (used as working asset by Requirements Prioritization with AHP). Artefacts are declared of a particular category. Examples are: Model (e.g., Integration-Oriented Ontology), File (e.g., Monitoring Data), DataSet (e.g., Project Schedule), Technology (e.g., Event Queue Endpoint) and Expression (e.g., Complex Event Pattern). Categories may be decomposed into subcategories at any level, for instance the Integration-Oriented Ontology belongs to the Ontology subcategory in Model. Artefacts may be related to each other by structural relationships: aggregation (a List of Requirements as aggregation of Requirement) and specialization (Prioritized List of Requirements as specialization of List of Requirements). Other relationships are possible, like constraint (e.g., a Release Plan constrains a Project Schedule) or in-sync, meaning that changes in one artifact imply changes in another (e.g., a Release Plan is in-sync with a Prioritized List of Requirements).

Roles are involved in activities either individually (e.g., Project Owner) or as a set of person (e.g. Set of Developers), which is expressed with an association class (SetOf).

Tools are also used in activities, and they produce and consume artefacts and involve roles. All these associations among the four type of process elements are not independent and some integrity constraints need to be declared (not included here for the sake of brevity), e.g. `context` Activity `inv`: self.tool->includesAll(self.input.-tool). Relationships among tools include connection (i.e., the result of one tool is used by another). Again this association is related with others through integrity constraints, e.g. if the RePlan tool (which produces a ReleasePlan) is connected to DMGame (which produces a Prioritized List of Requirements), then the corresponding artefacts need to be in-sync.

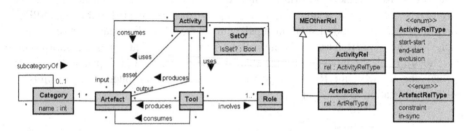

Fig. 3. Detail of the SUPERSEDE method elements

4.2   Definition of Method Chunks

Every activity in SUPERSEDE begets a method chunk for the catalogue. We consider that the description of the activity is the process part of the chunk, while the output of such activity (i.e., the list of produced artefacts) is the product part. The situation of the method chunk is given by the list of consumed artefacts, while the intention needs to be explicitly given in the form of a goal. We also include the definition of the chunk reuse context, which has been defined in [8] as a taxonomy of criteria (details are given in Sect. 5.2); not all the criteria apply to every chunk. Finally, we extend the definition of method chunks with the Role and Tool attributes representing the corresponding SUPERSEDE method elements and presented in the section above.

Figure 4 shows the metamodel corresponding to the definition of method chunk. For convenience, several derived roles are explicitly declared which can be computed

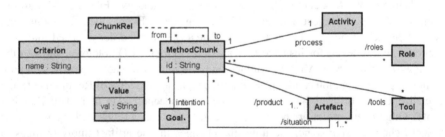

Fig. 4. Metamodel for method chunks

from roles appearing in Fig. 3. Also an associate association class is introduced as derived from the corresponding activity-binding association class appearing in Fig. 3.

5   A Catalogue of Method Chunks for SUPERSEDE

In this section, we address RQ2 (What are the criteria whose combination allows expressing the context in which these method chunks apply) and RQ3 (What are the most fundamental method chunks for data-driven evolution). To address RQ3, we present the method chunks of the SUPERSEDE method that we have developed following an ad-hoc approach and applying the method chunk metamodel shown in Fig. 4.

5.1   Context Criteria Applicable in the SUPERSEDE Method

As reported in Sects. 3.1 and 4.2, context definition using SME is implemented by the definition of context criteria that capture the factors that may influence in the selection of a method chunk for a particular instantiation of a method. In [8], a proposal is given resulting in two relevant outcomes: (1) a 3-tier context structure in which a few dimensions are decomposed into facets, and the facets into criteria; (2) a concrete proposal of this context structure for the information systems field by 2006.

Table 1. Criteria for SUPERSEDE reuse context

Criterion	Values
User involvement	Low, medium, high
Resources required	Few, fair, much
Size	Small, medium, large
Delivery strategy	Spare releases, frequent releases, continuous delivery
Type of end-user (*)	Citizen, organization, technician
Accuracy (*)	Low, medium, high
Motivation (*)	Low, medium, high
Other assumptions (*)	Minor, fair, major
Privacy	Very sensible, sensible, not sensible

We have evolved the original proposal by incorporating additional criteria that may be needed in a particular context in order to decide if one particular chunk has to be part of a customization of the SUPERSEDE method. Table 1 presents an excerpt of the result, focusing on the most relevant criteria for chunk selection (i.e., whose values affect a greater number of chunks). Criteria with an (*) are new, mostly related to the data-driven approach adopted in the SUPERSEDE method. Range of values adhere to the original proposal as much as possible; when subjective, value assignment relies in expert criteria. Definition of the relevant criteria follow:

- *User involvement.* Related to the participation of the user in the chunk's activity. Relevant for activities related to feedback gathering (end-user involvement) but also for technical ones (e.g., involvement of project managers in release planning).

- *Resources required* (evolution of *Means shortage* in [8]). To measure the complexity of an activity with respect to personnel, infrastructure, etc. It affects especially activities that require code development (e.g., new monitoring components).
- *Size* (of the project). Size may impact activities whose feasibility depends on the volume of work. A typical example would be prioritization techniques (e.g., AHP comparison) that do not scale well with a large number of requirements.
- *Delivery strategy.* The type of delivery impacts the later stages of evolution. Heavyweight prioritization techniques are cumbersome to use with frequent releases, and classical release planning does not apply in continuous delivery contexts.
- *Type of end-user.* Some users may be more educated than others from an IT consumer perspective. It may be expected that a technician using the feedback gathering mechanisms will be more accurate than a regular citizen.
- *Accuracy.* The level of accuracy sought impacts on the selection and customization of some techniques. For example, as we will see in Sect. 5.3, different prioritization techniques exhibit different accuracy levels.
- *Motivation.* In data-driven approaches, one strategy to get more data is to motivate stakeholders to make them willing to participate actively. Not only end-users, but also technical stakeholders may be involved using gamification techniques.
- *Other assumptions.* This criterion collects very specific conditions not belonging to the previous types that need to be fulfilled in order to apply the chunk. They may refer to the organization, adopted techniques, involved roles, etc.

5.2   A Catalogue of Method Chunks

We group the chunks into four categories. For all the categories, an initial chunk (Chu-XXX-01) represents an initial deployment activity, which is sometimes decomposed into subactivities. We omit these ones for brevity and present below the rest of chunks and their relationships (see Fig. 5). Being a data-driven approach, the chunks have emerged naturally by following the flow of data across the four phases of the control loop, with every significant data process converted into a process chunk.

Fig. 5. Relationships among SUPERSEDE evolution method chunks.

Chunks for Collection. Data enters the SUPERSEDE process in two different ways. First, end users may provide feedback using multi-modal mechanisms configured at design time but also at runtime (although runtime issues, belonging to the self-adaptation world, are not included in this paper) (Chu-Col-02: Feedback Collection). Second, SUPERSEDE may monitor data without users' explicit intervention (Chu-Col-03: Monitoring Data Collection). There are three different types of monitored data: quality of service, e.g. response time or availability (Chu-Col-03a: QoS

Monitoring); monitoring of social networks as Twitter (Chu-Col-03b: Social Network Monitoring); users' usage monitoring (Chu-Col-03c: Usage Monitoring). The collected data are sent for analysis following the SUPERSEDE loop.

Chunks for Analysis. The SUPERSEDE data-driven approach relies on the conceptualization of relevant ideas using a domain ontology defined by a data steward who mediates with domain experts from the organization (Chu-Ana-02: Domain Ontology Definition). This ontology is used as the basis to define the event recognition rules (Chu-Ana-03: Definition of Event Recognition Rules) that will capture needs for evolution as well as the logical definition of data coming from feedback and monitoring (Chu-Ana-04: Source Ontology Extraction). Collected data are processed in a last chunk (Chu-Ana-05: Event and Pattern Detection) that applies the recognition rules to the monitored data coming from Chu-Col-02 and Chu-Col-03 and produces the real needs in the form of events and patterns to be decided upon in the next phase.

Chunks for Decision. The alerts produced by Chu-Ana-05 are captured by a method chunk that converts them into a list of requirements by means of collaborative editing involving some selected requirements analysts (Chu-Dec-02: Requirements Collaborative Editing). Since they have been produced independently, it may be the case that requirements are overlapping or present other relationships, therefore a consolidation of the list is made right away (Chu-Dec-03: Requirements Similarity Check). This consolidated list is then prioritized involving again the appropriate (or available) set of stakeholders, to generate a prioritized list of requirements (Chu-Dec-04: Requirements Prioritization) to be processed by the enactment method chunks. There are several strategies for prioritizing requirements, and we present here three of them (Chu-Dec-04a, Chu-Dec-04b and Chu-Dec-04c, using AHP, Gamified AHP and Genetic Algorithms, respectively) which differ in some context criteria (see details in next subsection).

Chunks for Enactment. The prioritized list of requirements is processed by a release planning activity (Chu-Ena-02: Release Planning) that considers also the list of available resources as input, and then produces a release plan.

To ensure the correctness and adequacy of the aforementioned method chunks, we checked them against the requirements posed by the three companies mentioned in the introduction [5] ensuring that the catalogue of method chunks satisfied their needs. Furthermore, we conducted regular meetings to check that the different method chunks matched well with each other.

5.3   Example: Requirements Prioritization

For the sake of illustration, in this subsection we present a method chunk in detail, namely the chunk for requirements prioritization, together with a summary of its three current specializations.

Requirements prioritization elaborates on a set of requirements produced by a previous activity (Chu-Dec-03, which needs to be finished before starting this one), and applies a set of weighted criteria to the information elicited from several stakeholders, as to provide a list of prioritized requirements that serves as input of the release planning activities. The SUPERSEDE method applies a gamification approach to prioritization, and a software tool called DMGame has been developed with this purpose. Three types of stakeholders collaborate around this tool: a Supervisor supervising the process, a Negotiator mediating in conflicts and the Decision Providers that provide the necessary information for prioritizing. It is worth remarking that this chunk has not context criteria applicable, meaning that it will be instantiated in any possible situation. We expect this to be the usual situation in abstract method chunks, leaving the criteria to the specializations as shown below.

We have built three different specializations for this abstract chunk until now. Two of them are based on the AHP method, a third one on genetic algorithms. As a specialization, the information is inherited, thus only new information needs to be added. In the case of these three specializations, only the context is concerned. The context criteria that affect the selection of the specialization are shown in Table 3, with the values that apply to every specialization (Table 2).

Table 2. Method chunk for requirements prioritization (DM: Decision-Making)

Method chunk	Content
Id	Chu-Dec-04 [abstract]
Name	Requirements prioritization
Description	This activity applies some selected technique in order to prioritize a list of requirements involving several selected stakeholders
Context	–
Situation	• Set of (possibly interrelated) requirements • Set of weighted criteria for the prioritization
Intention	Prioritize requirements in a collaborative way
Process part	Description of the activity – not included for brevity reasons
Product part	LP: List of prioritized requirements
Roles	• DM Supervisor: Organizes the full setting of the prioritization and supervises the execution of the activity • DM Negotiator: Facilitates the resolution of conflicts among decision providers • Set of DM Decision Providers: Provides information useful for prioritization
Tools	DMGame: decision-making web-based tool
Related chunks	Chu-Dec-03 e-s Chu-Dec-04

As a short summary, AHP-based methods require more user involvement and thus more resources and also suffer from limitations on the number of requirements to handle and from the assumptions needed to ensure the accuracy of the method (which is higher than the case of genetic algorithms). Although not critical, there are some differences also in the type of end-users and the delivery strategy. Concerning the two AHP methods, the main difference among them is the higher motivation required for AHP.

6   Steps for Adopting SUPERSEDE: A Situational Approach

It is not sufficient to transform a method into a collection of method chunks to make it situational. Hence, in this section we address RQ4 (How can the different method chunks be combined in order to create a customization of the SUPERSEDE data-driven evolution method tailored to a particular context). To address this RQ, we need to provide guidance for creating situation-specific SUPERSEDE methods and allowing companies to tailor SUPERSEDE to their needs. For that, we follow the generic three-step assembly process mentioned in Sect. 3, that we specialise for SUPERSEDE as follows (see Fig. 6). The formalization of the process is just outlined; it cannot be developed in full due to space reasons.

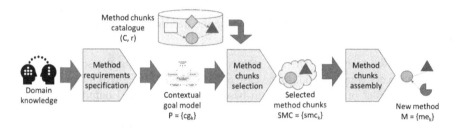

Fig. 6. The three-step assembly process for the SUPERSEDE method tailoring.

Method Requirements Specification. The first step consists in defining functional and contextual method requirements. Functional requirements capture a set of engineering intentions that shall be fulfilled by the new method, while the contextual ones reflect the situation of the project at hand in terms of assessed criteria from the reuse context.

We propose to use *i** goal models [20] and domain knowledge to extract method intentions and to assess context criteria. Indeed, a goal model interconnects the main actors via goal dependencies and allows to derive method intentions from these goals. We recommend, to construct first the as-is goal model reflecting the current situation of the organization. Then, the envisioned situation is designed as a to-be goal model. For the to-be situation, we propose to link intentional elements to relevant context information that allow identifying context criteria and, therefore, to derive not only method intentions but also contextual requirements. The result of this step can be stated then as a set of method requirements, namely contextual goals, $P = \{cg_k\}$ where every method requirement is a pair $cg_k = (g_k, C_k)$ of a goal and a set of conditions of the form $C_k = \{(c_i, v_i)\}$, being c_i a contextual criterion and v_i a valid value for such criterion. Note that C_k is a correspondence, allowing thus different values for a criterion.

Method Chunks Selection. In the second step, the method requirements are used to select method chunks by matching them with method chunk components. The selection query is the model P above which is compared against the catalogue (C, R) of method chunks, $C = \{mc_k\}$, and their relationships, $R = \{(mc1_i, mc2_i, tr_i)\}$, being $mc1_i$ and $mc2_i$ two method chunks and tr_i a type of relationship valid for them. For the selection process, we consider only as relevant information of method chunks the intention and the context, $mc_k = (int_k, cont_k)$, where $cont_k$ has a similar structure than C_k above.

The matching among P and (C, R) can then be defined as a set of selected method chunks SMC = {smc_k} where each smc_k = (int_k, $cont_k$) fulfils several conditions: (1) the intention of the chunk satisfies some functional requirement g_k of the query, $int_k \Rightarrow g_k$; (2) the context criteria of such functional requirements are satisfied by the chunk, $cont_k \Rightarrow C_k$; (3) the intention of the chunk does not violate any other functional requirement of the query, $\neg(int_k \Rightarrow \neg g_i)$. Please note that we do not consider relationships in the step, but in the next one.

Ideally, the selected method chunks should cover all functional method requirements and satisfy the context conditions. In case some functional requirement is not satisfied, other method sources will be explored and formalised as method chunks to fill the gap (see next step). On the contrary, if there are several method chunks satisfying the same requirement (i.e. producing the same outputs in different ways) two possibilities are to be considered: (1) selecting only one of these chunks, so the decision is taken by the method engineer, or (2) postpone the decision to method enactment time (see next step).

Method Chunks Assembly. The last step consists in assembling the selected method chunks into a coherent method. In the case of SUPERSEDE, that will mainly consist in defining the order of chunks execution based on their input and output artefacts and complete missing elements by ad hoc (eventually non-reusable) chunks.

Let M be the method corresponding to the SUPERSEDE instantiation under construction, M = me_k}, being me_k an instance of the MethodElement abstract class introduced in Fig. 2. The rules to be applied are:

1. Connect the selected method chunks according to the relationships in R: $\forall$(m1, m2) $\in$ SMC: (m1, m2, t) $\in$ R $\Rightarrow$ (m1, m2, t) $\in$ M. The inclusion in M takes care of including and connecting artefacts, roles and tools as specified in the metamodel.
2. For those contextual goals {cg_k} in P not covered by any method chunk in SMC, explore other method sources and formalise them as method chunks to fill the gap, applying the steps above as required.
3. For those contextual goals {cg_k} in P covered by more than one method chunk in SMC, providing arguments for the inclusion in M of either only one of them (if they are exclusive) or a subset (if they are complementary).

7   Example

We present an illustrative example of the application of the SUPERSEDE method to the SIEMENS project use case, which concerns the development and evolution of a smart-city platform (Eco Sys. Platform).

Method Requirements Specification. A SIEMENS department has adopted the SUPERSEDE method as shown in Fig. 7, which depicts an excerpt of the *i* to-be* goal-oriented model for this use case (see [5] for the full version). The Project Manager relies on the SUPERSEDE method for achieving key goals that will allow improving the way the department evolves and maintains the Eco Sys. Platform, in relation to feedback analysis, collaborative decision-making and release planning.

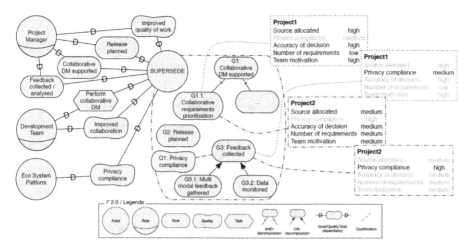

Fig. 7. To-be contextual goal model with information of two projects (simplified example)

The **Development Team** relies on **SUPERSEDE** method to improve collaborative decision-making. **Eco Sys. Platform** requires **SUPERSEDE** to ensure privacy compliance. The goal diagram of **SUPERSEDE** actor models the goals that the method has been delegated to achieve, which are refined using decomposition (just hinted in the figure).

The SIEMENS department can perform different, and possibly simultaneous projects. Each project is characterised by specific values of context properties (depicted as a simple list of item in the rectangular shapes in Fig. 7) which need to be mapped to contextual criteria (Table 1). For instance, in *Project1*, which concerns the implementation of a small set of new requirements to improve the platform reliability, privacy compliance is not critical, while in *Project2*, which deal with a larger set of requirements related to the management of usage logs, privacy compliance is highly relevant. The analysis of the resulting contextual goal model will lead to the identification of a set of method requirements, such as $cg_{1.1} = (G1.1, C_{1.1})$, where $C_{1.1} = \{$(user involvement, high) (size, small) (accuracy, high) (motivation, high)$\}$, and $cg_{1.2} = (G3, C_{1.2})$ where $C_{1.2} = \{$(privacy, sensible)$\}$ for the SUPERSEDE method configuration in *Project1*. Similarly, for *Project2*, method requirements include $cg_{2.1} = (G1.1, C_{2.1})$, where $C_{2.1} = \{$(user involvement, medium) (size, medium) (accuracy, fair) (motivation, medium)$\}$, and $cg_{2.2} = (G3, C_{2.2})$ where $C_{2.2} = \{$(privacy, very sensible)$\}$.

Method Chunks Selection. These method requirements drive the selection of method chunks. For instance, requirements $cg_{1.1}$ matches to the method chunk Chu-Dec-04 (for the goal part, $G1.1$), while for the selection of the method variant we need to match $C_{1.1}$ with context criteria (see Table 3), resulting ·in the selection of the AHP.variant (Chu-Dec-04a). Analogously, requirement $cg_{1.2}$ will lead to the identification of method chunks Chu-Col-2 (because privacy is not very sensible), Chu-Col-03a, and Chu-Col-03c. On its turn, for *Project2* the genetic algorithm-based prioritization technique Chu-Dec-04c will be selected, due to the increasing number of requirements). Concerning data collection, the analysis of context condition $C_{1.2}$ leads to exclude the

Table 3. Context criteria for the three specializations of requirements prioritization.

Criterion	AHP	Gamified AHP	Genetic
User involvement	High	High	Medium
Resources required	High	High	Medium
Project size	Small	Small	Medium
Delivery strategy	Frequent	Frequent	Continuous
Type of end-user	Technician	Technician	Mixed
Accuracy	High	High	Fair
Motivation	High	Medium	Medium
Assumptions	High	High	Low

use of both Chu-Col-02 and one specialization of Chu-Col-03, Chu-Col-03b, because neither feedback gathering nor usage monitoring are considered appropriate when privacy is a highly relevant issue (which is explicitly stated in the corresponding chunks). In both projects, Chu-Ena-02 (Release Planning) is selected due to the **Release Planned** goal.

Method Chunks Assembly. On the one hand, the selected method chunks are now assembled taking into account execution order (see Fig. 6) and input/output artefacts. This will create the assembly (Chu-Dec-04a, Chu-Ena-02) in *Project1* and (Chu-Dec-04c, Chu-Ena-02) in *Project2*. However, collection chunks cannot be assembled because the analysis chunks have not been selected. The reason is that the goal model does not include information enough as to realize that these chunks are needed. Therefore, they need to be selected in this phase and assembled correspondingly, e.g. (Chu-Col-02, Chu-Ana-05) in *Project1*. Figure 8 presents the resulting SUPERSEDE Method for *Project1,* with the main method chunks, the input-output artefacts and the roles involved (for the sake of brevity, method chunks required at design time are not depicted, namely, Chu-Ana-02, Chu-Ana-03 and Chu-Ana-04).

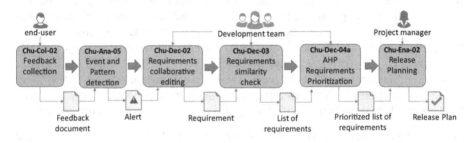

Fig. 8. Excerpt of the SUPERSEDE Method for Project

8   Conclusions and Future Work

In this paper, we have provided a systematic definition of the SUPERSEDE method for data-driven software evolution oriented towards its customization in particular contexts. We have defined a metamodel for method chunks built upon activities, artefacts, roles and tools (RQ1), then we have defined a set of context criteria to describe the context in which every chunk can be selected (RQ2), we have presented a catalogue of method chunks for SUPERSEDE-based software evolution compliant to such meta-model (RQ3) and we have presented an SME-based process to guide the definition of a tailoring of the SUPERSEDE method in a particular context (RQ4).

It is worth mentioning that the answer to RQ4 is a methodological contribution beyond the application to the SUPERSEDE method. It represents an evolution to recent SME works that we have undertaken [21, 22] given the inclusion of context criteria as part of the method requirements. The full development of the, e.g. the applicability of previously proposed context goal modelling approaches that support contextual annotations [23], is future work. Other aspects could be thought to be integrated in this extension of SME, as for instance adding existing guidelines for building methods based on metamodels from the literature about Domain Specific Modeling [24].

A point of discussion is the possible complexity of the method. As in any SME-based approach, we assume that a method engineer would lead the customization of the method with the help of key stakeholders, e.g. a requirements engineer for the construction of the goal model. We plan also to develop tool support as to assist the method engineer, e.g. by suggesting the missing chunks in the third step of the assembly

Future work spreads along several directions. First, we plan to conduct an exhaustive validation of the SUPERSEDE evolution method, which is partly required also due to the novelty of the topic and limitations that are inherent to data-driven decision support. Second, we aim at formalizing the method described in Sect. 6. Last, we aim at extending the catalogue of chunks for software evolution with adaptation and configuration artefacts obtaining then a holistic view of data-driven software engineering.

Acknowledgments. This work is a result of the SUPERSEDE project, funded by the EU's H2020 Programme under the agreement number 644018.

References

1. Wang, X., Guarino, N., Guizzardi, G., Mylopoulos, J.: Software as a social artifact: a management and evolution perspective. In: Yu, E., Dobbie, G., Jarke, M., Purao, S. (eds.) ER 2014. LNCS, vol. 8824, pp. 321–334. Springer, Cham (2014). https://doi.org/10.1007/978-3-319-12206-9_27
2. Mens, T., Serebrenik, A., Cleve, A. (eds.): Evolving Software Systems. Springer, Heidelberg (2014). https://doi.org/10.1007/978-3-642-45398-4
3. Lehman, M.M.: Programs, life cycles, and laws of software evolution. Proc. IEEE **68**(9), 1060–1076 (1980)
4. Henderson-Sellers, B., Ralyté, J., Ågerfalk, P., Rossi, M.: Situational Method Engineering. Springer, Heidelberg (2014). https://doi.org/10.1007/978-3-642-41467-1
5. Stade, M., et al.: D3.1: Requirements for Methods and Tools. SUPERSEDE project deliverable (2015). www.supersede.eu

6. Ralyté, J., Rolland, C.: An approach for method reengineering. In: S.Kunii, H., Jajodia, S., Sølvberg, A. (eds.) ER 2001. LNCS, vol. 2224, pp. 471–484. Springer, Heidelberg (2001). https://doi.org/10.1007/3-540-45581-7_35

7. Ralyté, J., Deneckère, R., Rolland, C.: Towards a generic model for situational method engineering. In: Eder, J., Missikoff, M. (eds.) CAiSE 2003. LNCS, vol. 2681, pp. 95–110. Springer, Heidelberg (2003). https://doi.org/10.1007/3-540-45017-3_9

8. Mirbel, I., Ralyté, J.: Situational method engineering: combining assembly-based and roadmap-driven approaches. Requir. Eng. J. **11**, 58–78 (2006)

9. Ralyté, J.: Towards situational methods for information systems development: engineering reusable method chunks. In: ISD 2004 (2004)

10. Brun, Y., et al.: Engineering self-adaptive systems through feedback loops. In: Cheng, B.H. C., de Lemos, R., Giese, H., Inverardi, P., Magee, J. (eds.) Software Engineering for Self-Adaptive Systems. LNCS, vol. 5525, pp. 48–70. Springer, Heidelberg (2009). https://doi.org/10.1007/978-3-642-02161-9_3

11. Stade, M., Fotrousi, F., Seyff, N., Albrecht, O.: Feedback gathering from an industrial point of view. In: RE 2017 (2017)

12. Nadal, S., et al.: A software reference architecture for semantic-aware big data systems. Inf. Softw. Technol. **90**, 75–92 (2017)

13. Morales-Ramirez, I., Kifetew, F.M., Perini, A.: Analysis of online discussions in support of requirements discovery. In: Dubois, E., Pohl, K. (eds.) CAiSE 2017. LNCS, vol. 10253, pp. 159–174. Springer, Cham (2017). https://doi.org/10.1007/978-3-319-59536-8_11

14. Guzmán, E., Alkadhi, R., Seyff, N.: An exploratory study of twitter messages about software applications. Requir. Eng. J. **22**(3), 387–412 (2017)

15. Busetta, P., et al: Tool-supported collaborative requirements prioritisation. In: COMPSAC 2017 (2017)

16. Ameller, D., et al.: Replan: a release planning tool. In: SANER 2017 (2017)

17. Ameller, D. et al.: Towards continuous software release planning. In: SANER 2017 (2017)

18. Object Management Group (OMG): Software & Systems Process Engineering Meta-Model Specification (SPEM), Version 2.0. Technical report, April 2008

19. Ribó, J.M., Franch, X.: A precedence-based approach for proactive control in software process modelling. In: SEKE 2002 (2002)

20. Dalpiaz, F., Franch, X., Horkoff, J.: iStar 2.0 Language Guide. https://arxiv.org/abs/1605.07767

21. López, L., Costal, D., Ralyté, J., Franch, X., Méndez, L., Annosi, M.C.: OSSAP – a situational method for defining open source software adoption processes. In: Nurcan, S., Soffer, P., Bajec, M., Eder, J. (eds.) CAiSE 2016. LNCS, vol. 9694, pp. 524–539. Springer, Cham (2016). https://doi.org/10.1007/978-3-319-39696-5_32

22. López, L., Behutiye, W., Karhapää, P., Ralyté, J., Franch, X., Oivo, M.: Agile quality requirements management best practices portfolio: a situational method engineering approach. In: Felderer, M., Méndez Fernández, D., Turhan, B., Kalinowski, M., Sarro, F., Winkler, D. (eds.) PROFES 2017. LNCS, vol. 10611, pp. 548–555. Springer, Cham (2017). https://doi.org/10.1007/978-3-319-69926-4_45

23. Ali, R., Dalpiaz, F., Giorgini, P.: A goal-based framework for contextual requirement modeling and analysis. Requir. Eng. J. **15**(4), 439–458 (2010)

24. Frank, U.: Domain-specific modeling languages: requirements analysis and design guidelines. In: Reinhartz-Berger, I., Sturm, A., Clark, T., Cohen, S., Bettin, J. (eds.) Domain Engineering. Springer, Berlin, Heidelberg (2013). https://doi.org/10.1007/978-3-642-36654-3_6

CAiSE 2018 Tutorials

CAiSE 2018 Tutorials

This section contains the abstracts of the tutorials accepted for presentation at the 30th International Conference on Advanced Information Systems Engineering (CAiSE 2018), held in Tallinn, Estonia, from 11–15 June 2018.

The objective of the tutorials is offering new insights, knowledge and skills to practitioners, professors, research fellows and students seeking to gain a better understanding of the state-of-the-art in Information Systems engineering. They are a good way to get a broad overview of a topic beyond a current paper presentation.

This year, 10 tutorial proposals were submitted for consideration at CAiSE 2018. They were evaluated according to several criteria, namely relevance to CAiSE, structure and content of the proposal, attractiveness, novelty of the topic, perceived importance in the field, methodology for the presentation, background of the speaker(s) and past experience.

As a result, 5 tutorials were selected for presentation at the conference:

- Teaching Conceptual Modelling: How can I improve? by Monique Snoeck, Estefanía Serral Asensio and Daria Bogdanova
- Fundamentals of Business Process Management: Fifty Years of BPM Teaching Distilled, by Marcello La Rosa and Jan Mendling
- Creating Quality User Stories to Generate Conceptual Models, by Sjaak Brinkkemper and Fabiano Dalpiaz
- Big Data Driven Software Reuse: Feature Models and Case-Based Reasoning, by Hermann Kaindl and Mike Mannion
- Using Fractal Enterprise Model for Business Model Innovation, by Ilia Bider and Erik Perjons

All the tutorials were assigned 90 minutes for the presentation and were included in the main conference program.

We would like to thank all the colleagues involved in the organizations of the event: the CAiSE 2018 General Chairs, Marlon Dumas and Andreas Opdahl, the CAiSE 2018 Program Chairs, John Krogstie and Hajo A. Reijers, and all the colleagues who submitted their tutorial proposal for consideration to the conference.

April 2018

Selmin Nurcan
Massimo Mecella

Teaching Conceptual Modelling:
How Can I Improve?

Monique Snoeck$^{(\boxtimes)}$, Estefanía Serral, and Daria Bogdanova

LIRIS, Research Centre for Management Informatics, KU Leuven,
Leuven, Belgium
{monique.snoeck, estefania.serral,
daria.bogdanova}@kuleuven.be

Abstract. CM fosters a good understanding of the problem domain, which is a key factor of IS development success. Conceptual modelers require a wide variety of skills involving "simple" skills, such as the ability to remember a modelling notation, and "complex" skills, such as the ability to translate a text description into a representative conceptual model. In order to build the "complex" skills, the educator must ensure the proper acquisition and integration of simpler skills by means of continuous support and timely feedback [1].

1 A Taxonomy of Learning Outcomes

While there is a general agreement in the CM community on what makes a good model, it is still unclear which learning outcomes are typically set by the educators, and how skill acquisition should be scaffolded. A careful review of the literature on CM education shows that no rubrics or taxonomies classifying learning outcomes exist for CM, though such rubrics have been recently adopted in other fields, such as histology [2], software engineering [3] and social studies [4].

Taking into account the fact that clearly set learning outcomes are essential for the learning process [5], and that evaluation is a core element of virtually any instructional design model [6], the lack of state-of-the-art accepted standards for such an important subject as CM is surprising and calls for action from the academic community.

This tutorial reviews the revised Bloom Taxonomy and presents the current state of practice of CM assessments, based on an analysis of: (a) four principal books from the fields of databases, CM with (E)ER, CM with UML, and requirements engineering, (b) three MOOCs and (c) four university courses on CM. Gaps in current practice are identified and an overview of assessment types for all learning outcomes is given.

2 Feedback Types and Their Effectiveness

Feedback gives students an opportunity to reflect on their progress, and is a means teachers use to scaffold the learning process. The timeliness of feedback, especially on task and process levels, is of crucial importance [7]. However, when educators have to teach large audiences that may reach hundreds in higher education or more than tens of

© Springer International Publishing AG, part of Springer Nature 2018
J. Krogstie and H. A. Reijers (Eds.): CAiSE 2018, LNCS 10816, pp. 621–622, 2018.
https://doi.org/10.1007/978-3-319-91563-0

thousands of people in case of MOOCs, it becomes impossible to give personal attention to every student and provide them with timely and personalized feedback. Consequently, university students are likely to be exposed to insufficient feedback, which influences their engagement and overall experience in a negative way [8]. The insufficiency of ongoing support and feedback becomes even more influential in the subjects that require high level of abstract thinking, such as CM.

Researching on feedback automation [9] exposed a lack of theoretical foundation clarifying the terminology and compiling what to considered in order to design and deliver high-quality feedback. This research resulted in a framework that identifies and clarifies the most important factors to consider for giving feedback and will be presented in the second part of the tutorial. It provides educators with a rich palette of types of feedback to choose from, including an indication in which context these should be used.

References

1. van Merriënboer, J.J.G., Kester, L., Paas, F.: Teaching complex rather than simple tasks: balancing intrinsic and germane load to enhance transfer of learning. Appl. Cogn. Psychol. **20**, 343–352 (2006)
2. Zaidi, N.B., Hwang, C., Scott, S., Stallard, S., Purkiss, J., Hortsch, M.: Climbing Bloom's taxonomy pyramid: lessons from a graduate histology course. Anat. Sci. Educ. (2017)
3. Dolog, P., Thomsen, L.L., Thomsen, B.: Assessing problem-based learning in a software engineering curriculum using bloom's taxonomy and the IEEE software engineering body of knowledge. ACM Trans. Comput. Educ. **16**, 1–41 (2016)
4. Gezer, M., Sunkur, M.O., Sah, F.: An evaluation of the exam questions of social studies course according to revized Bloom's taxonomy **2** (2014)
5. Harden, R.M.: Learning outcomes and instructional objectives: is there a difference? Med. Teach. **24**, 151–155 (2002)
6. Branch, R.M., Kopcha, T.J.: Instructional design models. In: Spector, J., Merrill, M., Elen, J., Bishop, M. (eds.) Handbook of Research on Educational Communications and Technology. Springer, New York (2014)
7. Hattie, J., Timperley, H.: The power of feedback. Rev. Educ. Res. **77**, 81–112 (2007)
8. Boud, D., Molloy, E.: Rethinking models of feedback for learning: the challenge of design. Assess. Eval. High. Educ. **38**, 698–712 (2013)
9. Serral, E., De Weerdt, J., Sedrakyan, G., Snoeck, M.: Automating immediate and personalized feedback taking conceptual modelling education to a next level. In: 2016 IEEE Tenth International Conference on Research Challenges in Information Science (RCIS), pp. 1–6. IEEE (2016)

Fundamentals of Business Process Management: Fifty Years of BPM Teaching Distilled

Marcello La Rosa[1(✉)] and Jan Mendling[2]

[1] University of Melbourne, Melbourne, Australia
marcello.larosa@unimelb.edu.au
[2] Vienna University of Economics and Business, Vienna, Austria
jan.mendling@wu.ac.at

Business Process Management (BPM) [1] is a well-established discipline for improving organisational performance. By using a *business process lens*, BPM enables organisations to systematically oversee how work is performed, in order to ensure consistent outcomes and take advantage of improvement opportunities.

Over time, the BPM discipline has become increasingly more complex and multi-faceted, as new technologies, business models and standards emerge on a continuous basis, and affect the way business processes are conceived, implemented and continuously monitored [2]. Businesses around the world are carrying out BPM initiatives with the aim to outperform their competitors or meet the demands of regulatory authorities. At the same time, a lively academic community is pushing the boundaries of the discipline: computer scientists, management scientists, and engineers add new ingredients to its repertoire, which are eagerly being picked up by practitioners.

Against this backdrop, the aim of this tutorial is to explain what has crystalized as the core body of knowledge of the BPM discipline, and how it can be organized for teaching at undergraduate and graduate levels. A particular focus will be on how four specific topics of BPM can be integrated in the teaching curriculum, namely: process redesign orbit, process implementation with standards, variants analysis with process mining, and BPM as an enterprise capability. The tutorial concludes with an outlook on future developments in BPM.

This tutorial is suited for educators who are interested in learning how BPM can be taught using a holistic approach. It is also interesting to a general audience who would like to gain an overall understanding of the BPM discipline.

References

1. Dumas, M., La Rosa, M., Mendling, J., Reijers, H.A.: Fundamentals of Business Process Management, 2nd edn. Springer, Heidelberg (2018)
2. Rosemann, M.: Proposals for future BPM research directions. In: Ouyang, C., Jung, J.-Y. (eds.) AP-BPM 2014. LNBIP, vol. 181, pp. 1–15. Springer, Cham (2014). https://doi.org/10.1007/978-3-319-08222-6_1

© Springer International Publishing AG, part of Springer Nature 2018
J. Krogstie and H. A. Reijers (Eds.): CAiSE 2018, LNCS 10816, p. 623, 2018.
https://doi.org/10.1007/978-3-319-91563-0

User Stories in Information Systems Engineering

Sjaak Brinkkemper(✉) ⓘ and Fabiano Dalpiaz ⓘ

Department of Information and Computing Sciences, Utrecht University,
Utrecht, The Netherlands
{S.Brinkkemper, F.Dalpiaz}@uu.nl

Abstract. The scope of this tutorial includes user stories in agile information systems engineering and generating conceptual models by relying on the inherent simplicity and structure of user stories. This tutorial presents the creation of high quality user stories based on state-of-the-art academic work and industrial best practices on the topic of user stories. By focusing on the context of user story formulation, this tutorial provides an overview of available literature and incorporates recent research results.

Keywords: User stories · Information systems engineering · Conceptual model

1 User Story Quality and Usage

90% of agile practitioners employ user stories for specifying information systems. Of these, 70% follow a simple template when creating user stories: **"As a <role>, I want to <action>, [so that <benefit>]"** [2]. For example: *"As an Administrator, I want to receive an email when a contact form is submitted, so that I can respond to it."* User stories' popularity among practitioners and simple yet strict structure make them ideal candidates for specification and the subsequent systems architecting and design ([3, 10, 11]). Unfortunately, however, 50% of real-world user stories contain easily preventable errors that sabotage their potential. To alleviate this problem, the presenters of this tutorial have created methods, theories and tools that support creating better user stories ([3–7, 9]).

In this tutorial, the following topics will be presented:

1. the basics of creating user stories;
2. the role of user stories in contemporary IS engineering methods;
3. incremental adoption of user story practices;
4. improving user story quality with the Quality User Story Framework and AQUSA tool;
5. generating conceptual models from high quality user stories with Visual Narrator;
6. insight in best practices of 20+ software companies employing user stories

This tutorial aims to provide the knowledge and resources to start working with user stories in IS engineering projects. A state-of-the-art overview of the theories behind user stories and their usage will be provided [1, 3, 5, 8]. Participants are

J. Krogstie and H. A. Reijers (Eds.): CAiSE 2018, LNCS 10816, pp. 624–625, 2018.
https://doi.org/10.1007/978-3-319-91563-0

encouraged to bring existing sets of user stories available from projects they or someone else in their organization participated in.

References

1. Dimitrijević, S., Jovanović, J., Devedžić, V.: A comparative study of software tools for user story management. Inf. Softw. Technol. **57**, 352–368 (2015)
2. Lucassen, G., Dalpiaz, F., van der Werf, J.M.E.M., Brinkkemper, S.: The use and effectiveness of user stories in practice. In: Daneva, M., Pastor, O. (eds.) REFSQ 2016. LNCS, vol. 9619, pp. 205–222. Springer, Cham (2016). https://doi.org/10.1007/978-3-319-30282-9_14
3. Lucassen, G., Dalpiaz, F., van der Werf, J.M., Brinkkemper, S.: Improving agile requirements: the quality user story framework and tool. Requirements Eng. **15**(1), 1–21 (2016)
4. Lucassen, G., Dalpiaz, F., van der Werf, J.M.E.M., Brinkkemper, S.: Visualizing user story requirements at multiple granularity levels via semantic relatedness. In: Comyn-Wattiau, I., Tanaka, K., Song, I.-Y., Yamamoto, S., Saeki, M. (eds.) ER 2016. LNCS, vol. 9974, pp. 463–478. Springer, Cham (2016). https://doi.org/10.1007/978-3-319-46397-1_35
5. Lucassen, G., Robeer, M., Dalpiaz, F., van der Werf, J.M., Brinkkemper, S.: Extracting conceptual models from user stories with Visual Narrator. Requir. Eng. **22**(3), 339–358 (2017)
6. Lucassen, G., van de Keuken, M., Dalpiaz, F., Brinkkemper, S., Sloof, G.W., Schlingmann, J.: Jobs-to-be-done oriented requirements engineering: a method for defining job stories. In: Kamsties, E., Horkoff, J., Dalpiaz, F. (eds.) REFSQ 2018. LNCS, vol. 10753, pp. 227–243. Springer, Cham (2018). https://doi.org/10.1007/978-3-319-77243-1_14
7. van de Weerd, I., Brinkkemper, S., Versendaal, J.: Concepts for incremental method evolution: empirical exploration and validation in requirements management. In: Krogstie, J., Opdahl, A., Sindre, G. (eds.) CAiSE 2007. LNCS, vol. 4495, pp. 469–484. Springer, Heidelberg (2007). https://doi.org/10.1007/978-3-540-72988-4_33
8. Kassab, M.: The changing landscape of requirements engineering practices over the past decade. In: Proceedings of EmpiRE, pp. 1–8. IEEE (2015)
9. Vlaanderen, K., Dalpiaz, F., Brinkkemper, S.: Finding optimal plans for incremental method engineering. In: Jarke, M., Mylopoulos, J., Quix, C., Rolland, C., Manolopoulos, Y., Mouratidis, H., Horkoff, J. (eds.) CAiSE 2014. LNCS, vol. 8484, pp. 640–655. Springer, Cham (2014). https://doi.org/10.1007/978-3-319-07881-6_43
10. Liskin, O., Pham, R., Kiesling, S., Schneider, K.: Why we need a granularity concept for user stories. In: Cantone, G., Marchesi, M. (eds.) XP 2014. LNBIP, vol. 179, pp. 110–125. Springer, Cham (2014). https://doi.org/10.1007/978-3-319-06862-6_8
11. Cohn, M.: User Stories Applied: for Agile Software Development. Addison Wesley Longman Publishing Co., Inc., Redwood City (2004)

Big Data Driven Software Reuse: Feature Models and Case-Based Reasoning

Hermann Kaindl[1(✉)] and Mike Mannion[2(✉)]

[1] TU Wien, ICT, Gußhausstr 27–29, 1040 Vienna, Austria
kaindl@ict.tuwien.ac.at
[2] Glasgow Caledonian University, 70 Cowcaddens Road, Glasgow G4 0BA, UK
m.a.g.mannion@gcu.ac.uk

Abstract. A range of socio-economic trends are driving customer demands towards personalization and suppliers toward mass customization. At the same time the increasing development of big data driven cyber-physical systems (Industry 4.0) is coalescing ideas from software product line engineering and from flexible manufacturing. One significant implication is the ability to manage the process of identification, selection and deployment of reusable features on a large scale, regardless of whether the features are to be implemented in software or hardware. Such reuse ranges from operational, ad-hoc and short-term to strategic, planned and long-term. This tutorial presents and compares two different feature-led approaches.

The first approach deals with feature reuse and reusability in the context of *product line engineering* and platform eco-system development. The second approach deals with feature reuse and reusability in the context of *case-based reasoning*. Both approaches have different key properties and trade-offs between the costs of making software artefacts *reusable* and the benefits of *reusing* them. To aid large-scale development we have proposed a *Feature-Similarity Model*, which draws on both approaches to facilitate discovering features relationships using similarity metrics. A Feature-Similarity Model also helps with the evolution of a product line, since new features can be introduced first into a case base and then gradually included into a product line representation. This tutorial focuses on effective feature reuse to reduce the effort in developing features whilst maintaining the level of precision and quality and risk mitigation that has been worked through on previous projects.

Keywords: Reuse · Feature model · Case-based reasoning

J. Krogstie and H. A. Reijers (Eds.): CAiSE 2018, LNCS 10816, p. 626, 2018.
https://doi.org/10.1007/978-3-319-91563-0

Using Fractal Enterprise Model for Business Model Innovation

Ilia Bider[(⊠)] and Erik Perjons

DSV - Stockholm University, Stockholm, Sweden
{ilia,perjons}@dsv.su.se

Abstract. The pace of changes in the business environment in which a modern enterprise operates requires the enterprise to constantly review its business models in order to survive and prosper in the dynamic world. The tutorial introduces an approach to Business Model Innovation (BMI) based on a new type of enterprise models called Fractal Enterprise Model (FEM). The model express the relationship between the enterprise assets and its business processes. The innovation consists of finding a new way of using enterprise assets to produce value for a new group of customers. The tutorial introduces FEM and shows how it could be used for BMI. The introduction is followed by an exercise in which the participants will apply the approach and invent a new business model for their own organizations.

Keywords: Enterprise model · Business model innovation
Business transformation

1 Overview

In the dynamic world of today, enterprises need to be innovative not only in the line of products and services they offer, but also in who they are and what they do, i.e. under which business models they operate. For example, in the future, a traditional manufacturing company may not be able to continue its business as usual, i.e. designing and manufacturing their own products, due to the emergence of mature 3-D printing. Instead, the company may need to change its business model, for example, becoming a designer while letting the customer 'print' the design in the place convenient for the customer, or becoming a manufacturer, providing the customer with a service to 'print' somebody else's design (having both alternative could do as well). This change could be more radical than adding a new product or service to the company's offerings.

The tutorial introduces an approach to Business Model Innovation (BMI) based on a new type of enterprise models called Fractal Enterprise Model (FEM). FEM has a form of a directed graph with two types of nodes *Processes* and *Assets*, where the arrows (edges) from assets to processes show which assets are utilized by which processes and arrows from processes to assets show which processes help to have specific assets in healthy and working order. The arrows are labeled with meta-tags that show in what way a given asset is utilized, e.g., as *workforce, reputation, infrastructure*, etc., or in what way a given process helps to have the given assets "in order", i.e., *acquire, maintain* or *retire* the assets. Building a FEM is supported by a set of

© Springer International Publishing AG, part of Springer Nature 2018
J. Krogstie and H. A. Reijers (Eds.): CAiSE 2018, LNCS 10816, pp. 627–628, 2018.
https://doi.org/10.1007/978-3-319-91563-0

archetypes that show what kinds of assets are needed for particular process types and which assets they can help to acquire, maintain and retire. An archetype can be generic – applicable to all processes, or specific for some class of processes, e.g., acquiring stakeholders.

The enterprise processes are divided into two categories – main processes and supporting processes. A main process delivers value to the customer, while a supporting process manages assets needed for other processes to function problem free. The approach, BMI with the help FEM, aims at designing a new main process using assets that already exist in the enterprise.

BMI with FEM consists of two steps: (1) generating hypotheses, and (2) assessing promising hypotheses. The first step is based on analyzing which asset(s) should be used in a new business activity (new main process). The second step consists of comparing a FEM for a new business activity with the FEM for already existing one(s), and assessing the differences, in order to understand which assets and processes can be reused and which new ones need to be acquired.

BMI with FEM is introduced with the help of an example, inspired by Amazon, which while continue its primary business of selling books over the internet, added a completely new business of providing a general purpose IT platform.

2 Tutorial Agenda

The tutorial will follow the following plan:

1. Presentation of basics of FEM and how to use it for BMI by tutorial instructors
2. Attendees work on innovating a business model of their current company/ organization (which might be an academic institution). Dependent on the number of participants, this can be done in groups. Instructors go around and help.
3. Attendees present the results of their work to each other
4. General discussion on the applicability of the approach to BMI with FEM.

Reading Material

1. Bider, I., Perjons, E., Elias, M., Johannesson, P.: A fractal enterprise model and its application for business development. Softw Syst Model (2016)
2. Henkel, M., Bider, I., Perjons, E.: Capability-based business model transformation. In: Advanced Information Systems Engineering Workshops, LNBIP, vol. 178, pp. 88–99. Thesaloniki, Greece (2014)
3. Elias, M., Bider, I., Johannesson, P.: Using fractal process-asset model to design the process architecture of an enterprise: experience report. In: Bider, I., Gaaloul, K., Krogstie, J., Nurcan, S., Proper, H.A., Schmidt, R., Soffer, P. (eds.) BPMDS/EMMSAD - 2014. LNBIP, vol. 175, pp. 287–301. Springer, Heidelberg (2014). https://doi.org/10.1007/978-3-662-43745-2_20
4. Josefsson, M., Widman, K., Bider, I.: Using the process-assets framework for creating a holistic view over process documentation. In: Gaaloul, K., Schmidt, R., Nurcan, S., Guerreiro, S., Ma, Q. (eds.) CAISE 2015. LNBIP, vol. 214, pp. 169–183. Springer, Cham (2015). https://doi.org/10.1007/978-3-319-19237-6_11

Author Index

Printed in the United States
By Bookmasters